Antimutagenesis and Anticarcinogenesis Mechanisms II

BASIC LIFE SCIENCES

Ernest H. Y. Chu, Series Editor

The University of Michigan Medical School
Ann Arbor, Michigan

Alexander Hollaender, Founding Editor

Recent volumes in the series:

A Continuation Order Plan is available for this series. A continuation order will bring
delivery of each new volume immediately upon publication. Volumes are billed only upon
actual shipment. For further information please contact the publisher.

Antimutagenesis and Anticarcinogenesis Mechanisms II

Edited by

Yukiaki Kuroda
National Institute of Genetics
Mishima, Japan

Delbert M. Shankel
University of Kansas
Lawrence, Kansas

and

Michael D. Waters
United States Environmental Protection Agency
Research Triangle Park, North Carolina

Technical Editor
Claire Wilson & Associates
Washington, D.C.

PLENUM PRESS • NEW YORK AND LONDON

Proceedings of the Second International Conference on
Mechanisms of Antimutagenesis and Anticarcinogenesis,
held December 4–9, 1988, in Ohito, Japan

ISBN-13: 978-1-4615-9563-2   e-ISBN-13: 978-1-4615-9561-8
DOI: 10.1007/978-1-4615-9561-8

DEDICATION
IN MEMORY OF

Dr. Tsuneo Kada   Dr. Alexander Hollaender

    With respect and affection, the editors dedicate this volume to Dr. Tsuneo Kada and Dr. Alexander Hollaender, each of whom played a significant and important role in the establishment and continuation of these Conferences.

    Tsuneo Kada was born in Japan, where he lived and worked throughout most of his scientific career. He received his undergraduate degree from the University of Tokyo and his doctorate in Natural Sciences from the University of Paris in 1963 for studies on the mechanism of mutation induction by X-rays and ultraviolet light in E. coli. He also studied in the United States, and was a frequent participant in international conferences and meetings. Dr. Kada was a pioneer in the area of studies on antimutagenesis, and it was his suggestion that led to the first International Conference on Mechanisms of Antimutagenesis and Anticarcinogenesis, for which he served as co-organizer and in which he participated vigorously. He will long be remembered for his numerous scientific contributions, including the development of the widely used "rec assay," and for his infectious enthusiasm which stimulated his colleagues and students. He was Professor and Head of the Department of Molecular Genetics at the National Institute of Genetics in Mishima, leading the planning for this Conference at the time of his untimely death in November of 1986.

    "Alex" Hollaender was born in Germany, but in 1921 moved to the United States, where he pursued his educational and scientific careers. He received all of his degrees from the University of Wisconsin, where he completed his Ph.D. in 1931. He did pioneering experimental research in radiation biology, and was a driving force behind the organization of the Radiation Research Society and the Environmental Mutagen Society, and their international counterparts. He was also instrumental in founding the Biology Division of the Oak Ridge National Laboratory and served as

its Director for 20 years.   From 1973 until his death in 1986, he operated
the Council for Research Planning in Biological Sciences in Washington,
DC, which played a strong advocacy role for science and scientific communication.   He was a member of the National Academy of Sciences and
holder of innumerable prestigious awards.   He was enthusiastic, persis-
tent, helpful, and loved and respected.   At the time of his death in
December of 1986, he was deeply involved in the planning for this Conference.

   For their pioneering scientific contributions, and for their key roles
in stimulating and organizing these Conferences, we dedicate this volume
to our departed friends and colleagues.

Yukiaki Kuroda
Delbert M. Shankel
Michael D. Waters

ACKNOWLEDGEMENTS

Antimutagenesis and anticarcinogenesis and the intricate relationships between them are of clear significance in cancer, hereditary illnesses, and perhaps even in aging. The elaboration of the mechanisms involved in these processes can confer substantial benefits on this and future generations. It is our hope that bringing together the basic, clinical, and applied scientists who generously shared their data and their thoughts, both formally and informally, throughout this Conference will increase our understanding and lead to those future benefits.

The success of a conference depends upon the combined efforts, skills, and support of many individuals and organizations. Listed below you will find the membership of the International Advisory Board, the Japanese Advisory Board, the Organizing Committee, and the Local Committee. Their contributions were all important. Also listed are those local and federal agencies and companies in Japan and the United States which contributed financial support for the meeting. Without their support, the Conference could not have been successful; and we are deeply grateful for that support. Less obvious, but also important, were the contributions of the excellent secretarial and logistic support provided by the Conference Secretariat Staff at the National Institute of Genetics and by the staff of the Ohito Hotel at the conference site.

We are also indebted, of course, to each of the major speakers for their excellent presentations, and to all of those who presented short oral papers or posters and participated in the intellectual excitement of the meeting.

We are doubly grateful to those who have prepared their manuscripts for inclusion in these proceedings. The final success of any meeting depends upon the quality of the speakers and papers--and that quality was very high at this meeting, as evidenced by these papers. We express also our deep gratitude to Ms. Claire Wilson and the staff of Claire Wilson & Associates in Washington, DC, for technical editing and typing, and to the staff at Plenum Press who joined with Ms. Wilson to assure that these Proceedings could be published rapidly and efficiently.

    Funding for the Conference was provided by the following organiza-
tions and companies:

The Agricultural Chemical Society of Japan
The Commemoration Association for the Japan World Exposition
The Environmental Mutagen Society of Japan
The Foundation for Advancement of International Science
The Genetics Society of Japan
The International Association of Environmental Mutagen Societies
The National Cancer Institute/National Institute of Environmental
    Health Sciences (USA, Grant #1R13CA49398-01)
The National Institute for Environmental Health Sciences (USA,
    Award #59-32U4-8-40)
The Pharmaceutical Society of Japan
The United States Department of Agriculture (Grant #59-32u4-8-40)
The Schering-Plough Corporation

    The members of the International Advisory Board were Drs. Sherry
Ansher (USA), Bryn Bridges (England), David Brusick (USA), Silvio de
Flora (Italy), Barry Glickman (Canada), Philip Hartman (USA), Nicola
Loprieno (Italy), Donald MacPhee (Australia), E. Moustacchi (France),
Earle Nestmann (Canada), Claes Ramel (Sweden), Frederic Sobels (The
Netherlands), R.C. "Jack" von Borstel (Canada), Graham Walker (USA),
Michael D. Waters (USA), and Friederich Wurgler (Switzerland).

    The members of the Japanese Advisory Board were Drs. T. Iino
(Tokyo), K. Imamura (Tsukuba), T. Matsushima (Tokyo), T. Sugimura
(Tokyo), and Y. Tazima (Ibaraki).

    In addition to Drs. Y. Kuroda (Chair) and D.M. Shankel (Co-chair),
the other members of the Organizing Committee were Drs. Y. Shirasu
(also Co-chair), Y. Sadaie (Secretary), H. Tezuka (Treasurer), and I.
Tomita (Program).

    The members of the Local Organization Committee were Drs.
H. Fujiki (Tokyo), H. Hayatsu (Okayama), T. Inoue (Fujisawa),
Y. Nakamura (Shizuoka), M. Namiki (Nagoya), H. Nishioka (Kyoto), T.
Ohta (Tokyo), T. Ohsawa (Nagoya), T. Seno (Mishima), K. Shimoi
(Shizuoka), and K. Tutikawa (Mishima).

    To all of the above, and to all of those who participated, we express
our sincere thanks.

                              Yukiaki Kuroda
                              Delbert M. Shankel
                              Michael D. Waters

CONTENTS

MOLECULAR ASPECTS OF
MUTAGENESIS AND ANTIMUTAGENESIS

ONCOGENES AND ANTIONCOGENES

SHORT PAPERS

ANTIMUTAGENESIS STUDIES IN JAPAN

Yukiaki Kuroda

Department of Ontogenetics
National Institute of Genetics
Mishima, Shizuoka 411, Japan

INTRODUCTION

Studies on antimutagenic factors were initially carried out in the 1950s in the field of microbial genetics [for review, see Clarke and Shankel (3)]. During the past ten years, a wide variety of chemical mutagens and carcinogens has been detected in foods, medicines, cosmetics, insecticides, and even in the atmosphere and water which we utilize daily. Some mutagens act directly on cells to produce mutations, and others act following their modification by other factors. Some of these indirect mutagens are metabolically activated by enzymes in organs or tissues, and others may be inactivated and inhibited by some dietary foods and by components of our cells.

In Japan, Tsuneo Kada (10,20), a pioneer in antimutagenesis studies, found effective factors in vegetables and fruits which inactivate the mutagenic action of amino acid pyrolysis products. He has proposed a new word, "desmutagens," for factors which act directly on mutagens or their precursors and inactivate them (20). It has been proposed that other factors which act on the processes of mutagenesis or repair DNA damage to result in decreasing mutation frequency should be called "bio-antimutagens" (11). Figure 1 indicates the actions of desmutagens and bio-antimutagens in the process of mutation induction.

So we can divide antimutagenesis into two different processes: desmutagenesis and bio-antimutagenesis. This article reviews antimutagenesis studies in Japan.

DESMUTAGENESIS

The mechanisms of desmutagenesis involve the following processes:

1. Chemical inactivation of mutagens.

2. Enzymatic inactivation of mutagens.

1

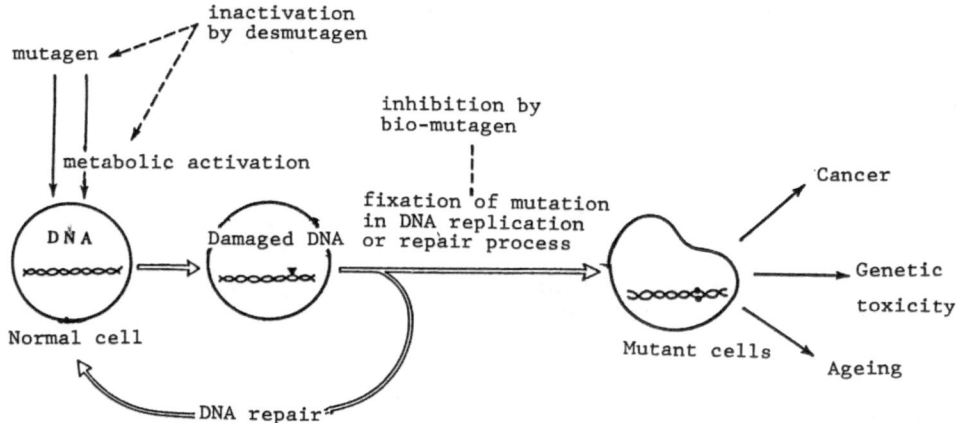

Fig. 1.   Schematic representation showing the process of antimutagenesis
          (11).

3. Inhibition of metabolic activation of promutagens.

4. Inactivation of activated mutagens, including scavenging.

Chemical Inactivation of Mutagens

     Most mutagens are chemically-active substances which act not only on
DNA but also on proteins and enzymes. This suggests that mutagens may
frequently be inactivated by other factors. For example, some mutagenic
pesticides, such as captan, captafol, folpet, and 2,4-dinitrophenyl
thiocyanate (NBT), are extremely sensitive to compounds containing the
sulfhydryl group, such as cysteine, rat-liver homogenate, and blood
(21). The numbers of revertants induced by these pesticides in Escherichia coli and Salmonella typhimurium were reduced by cysteine, liver
homogenate, or blood. These characteristics should be considered in any
explanation of the lack of carcinogenicity in the chronic feeding studies
on captan, captafol, and folpet. Various factors which inactivate some
direct mutagens have been extensively surveyed by Onitsuka et al. (32),
as shown in Tab. 1. It is known that some food additives react with
each other and produce new mutagens. Sorbic acid (2,4-hexadienoic
acid) is a widely used food preservative. Nitrite is also a food additive
used as a preservative of meat products, and often co-exists with sorbic
acid. These two food additives, sorbic acid and sodium nitrite, can react
to produce a mutagenic nitroso-compound (Y substance). It was found
that vitamin C has the ability to react with Y substance and produce a
nonmutagenic compound (33). Sodium nitrite also reacts with dimethylamine contained in various foods and produces a strong carcinogen, nitros-
oamine. Proline has a competitive inhibitory action with the reaction
between sodium nitrite and dimethylamine. Proline reacts with sodium
nitrite and produces noncarcinogenic nitrosoproline, which is excreted to
the outside of the body by urine (44).

     The pyrolysis products of amino acids and of certain foods have a
strong mutagenic activity (22,42). Negishi and Hayatsu (23) found that
cysteine and its derivatives have an enhancing effect on the mutagenicity

Tab. 1.   Inactivating factors of some chemical mutagens (32).

Factors \ Mutagens	Dexon	Captan	MC	4NQO	MNNG
Control	-	-	-	-	-
Vitamin C	++	++	++	-	++
D-Erythrobin	±	++	++	-	++
BHT	-	+	-	±	-
Gallic acid	++	±	++	-	-
n-Propyl gallate	-	±	+	-	-
Vitamin E	-	+	-	-	-
Glutathione	++	++	++	-	++
L-Cysteine	++	++	++	+	++
Cysteamine	+	++	+	+	++
Thiourea	-	++	-	-	-
AET	-	++	++	-	-

Test system : Rec-assay

of tryptophan pyrolysis products, Trp-P-1 and Trp-P-2.   On the contrary, Arimoto et al. (1) in the same group demonstrated that hemin and other biological pyrrole pigments have an inhibitory effect on the mutagenicity of tryptophan and glutamic acid pyrolysis products.   Hemin, biliverdin, and chlorophyllin produced inhibitions of all six mutagens tested, and protoporphyrin inhibited three of them (Tab. 2).   Hemin was the most effective among these pigments.   It was suggested that hemin may interact with the metabolically-activated form of amino acid pyrolysis products and as a result may inhibit mutagenicity.

Tab. 2.   Inactivation of the mutagenicity of amino acid pyrolysis products by pyrrole pigments (1).

Mutagen	Dose of pigments required for 50% inhibition (nmole/plate)			
	Hemin	Chlorophyllin	Biliverdin	Protoporphyrin
Trp-P-1	3	100	200	50
Trp-P-2	20	200	500	100
Glu-P-1	75	100	300	(-)
Glu-P-2	40	150	200	(-)
Amino-α-carboline	30	150	500	(-)
Aminomethyl-α-carboline	25	200	500	50

Test system : TA98 + S9 Mix

Enzymatic Inactivation of Mutagens

Environmental mutagens, especially natural mutagens, are often in-activated by various enzymes contained in the organs and tissues. Kada et al. (9) found that juices of vegetables such as cabbage, turnip, radish, or ginger contained antimutagenic factor(s) which acted on tryp-tophan pyrolysis products in Salmonella TA98 with rat liver homogenate activating mixture (S-9 mix) (Tab. 3). This antimutagenic effect was abolished by prior heating of the juice at 100°C. This indicates that the antimutagenic factor(s) may be a desmutagen(s) and is a heat-sensitive substance(s).

The desmutagenic factors of cabbage juice were purified by proce-dures shown in Tab. 4 (4-6). The mutagenicities of Trp-P-1 and Trp-P-2 were abolished by this factor. Mutagenic activities of ethidium bromide and 2-aminoanthracene (2AA) were also susceptible to the factor. However, the factor had no effect on the mutagenicity of ICR-170 or furyl-furamide (AF2). The purified factor had a molecular weight of 43,000 and had 54.2 µg of sugar per mg protein. The factor exhibited a hemoprotein-like absorption spectrum with a Soret band at 404 nm, and α and β bands at 640 nm and 497 nm, respectively. The Soret band shift resulting from treatment with sodium hydrosulfite and cyanide indicated that the factor had very similar properties to horseradish peroxidase with NADPH-oxidase activity. The factor had both peroxidase and NADPH-oxi-dase activities. This suggests that the desmutagenic activity of vegetable juice may be due to the inactivation of amino acid pyrolysis products by these enzymes.

Myeloperoxidase that was extracted from human promyelocytic leu-kemia HL-60 cells also has the ability to degradate pyrolysis products of tryptophan, glutamic acid, and globulin (46). This is another example of enzymatic inactivation of mutagens, which may be a general phenomenon. Enzymatic detoxification in the liver has been extensively examined during studies on drug metabolism.

Inhibition of Metabolic Activation

It is known that many procarcinogens are changed into ultimate car-cinogens by metabolic activation in cells. In the short-term assay with S. typhimurium, used for detecting mutagens and carcinogens for higher

Tab. 3.   Effects of various vegetable juices on the mutagenicity of Trp-P (10).

Conc. of Trp-P (µg/plate)	Vegetable juice	Mean No. of His[+] revertants/plate
200	None	382
0	None	25
200	Spinach	360
200	Celery	286
200	Cabbage	106
200	Lettuce	327
200	Radish	212
200	Turnip	126
200	Sprouts	357
200	Ginger	104

Test system : TA98 + S9 Mix

animals, including humans, it has become clearer that the process of metabolic activation is one of the important procedures that should be included in this assay system. In the process of metabolic detoxification in the living body, the liver plays an important role. In many cases, external toxic substances are expected to be metabolized into hydrophobic ones which are excreted to the outside from the body. During metabolic changes of chemical compounds, they are sometimes converted to mutagenic compounds. Some inhibitors can block these metabolic activations of promutagens and procarcinogens, thus suppressing mutagenesis and carcinogenesis.

The desmutagenic factor with NADPH-oxidase activity, purified from cabbage juice, also has an inhibitory effect on the metabolic activation of mutagens by the S-9 fraction from rat liver homogenate.

In general, it is very difficult to distinguish whether desmutagenic activity of a certain substance is due to inhibition of metabolic activation or to direct inactivation of mutagens. For mutagens acting without metabolic activation by S-9 mix, the possibility that their inhibition is due to blocking of metabolic activation can be excluded.

Inactivation of Activated Mutagens, Including Scavenging

When promutagens are activated by liver homogenates, they may become highly active mutagens. The effective screening of desmutagens may be possible by using activated mutagens.

Some desmutagens have a scavenging or binding activity for mutagens. Burdock juice has inhibitory effects on the mutagenicity of Trp-P-1, Trp-P-2, 2AA, and ethidium bromide. This desmutagenic activity of burdock juice is not affected by heating at 100°C for 15 min. The active principle of burdock juice was partially purified. The factor was absorbed to DEAE-cellulose but not to CM-cellulose, indicating that it may be a strong anionic polyelectrolyte. This suggests that the desmutagenic activity of burdock juice may be due to the ability of burdock fibers to adsorb mutagens.

Purified fibers were prepared from various vegetables (43). These fibers were added to water solutions of Trp-P-1, Trp-P-2, or Glu-P-1 at concentrations of 1-5% (w/w), and the mixtures were kept at room temperature for several hours. The mutagenic activities of pyrolysate

Tab. 4. Purification of the desmutagenic factor from cabbage leaves (6).

Purification	Volume (ml)	Activity	Protein (mg/ml)	Specific activity/mg
I. Dialyzed 9,000 x g sup.	3,900	7,100	7.86	0.24
II. DEAE-cellulose	5,000	10,600	4.16	0.51
III. CM-cellulose	1.4	2,300	33.6	49.5
IV. Sephacyl S200	8.3	1.980	2.24	106.3
V. DEAE-Sephadex A25	3.6	2,050	2.08	123.3
VI. Second CM-cellulose	3.6	1,522	3.6	87

mutagens were assayed. The fibers from cabbage, burdock, radish, car-
rots, and green pepper showed a strong absorption of pyrolysate mutagens (12). Figure 2 shows the absorption of mutagens by burdock
fibers. The mutagens absorbed to the vegetable fibers were recovered
by treatment with special organic solvents.

Nishioka et al. (24) found that mutagenicities of AF2, N-methyl-N'-
nitro-N-nitrosoguanidine (MNNG), 4-nitroquinoline-1-oxide (4NQO), aflatoxin B, benzo(a)pyrene, and Trp-P-1, with or without metabolic activation, were inactivated by treatment with human saliva using S. typhimurium TA98 and TA100 strains. Mutagenic activities of quercetin, pyrolysates of beef, salmon, and sodium glutamate, and condensate of cigarette smoke were also decreased by saliva treatment, but no significant
effect was found on the activity of methyl methanesulfonate (MMS) or
pyrolysate of polypeptone. Boiled saliva was still active in TA98, but not
in TA100 cultures. Complex mechanisms may be involved in the inactivation of mutagenicity of carcinogens by saliva, including biochemical reactions with enzymes, vitamins, and/or adsorption.

BIO-ANTIMUTAGENESIS

Mutation Induction and Repair Process

Bio-antimutagens act on repair and replication processes of the damaged DNA produced by mutagens, resulting in decreases in mutation frequency. Figure 3 indicates the production of mutations by various
mutagens and the DNA repair processes in E. coli (45). Mutations are
originally produced by modifications induced in nucleotide bases of DNA
by mutagens. The base modifications may produce errors in base pairing
followed by base exchanges, insertions, or deletions which result in shifts
of the codon frame (frame-shift).

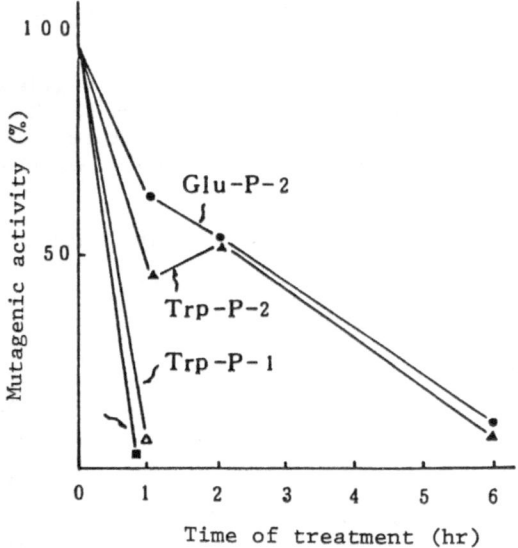

Fig. 2.   Absorption of mutagens by vegetable fiber (12).

For these types of damage occurring in DNA, various repair systems present the cells with an opportunity to normalize DNA. The DNA damage induced by ultraviolet (UV), 4NQO, or AF2 are repaired by one of three major systems. Minor damage to DNA is repaired by the excision repair system controlled by the uvrA, uvrB, and uvrC genes or by the recombination repair system controlled by the recA, recB, and recF genes, which are normally present in the cells. Larger damage which does not permit DNA replication is repaired by the SOS repair system which operates to repair the DNA damage in an "emergency." The SOS repair system, which is controlled by the umuC, umuD, and recA genes, frequently produces errors in the process of DNA repair, resulting in the production of mutations at high frequency. Therefore, in the SOS repair-deficient mutant (umuC⁻) strain, mutations are not induced by UV, 4NQO, or AF2.

On the other hand, DNA damage induced by alkylating agents such as ethyl methanesulfonate (EMS) and MNNG are produced in the SOS repair-deficient mutant strain as well as in the wild-type strain. For alkylated DNA bases, the glycosylase excision repair system controlled by the tag and alkA genes releases the alkylated bases. The methyl transferase controlled by the ada gene also operates to transfer the methylated bases. DNA repair carried out by these enzymes does not produce errors. Therefore, most of the mutations induced by alkylating agents may be produced by errors which occur during replication of alkylated bases and remain as not-repaired bases.

Assay Systems for Detecting Bio-Antimutagens

To detect bio-antimutagens, the assay systems which exclude desmutagens should be used. One procedure for detecting bio-antimutagens is the use of radiation such as UV or γ-rays as mutagens. When chemical mutagens are used as mutagens, the treatment with antimutagens should be carried out at a different time from that for the treatment with mutagens.

Another possible procedure for detecting bio-antimutagens is the use of mutator mutants, in which the effect of antimutagens on the frequency of spontaneous mutations is determined. In both procedures, effects of bio-antimutagens can be detected without chemical mutagens.

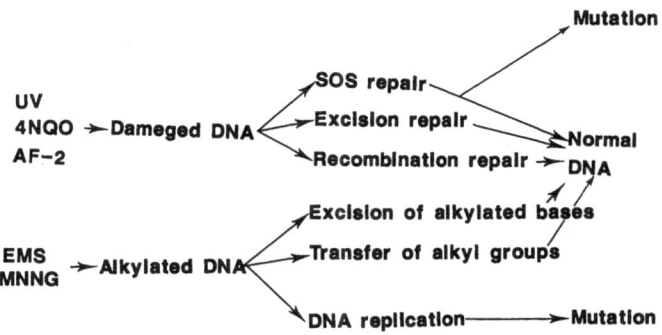

Fig. 3. Mutation induction and repair process (45).

Mechanisms of Bio-Antimutagenesis

As suggested from the presence of various systems to repair DNA damage, bio-antimutagens may act on various repair processes of damaged DNA. Some of the mechanisms of bio-antimutagenesis are as follows:

  a. Inhibition of induction of SOS repair.

  b. Normalization of proofreading in SOS repair.

  c. Acceleration of recombination-SOS repair.

  a. Inhibition of induction of SOS repair. The induction of the SOS repair system is produced as shown in Fig. 4 (18). In this model, the LexA protein is a repressor of a number of unlinked genes that play roles in the SOS response. During exponential growth, the LexA protein acts to repress its target genes. When DNA is damaged, or when replication is inhibited in various ways, an inducing signal is produced. The inducing signal reversibly activates a specific protease activity of the RecA protein, and this protease cleaves and inactivates the LexA repressor. In consequence, the products of the SOS target genes are expressed at much higher levels, and the secondary SOS functions are expressed. Finally, when DNA damage is repaired, the level of the signal molecule drops, the protease activity of the RecA protein disappears, and the LexA repressor can accumulate and express the SOS genes once again.

It has been found that "instant" coffee suppresses strongly the SOS-inducing activity of UV or chemical mutagens such as AF2, 4NQO, and MNNG in S. typhimurium TA1535/p SK 1002 (26). As decaffeinated instant coffee shows a similarly strong suppressive effect, it would seem that caffeine, a known inhibitor of SOS responses, is not responsible for the effect observed. The suppression was also shown by freshly brewed coffee extracts (Fig. 5).

In this figure, the expression of the SOS repair gene is represented by β-galactosidase activity controlled by the lacZ gene, which is fused with the sulA gene controlled by the same repressor protein. The suppression of SOS-inducing activity was not shown by green coffee-bean

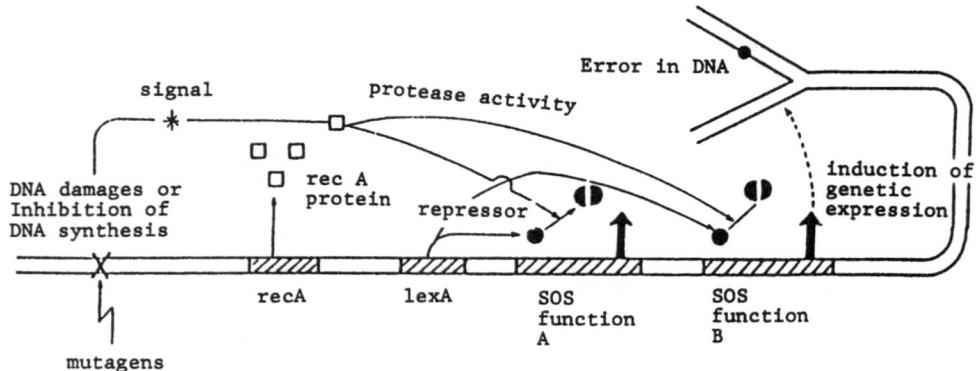

Fig. 4. Model for the SOS regulatory system (18).

extracts.   These results suggest that coffee contains some substance(s) apart from caffeine which suppresses the SOS-inducing activity of UV or chemical mutagens, and that the suppressive substance(s) is produced by roasting coffee beans.

Ohta et al. (30) have screened inhibitors of UV induction of the SOS function.   A log-phase culture of E. coli PQ37 (sulA::lacZ, rfa, uvrA, Pho^c) was irradiated with UV and then immediately grown for 2 hr in a liquid Luria-Bertani medium (LB) medium containing each test compound. Expression of the SOS gene (sulA) was assayed by monitoring the levels of β-galactosidase.   The total number of compounds tested was 233, including 44 food and feed additives, 23 naturally-occurring compounds and derivatives, 21 antibiotics, 61 pesticides, 33 inorganics, and 51 other chemicals.   As a result, 5-fluorouracil (5FU) and 5-fluorodeoxyuridine (FUDR) were found to inhibit considerably the UV induction of the SOS gene without any inhibition of protein synthesis (Tab. 5).   Mutagenesis induced by UV irradiation was depressed by the addition of either compound at nontoxic concentrations.   Since 5FU and FUDR are pyrimidine derivatives and their inhibitions of the UV induction of the SOS gene are reversible, it is suggested that they may competitively inhibit the reaction activating the protease activity of the RecA protein.

Nunoshiba and Nishioka (25) have found that sodium arsenite at a nontoxic   concentration   strongly   inhibits   mutagenesis   induced   by UV, 4NQO, AF2, and MMS as well as spontaneous mutation in the reversion assay of E. coli WP2 uvr A/pKM101.   The effect was not seen in

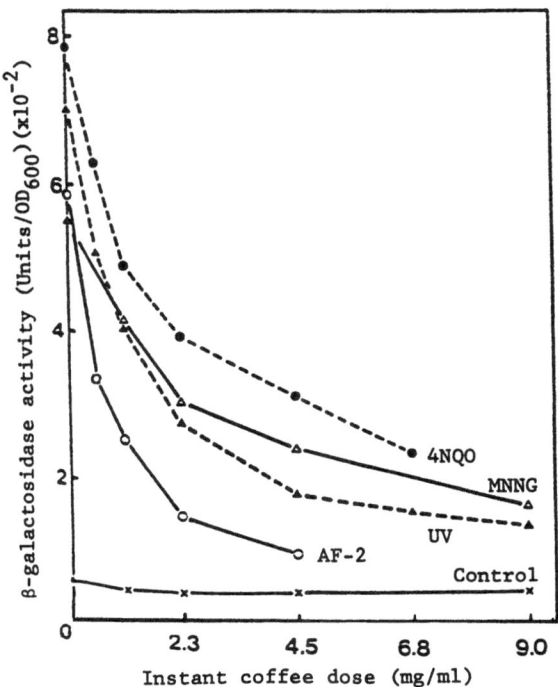

Fig. 5.   Effect of coffee extract on SOS response (26).

Tab. 5.   Inhibitory effect of 5FU and FUDR on UV induction of the SOS
           gene (30).

UV dose (J/cm^2)	Compound added	Cell Growth	β-Galactosidase	Alkaline phosphatase (Unit)
0	None	0.502	82	289
1.5		0.474	588	305
1.5	5-FU	0.426	186	334
1.5	FUDR	0.429	184	327

Test system : E. coli PQ 37

mutagenesis induced by MNNG (Tab. 6). In order to elucidate the mechanism of the mutation-inhibitory effect of sodium arsenite, its action on
umuC gene expression and DNA-repair systems was investigated. It was
found that sodium arsenite depressed β-galactosidase induction, corresponding to umuC gene expression. Nunoshiba and Nishioka assumed that
sodium arsenite may have at least two roles in antimutagenesis: as an
inhibitor of umuC gene expression, and as an enhancer of the error-free
repair depending on the uvrA and recA genes.

       b. Normalization of proofreading in SOS repair. MNNG is a typical
alkylating agent inducing base-change type mutations. Kada and
Kanematsu (9) found that MNNG-induced mutations were remarkably reduced in E. coli B/r WP2 by the presence of cobalt chloride. It was
reported that cobalt chloride inhibited spontaneous mutations in Bacillus
subtilis mutator strain NIG 1125 (his met mut-1) (7). It has also been
shown that the mut-1 cells produced an altered DNA polymerase III which
is the DNA-replicating enzyme of this bacterium (2). An error-prone
DNA replicating enzyme(s) is responsible for frequent spontaneous mutations in mutator strains of E. coli (38) and coli phage T4 (41). Therefore, it is believed that cobalt chloride may correct the error-prone-
ness of the DNA replicating enzyme(s) by improving its fidelity in DNA

Tab. 6.   Antimutagenic effect of sodium arsenite on spontaneous and in-
           duced mutations (25).

Mutagen	Sodium arsenite (µg/ml)	Revertants /viable cells (%)
Control	0	100
	250	43
UV	250	13
4NQO	250	16
MNNG	250	120
AF-2	250	31
MMS	250	52

Test system : E. coli WP2

synthesis. This indicates that mutation frequency could be reduced by inhibiting the factors suppressing the proofreading function in SOS repair. This would allow the proofreading function to remain in the normal state even after the RecA protein cleaves the repressor and the SOS repair functions are expressed.

Mochizuki and Kada (19) found that placental extracts from human, monkey, dog, rat, and mouse had strong antimutagenic activities for reversion mutations induced in E. coli B/r WP2 trp by UV irradiation, γ-rays, or MNNG (Tab. 7).

Preliminary studies on the purification and characterization of the antimutagenic factor(s) from human placental extracts indicate that it is stable to heating and to pronase treatment. In addition, the factor(s) is not lost by dialysis and is not absorbed by ion-exchange resin (Amberlite IR-120B, IRA-410). These results suggest that one of the major factors is not an amino acid or protein and that its molecular weight is greater than 2,000. The other factor was cobaltous metal ion (15).

It was also shown that the placental extracts worked as a potent antimutagen against spontaneous mutagenesis in E. coli strain DM1187 (SOS-repair-constitutive).

Kada et al. (13) systematically surveyed antimutagens in a number of metal compounds and found that germanium oxide also works as a potent antimutagen on frame-shift-type reversion mutations induced by Trp-P-2 in Salmonella (Fig. 6).

Germanium oxide seems to work as an antimutagen in strains TA98 and TA1538. This suggests that this antimutagen might antagonize errors in the SOS repair function residing in the cell but not those endowed by the plasmid.

Germanium, widely used in the field of electronics, is now being applied to the field of medicine in the form of organometallic compounds or organic chelating agents, and the antitumor effect of an organogermanium compound in mouse tumors has been reported (37).

Tab. 7.   Effect of mammalian placental extract on MNNG-induced mutations (19).

Animal species	MNNG (μg/ml)	Placental extract (ml)	Surviving cells (x 10^6)	Revertants /10^8 cells
Human	0	0	292	-
	60	0	231	868
	60	0.3	213	5
Monkey	60	0.3	202	14
Dog	60	0.3	227	7
Rat	60	0.3	231	10
Mouse	60	0.3	225	8

Test system : E. coli B/r WP2

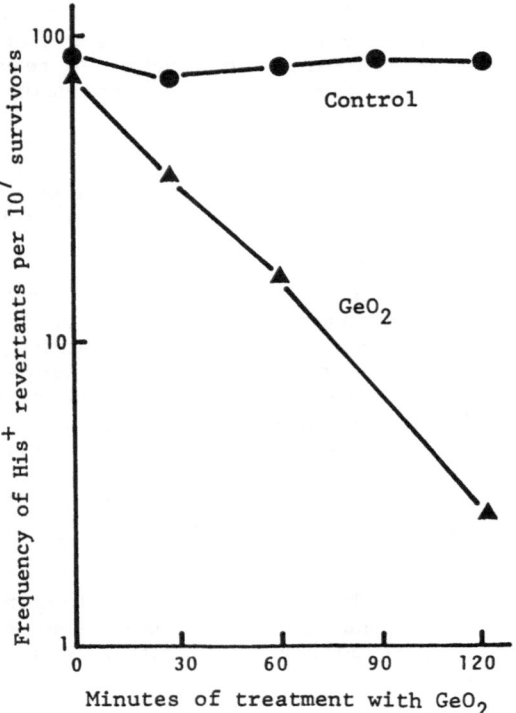

Fig. 6. Effects of germanium oxide on Trp-P-2-induced mutations (13).

c. Acceleration of recombination-repair activity. Some of the damaged DNA may be repaired by the recombination repair system, resulting
in normal DNA. When the damaged DNA remains without repair, it leads
to mutations. If some substances accelerate the recombination-repair
activity and the mutation frequency is reduced, these substances may
seem to be bio-antimutagens. This type of bio-antimutagen is:

1. Effective in UV-, 4NQO-, and AF2-induced mutations.

2. Noneffective in MNNG-induced mutations.

3. Noneffective in improving survival of recombination-deficient
   mutants.

4. An enhancer of recombination frequency in plasmids.

Ohta et al. (27) have examined the antimutagenic effects of cinnamaldehyde on mutagenesis by chemical agents in E. coli WP2 uvrA⁻ trpE⁻
(Fig. 7). Cinnamaldehyde greatly reduces the number of Trp⁺ revertants
induced by 4NQO (Tab. 8). This antimutagenic effect could not be explained by inactivation of 4NQO. Mutagenesis by AF2 was also suppressed significantly. Mutations induced by MMS and EMS were slightly
inhibited. However, cinnamaldehyde was not at all effective against mutagenesis of MNNG. Two derivatives of cinnamaldehyde, cinnamyl alcohol,
and trans-cinnamic acid, did not have as strong antimutagenic effects on

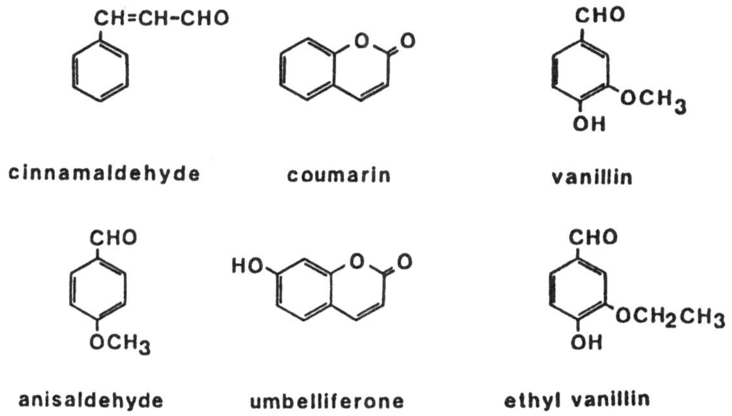

CH=CH-CHO

cinnamaldehyde   coumarin   vanillin

anisaldehyde   umbelliferone   ethyl vanillin

Fig. 7. The chemical structures of flavorings.

4NQO mutagenesis. These results suggest that cinnamaldehyde might act by interfering with an inducible error-prone DNA repair pathway.

The frequency of mutation induction by ten kinds of chemical mutagens in an SOS repair-deficient (umuC⁻) strain was compared with that in a wild-type strain. Cinnamaldehyde greatly suppressed umuC-dependent mutagenesis induced by 4NQO, AF2, or captan. However, cinnamaldehyde was less effective against umuC-independent mutagenesis by alkylating agents such as MNNG and EMS (28). These results suggest that cinnamaldehyde does not prevent the induction of the SOS functions. A remarkable increase was observed in the survival of 4NQO-treated cells after exposure to cinnamaldehyde. This increase in survival suggests the promotion of some DNA repair system by cinnamaldehyde. Therefore, it is assumed that cinnamaldehyde may enhance an error-free recombinational repair system by acting on recA-enzyme activity.

Ohta et al. (29) further screened 18 compounds that are structurally related to cinnnamaldehyde for their antimutagenic effects on mutagenesis induced by 4NQO or UV irradiation. Among the 18 test chemicals, coumarin and umbelliferone, in addition to cinnamaldehyde, showed clear antimutagenic activities (Tab. 9). Cinnamaldehyde is used as a flavoring

Tab. 8. Effects of cinnamaldehyde on 4NQO-induced mutations (27).

Conc. of cinnamaldehyle (μg/ml)	Viable cells/plate	Revertants /plate
0	127	816
20	140	358
40	153	122
50	144	113

Test system : E. coli WP2

Tab. 9.   Effects of cinnamaldehyde, coumarin, and umbelliferone on
4NQO-induced mutations (28).

Anti-mutagen	Viable cells/plate	Revetants /plate
None	124	511
Cinnamaldehyde	131	37
Cumarin	127	78
Umbelliferone	137	62

Test system : E. coli WP2

agent in foodstuffs.   Coumarin is found in lavender, woodruff, and sweet
clover and is used as a perfume.   Umbelliferone, which occurs in many
umbelliferate plants, is used as a fluorescent dye in cosmetics.

Acceleration of Excision-Repair Activity

Shimoi et al. (39) screened about 150 kinds of plants, which were
selected mainly from medicinal plants, to detect the factors which show
bio-antimutagenic effects on UV- or MNNG-induced mutagenesis.   They
used an E. coli B/r WP2 trp⁻ strain.   They found that tannic acid (m-
galoyl gallic acid) suppresses mutagenesis induced by UV or 4NQO, but
not that induced by γ-rays or MNNG.   The depression of mutations induced by UV was most remarkable in the DNA repair-proficient strain
(WP2).   Tannic acid, however, showed no bio-antimutagenic effect in the
excision repair-deficient strain (WP2s uvrA or ZA159 uvrB) under the
best conditions, where no cellular toxicity was observed (Fig. 8).   The
inhibition of the expression of Trp^+ phenotype and the delay of the first
cell division after UV irradiation were not observed in the presence of
tannic acid.   From these results, it was concluded that tannic acid may
enhance the excision-repair system, probably by activating the repair
enzymes or by interacting with DNA.

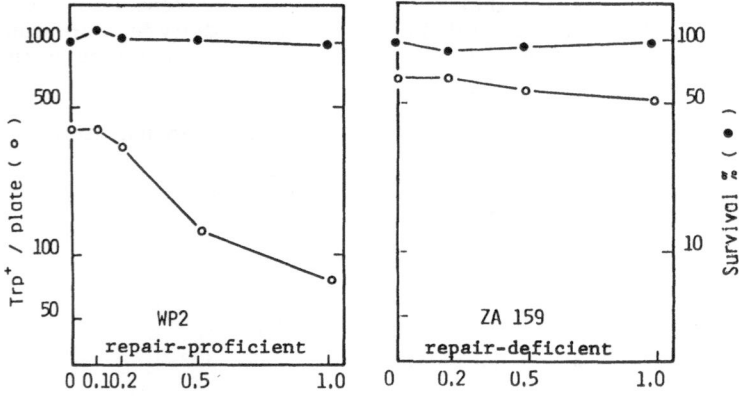

Fig. 8.   Effects of tannic acid on UV-induced mutagensis (39).

Kada et al. (14) carried out a screening for bio-antimutagens in several hundred plant specimens. They found that the homogenate of Japanese green tea (Camellia sinensis) gave the highest bio-antimutagenic activity in the B. subtilis NIG 1125 strain. They then tried to identify chemically the active principles. From 12 g of the crude material, they obtained 0.85 g of epicatechin, 1.44 g of epigallocatechin, 1.24 g of epicatechin gallate, and 4.87 g of epigallocatechin gallate. As a result of the mutagenicity assays on the four catechins, it was found that only the epigallocatechin gallate reduced the frequency of spontaneous mutations due to altered DNA polymerase III in a mutator stain of B. subtilis strain NIG 1125 (Fig. 9).

In China, green tea has been considered to be a crude medicine for 4,000 years. Different kinds of pharmaceutical effects, such as protection of blood vessels, suppression of cancer, and prolongation of life span, were reported. All four catechins isolated have antioxidative characteristics. It is interesting that the major component, epigallocatechin gallate, has a strong bio-antimutagenic activity on spontaneous mutations.

Shimoi et al. (40) also screened plant components with bio-antimutagenic activity against UV-induced mutagenesis using E. coli B/r WP2. The components with a pyrogallol moiety, including gallic acid, (-)-epicatechin gallate, (-)-epigallocatechin, and (-)-epigallocatechin gallate, reduced mutation induction, but other components such as caffeic acid, chlorogenic acid, and quercetin did not (Tab. 10). The above compounds with a pyrogallol moiety were effective on UV-induced mutagenesis, while they showed little effect on MNNG-induced mutagenesis. As this bio-antimutagenic effect was not seen in the DNA excision-repair-deficient strains

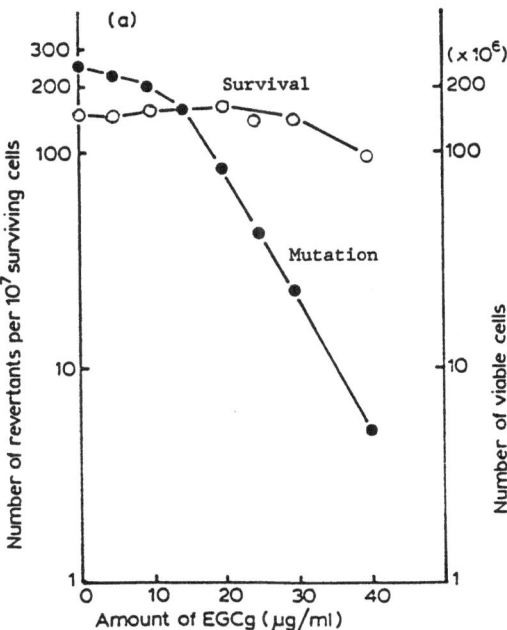

Fig. 9. Effect of epigallocatechin gallate on spontaneous mutations (14).

Tab. 10. Chemical structures of plant components and their bio-antimuta-
          genic activity (40).

Structural formula	Name	Bio-antimutagenic effect on. UV-induced mutagenesis
	catechol	+
	gallic acid	+
	tannic acid (*m*-galoyl gallic acid)	+
	(−)-epicatechin gallate (ECG)	+
	(−)-epigallocatechin (EGC)	+
	(−)-epigallocatechin gallate (EGCG)	+

R, 3,4,5-trihydroxybenzoyl.

WP2 and ZA159, the activity of the above plant components might be based
on the stimulation of the excision-repair system in E. coli B/r WF⌐
(Fig. 10).

ANTIMUTAGENESIS IN MAMMALIAN CELL SYSTEMS

The mechanisms of antimutagenesis described above have been studied on mutation induction and DNA-repair systems in microbial cells.

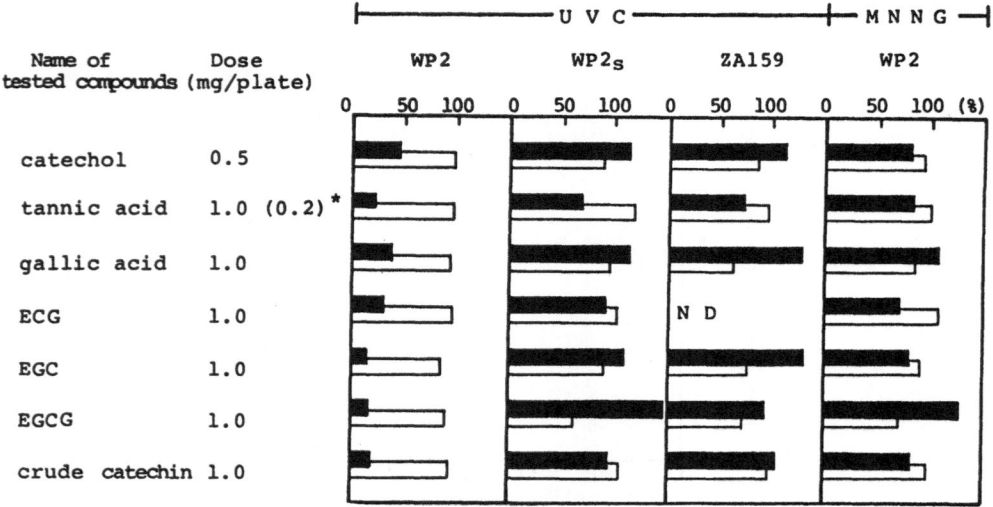

Fig. 10. Effects of plant components on UV- and MNNG-induced muta-
          tions (40).

However, higher animal cells, including human cells, may have more complex characteristics and DNA repair systems which are different from those in bacterial cells.

It has been found that vanillin acts as an antimutagenic factor, especially against mutagenesis induced by 4NQO or UV irradiation (31). Vanillin occurs naturally in vanilla beans. It has been widely used as a flavoring agent in confectioneries, beverages, and food. Sasaki et al. (34) examined the effects of vanillin on the induction of sister-chromatid exchanges (SCEs) and structural chromosome aberrations produced by mitomycin C (MMC) in cultured Chinese hamster ovary cells. Although vanillin enhanced the frequency of SCEs in MMC-treated cells, the frequency of chromosome aberrations was significantly decreased by posttreatment with vanillin at G2 phase (Fig. 11). Among various types of chromosome aberrations, the frequencies of breakage-types (chromatid gaps and breaks, acentric fragments, and fragmentations) were significantly decreased. However, the frequencies of exchange-types (chromatid exchanges and rings) were not suppressed by vanillin.

Sasaki et al. (35) also investigated effects of six flavorings (Fig. 7), including vanillin, on the induction of SCEs by MMC in cultured Chinese hamster ovary cells. All these flavorings, such as vanillin, ethylvanillin, anisaldehyde, cinnamaldehyde, coumarin, and umbelliferone,

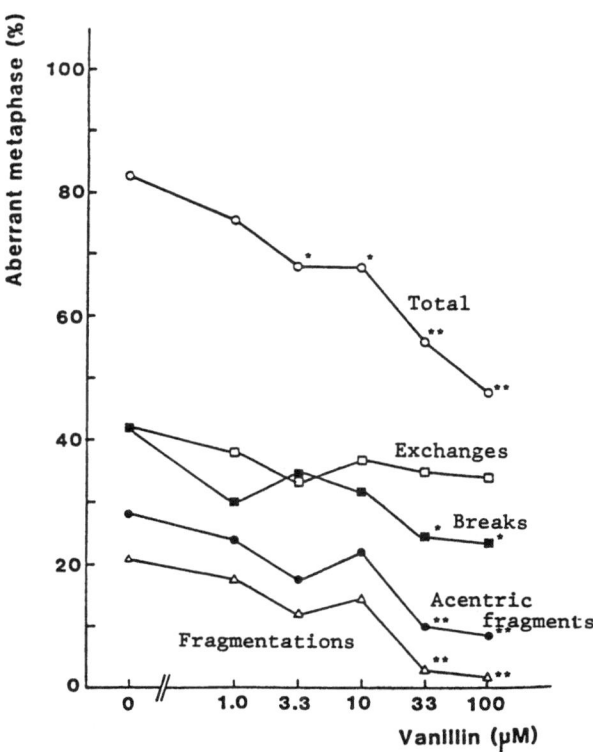

Fig. 11.   Effect of vanillin on MMC-induced chromosome aberrations in Chinese hamster CHO cells (34).

enhanced the frequencies of SCEs in MMC–pretreated cells (Fig. 12). These compounds have either an α,β–unsaturated carbonyl group, or a carbonyl functionality neighboring the phenyl group, which may react with an enzyme SH–group and cause higher-order structure changes. SCE-enhancing effects of vanillin were also observed in cells treated with EMS, N-ethyl-N'-nitro-N-nitrosoguanidine (ENNG), N-ethyl-N-nitrosourea (ENU), or N-methyl-N-nitrosourea (MNU), but not observed in cells treated with MMS or MNNG. SCE-enhancing effects of vanillin seemed to be dependent on the quality of lesions in DNA. It has been reported that vanillin also suppressed MMC-induced micronuclei in mouse bone marrow cells (8).

On the other hand, tannic acid suppressed UV-induced structural chromosome aberrations in cultured Chinese hamster ovary cells (36) (Fig. 13). The suppression of chromosome aberrations was also found in cells treated with MMC or MMS, but not with X-rays or bleomycin. The effects of tannic acid were dependent on the DNA prereplicational G1 phase of the cell cycle. Tannic acid enhanced the liquid-holding effect on UV-irradiated cells at G1 phase. The number of surviving UV-irradiated cells increased in the presence of tannic acid. These results suggest the possibility of promotion by tannic acid of an excision repair system of mammalian cells.

The author examined the antimutagenic effects of vitamins on gene mutations in Chinese hamster V79 cells. Vitamin C (L-ascorbic acid) was highly effective in decreasing 6-thioguanine (6TG)-resistant mutations

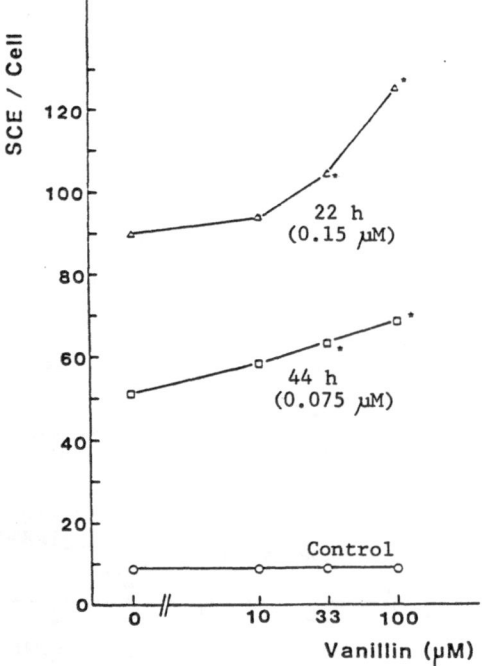

Fig. 12.   Effect of post-treatment with vanillin on MMC-induced SCEs in Chinese hamster CHO cells (34).

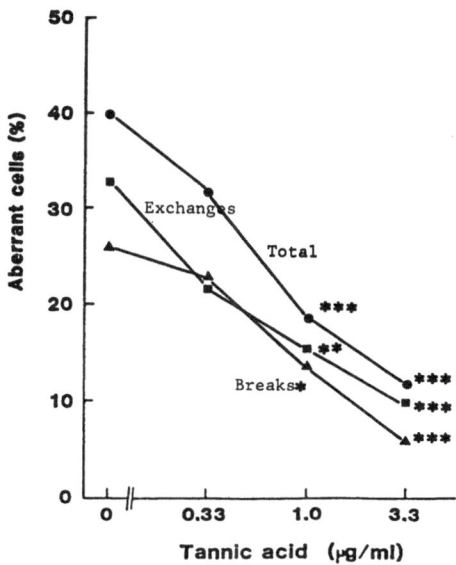

Fig. 13. Suppressing effect of tannic acid on UV-induced chromosome aberrations in cultured Chinese hamster ovary cells (36).

induced by EMS in V79 cells (16). In the presence of vitamin C, EMS-induced mutations were reduced to one-third to one-quarter. Dehydro-vitamin C and iso-vitamin C also decreased the EMS-induced mutations by one-half to one-third. Pretreatment with vitamin C was also effective in reducing EMS-induced mutations. Preincubation of a mixture of vitamin C and EMS has a strong effect in reducing mutations. This suggests that vitamin C was desmutagenic rather than bio-antimutagenic against induction of mutations by EMS. Vitamin A also reduced EMS-induced mutations by one-half to two-thirds. On the contrary, vitamin E enhanced three times the frequency of EMS-induced mutations (17). More detailed descriptions are given in another article by the author in this book.

CONCLUSION

Studies on antimutagenic factors, both desmutagens and bio-anti-mutagens, and their action mechanisms have been almost always carried out in microbial systems. Studies using cultured mammalian cell systems have been initiated only very recently, and there are still very few in Japan. Desmutagenic activity of vitamins and flavoring substances in mammalian cell systems has been detected as described here. Studies on the bio-antimutagenic activity of tannic acid were also carried out by using some repair-deficient mutant human cells.

The molecular and cellular mechanisms of antimutagenesis and anti-carcinogenesis will be more extensively examined in the future with these higher animal cells, including human cells. The common and extensive use of antimutagens and anticarcinogens in daily life will be the most effective procedure for preventing human cancer and genetic diseases.

ACKNOWLEDGEMENTS

    This review article was written as a memorial lecture in honor of my
colleague, Dr. Tsuneo Kada, Professor of the National Institute of Genetics, who initiated antimutagenesis studies in Japan. This review was
based on presentations and discussions in several scientific meetings held
by the Japanese Organizing Committee of the Second International Conference on Mechanisms of Antimutagenesis and Anticarcinogenesis. The
author expresses his deep gratitude to all members of the Committee.
This is contribution No. 1786 from the National Institute of Genetics,
Mishima, Japan.

REFERENCES

1. Arimoto, S., Y. Ohara, T. Namba, T. Negishi, and H. Hayatsu
    (1980) Inhibition of the mutagenicity of amino acid pyrolysis products
    by hemin and other biological pyrrole pigments. Biochem. Biophys.
    Res. Commun. 92:662–668.
2. Bazill, G.W., and J.D. Gross (1973) Mutagenic DNA polymerase in
    B. subtilis. Nature 243:241–243.
3. Clarke, C.H., and D.M. Shankel (1975) Antimutagenesis in microbial
    systems. Bacteriol. Rev. 39:33–53.
4. Inoue, T., K. Morita, and T. Kada (1978) Purification and proper-
    ties of a desmutagenic factor from plant (Brassica oleracea) for
    mutagenic principle of tryptophan pyrolysate. Ann. Rep. Natl.
    Inst. Genet. 28:79.
5. Inoue, T., and T. Kada (1979) A desmutagenic factor from plant
    (Brassica oleracea) for mutagenic principle of tryptophan pyrolysate
    has peroxidase and NADPH-oxidase activities. Ann. Rep. Natl.
    Inst. Genet. 29:66.
6. Inoue, T., K. Morita, and T. Kada (1981) Purification and proper-
    ties of a plant desmutagenic factor for the mutagenic principle of
    tryptophan pyrolysate. Agric. Biol. Chem. 45:345–353.
7. Inoue, T., Y. Ohta, Y. Sadaie, and T. Kada (1981) Effect of co-
    baltous chloride on spontaneous mutation induction in a Bacillus
    subtilis mutator strain. Mutat. Res. 91:41–45.
8. Inouye, T., Y.F. Sasaki, H. Imanishi, M. Watanabe, T. Ohta, and
    Y. Shirasu (1988) Suppression of mitomycin C-induced micronuclei in
    mouse bone marrow cells by post-treatment with vanillin. Mutat.
    Res. 202:93–95.
9. Kada, T., and N. Kanematsu (1978) Reduction of N-methyl-N'-nitro-
    N-nitrosoguanidine-induced mutations by cobalt chloride in Es-
    cherichia coli. Proc. Japan Acad. 54(Ser. B):234–237.
10. Kada, T., K. Morita, and T. Inoue (1978) Antimutagenic action of
    vegetable factor(s) on the mutagenic principle of tryptophan pyro-
    lysate. Mutat. Res. 53:351–353.
11. Kada. T. (1983) Environmental and biological factors suppressing in-
    duction of mutations (in Japanese). Toxicology Forum 6:580–589.
12. Kada, T., M. Kato, K. Aikawa, and S. Kiriyama (1984) Adsorption
    of pyrolysate by vegetable fibers. Mutat. Res. 141:149–152.
13. Kada, T., H. Mochizuki, and K. Miyao (1984) Antimutagenic effects
    of germanium oxide on Trp-P-2-induced frameshift mutations in Sal-
    monella typhimurium TA98 and TA1538. Mutat. Res. 125:145–151.
14. Kada, T., K. Kaneko, S. Matsuzaki, T. Matsuzaki, and Y. Hara
    (1985) Detection and chemical identification of natural bio-antimuta-
    gens. A case of the green tea factor. Mutat. Res. 150:127–132.

15. Komura, H., H. Minakata, K. Nakanishi, H. Mochizuki, and T. Kada (1981) Chemical features of antimutagenic factor in human placenta. Third International Conference on Environmental Mutagens (abstract), 3B 18, p. 83.

16. Kuroda, Y. (1987) Antimutagenic mechanism of vitamin C and its derivatives in mammalian cells in culture (abstract). Mutat. Res. 182:365.

17. Kuroda, Y. (1988) Mutagen-modifying effects of vitamin A and vitamin E in mammalian cells in culture (abstract). Mutat. Res. 203:377.

18. Little, J.W., and D.W. Mount (1982) The SOS regulatory system of Escherichia coli. Cell 29:11-22.

19. Mochizuki, H., and T. Kada (1982) Antimutagenic action of mammalian placental extracts on mutations induced in Escherichia coli by UV irradiation, γ-rays and N-methyl-N'-nitro-N-nitrosoguanidine. Mutat. Res. 95:457-474.

20. Morita, K., M. Hara, and T. Kada (1978) Studies on natural desmutagens: Screening for vegetable and fruit factors active in inactivation of mutagenic pyrolysis products from amino acids. Agric. Biol. Chem. 42:1235-1238.

21. Moriya, M., K. Kato, and Y. Shirasu (1978) Effects of cysteine and a liver metabolic activation system on the activities of mutagenic pesticides. Mutat. Res. 57:259-263.

22. Nagao, M., M. Honda, Y. Seino, T. Yahagi, T. Kawachi, and T. Sugimura (1977) Mutagenicities of protein pyrolysates. Cancer Lett. 2:335-340.

23. Negishi, T., and H. Hayatsu (1979) The enhancing effect of cysteine and its derivatives on the mutagenic activities of the tryptophan-pyrolysis products, Trp-P-1 and Trp-P-2. Biochem. Biophys. Res. Commun. 88:97-102.

24. Nishioka, H., K. Nishi, and K. Kyokane (1981) Human saliva inactivates mutagenicity of carcinogens. Mutat. Res. 85:323-333.

25. Nunoshiba, T., and H. Nishioka (1987) Sodium arsenite inhibits spontaneous and induced mutations in Escherichia coli. Mutat. Res. 184:99-105.

26. Obana, H., S. Nakamura, and R. Tanaka (1986) Suppressive effects of coffee on the SOS responses induced by UV and chemical mutagens. Mutat. Res. 175:47-50.

27. Ohta, T., K. Watanabe, M. Moriya, and Y. Shirasu (1983) Antimutagenic effects of cinnamaldehyde on chemical mutagenesis in Escherichia coli. Mutat. Res. 107:219-227.

28. Ohta, T., K. Watanabe, M. Moriya, Y. Shirasu, and T. Kada (1983) Analysis of the antimutagenic effect of cinnamaldehyde on chemically induced mutagenesis in Escherichia coli. Molec. Gen. Genet. 192:309-315.

29. Ohta, T., K. Watanabe, M. Moriya, Y. Shirasu, and T. Kada (1983) Anti-mutagenic effects of coumarin and umbelliferone on mutagenesis induced by 4-nitroquinoline 1-oxide or UV irradiation in E. coli. Mutat. Res. 117:135-138.

30. Ohta, T., M. Watanabe, R. Tsukamoto, Y. Shirasu, and T. Kada (1986) Antimutagenic effects of 5-fluorouracil and 5-fluorodeoxyuridine on UV-induced mutagenesis in Escherichia coli. Mutat. Res. 173:19-24.

31. Ohta, T., M. Watanabe, K. Watanabe, Y. Shirasu, and T. Kada (1986) Inhibitory effects of flavourings on mutagenesis induced by chemicals in bacteria. Fd. Chem. Toxicol. 24:51-54.

32. Onitsuka, S., N.V. Chanh, S. Murakawa, and T. Takahashi (1978) Desmutagenicity of several chemical compounds and vegetables on

some mutagens. Report of Special Research Projects (Environmental Sciences) of the Ministry of Education, Tokyo, Japan, A-2.

33. Osawa, T., H. Ishibashi, N. Namiki, and T. Kada (1980) Desmutagenic actions of ascorbic acid and cysteine on a new pyrrol mutagen formed by reaction between food additives: Ascorbic acid and sodium nitrite. Biochem. Biophys. Res. Commun. 95:835-841.

34. Sasaki, Y.F., H. Imanishi, T. Ohta, and Y. Shirasu (1987) Effects of vanillin on sister-chromatid exchanges and chromosome aberrations by mitomycin C in cultured Chinese hamster ovary cells. Mutat. Res. 191:193-200.

35. Sasaki, Y.F., H. Imanishi, T. Ohta, and Y. Shirasu (1987) Effects of antimutagenic flavourings on SCEs induced by chemical mutagens in cultured Chinese hamster cells. Mutat. Res. 189:313-318.

36. Sasaki, Y.F., H. Imanishi, T. Ohta, M. Watanabe, K. Matsumoto, and Y. Shirasu (1988) Suppressing effect of tannic acid on UV and chemically induced chromosome aberrations in cultured mammalian cells. Agric. Biol. Chem. 52:2423-2428.

37. Satoh, H., and T. Iwaguchi (1979) Antitumor activity of new novel organogermanium compound, Ge-132. Cancer and Chemotherapy Publishers, Tokyo, Japan, pp.79-83.

38. Sevastopoulos, C.G., and D.A. Glaser (1977) Mutator action by Escherichia coli strains carrying dnaE mutations. Proc. Natl. Acad. Sci., USA 74:3947-3950.

39. Shimoi, K., Y. Nakamura, I. Tomita, and T. Kada (1985) Bio-antimutagenic effects of tannic acid on UV and chemically induced mutagenesis in Escherichia coli B/r. Mutat. Res. 149:17-23.

40. Shimoi, K., Y. Nakamura, I. Tomita, Y. Hara, and T. Kada (1986) The pyrogallol related compounds reduce UV-induced mutations in Escherichia coli B/r WP2. Mutat. Res. 173:239-244.

41. Speyer, J.F. (1965) Mutagenic DNA polymerase. Biochem. Biophys. Res. Commun. 21:6-8.

42. Sugimura, T., T. Kawachi, M. Nagao, T. Yahagi, Y. Seino, T. Okamoto, K. Shudo, T. Kosuge, K. Tsuji, K. Wakabayashi, Y. Iitaka, and A. Itai (1977) Mutagenic principle(s) in tryptophan and phenylalanine pyrolysis products. Proc. Japan Acad. 53:58-61.

43. Takeda, H., and S. Kiriyama (1979) Correlation between the physical properties of dietary fibers and their protective activity against amaranth toxicity in rats. J. Nutrition 109:388-396.

44. Tokuue, N., Y. Kumai, H. Miyazaki, T. Yamaguchi, and S. Kusunoki (1986) The possibility of protecting carcinogenesis by proline. Environ. Mutagen. Res. Commun. 8(3):49.

45. Walker, C.C. (1984) Mutagenesis and inducible responses to deoxyribonucleic acid damage in Escherichia coli. Microbiol. Rev. 48:60-93.

46. Yamada, M., M. Tsuda, M. Nagao, M. Mori, and T. Sugimura (1979) Degradation of mutagens from pyrolysates of tryptophan, glutamic acid and globulin by myeloperoxidase. Biochem. Biophys. Res. Commun. 90:769-776.

CANCER PREVENTION: UNDERLYING PRINCIPLES

AND PRACTICAL PROPOSALS

Takashi Sugimura

National Cancer Center
Tsukiji, Chuo-ku, Tokyo, Japan

INTRODUCTION

The aim of this symposium on antimutagenesis and anticarcinogenesis is to obtain information on the problem of prevention of human cancer. Carcinogenesis is associated with genetic changes in somatic cells produced by mutagens and carcinogens. The molecular mechanisms of human carcinogenesis are very complicated: multiple genetic changes are often observed in established cancer cell lines, and these changes are generally more numerous in more malignant cancer cells. Completion of the carcinogenic process requires a long time, and precancerous changes often persist for many years.

This chapter is concerned with the importance of primary prevention of cancers. As necessary background information for considering ways of primary prevention of cancer, recent findings on the complex multiple genetic alterations observed in human cancers, new genotoxic substances in our daily life, new tumor promoters, and human precancerous changes are reported. Recommendations for improvement of lifestyle for prevention of cancer can be proposed on the basis of this information. Data given in this article were obtained mainly at the National Cancer Center.

BOTH PRIMARY AND SECONDARY PREVENTIONS OF CANCER ARE IMPORTANT

Although the object of this symposium is to consider ways of preventing development of cancer in humans, i.e., primary prevention of cancer, secondary prevention of cancer, meaning early diagnosis and treatment, is as important as primary prevention. The cancers occurring today result from exposures to environmental mutagens and carcinogens and from conditions under which the carcinogenic process has been promoted for several decades. But measures taken today for primary prevention of cancer may not be effective for suppressing the appearance of cancers tomorrow. Thus, early diagnosis is still very necessary for saving lives. As shown in Tab. 1, the success rate in treatment of early stomach cancer in Japan is almost 100%; but, of course, at progressive

Tab. 1.   Relationship of stage of diagnosis of stomach cancer with the
          cure rate.

Stage	Cure rate %
1	96.6
2	63.7
3	40.5
4	14.3

stages of cancer growth and invasion, the success rates in stomach cancer treatment decrease markedly (5).   An effective method for early diagnosis is screening persons without subjective symptoms by experienced
physicians.   In countries such as Japan, where the risk of stomach cancer is very high, examination of subjects who are over 40 years old by
double-contrast radiography is recommended.   Radiation exposure is being
minimized by using a sophisticated device such as digital radiography.
Early detections of breast cancers and uterine cervix cancers have also
proved very effective in saving lives.

Unfortunately, early diagnosis of some cancers such as pancreatic
cancers is practically impossible.   Therefore, primary prevention of cancer is very important, especially now, when second and third cancers
that are different from recurrence and metastasis of a first cancer often
develop anew in survivors from a first cancer.   The medical costs to be
saved by primary prevention of cancer are also enormous.

EPIDEMIOLOGICAL STUDIES SUGGEST THE IMPORTANCE OF ENVIRONMENTAL FACTORS AS CAUSES OF CANCER

The predominating cancers vary greatly in different parts of the
world: stomach cancer is most common in Japan, liver cancer in South-
east Asian countries, nasopharyngeal cancer in the southern part of
China, cancer of the oral cavity in India, and urinary bladder cancer in
Egypt.   These cancers are suspected to be caused by conditions related
to particular lifestyles or endemic diseases in these regions.   For instance, urinary bladder cancer in Egypt is related to infection with
Schistosoma mansoni, and cancer of the oral cavity in India is related to
the habit of chewing betel nut tobacco.   Moreover, in western countries,
the high incidences of breast and colon cancers are probably related to
excess intake of animal fat (2).

The greater importance of environmental factors than of genetic
background on the prevelance of different types of cancer is indicated by
the finding that the incidence of stomach cancer tends to be lower, and
that of colon cancer to be higher, in Japanese immigrants in Hawaii than
in Japanese residents in Japan (7).   Moreover, these changes in the patterns of incidence of cancer in Hawaii are greater in Japanese immigrants
of the second and third generations.   These changes result from a shift
from the traditional Japanese lifestyle to the western lifestyle.   Very
interesting data on occult and overt cancers of the prostate among

Japanese in Japan and in Hawaii are shown in Tab. 2 (22). The prevalence of latent prostate cancers is similar in the two populations, but the
incidence of overt and clinically problematic prostate cancers is 18 times
higher in Hawaii than in Japan. These data suggest that the lifestyle in
Hawaii favors the final step for manifestation of clinically active prostate
cancer.

Other epidemiological studies have also provided much important information on environmental carcinogenic and anticarcinogenic factors,
which should be considered in taking measures for preventing primary
cancer.

COMPLEXITY OF GENETIC ALTERATIONS FOUND IN HUMAN CANCERS

The genetic alterations found in cancer cells are listed in Tab. 3.
Several of these changes may be observed concurrently in monoclonal cancer cells. These changes do not necessarily occur in a fixed order.
Heritable genetic alteration leading to cancer at a later stage of life
should be present in all cells at birth, whereas genetic alteration produced by virus infection should occur immediately or some time after infection in an early period of life. Gene amplifications are frequently
observed in the advanced stage of malignant growth and are found in
metastatic foci (23). Increased cell divisions after cell death caused by
viral or bacterial infection and repeated tissue damage, and during tumor
promotion and chronic inflammation, may enhance the chance of genetic alterations caused by errors in replication of the genetic apparatus.

Activation of c-Ki-ras with conversion of the 12th amino acid glycine
to arginine was demonstrated in a case of highly malignant pancreatic
cancer which we reported in 1986 (21). This change was caused by conversion of the codon for glycine, GGT, to CGT. In this case, the mutated c-Ki-ras gene was also amplified about ten times, and in addition, the
c-myc gene was amplified more than ten times. Thus, in this cancer,
there were at least three genetic alterations (21). Recently, a c-Ki-ras
mutation at the 12th codon was demonstrated by Perucho and his colleagues in 21 of 22 cases of pancreatic cancer by the polymerase chain
reaction (1). Activation of c-Ki-ras may be necessary, but additional
multiple genetic changes may provide more growth advantages for pancreatic cancers.

Tab. 2.   Occult and overt prostate cancer among Japanese in Japan and
          Hawaii.

Country	Occult[a]	Overt[b]
Japan	20.5	2.3
Hawaii	25.6	40.9

a.   % of cases of over 50 years old (3)

b.   Incidence per 100,000 head of population according to WHO
     data in 1982.

Tab. 3. Genetic alterations found in human cancers.

```
Oncogene activation
  Point mutations
  Amplifications
  Insertions
  Translocations

Tumor suppressor genes
  Point mutations
  Deletions

Oncogenic viruses
  Integration
```

Gene amplifications are often found with other oncogenes besides Ki-ras and c-myc. The oncogenes amplified depend greatly on the organ in which cancer originates and on the histopathological type of cancer. We cloned a unique amplified oncogene from human stomach cancer by an in-gel renaturation method. The gene, which encodes a new receptor type of protein kinase, was designated as sam ("stomach cancer and amplified") (11). As shown in Tab. 4, amplification of sam was observed in many poorly-differentiated stomach cancers, while amplification of c-erbB-2 (neu gene) was observed in many well-differentiated stomach cancers (25).

The hst1 gene is a newly identified oncogene isolated from transformants of a mouse NIH 3T3 cell line after transfection of human stomach cancer DNA by Terada's group at the National Cancer Center. The hst1 gene product can bind heparin and is a growth factor for human fibroblasts and endothelial cells of umbilical cord blood vessels. The hst1 gene is located on chromosome band 11q13, about 50 kb from another oncogene, int2 (28). Co-amplifications of hst1 and int2 genes were observed in more than half the cases of esophageal cancer examined (Tab. 5), but were seldom found in other cancers, including stomach cancers.

The integration of hepatitis B virus into the cellular genome is responsible for the development of some hepatocellular carcinomas (20). Similarly, the integration of human papilloma virus may be related to the development of uterine cervix cancer (3). The integration of human papilloma virus, type 16 or 18, was detected in over half the Japanese

Tab. 4. Amplifications of sam and c-erbB-2 genes in stomach cancers.

	Poorly differentiated		Well differentiated	
	sam	c-erbB-2	sam	c-erbB-2
Fresh tumor	3/19	0/24	0/7	4/10
Xenografts	2/6	-	0/7	-
Cell lines	1/4	0/3	0/3	1/3
Total	6/29	0/27	0/17	5/13

Tab. 5. Co-amplification of hst1 and int2 genes in alimentary tract cancers.

	Frequency
Oesophageal squamous carcinomas	19/36 (52%)
Gastric adenocarcinomas	0/42 (0%)
Colorectal adenocarcinomas	0/52 (0%)

cases of uterine cervix cancer examined. In addition, loss of heterozygosity detected using probes of restriction enzyme fragment-length polymorphism (RFLP) was found exclusively on chromosome 3p in uterine cervix cancers, in which gene amplification of c-myc, c-Ki-ras, and c-Ha-ras (26) was rarely observed, as shown in Tab. 6.

In the cases of small-cell lung carcinoma examined, loss of heterozygosity was observed on chromosome 3p in all 7 cases, on 13q in 10 of 11 cases, and on 17p in all 5 cases examined (Tab. 7) (24). The 13q deletion was found to be in a range covering the RB (retinoblastoma) gene. Studies by Yokota's group at the National Cancer Center using a monoclonal antibody to a synthetic peptide of the carboxy-terminal region of the RB protein showed that this protein was always missing in small-cell lung carcinomas (27). Genetic alteration of the RB gene in somatic cells may be related to development of at least some common types of cancer.

From these findings it is evident that various types of cancers show a wide variety of genetic alterations involving oncogenes and antioncogenes.

NEW ENVIRONMENTAL MUTAGENS AND CARCINOGENS

The finding of the multiple genetic changes described above suggests

Tab. 6. Genetic alterations in uterine cervix cancers.

	Frequency

Integration of
  human papilloma virus
      type 16
      type 18
	
	11/30 (37%)
	6/30 (20%)

Gene amplification
  c-myc
  Ha-ras
	
	1/30 (3%)
	0/18 (0%)

Chromosome deletion
  3p
  11p
  11q
	
	9/9 (100%)
	0/8 (0%)
	0/15 (0%)

Tab. 7. Loss of heterozygosity on chromosomes 3p, 13q, and 17p loci in
         human lung cancer.

Type of tumor	Loss of heterozygosity		
	3p	13q	17p
Small cell carcinoma	7/7	10/11	5/5
Adenocarcinoma	5/6	4/15	2/7
Squamous cell carcinoma	0/1	4/11	1/4
Large cell carcinoma	0/1	1/3	0/0

the existence of numerous, as yet unidentified environmental mutagens
and carcinogens. In fact, we have discovered a series of heterocyclic
amines produced in meat and fish during cooking, especially broiling and
prolonged heating. Some examples are given in Fig. 1 (15). These imi-
dazoquinoline (IQ) and imidazoquinoxaline (IQx) derivatives are produced
from creatinine, sugar, and amino acids in meat. Their production involves a Maillard reaction, and creatinine provides the imidazo-moiety of
IQ and IQx derivatives (9). All these mutagenic heterocyclic amines were
shown to be carcinogenic in long-term animal experiments (14).
Phenylimidazopyridine (PhIP), which was identified by Felton et al. (4),
is present at a higher concentration than other heterocyclic amines in

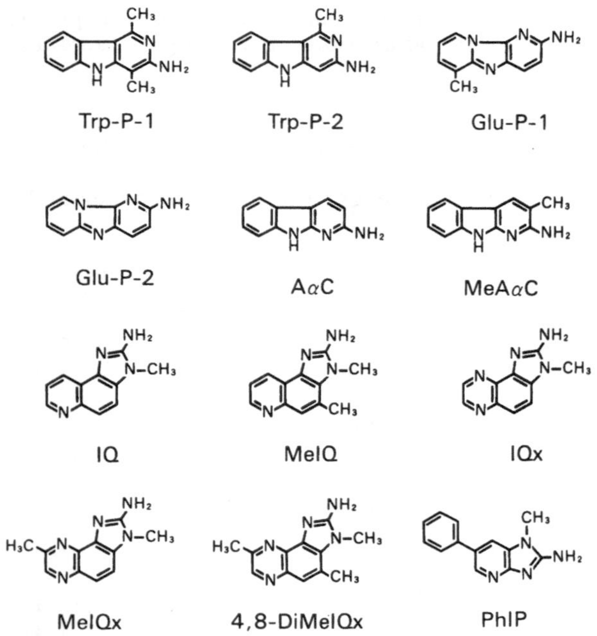

Fig. 1. Structure of mutagenic heterocyclic amines produced during
         cooking of meat and fish.

cooked food. When given in the diet of rats, PhIP formed adducts most efficiently with DNA of the pancreas, suggesting that it is a pancreatic carcinogen (19).

In addition to chemical carcinogens, active oxygen molecules are formed continuously in the human body. Oxygen radicals produce strand scissions of DNA, and thymine glycol and hydroxymethyldeoxycytidine are known to be oxidative products of DNA bases. Recently, Nishimura's group at the National Cancer Center demonstrated the formation of 8-hydroxydeoxyguanosine in DNA by a variety of oxidative agents. A polydeoxyribonucleotide template containing 8-hydroxydeoxyguanosine in place of a deoxyguanosine residue was synthesized. The fidelity of DNA replication catalyzed by the Klenow fragment of DNA polymerase of Escherichia coli was found to be lost at the locus of the 8-hydroxydeoxyguanosine and at the 3' and 5' adjacent positions, resulting in incorporation of four nucleotides with almost equal frequency (10).

The above chemicals are only two examples of many mutagens and carcinogens occurring in our daily life.

NEW TUMOR PROMOTERS

Tumor promoters are understood to be important In the development of human cancers. Initiation and subsequent promotion steps may be essential, if not sufficient, for development of true malignant tumors. Progression steps, induced by mutagens and carcinogens, are also very important for full transformation to cancer cells. Cancer cells can be converted further to cells with more malignancy, more resistance to cancer drugs, and higher metastatic potentials. For the last 30 years, 12-O-tetradecanoylphorbol-13-acetate (TPA) has been the only tumor promoter available. But by use of the mouse ear irritation test, as well as tests on induction of ornithine decarboxylase on mouse skin and induction of adhesion of cultured HL-60 cells as screening tests for tumor promoters, new tumor promoters were isolated by Fujiki's group at the National Cancer Center (6). These promoters were teleocidines A and B from Streptomyces, and lyngbyatoxin A and aplysiatoxins from the marine blue-green alga, Lyngbya majuscula. Aplysiatoxin is also stored in the midgut gland of the sea-hare, Stylocheilus longicauda.

The tumor-promoting potencies of these compounds in two-step carcinogenesis in mouse skin are the same as that of TPA. These compounds bind to the same receptors as TPA and, like TPA, activate protein kinase C (6). Their structures are shown in Fig. 2. Although two-dimensional structures of these compounds differ, computer-graphic analysis showed that they had some common stereostructural characteristics (8).

Recently, okadaic acid (OA) was isolated from a black sponge, Halichondria okadae. The structure of OA is shown in Fig. 3. Figure 4 summarizes data obtained in a two-stage carcinogenesis experiment in which limited amounts of 7,12-dimethylbenz(a)anthracene (DMBA) and OA were painted on mouse skin. DMBA or OA alone produced no tumors, but a combination of DMBA and OA produced many tumors. OA cannot bind to phorbol ester receptors or activate protein kinase C (13), but it was reported to inhibit protein phosphatase (17). Thus, OA may cause tumor promotion by inhibiting protein-serine-phosphatase, with a consequent increase in phosphorylation of serine of key proteins.

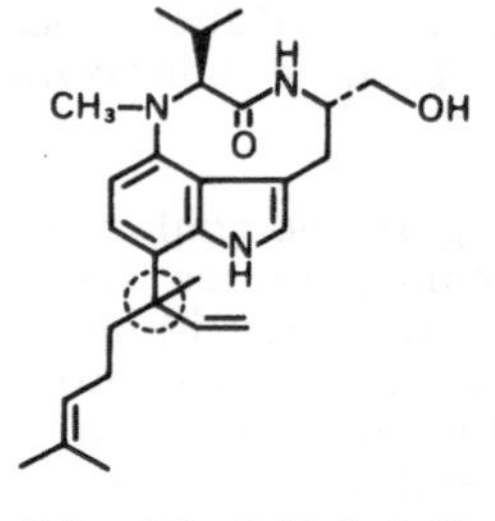

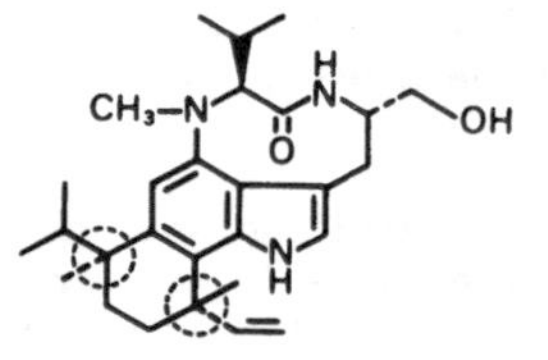

Teleocidin B
(B-1,B-2,B-3,B-4)

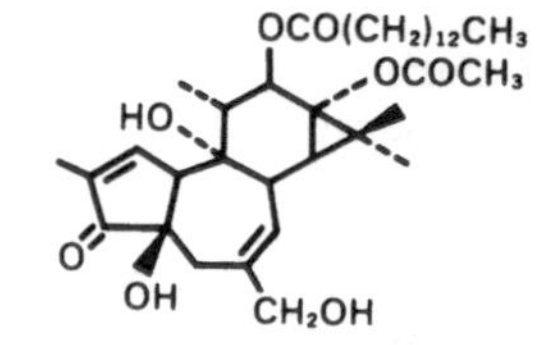

TPA

Teleocidin A (A-1,A-2)
Lyngbyatoxin A

Fig. 2.   Structures of tumor promoters, teleocidines and TPA.

Many other tumor promoters are present in the human environment. Sodium chloride is suspected to be a tumor promoter for stomach carcinogenesis (16). Excess fat intake is positively correlated with production of colon cancers, breast cancers, and pancreatic cancers. Excess fat intake may promote tumor development by increasing bile formation and/or altering hormonal conditions. Cigarette smoke contains many types of tumor promoters and initiators, so stopping smoking results in an immediate decrease in lung cancer development. Promotion or progression steps are probably the most realistic targets to consider in developing strategies for primary cancer prevention.

RECOMMENDATIONS FOR IMPROVEMENTS IN LIFESTYLE
FOR PRIMARY PREVENTION OF CANCER

The levels of most environmental carcinogens, including heterocyclic amines in ordinarily cooked foods, are much lower than those required to induce cancers in experimental animals. However, as discussed above, the carcinogenic process is very complicated, and cancer development is the final outcome of multiple, sequential genetic changes.

In the human body, there are many precancerous states such as leukoplakia of the tongue and oral mucosa membrane induced by chronic mechanical irritation, liver cirrhosis caused by hepatitis B virus infection and intake of excess alcohol, and intestinalization of gastric mucosa resulting from recurrent gastritis. Therefore, it is very hard to determine the actual risks of carcinogenic factors to humans from results obtained in experiments using healthy young rats or mice which lack any pathological condition. Many methods for estimating these risks have been proposed, but some of them are too simple, such as, for instance, the ratio of the TD_{50} value in rats to the daily intake in humans. Thus, at the moment,

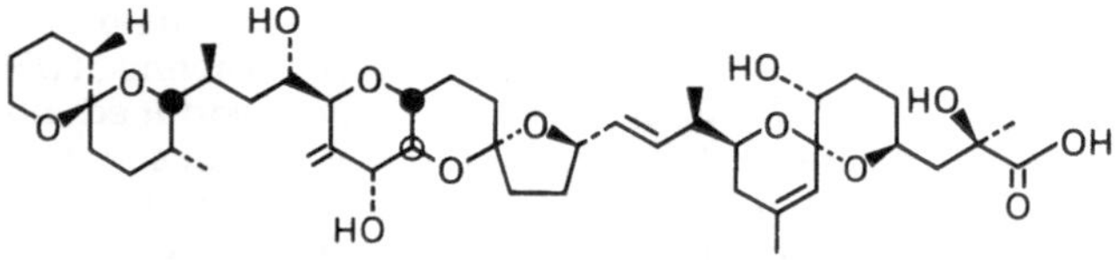

Fig. 3.   Structure of okadaic acid.

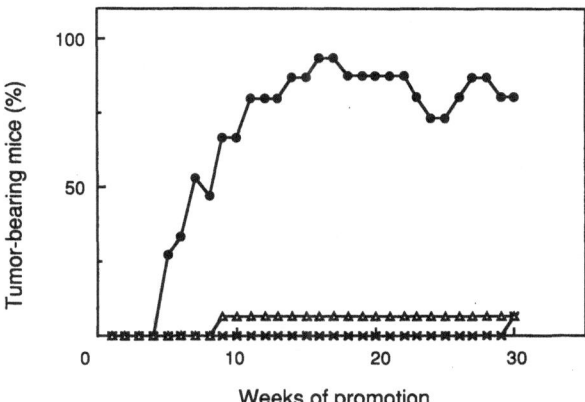

Fig. 4.   Tumor-promoting activities of okadaic acid and teleocidine. Mice
were treated with DMBA and okadaic acid (●), with DMBA alone
(X), and with okadaic acid alone (Δ). To the skin of the back
of 8-wk-old female CD-1 mice, 100 μg of DMBA in 0.1 ml of acetone was applied. After 1 wk, 10 μg of okadaic acid in 0.1 ml
of acetone was applied twice a week until week 30.

quantitative estimations of the carcinogenic risks of chemicals and factors
are difficult. Thus it seems more reasonable and practical to suggest
that exposure to any kind of mutagen or carcinogen at any level and to
any condition that favors carcinogenesis should be avoided, if this is possible without serious disturbance of daily life.

Much still needs to be understood about cancer prevention. However, we are seeing changes in cancer patterns as a function of changes
in lifestyle. It should be noted that some types of cancers have already
disappeared. Iatrogenic cancers due to radiation exposure, which were
found several decades ago among X-ray workers in hospitals, have disappeared. Cleaner and more hygienic conditions have almost eradicated
penile cancers in Japan. Moreover, the age-adjusted mortality from stomach cancer in Japan is declining very quickly, probably as the result of
changes in dietary conditions and progress in early diagnosis and treatment of this type of cancer (5). Thus the incidences of some cancers are
diminishing.

The 12 points for primary cancer prevention that we have recommended previously are given in Tab. 8 (14). After collecting and considering available information obtained in laboratory and epidemiology studies, we tried to propose modest recommendations that could be adopted by
many people without provoking serious antagonistic reactions. Of these
proposals, stopping smoking should have the highest impact in cancer
prevention. These proposals are in accordance with those made by the
American Cancer Society and other organizations.

Finally, it should be pointed out that these measures would also reduce hypertension/strokes resulting from less intake of sodium chloride,
and cardiovascular diseases/heart attacks resulting from less intake of
animal fat. Many environmental mutagens can cause other changes
besides malignant cancer formation. Even some of the newly isolated

Tab. 8.   Twelve points for cancer prevention.

---

(1) Have a nutritionally balanced diet.

(2) Eat a variety of types of food.

(3) Avoid excess calories, especially as fat.

(4) Avoid excessive drinking of alcohol.

(5) Smoke as little as possible.

(6) Take vitamins in appropriate amounts and eat fiber and green
    and yellow vegetables rich in carotene.

(7) Avoid drinking fluids that are too hot and eating foods
    that are too salty.

(8) Avoid the charred parts of cooked food.

(9) Avoid food possibly contaminated by fungal toxins.

(10) Avoid over-exposure to sunlight.

(11) Have an exercise program matched to the individual's
     condition.

(12) Keep the body clean.

---

heterocyclic amines can produce atrophy of the salivary glands and pancreas (18).   Dimethylbenz(a)anthracene can produce the growth of smooth
muscle cells beneath the endothelium of the aorta (12).   Decreased exposure to environmental mutagens/carcinogens and greater exposure to anti-
mutagens/anticarcinogens should be beneficial for reducing age-related
deleterious changes as well as cancers.

ACKNOWLEDGEMENT

The author is very much indebted to the staff of the National Cancer
Center for their cooperation.

REFERENCES

1. Almoguera, C., D. Shibata, K. Forrester, J. Martin, N. Arnheim,
    and M. Perucho (1988) Most human carcinomas of the exocrine pan-
    creas contain mutant c-K-ras genes.   Cell 53:549-554.
2. Doll, R. (1977) Strategy for detection of cancer hazards to man.
    Nature (London) 265:589-596.
3. Durst, M., E. Schwarz, and L. Gissmann (1986) Integration and
    persistence of human papilloma virus DNA in genital tumors.   In
    Viral Etiology of Cervical Cancer, Banbury Report 21, R. Peto and
    H. zur Hausen, eds.   Cold Spring Harbor Laboratory, Cold Spring
    Harbor, New York, pp. 273-280.
4. Felton, J.S., M.G. Knize, N.H. Shen, P.R. Lewis, B.D. Andresen,
    J. Happe, and F.T. Hatch (1986) The isolation and identification of
    a new mutagen from fried ground beef: 2-Amino-1-methyl-6-phenyl-
    imidazo[4,5-b]pyridine (PhIP).   Carcinogenesis 7:1081-1086.

5. Foundation for Promotion of Cancer Research (1987) Figures on Cancer in Japan--1987, FPCR, Tokyo, 10 pp.
6. Fujiki, H., and T. Sugimura (1987) New classes of tumor promoters: Teleocidine, aplysiatoxin, and palytoxin. Adv. Cancer Res. 49:223-264.
7. Hirohata, T. (1979) Shift in cancer mortality from 1920 to 1970 among various ethnic groups in Hawaii. In Genetic and Environmental Factors in Experimental and Human Cancer, H.V. Gelboin et al., eds. Japan Science Societies Press, Tokyo, pp. 341-350.
8. Itai, A., Y. Kato, N. Tomioka, Y. Iitaka, Y. Endo, M. Hasegawa, K. Shudo, H. Fujiki, and S. Sakai (1988) A receptor model for tumor promoters: Rational superposition of teleocidines and phorbol esters. Proc. Natl. Acad. Sci., USA 85:3688-3692.
9. Jagerstad, M., K. Olsson, S. Grivas, C. Negishi, K. Wakabayashi, M. Tsuda, S. Sato, and T. Sugimura (1984) Formation of 2-amino-3,8-dimethylimidazo[4,5-f]quinoxaline in a model system by heating creatinine, glycine and glucose. Mutat. Res. 126:239-244.
10. Kuchino, Y., F. Mori, H. Kasai, H. Inoue, S. Iwai, K. Miura, E. Ohtsuka, and S. Nishimura (1987) Misreading of DNA templates containing 8-hydroxydeoxyguanosine at the modified base and at adjacent residues. Nature 327:77-79.
11. Nakatani, H., E. Tahara, T. Yoshida, H. Sakamoto, T. Suzuki, H. Watanabe, M. Sekiguchi, Y. Kaneko, M. Sakurai, M. Terada, and T. Sugimura (1986) Detection of amplified DNA sequences in gastric cancers by a DNA renaturation method in gel. Jpn. J. Cancer Res. (Gann) 77:849-853.
12. Penn, A., G. Batastini, J. Soloman, F. Burnsand, and R. Albert (1981) Dose-dependent size increases of aortic lesions following chronic exposure to 7,12-dimethylbenz(a)anthracene. Cancer Res. 41:588-592.
13. Suganuma, M., H. Fujiki, H. Suguri, S. Yoshizawa, M. Hirota, M. Nakayasu, M. Ojika, K. Wakamatsu, K. Yamada, and T. Sugimura (1988) Okadaic acid: An additional non-phorbol-12-tetradecanoate-13-acetate-type tumor promoter. Proc. Natl. Acad. Sci. USA 85:1768-1771.
14. Sugimura, T. (1986) Studies on environmental chemical carcinogenesis in Japan. Science 233:312-318.
15. Sugimura, T. (1988) Successful use of short-term tests for academic purpose: Their use in identification of new environmental carcinogens with possible risk for humans. Mutat. Res. 205:33-39.
16. Takahashi, M., and R. Hasegawa (1986) Enhancing effects of dietary salt on both initiation and promotion stages of rat gastric carcinogenesis. In Diet, Nutrition and Cancer, Y. Hayashi et al., eds. Japan Science Societies Press, Tokyo/VNU Science Press, Utrecht, pp. 169-182.
17. Takai, A., C. Bialojan, M. Troschka, and J.C. Ruegg (1987) Smooth muscle myosin phosphatase inhibition and force enhancement by black sponge toxin. FEBS Lett. 217:81-84.
18. Takayama, S., Y. Nakatsuru, H. Ohgaki, S. Sato, and T. Sugimura (1985) Atrophy of salivary glands and pancreas of rats fed on diet with amino-methyl-α-carboline. Proc. Japan Acad. 61(B):277-280.
19. Takayama, K., K. Yamashita, K. Wakabayashi, M. Nagao, and T. Sugimura (1989) Formation of PhIP-DNA adduct in F344 rats. Mutat. Res. (in press).
20. Tillaris, P., C. Pourcel, and A. Dejean (1985) The hepatitis B virus. Nature 317:489-495.

21. Yamada, H., H. Sakamoto, M. Taira, S. Nishimura, Y. Shimosato, M. Terada, and T. Sugimura (1986) Amplifications of both c-Ki-ras with a point mutation and c-myc in primary pancreatic cancer and its metastatic tumors in lymph nodes. Jpn. J. Cancer Res. (Gann) 77:370-375.

22. Yatani, R., I. Chigusa, K. Akazaki, G.N. Stemmerman, R.A. Welsh, and P. Correa (1982) Geographic pathology of latent prostatic cancer. Int. J. Cancer 29:611-616.

23. Yokota, J., Y. Tsunetsugu-Yokota, H. Battifora, C. LeFevre, and M.J. Cline (1986) Alterations of myc, myb and ras Ha proto-oncogenes in cancers are frequent and show clinical correlation. Science 231:261-265.

24. Yokota, J., M. Wada, Y. Shimosato, M. Terada, and T. Sugimura (1987) Loss of heterozygosity on chromosomes 3, 13 and 17 in small-cell carcinoma and on chromosome 3 in adenocarcinoma of the lung. Proc. Natl. Acad. Sci., USA 84:9252-9256.

25. Yokota, J., T. Yamamoto, N. Miyajima, K. Toyoshima, N. Nomura, H. Sakamoto, T. Yoshida, M. Terada, and T. Sugimura (1988) Genetic alterations of the c-erbB-2 oncogene occur frequently in tubular adenocarcinoma of the stomach and are often accompanied by amplification of the v-erbA homologue. Oncogene 2:283-287.

26. Yokota, J., Y. Tsukada, T. Nakajima, M. Gotoh, Y. Shimosato, N. Mori, Y. Tsunokawa, T. Sugimura, and M. Terada (1989) Loss of heterozygosity on the short arm of chromosome 3 in carcinoma of the uterine cervix. Cancer Res. (in press).

27. Yokota, J., T. Akiyama, Y.-K. Fung, W.F. Benedict, Y. Namba, M. Hanaoka, M. Wada, T. Terasaki, Y. Shimosato, T. Sugimura, and M. Terada (1988) Altered expression of the retinoblastoma (RB) gene in small-cell carcinoma of the lung. Oncogene 3:471-475.

28. Yoshida, M.C., M. Wada, H. Satoh, T. Yoshida, H. Sakamoto, K. Miyagawa, J. Yokota, M. Kakinuma, T. Sugimura, and M. Terada (1987) Human HST1 (HSTF1) gene maps to chromosome band 11q13 and coamplifies with the INT2 gene in human cancer. Proc. Natl. Acad. Sci., USA 84:2980-2984.

ANTICLASTOGENIC DIETARY FACTORS ASSESSED

IN MAMMALIAN ASSAYS

Heinz W. Renner

Federal Research Centre for Nutrition
Institute of Hygiene and Toxicology
D-7500 Karlsruhe, Federal Republic of Germany

INTRODUCTION

Epidemiological investigations resulted in the formulation of a hypothesis that could be tested in the laboratory, namely, high levels of dietary fat increase the risk of breast cancer. Dietary guidelines with respect to fat and cancer are usually expressed in terms of the percentage of total calories derived from fat. In this way we obtain a positive correlation for the incidence of breast cancer deaths when we compare the daily fat intake in the population of different countries. In the diet of most western industrialized countries, fat consumption comprises about 40% of total daily calories, whereas dietary guidelines recommend only 30%. When, instead of total fat intake, only the percentage of calories from polyunsaturated fatty acids is taken into consideration, no correlation between age-adjusted mortality from breast cancer and polyunsaturated fatty acids is apparent (3). On the other hand, animal experiments suggest that polyunsaturated fatty acids under a low fat regime contribute more to tumorigenesis than do saturated fatty acids (7).

Results of further epidemiological studies have also shown a connection between dietary fat and the cancer frequency in other organs, particularly the prostate and colon. In general, it is not possible in this way to detect which specific constituents of fat are clearly responsible for the observed effects. Is it actually a question of fat itself or more a question of calories from fat? While the mechanisms which lead to the development of tumors are still quite unclear, most of the data seem to indicate that dietary fat exerts a promotor effect rather than an initiation effect. Despite the apparent discrepancy between epidemiological and experimental data, a connection between fat consumption and cancer risk is considered very probable.

The studies on which I want to report here were designed to examine the question of possible relationships between fatty acids and mutations. Where relationships between fat and cancer are quite apparent, it should be possible to demonstrate a connection between fat or fatty acids and mutagenicity. Why has so little importance been attached to this aspect until now? I would like to quote some figures here: a catalogue of publications in the Oak Ridge Center, from the years 1981-1987, in the area of mutagenesis- carcinogenesis listed about 30,000 titles; 4% of these could be said to deal with anticarcinogenesis and 0.5% only with antimutagenesis.

The subject of fat mutagenesis has been studied by only a few authors, especially when experiments under in vivo conditions are considered. Surprisingly, however, these authors have reported on antimutagenic properties of fatty acids or plant oils on mutagens. These authors (5,6,17) used the micronucleus test with mice.

ANTICLASTOGENIC EFFECTS OF FATTY ACIDS

We directed our attention first of all to the biologically important fatty acids, saturated or unsaturated, or polyunsaturated fatty acids. We administered fatty acids in their methyl ester form via stomach tube to Chinese hamsters, using liquid paraffin as a solvent. To test that the fatty acids themselves have no effect, we took bone marrow cells after various treatment times from the animals and analyzed the chromosomes. They were in no way different from those of the untreated control. Immediately after the administration of the fatty acid methyl ester we applied a powerful mutagen and tested whether any modulating effects could be observed. The mutagen was busulfan (1,4-butandiolbismethane sulfonate). When both substances--the fatty acid ester and the mutagen--affected the organism for 30 hours after oral application, the following cytogenetic result occurred: all types of structural aberrations in-

Tab. 1.   Aberration spectrum for clastogenic (busulfan) and anticlastogenic activity (Chinese hamsters, bone marrow cells).

Treatment (mg/kg p.o.)	Gaps	Chromatid breaks	Chromosome breaks	Translocations	Multiple breaks
50 mg of busulfan	33.5	6.3	1.8	1.0	0.3
50 mg of busulfan + 100 mg of linoleic fatty acid ester	24.5	2.0	1.0	0.2	0
Untreated control	0.2	0.3	0	0	0

Numbers represent % of aberrant metaphases

duced by the mutagen were reduced by the methyl ester of the fatty acid linoleic acid. The percentage of damaged cells dropped from about 9% to 3% (Tab. 1) (18).

In such bioassays the question of suitable timing arises. When should we check for antimutagenic actions? The primary requirement is to establish the time of maximum response of the mutagen and then to follow up the anticlastogenic action. In the example with busulfan and linoleic acid ester the highest response of the mutagen and the strongest action of the antimutagen coincided at 30 hours treatment time prior to cell sampling.

The next question is whether a dose relationship exists for the anticlastogenic effect of the fatty acid ester. We conducted experiments with the mutagen busulfan and linoleic acid ester using doses of fatty acids from 5-5,000 mg/kg, so it could be demonstrated that a threshold of between 5 and 10 mg/kg body weight is necessary to achieve the anticlastogenic effect. This means less than 1 g of linoleic acid per man and day, while about 10 g of this essential fatty acid are normally ingested per day (18).

We assume that these antimutagenic effects originate from the fatty acids themselves. Dietary fats are hydrolyzed and resorbed in the wall of the small intestine. Behind the intestinal wall the organism creates new compounds from the fatty acids. To verify this hypothesis, we administered the fatty acids to the animals in various forms: (a) as pure fatty acids; (b) as methyl or ethyl ester; (c) as sodium salt; and (d) as triglyceride ester. In all these tests an anticlastogenic action on the mutagen busulfan was evident, when appropriate treatment times are considered.

Do these experiments demonstrate that fatty acids, such as those consumed by us daily in plant oils, have the same modulating effects on our human organism?

In further experiments we administered corn oil along with the mutagen, as was done in the previous example with the methyl ester of the fatty acids. We then determined the proportion of aberrant chromosomes. It was as high as in the positive control. We then lengthened the time gap between corn oil treatment and mutagen treatment by 1 day at a time. When the corn oil pretreatment was at least 3, 4, or 5 days, then again the anticlastogenic effect was clearly recognizable (Fig. 1). We therefore believe that we are quite justified in stating that corn oil--the main constituent of which is linoleic acid--is a true dietary antimutagen. The critics may claim that this is merely a specific interaction between the fatty acids and busulfan. We can, however, prove that fatty acids such as linoleic acid or alpha-linolenic acid are qualified for a considerable reduction of chromosomal damage induced by thiotepa as well as by the indirectly-acting mutagen cyclophosphamide.

What is the reason behind these findings? What are the possible mechanisms which give rise to this phenomenon? A number of authors (13,20,22) provided evidence that dietary lipid is essential for maximum activity of detoxifying enzymes in the liver. But comparable information on short-term treatment with fatty acids is not available.

At this point I should like to emphasize once again that we were able to demonstrate in animal experiments the anticlastogenic action with a

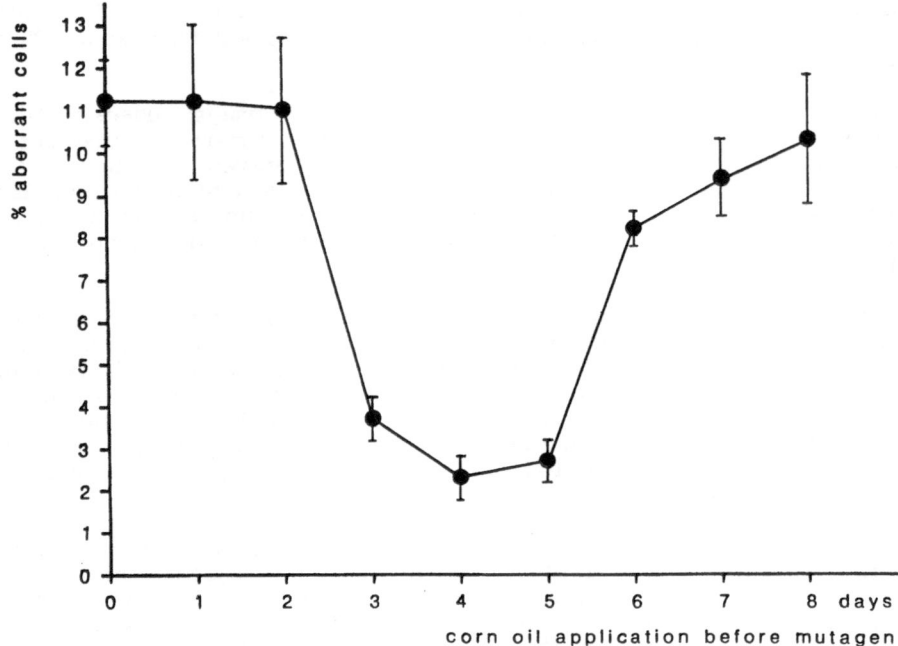

Fig. 1.   Anticlastogenic effect of corn oil on busulfan-induced chromo-
          some aberrations (50 mg busulfan/kg; 5 ml corn oil/kg p.o.)
          (Chinese hamsters, bone marrow cells).

series of biologically important fatty acids, not only with the linoleic acid.
I would like to remind you that the nutritional sciences describe linoleic
acid and alpha-linolenic acid as essential fatty acids.   These two fatty
acids are the parent fatty acids of the omega-6 and the omega-3 fatty
acid family.   This means that both fatty acids form derivatives which
are important for a large number of functions in the human or mammalian
organism.

     Using the same experimental design as described above, this time we
tested the derivatives of linoleic acid and of alpha-linolenic acid (Tab 2).
The anticlastogenic action of the parent fatty acid on busulfan was known
to us.   This same effect is apparent with all the polyunsaturated fatty
acids derived from alpha-linolenic acid (omega-3 fatty acids).   On the
other hand, antimutagenic action is not achieved with any derivative of
linoleic acid (omega-6 fatty acids) (19).

EVALUATION AND INTERPRETATION OF THE ANTICLASTOGENIC
EFFECTS OF FATTY ACIDS

     The presented data suggest that some biologically important fatty
acids bear a remarkable antimutagenic potential.   The detection of specific
activities of the omega-3 fatty acids calls attention to the eicosanoids.
These result from the parent fatty acids linoleic- and alpha-linolenic acid
by way of desaturation and chain elongation.   A very important 20 C-

compound in the omega-6 family is the arachidonic acid. The enzyme cyclooxygenase catalyzes the first reaction steps of the arachidonic acid toward the biologically highly effective prostaglandins as well as thromboxanes and leucotrienes—substances with manifold effects in various organs of the body. Arachidonic acid has a counterpart in the omega-3 family, namely eicosapentaenoic acid (EPA).

Figure 2 illustrates that in the omega-6 group particular prostanoids are formed, namely the series 1 and series 2 of prostaglandins, while in the omega-3 group the series 3 of prostaglandins from the EPA are created.

Relationships between prostaglandins and cancer have often been observed. Increased prostaglandin levels have been found in the blood and urine of animals carrying neoplasma as well as in transformed cells growing in tissue culture. The prostaglandins of series 2 are mainly thought to be implicated in tumorigenesis. For the conversion of arachidonic acid to prostaglandins of series 2, EPA is an effective antagonist just as the other important omega-3 fatty acid, the docosahexaenoic acid (DHA) is a strong inhibitor (4,11).

If the organism receives EPA and DHA along with the diet, a switch of the balance in prostaglandin metabolism takes place, with the result that fewer eicosanoids from the arachidonic acid are formed. For example, in the organism the arachidonic acid-derived prostaglandins are displaced in the liver and spleen phospholipids of the cell membrane of mice and rats by EPA and DHA (23,24). Transplanted mammary tumors and

Tab. 2.   Anticlastogenic action of linoleic- and α-linolenic acid methyl esters and their derivatives on busulfan-induced chromosome aberrations (Chinese hamsters, bone marrow cells).

	Aberrant metaphases % ± S.D.		Aberrant metaphases* % ± S.D.
Linoleic acid 18:2, n-6 $\Delta^{9,12}$-octadecadienoic acid	3.2 ± 0.8	α-Linoleic acid 18:3, n-3 $\Delta^{9,12,15}$-octadecatrienoic acid	3.5 ± 1.1
γ-Linolenic acid 18:3, n-6 $\Delta^{6,9,12}$-octadecatrienoic acid	7.0 ± 3.2	18:4, n-3 $\Delta^{6,9,12,15}$-octadecatetraenoic acid	2.8 ± 0.8
20:2, n-6 $\Delta^{11,14}$-Eicosadienoic acid	6.0 ± 2.8	20:3, n-3 $\Delta^{11,14,17}$-eicosatrienoic acid	3.2 ± 0.8
Dihomo-γ-linolenic acid 20:3, n-6 $\Delta^{8,11,14}$-eicosatrienoic acid	9.8 ± 0.4	20:4, n-3 $\Delta^{8,11,14,17}$-eicosatetraenoic acid	NA**
Arachidonic acid 20:4, n-6 $\Delta^{5,8,11,14}$-eicosatetraenoic acid	9.0 ± 0.9	20:5, n-3 $\Delta^{5,8,11,14,17}$-eicosapentaenoic acid	3.0 ± 1.4
22:4, n-6 $\Delta^{7,10,13,16}$-docosatetraenoic acid	10.2 ± 1.8	22:6, n-3 $\Delta^{4,7,10,13,16,19}$-docosahexaenoic acid	3.3 ± 0.5
Untreated control	0.4		
Positive control	9.4 ± 1.0		

*Excluding gaps.
**Not available.

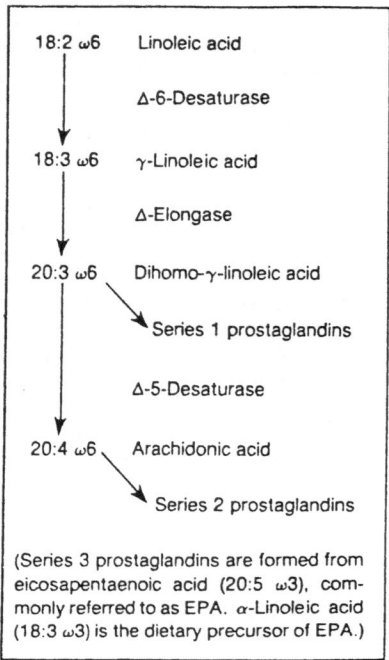

Fig. 2. Origin of different prostaglandins (from Ref. 12).

7,12-dimethylbenzanthracene-induced mammary tumors in rats were reduced in development and growth by EPA and DHA (9,10); it was also
reported that transplantable prostatic tumors were reduced in nude mice
and intestinal carcinogenesis in rats (9,16).   The authors of these investigations stated that the mechanism underlying inhibition of mammary
tumorigenesis may be linked to the inhibitory effect of EPA and DHA on
arachidonic acid metabolism.   Similar effects could be shown in rats with
indomethacin, an inhibitor of the cyclooxygenase pathway of arachidonic
acid, after dietary supplementation of fish oil rich in EPA and DHA.
Furthermore, dietary EPA exerted inhibitory effects on azoxymethane-
induced colon tumors in rats by modulating lipid metabolism and inhibiting
prostaglandin E 2 synthesis (14).

When considering the corresponding in vitro studies, particular attention should be given to investigations in which the ability of certain
polyunsaturated fatty acids to kill human carcinoma cells and leave normal
cells unharmed is demonstrated.   Both omega-6 fatty acids and omega-3
fatty acids exhibited this phenomenon without any discrimination.   It was
suggested that the effectiveness of a given fatty acid in killing cancer
cells correlated with the extent of lipid peroxidation of the added fatty
acid in the cells (2).

Tab. 3. Differences in morbidity from chronic diseases between Greenland Eskimos and Scandinavians (from Ref. 21).

Disease	Eskimos/ Scandinavians
Stroke	2/1
Acute myocardial infarction	1/10
Psoriasis	1/20
Diabetes	Rare
Bronchial asthma	1/25
Thyrotoxicosis	Rare
Multiple sclerosis	0
Epilepsy	2/1
Polyarthritis	Low

Possible Connections with Epidemiological Facts

Our investigations have revealed the ability of the living organism to differentiate between the anticlastogenic actions of omega-6 and omega-3 fatty acids. Possible connections between the inhibition of carcinogenesis in different organs by omega-3 fatty acids and our results are considered likely. I should like, therefore, to draw attention to some epidemiological data. Such data have revealed a very low rate of breast cancer in the Eskimo women of Greenland in contrast to those in Denmark--although the total fat intake is approximately equal in these two populations. Differences in the composition of dietary fat in Greenland and in Denmark are, however, obvious and could be responsible for the reported phenomenon: the amount of fish oil and therefore of omega-3 polyunsaturated fatty acids in Eskimo nutrition is high in contrast to Danish nutrition, with its high content of saturated dairy fats (1,15).

Similar dietary differences between the inhabitants of Japanese fishing villages and Japanese farming villages were described in 1980 (8). The authors drew a parallel between these differences and the relatively low incidence of cardiovascular thrombotic disease in Japanese fishermen. The ratio of eicosapentaenoic acid to arachidonic acid in blood plasma is 0.5 for Japanese, 0.1 for Europeans, and 7.0 for Eskimos.

The compilation shown in Tab. 3 reveals evident differences between Eskimos and Scandinavians in morbidity resulting from chronic diseases. The low prevalence of a series of chronic illnesses in Eskimos is considered to be related to their high consumption of fish, a rich source of omega-3 fatty acids.

This hypothesis should stimulate further intensive research into the antimutagenic and anticarcinogenic properties of alpha-linoleic acid and its derivatives.

REFERENCES

1. Bang, H.O., J. Dyerberg, and N. Hjorne (1976) The composition of food consumed by Greenland eskimos. Acta Med. Scand. 200:69-72.

2.   Begin, M.E., G. Ells, and D.F. Horrobin (1988) Polyunsaturated
     fatty acid-induced cytotoxicity against tumor cells and its relation-
     ship to lipid peroxidation. J. Natl. Cancer Inst. 80:188-194
3.   Carroll, K.K. (1986) Experimental studies on dietary fat and cancer
     in relation to epidemiological data. In Dietary Fat and Cancer, C.
     Ip, D.F. Birt, A.E. Rogers, and C. Mettlin, eds. A.R. Liss, Inc.,
     New York, pp. 231-248.
4.   Corey, R.J., C. Shih, and J.R. Cashman (1983) Docosahexaenoid-
     acid acid is a strong inhibitor of prostaglandin but not leucotriene
     biosynthesis. Proc. Natl. Acad. Sci., USA 80:3581-3584.
5.   Das, U.N., G.R. Devi, K.P. Rao, and M.S. Rao (1983) Prevention
     of radiation induced polychromatophilia by prostaglandin E and col-
     chicine. IRCS Med. Sci. 11:863-864.
6.   Das, U.N., G.R. Devi, K.P. Rao, and M.S. Rao (1985) Benzo[a]-
     pyrene and γ-radiation-induced genetic damage in mice can be pre-
     vented by γ-linolenic acid but not by arachidonic acid. Nutrition
     Res. 5:101-105.
7.   Diet, Nutrition and Cancer (1982) Report of the Committee of the
     Assembly of Life Sciences, National Academy Press, Washington,
     D.C.
8.   Hirai, A., T. Hamazaki, T. Terano (1980) Eicosapentaenoic acid and
     platelet function in Japanese. Lancet 2:1132-1136.
9.   Karmali, R.A., J. March, and C. Fuchs (1984) Effect of omega-3
     fatty acids on growth of a rat mammary tumor. J. Natl. Cancer
     Inst. 73:457-461.
10. Kort, W.J., F.M. Weijma, A.M. Bijma, W.P. von Schalkwijk, A.J.
     Vergroesen, and D.L. Westbroek (1987) Omega-3 fatty acids inhibit-
     ing the growth of a transplantable rat mammary adenocarcinoma. J.
     Natl. Cancer Inst. 79:593-599.
11. Lands, W.E.M. (1986) Renewed questions about polyunsaturated fatty
     acids. Nutr. Rev. 44:189-195.
12. Marshall, M.M. (1987) The nutritional importance of biotin-an up-
     date. Nutrition Today Nov./Dec. 26-30.
13. Michalik, H., R. Carmine, and G.L. Gatty (1975) The effect of die-
     tary rape-seed oil on hepatic hexobarbital metabolism in mice. Nutr.
     Metabol. 18:272-282.
14. Minoura, T., T. Takata, M. Sakaguchi, H. Takada, M. Yamamura,
     K. Hioki, and M. Yamamoto (1988) Effect of dietary eicosapentaenoic
     acid on azoxymethane--induced colon carcinogenesis in rats. Cancer
     Res. 48:4790-4794.
15. Nielsen, N.H., and J.P. Hansen (1980) Breast cancer in Greenland--
     Selected epidemiological and historical features. J. Cancer Res.
     Clin. Oncol. 98:287-299.
16. Nigro, N.D., A.W. Bull, and M.E. Boyd (1986) Inhibition of intes-
     tinal carcinogenesis in rats: Effect of difluoromethylornithine with
     piroxicam or fish oil. J. Natl. Cancer Inst. 77:1309-1313.
17. Raj, A.S., and M. Katz (1984) Corn oil and its minor constituents as
     inhibitors of DMBA-induced chromosomal breaks in vivo. Mutation
     Res. 136:247-253.
18. Renner, H.W. (1986) The anticlastogenic potential of fatty acid meth-
     yl esters. Mutation Res. 172:265-269.
19. Renner, H.W., and H. Delincee (1988) Different antimutagenic ac-
     tions of linoleic- and linolenic acid derivatives on busulfan-induced
     genotoxicity in Chinese hamsters. Nutrition Res. 8:635-642.
20. Rowe, L., and E.D. Wills (1976) The effect of dietary lipids and
     vitamin E on lipid peroxide formation, cytochrome P-450 and oxida-
     tive de-methylation in the endoplasmatic reticulum. Biochem.
     Pharmacol. 25175-179.

21. Simopoulos, A.P. (1988) Omega-3 fatty acids in growth and develop-
     ment and in health and disease, Part II. Nutrition Today May/June,
     13-18.
22. Wade, A.E., and W.P. Norred (1976) Effect of dietary lipid on drug-
     metabolizing enzymes. Fed. Proc. 35:2475-2479.
23. Zuniga, M.E., B.R. Lokesh, and J.E. Kinsella (1987) Dietary n-3
     polyunsaturated fatty acids alter lipid composition and decrease pros-
     taglandin synthesis in rat spleen. Nutrition Res. 7:299-306.
24. Zuniga, M.E., B.R. Lokesh, and J.E. Kinsella (1988) Effect of die-
     tary n-6 and n-3 polyunsaturated fatty acids on composition and en-
     zyme activities in liver plasma membrane of mice. Nutrition Res.
     8:1051-1059.

ANTIGENOTOXIC ACTIVITY OF CAROTENOIDS
IN CARCINOGEN-EXPOSED POPULATIONS

Miriam P. Rosin

Environmental Carcinogenesis Unit
British Columbia Cancer Research Centre

and

Occupational Health Unit, School of Kinesiology
Simon Fraser University
Vancouver, British Columbia V5Z 1L3, Canada

ABSTRACT

Although epidemiological studies suggest the presence of anticarcino-genic agents in the diet, it is difficult to obtain actual proof for the activity of such agents in humans. One approach is to develop and vali-date potential quantifiable indicators of antigenotoxic/anticarcinogenic agents which can be used in humans belonging to populations at elevated risk for cancer. This paper provides evidence that the exfoliated cell micronucleus test (MEC test) can be used (i) to provide a quantifiable marker for the amount of chromosomal breakage occurring in target tis-sues of carcinogen-exposed populations; (ii) to indicate the capacity of beta-carotene, alone or in combination with vitamin A, to prevent such damage; and (iii) to predict the response of other biological indicators of cancer risk, such as oral leukoplakias, in individuals receiving oral sup-plementation with beta-carotene/vitamin A (although the dose and time to response may differ for these endpoints). Future extensions of this approach include establishing the levels of beta-carotene required for antigenotoxic activity in a carcinogen's target tissue by concurrently measuring MEC frequencies and beta-carotene levels in exfoliated cells. In summary, early indications are that the MEC assay is an effective indi-cator for antigenotoxic agents in carcinogen-exposed individuals and that beta-carotene and vitamin A can suppress such genotoxic activity in at least some populations.

INTRODUCTION

The hypothesis that dietary anticarcinogens play a role in reducing the incidence of human cancers is primarily supported by epidemiological analyses of high- and low-risk population groups. Such studies have shown an inverse relationship between dietary intake and/or blood level of vitamin A and human cancer risk (2,17,18). More specifically, many of these studies have implicated the dietary carotenoids rather than pre-formed retinol or retinol esters as the putative protective agent(s) (2, 9,48). Of the carotenoids, attention has centered on beta-carotene, a compound with antioxidant capacity resulting from its carotenoid struc-ture, and in addition, capable of being converted in the body to retinoids (provitamin A activity).

Laboratory studies have confirmed the protective effect of beta-caro-tene and the retinoids (term includes both natural and synthetic retinol analogues; see Ref. 2) in suppressing tumor development in animal models and carcinogen/promoter activity in in vitro assays (2,15,19,30). How-ever, there is still little direct evidence supporting the role of these agents in decreasing human cancers. One approach to obtaining such in-formation is through chemoprevention trials. To date, completed trials have focused on the retinoids and have studied individuals with premalig-nant conditions, e.g., cervical dysplasia, bronchial metaplasia, leuko-plakia, rectal polyps (reviewed in Ref. 2). These studies have shown retinol and various synthetic retinoids to be effective in causing remission of early lesions or in preventing lesion development. Whether beta-caro-tene will have a similar chemopreventive effect is unknown, since trials with beta-carotene are not yet complete.

Since clinical trials are expensive and time-consuming, alternative approaches to obtaining information on protective effects of dietary agents in humans need to be explored. One such approach is to develop and validate biological markers which can be used to screen for the activity of such agents in individuals at elevated risk for cancer. These markers must have some biological relevance with respect to development of can-cer, be quantifiable, and be noninvasive. The objective of this paper is to summarize evidence supporting the role of one assay, the micronucleus test on exfoliated cells (MEC test), as such a biomarker.

BIOLOGICAL RELEVANCE OF MICRONUCLEI

Micronuclei are formed from chromosomal fragments (and less fre-quently, from entire chromosomes) which lack connection to the spindle apparatus and lag behind when the cell divides. Such fragments are ex-cluded from the main nucleus when it reforms and are visible as feulgen-positive extranuclear bodies in the daughter cells. The use of micro-nucleus frequencies as a quantifiable indicator of chromosomal breakage has been validated on more than 100 genotoxic and/or carcinogenic chemi-cals using either cultured fibroblasts, hepatocytes, human lymphocytes, myelo- and erythroblasts, or animals (8,28,46). The approach discussed in this paper involves a modification of the micronucleus test to use on exfoliated human cells (MEC test). Micronuclei present in these cells represent chromosomal breakage occurring in the dividing cell populations of the basal epithelial layers. The daughter cells containing the micro-nuclei migrate up through the epithelium and are exfoliated.

The biological relevance of employing the MEC test to study carcino-gen-exposed populations is that the approach serves as an "endogenous dosimeter" of genotoxic damage directly in the tissue that is the target for the carcinogens and from which tumors will later arise. Furthermore,

the fact that the assay provides an estimate of the frequency of chromosomal breakage is itself significant. Evidence suggests that chromosomal
changes may be intrinsically linked to cancer development. Chromosomal
instability is characteristic of dysplasias and many premalignant conditions, and specific chromosomal aberrations appear to be associated with
many types of cancer (14). Furthermore, individuals belonging to genetic
syndromes in which chromosomal breakage rates are elevated, such as the
chromosomal breakage syndromes Bloom syndrome, ataxia-telangiectasia,
Fanconi anemia, and xeroderma pigmentosum, are also characterized by an
increase in risk for cancer (7,21). Finally, a role for chromosomal
breakage, translocation, or loss is implicated in the sequence of events
leading to development of neoplasia, since such changes can activate
oncogenes or result in the loss of tumor "antioncogenes" or "suppressor"
genes (4,5,11). Although the MEC test will not indicate whether a specific required chromosome change has occurred in a carcinogen-exposed
tissue, a chronic increase in the level of MEC in a tissue may suggest an
increase in the probability of the necessary chromosomal change occurring.

VALIDATION OF THE MEC TEST AS A TISSUE-SPECIFIC DOSIMETER
OF CHROMOSOMAL BREAKAGE IN HUMANS

Experimental evidence with animal models suggests that the protective effects of chemopreventive agents will show both carcinogen and or-
gan/tissue specificity (2,15). Thus, biomarkers will be needed which can
be used to monitor responses in different tissues. Currently, it is possible to compare the frequency of micronuclei in exfoliated cells of the
following tissues: mucosa of the oral cavity (swab), urinary bladder
(centrifugation of urine samples), cervix (swabs), esophagus (scrapings
or biopsies), and bronchi (biopsies or washings). The majority of research from this laboratory has focused on the oral cavity and the validation of the MEC assay in populations at elevated risk for oral cancer.
Hence, the following discussion will focus on oral cancers.

The evidence in support of the MEC test being a dosimeter of chromosomal breakage with relevance to oral cancer is the following:

(1) MEC frequencies are significantly elevated in oral smears of
individuals with lifestyle habits strongly associated with an
increase in risk for oral cancer (e.g., oral tobacco and/or betel
quid users; for review, see Ref. 37).

(2) In contrast, MEC frequencies have been consistently low in oral
smears from individuals with no known carcinogen exposure (for
review, see Ref. 37). These individuals were chosen as controls for the above-mentioned carcinogen-exposed populations,
and matched for age, sex, and dietary habits.

(3) The elevation in MEC frequency is dependent on the dose of
carcinogen exposure. For example, a dose-dependent increase
in MEC frequencies is observed throughout the course of radiotherapy in samples from the irradiated buccal cavity of head
and neck cancer patients (27,35,42). Dose response relationships have also been demonstrated for MEC frequencies among
betel quid users (42) and among alcohol drinkers who smoke
(36).

(4) Site-specific variations in the MEC frequency within the oral cavity occur in response to different levels of exposure of particular sites to carcinogens. The best example involves a population of "reverse-smokers" in India (37). These individuals routinely hold the lit end of the cigarette inside the mouth, with the cigarette resting on the tongue and often touching the palate. The MEC frequency is most elevated in samples from the dorsum of the tongue and from the palate; these are also the sites at which carcinomas most often develop (over 73% of all oral cancers arise in the palate; Ref. 22).

(5) A temporal relationship exists between an observed elevation in MEC frequencies of buccal smears and exposure to a carcinogenic agent. Patients receiving radiotherapy to the head and neck region show an elevation in MEC frequencies at 7-10 days after commencement of radiotherapy. Since chromosome damage occurs in the dividing basal cells of the epithelium, the indicated time is required for the acentric fragments to form micronuclei in the daughter cells, and for the migration of these daughter cells up through the epithelium. When therapy ceases, the MEC frequency decreases, and returns to basal, pretreatment levels within 2-3 wk (27,35,42).

(6) Additive and synergistic effects on MEC frequencies in the oral cavity have been observed in response to the exposure of an individual to combinations of carcinogens and co-carcinogens. For example, the combination of both alcohol and cigarette consumption in an individual produces a synergistic effect on the level of MEC in smears of the oral cavity (36). A synergistic interaction between these two lifestyle habits in the induction of oral cancers has been noted in epidemiological studies (49).

(7) MEC frequencies are elevated in oral smears of patients with Bloom syndrome (BS) or ataxia-telangiectasia (AT), chromosomal instability syndromes which are characterized by an elevated level of spontaneous chromosomal breakage (occurs in cultured cells from these individuals without the addition of a carcinogenic/mutagenic agent) and a significantly increased risk for cancer (Fig. 1; Ref. 7, 24, 25, 26). Of significance is the increased MEC frequencies observed in 60% of the examined parents of AT patients who are heterozygous for the AT gene. Current estimates are that these heterozygotes may comprise up to 2.8% of the general population (44), and may represent up to 5% of individuals dying of cancer before age 45 and from 8.8-20% of breast cancers (6,45). In contrast to the results with the AT heterozygotes, MEC frequencies are not elevated among the parents of BS patients; nor is there any evidence for an increase in risk for cancer for these parents.

(8) Indications are that among individuals with a genetic defect in DNA repair of certain DNA lesions, exposure to an agent producing such damage will result in an elevation in MEC frequency specific to the exposed site. Results from xeroderma pigmentosum patients show an unequal distribution of MEC frequencies throughout the oral cavity (Tab. 1). MEC frequencies lie within control values for samples from the right buccal mucosa (with the exception of patient X7); the largest increase

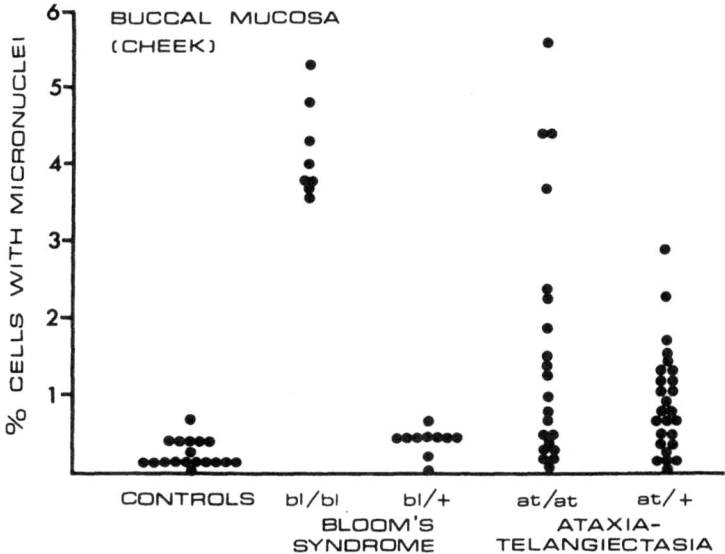

Fig. 1. Frequency of micronucleated cells in exfoliated cell samples taken from the buccal mucosa of controls (individuals with no known carcinogen exposure and no known genetic chromosomal instability), individuals homozygous (bl/bl) and heterozygous (bl/+) for Bloom's syndrome, and individuals homozygous (at/at) or heterozygous (at/+) for ataxia-telangiectasia. The heterozygous individuals were parents of the patients sampled.

Tab. 1. Unequal distribution of micronucleated cells in the oral mucosa of xeroderma pigmentosum patients.

Subject No.	MEC frequency[a]			
	Right Buccal Mucosa	Tongue	Palate	Urogenital Tract
X1	0.5	-	-	-
X2	0.5	-	-	-
X3	0.5	-	-	-
X4	0.7	-	-	-
X5	0.4	-	-	-
X6	0.3	2.6	1.3	1.0
X7	1.3	9.1	-	0.9

[a]Historical range of MEC frequencies for controls (no known carcinogen exposure, no known genetic chromosomal instability) is from 0.0 - 0.9.

in MEC frequencies occurs at the dorsal tip of the tongue, and
this increase is present in both XP patients sampled at this
site. This observation may have some significance, since xeroderma patients have a defect in excision repair of UV-induced
DNA damage and this site in the oral cavity would receive the
greatest sunlight exposure. Carcinomas on the tongue do occur
among XP patients in Egypt (J. German, pers. comm.).

(9) Points 4 and 8 have referred to a site-specificity of elevation in
MEC frequencies within a target organ. There is also considerable evidence supporting a site-specificity for this assay with
respect to the tissue/organ showing the response (Tab. 2).
For example, oral cavity MEC frequencies are not elevated in
smokers who do not drink alcohol; yet a significant elevation in
MEC frequencies occurs in cells isolated from urine voids. An
elevation of MEC frequencies in the oral cavity requires the
presence of alcohol (36). Similarly, individuals with Schistosoma haematobium infection in the urinary bladder, a parasite
infection associated with an elevated risk for bladder cancer,
have an increase in MEC frequencies of cells from the urinary
bladder and not from the oral cavity (Ref. 20; H.F. Stich,

Tab. 2. Site specificity of elevation in micronucleus frequency.

| Group studied | Number of subjects | Sample site of exfoliated cells | | | |
| | | Oral cavity | | Urogenital tract | |
		%MEC(average)[a]	R.I.[b]	%MEC(average)	R.I.
Cigarette Smokers[c]	20	0.5 (0.1-1.0)	1.0	3.3 (1.6-6.5)	5.5
Cigarette Smokers[d] + Alcohol	24	1.7 (0.5-3.5)	3.4	-	
Schistosoma haemotobium infection	4	0.6 (0.3-0.8)	1.2	8.6 (5.5-10.5)	14.3
Ataxia-telangiectasia	6	3.4 (1.5-5.2)	6.8	2.3 (1.6-3.2)	3.8
Bloom syndrome	8	4.2 (3.7-4.8)	8.4	4.4 (3.0-5.8)	7.3

[a] Average values for %MEC of individuals belonging to indicated population. Range of MEC
values is shown in parentheses.

[b] R.I.= relative increase in %MEC above values observed in controls (non-smokers, non-alcohol
drinkers, no known chromosomal instability, no bladder infection). %MEC for these controls
was 0.5 for the oral cavity (range, 0.1-0.7); urogenital tract 0.6 (range, 0.0-0.8). All increases are
significant at P<0.01 with the exception of values for oral cavity of cigarette smokers (who abstain
from alcohol) and S. haematobium infection

[c] 40-59 cigarettes per day, no alcohol

[d] 40-59 cigarettes per day, >150 ml ethanol per day

pers. comm.). In contrast, MEC frequencies in AT and BS patients are significantly increased in exfoliated cells from both sites (24,25).

PREVENTION OF CHROMOSOMAL DAMAGE IN TARGET TISSUES WITH CAROTENOIDS/RETINOIDS

Table 3 shows the results of four pilot studies conducted on tobacco- and/or betel quid-chewing populations to determine whether supplementation with various chemopreventive regimes would alter MEC frequencies in the oral cavity. Three of the groups (groups 1, 2, and 4 of Tab. 1; populations described in Ref. 31, 34, 39, 40, 42) consist of chronic users of a betel quid made of sun-dried tobacco, betel leaf (Piper betle L.), areca nut (Areca catechu L.), and slaked lime from marine shells (calcium hydroxide). The fourth group (group 3 of Tab. 1) is a population of Native Inuit snuff-users, from the Northwest Territories of Canada, who hold a finely-powdered tobacco in the lower gingival groove of the oral cavity. In each population, the tobacco habit results in a significant elevation in the frequencies of micronucleated cells in the oral mucosa compared with levels observed among nonusers. Oral tobacco and betel quid use are strongly associated with an elevation in oral cancer (10).

Tab. 3. The effect of supplementation with beta-carotene (B) and/or vitamin A (A) on MEC frequencies in the buccal mucosa of tobacco/betel quid users.

| Population | No. Subjects | Treatment[a] | Average % MEC | | % Subjects with decrease in MEC |
			Before	After[b]	
1.Betel quid chewers (Philippians)	40	B & A	4.2	1.4 (33.3)[c]	92.5
	11	-	4.3	4.8 (111.6)	0
2.Betel quid chewers (Philippians)	25	B	3.4	1.2 (35.3)[c]	88.0
	26	A	4.0	1.7 (42.5)[c]	92.3
	18	-	3.4	3.3 (103.0)	5.6
	20	C	3.4	3.4 (100.0)	0
3.Snuff users (NWT, Canada)	23	B	1.9	0.7 (36.8)[c]	78.3
	31	-	2.0	2.0 (100.0)	6.5
4.Betel quid chewers (India)	31	B	4.1	1.0 (24.4)[c]	96.8
	51	B & A	4.2	1.2 (28.6)[c]	94.1
	30	-	4.1	3.8 (92.7)	0

[a]Treatments consisted of twice-weekly oral supplementation with the following agents for a duration of 3 months (doses are total per week): B & A, beta-carotene and retinyl palmitate (beta-carotene 180 mg/week, equivalent to 300,000 IU/week and retinyl palmitate 100,000 IU/week);-, dextrose-containing placebo; B, beta-carotene (180 mg/week); A, retinyl palmitate (150,000 IU/week); C, canthaxanthin (180 mg/week).

[b]Values in parentheses represent the MEC frequency at end of treatment expressed as % of initial value. Mean MEC frequencies for non-chewers in these populations were the following: 1) 0.50 (n=10), 2.) 0.51 (n=52), 3.) 0.47 (n=12), and 4.) 0.4 (n=34).

[c]Significant difference between MEC frequencies before and after treatment, P< 0.001

The following points of significance can be drawn from these studies:

(1)   In each of the four populations, MEC frequencies of tobacco/-
      quid users receiving placebo did not significantly change during
      the course of the treatment.

(2)   In contrast, several of the chemopreventive regimes resulted in
      a significant reduction in average MEC frequencies. Moreover,
      a reduction in MEC frequencies occurred in the majority of indi-
      viduals in these responding groups (from 88 to 96%). This de-
      crease in MEC frequencies was not due to reduction in carcino-
      gen exposure, since questionnaires established that the chewing
      habit had not altered during the treatment period.

(3)   It is possible to see a chemopreventive effect with this endpoint
      within 3 mo. Indeed, in individuals sampled throughout the
      treatment (37), a decrease in MEC frequencies was visible with-
      in 1 mo of treatment. This point is significant from the
      standpoint of devising a preliminary screen of promising chemo-
      preventive agents and regimes prior to undertaking long-term
      trials.

(4)   Not all chemopreventive regimes were effective. Treatment with
      beta- carotene, vitamin A (retinyl palmitate), or a combination
      of these two agents caused a decrease in MEC frequencies
      among betel quid chewers. Canthaxanthin was ineffective (43).
      Canthaxanthin is a carotenoid which possesses similar antioxi-
      dant properties to beta-carotene (1) but lacks the capacity to
      be converted to a retinoid. Although this observation suggests
      that the protective effect of beta-carotene in this test system is
      due to its retinoid activity, the data are not conclusive since
      repetition is required.

    With the exception of one of the above studies (group 3), each of
the population groups examined was marginally deficient in both beta-
carotene and vitamin A. One question which arises is whether beta-
carotene or vitamin A will have a similar capacity to prevent carcinogen-
induced damage in European or North American populations. From this
standpoint, the protection observed against carcinogen damage in the
Inuit population is somewhat reassuring, since these individuals had normal levels of vitamin A but a decreased beta-carotene status, a situation
which may more closely resemble populations at elevated risk in western
cultures.

MEC FREQUENCIES AND PREMALIGNANT LESIONS

    Oral leukoplakias occur frequently among fishermen in Kerala, Tri-
vandrum, India who are betel quid chewers (30-50% of chewers have
lesions). Thus, it is possible to concurrently monitor the effect of
beta-carotene/vitamin A supplementation on two endpoints: the frequency
of MEC cells in the buccal mucosa and the presence of oral leukoplakias.
Table 4 presents the data obtained for the response of these two endpoints to supplementation with beta-carotene, beta-carotene plus vitamin
A, and vitamin A alone. A determination was made of the effect of the
treatments on lesions present at the beginning of the trial (expressed as
the percentage of individuals who showed a remission of leukoplakia) and

Tab. 4. Response of MEC frequencies and oral leukoplakias in betel quid/tobacco chewers to supplementation with beta-carotene/vitamin A.

Endpoint	%Individuals with indicated response			
	Treatment[a]			
	Placebo	Beta-carotene	Beta-carotene plus Vitamin A	Vitamin A
Decrease in MEC frequency	0	96.8***	94.1***	ND[c]
Remission of "old" leukoplakia	3.0	14.8	27.5***	57.1***
% individuals with new leukoplakias	21.2	14.8	7.8*	0**

[a]Twice weekly treatment for respective groups: beta-carotene group (180 mg/week, equivalent to 300,000 IU), beta-carotene plus vitamin A group (180 mg/week beta-carotene plus 100,000 IU/week retinyl palmitate, equivalent to total of 400,000 IU/week retinol equivalents), and vitamin A group (200,000 IU retinyl palmitate/week).

[b]Test for significance with Fisher's exact test, P values (2-sided): * P=0.02, ** P=0.08 and *** P<0.01 for comparison with value observed for indicated endpoint in placebo

[c]ND=not done

on the ability of the treatment to prevent the development of new lesions (expressed as the percentage of individuals with new leukoplakias). The corresponding values of MEC frequencies in smears taken at 3 mo in each treatment group are also shown. No values are available for the group receiving vitamin A alone.

The results indicate that the two endpoints respond differently to the chemopreventive regimes. Supplementation with beta-carotene or a combination of beta-carotene and vitamin A caused a similar decrease in MEC values, with 94-96% of the treated individuals responding. Although MEC frequencies in the group receiving vitamin A alone were not obtained, previous studies with another betel quid-chewing group in which oral supplementation with less vitamin A was used (150,000 IU/wk compared with 200,000 IU/wk in this trial) resulted in a reduction in MEC frequencies in 92.3% of individuals (group 2, Tab. 3). In contrast to the results observed with MEC frequencies, the remission of leukoplakias and the prevention of development of new leukoplakias differed significantly in the three treatment groups, with the vitamin A treatment being the most effective. Furthermore, a larger portion of the treated individuals showed a reduction in MEC frequency than showed a response in lesions. Finally, the rate of response to the treatments differs for the two endpoints. Decreases in MEC frequencies are observed after treatment for 1 mo (37), whereas the effect on the lesions developed gradually over 6 mo, a slight effect becoming apparent after treatment for 3 mo (39) but significant only at 6 mo.

These results indicate that a chemopreventive regime may need to be altered depending upon the biological response to be modified. Treatments to cause regression of lesions, or to block the development of such lesions from initiated cells, may differ in may respects from regimes which result in the scavenging of electrophiles to prevent carcinogen damage: the effective dose of agent, time to response, and even the type of agent which is protective may each need to be varied. In this study, vitamin A alone, or in combination with beta-carotene, was effective in producing a protective effect in two endpoints. It is not known what effect an increase in beta-carotene intake or in the length of the supplementation period would have on the ability of this agent to affect the leukoplakias. In contrast to these results with betel quid chewers, oral supplementation of a group of "reverse smokers" from the Philippines for 6 mo with a combination of beta-carotene (180 mg/wk) and vitamin A (150,000 IU retinyl palmitate/wk) did not result in a response on either the MEC frequencies in the oral cavity or the clinical lesions which were present. This latter study indicates that population characteristics such as the pattern of carcinogen exposure (betel quid containing tobacco versus cigarette smoke, and damage to the tissue by extreme heat) can have a significant impact on the success of chemoprevention studies.

THE IMPORTANCE OF TISSUE LEVELS OF BETA-CAROTENE

One of the characteristics of beta-carotene which makes it particularly attractive as a potential chemopreventive agent is its ability to accumulate in epithelial tissues. This property could yield a reservoir of protective agents, with cells in the tissue having ready access to beta-carotene and also, perhaps, to retinoids. Cytosolic homogenates of rat intestine, kidney, lung, testes, and liver all have the capacity to synthesize retinoic acid from beta-carotene (16). These data suggest that in situ generation of retinoic acid from beta-carotene in epithelial tissues could represent an alternate source of retinoids apart from the mobilization of retinyl esters from the liver, a source that is under tight homeostatic control.

One approach to obtaining information on the accumulation of beta-carotene in epithelial tissues in humans is to use the exfoliated cells collected from such tissues as indicators. Cells can be collected from the oral cavity by brushing the mucosa and from urine voids by centrifugation. These cells are extracted and are assayed with high-pressure liquid chromatography procedures (32). Not only does this approach provide information on basal (unsupplemented) levels of incorporated beta-carotene in different populations such as alcoholics, vegetarians, non-vegetarians, etc. (29,32), but it yields data on the change in such levels in response to oral supplementation with this agent (5,23,33). Table 5 shows data obtained in a small group of individuals receiving a short-term oral supplementation with beta-carotene. Levels of beta-carotene increased significantly in exfoliated cells collected from the oral cavity and from urine voids. This change was not apparent in placebo-treated individuals.

Tab. 5. Beta-carotene levels in exfoliated urogenital tract and oral mucosa cells following supplementation with beta-carotene.

		Beta-carotene content (ng/10^6 cells)		
Site	Treatment[a]	Before	After	Relative increase
Urogenital tract	Placebo	0.5	0.6[b]	1.2
	Beta-carotene	0.6	4.1[c]	6.8[d]
Oral cavity	Placebo	1.2	1.7[b]	1.4
	Beta-carotene	2.0	9.9[c]	5.0[d]

[a] Study consisted of two treatment groups, one of which received placebo (n=7) and the other, 90 mg beta-carotene, twice weekly for four weeks (n=10). Exfoliated cells were collected from the oral cavity by brushing the mucosa with a toothbrush, and from the urogenital tract by centrifuging urine voids. All samples were from females.
[b] Comparison of beta-carotene levels in the two cell types in individuals receiving placebo, P=0.05.
[c] Comparison of beta-carotene levels in the two cell types in individuals receiving beta-carotene P<0.02.
[d] Relative increase before vs after supplementation, P<0.02.

The data in Tab. 5 also suggest that the tissues from which these cells were obtained may be accumulating beta-carotene with different kinetics. Beta-carotene levels were significantly different between the two cell isolates, and this difference was apparent before and after supplementation. Results from animal studies indicate that tissues differ in their extent of accumulation, rate of accumulation, time to saturation, and turnover time of beta-carotene (13,29). What effect this difference in accumulation kinetics in various tissues could have on the risk for cancers at specific sites is unknown. The possibility exists that the level of beta-carotene accumulated in the tissue at risk for development of cancer may be more closely correlated with cancer incidence than are either intake or blood levels. Animal studies suggest that although blood beta-carotene levels may yield information on recent intake, absorption, and transport of beta-carotene, they do not indicate tissue levels (29).

CONCLUSIONS

This article has reviewed the benefits of using exfoliated cells to obtain information at a tissue level in individuals belonging to population groups at elevated risk for cancer. Micronucleated exfoliated cells can serve as quantitative indicators of chromosomal damage resulting from exposure to carcinogenic lifestyle habits. In addition, the assay can be used to study the interplay of carcinogens, co-carcinogens, genetics, and diet in a biological response with relevance to cancer, i.e., chromosome breakage in target epithelial sites.

The observation that exfoliated cells can also serve as dosimeters of beta-carotene accumulation suggests that it may be feasible to determine tissue levels of beta-carotene which result in changes in various biological indicators, such as a decrease in MEC frequencies, remission of premalignant lesions, or the prevention of development of lesions. An additional possibility is that this approach may be of use in establishing supplementation regimes capable of maintaining protective effects. MEC frequencies return to presupplemented levels in carcinogen-exposed populations upon withdrawal of beta-carotene (37). Similarly, leukoplakias and premalignant lesions recur when chemopreventive treatment ceases (12).

Scientists working in this area are faced with an ethical dilemma when a population at elevated risk for cancer responds positively to a chemopreventive regime. The question arises as to whether an obligation to continue treatment in the population exists, especially if the treatment produces no detrimental effects, as is the case with beta-carotene. In such a situation, the possibility exists that "booster treatments" at fixed intervals with the chemopreventive agent could be used to maintain tissues at a "chemopreventive level."

ACKNOWLEDGEMENTS

These studies were supported by the Natural Sciences and Engineering Research Council of Canada.

REFERENCES

1. Anderson, S.M., and N.I. Krinsky (1973) Protective action of carotenoid pigments against photodynamic damage to liposomes. Photochem. Photobiol. 18:403-408.
2. Bertram, J.S., L.N. Kolonel, and F.L. Meyskens (1987) Rationale and strategies for chemoprevention of cancer in humans. Cancer Res. 47:3012-3031.
3. Cameron, L.M., M.P. Rosin, and H.F. Stich (1989) Use of exfoliated cells to study tissue-specific levels of beta-carotene in humans. Cancer Lett. (submitted for publication).
4. Cooper, G.M. (1984) Activation of transforming genes in neoplasms. Br. J. Cancer 50:137-142.
5. Flier, J.S., L.H. Underhill, S.H. Friend, T.P. Dryja, and R.A Weinberg (1988) Oncogenes and tumor-suppressing genes. New Engl. J. Med. 318:618-622.
6. Gatti, R.A., I. Berkel, E. Boder, et al. (1988) Localization of an ataxia-telangiectasia gene to chromosome 11q22-23. Nature 336:577-580.
7. German, J. (1983) Patterns of neoplasia associated with the chromosome-breakage syndromes. In Chromosome Mutation and Neoplasia, J. German, ed. Alan R. Liss, Inc., New York, pp. 97-134.
8. Heddle, J.A., R.D. Benz, and P.I. Countryman (1978) Measurement of chromosomal breakage in cultured cells by the micronucleus technique. In Mutagen-Induced Chromosome Damage in Man, H.J. Evans and D.C. Lloyd, eds. Edinburgh University Press, Edinburgh, p. 191.
9. Hinds, M.W., L.N. Kolonel, J.H. Hankin, and J. Lee (1984) Dietary vitamin A, carotene, vitamin C and risk of lung cancer in Hawaii. Am. J. Epidemiol. 119:227-237.
10. IARC (1985) Monographs on the Evaluation of the Carcinogenic Risk of Chemicals to Humans. 37. Tobacco Habits Other Than Smoking: Betel-nut and Areca-nut Chewing: and Some Related Nitrosamines, International Agency for Research on Cancer, Lyon.

11. Knudson, A.G. (1985) Hereditary cancer, oncogenes, and antionco-
    genes. Cancer Res. 45:1437-1443.
12. Kraemer, K.H., J.J. DiGiovanna, A.N. Moshell, R.E. Tarone, and
    G.L. Peck (1988) Prevention of skin cancer in xeroderma pigmen-
    tosum with the use of oral isotretinoin. New Engl. J. Med.
    318:1533-1637.
13. Mathews-Roth, M.M., D. Hummel, and C. Crean (1977) The carot-
    enoid content of various organs of animals administered large
    amounts of beta-carotene. Nutr. Rep. Intl. 16:419-423.
14. Mitelman, F., and S. Heim (1988) Consistent involvement of only 71
    of the 329 chromosomal bands of the human genome in primary neo-
    plasia-associated rearrangements. Cancer Res. 48:7115-7119.
15. Moon, R.C., and L.M. Itri (1984) Retinoids and cancer. In The
    Retinoids, Vol. 2, M.B. Sporn, A.B. Roberts, and D.S. Goodman,
    eds. Academic Press, New York, pp. 327-371.
16. Napoli, J.L., and K.R. Race (1988) Biogenesis of retinoic acid from
    beta-carotene--Differences between the metabolism of beta-carotene
    and retinal. J. Biol. Chem. 263:17372-17377.
17. Peto, R. (1983) The marked differences between carotenoids and
    retinoids: Methodological implications for biochemical epidemiology.
    Cancer Surv. 2:327-340.
18. Peto, R., R. Doll, J.D. Buckley, and M.B. Sporn (1981) Can die-
    tary beta-carotene materially reduce human cancer rates? Nature
    290:201-208.
19. Pung, A., J.E. Rundhaug, C.N. Yoshizawa, and J.S. Bertram
    (1988) Beta-carotene and canthaxanthin inhibit chemically- and phys-
    ically-induced neoplastic transformation in 10T1/2 cells. Carcino-
    genesis 9:1533-1539.
20. Raafat, M., S. El-Gerzawi, and H.F. Stich (1984) Detection of muta-
    genicity in urothelial cells of bilharzial patients by "the micronucleus
    test." J. Egypt. Natl. Cancer Inst. 1(3):63-73.
21. Ray, J.H., and J. German (1983) The cytogenetics of the "chromo-
    somal breakage syndromes." In Chromosome Mutation and Neoplasia,
    J. German, ed. Alan R. Liss, Inc., New York, pp. 135-167.
22. Reddy, C.R.R.M. (1974) Carcinoma of hard palate in India in rela-
    tion to reverse smoking of chuttas. J. Natl. Cancer Inst. 53:615-
    619.
23. Rosin, M.P., B.P. Dunn, and H.F. Stich (1987) Use of intermediate
    end-points in quantitating the response of precancerous lesions to
    chemopreventive agents. Can. J. Physiol. Pharmacol. 65:483-487.
24. Rosin, M.P., and J. German (1985) Evidence for chromosomal insta-
    bility in vivo in Bloom syndrome: Increased number of micronuclei
    in exfoliated cells. Hum. Genet. 71:187-191.
25. Rosin, M.P., and H.D. Ochs (1986) In vivo chromosomal instability
    in ataxia telangiectasia homozygotes and heterozygotes. Hum. Genet.
    74:335-340.
26. Rosin, M.P., H.D. Ochs, and E. Boder (1989) Heterogeneity of lev-
    els of chromosomal breakage in epithelial tissue of ataxia-telangi-
    ectasia homozygotes and heterozygotes. Hum. Genet. (submitted for
    publication).
27. Rosin, M.P., and H.F. Stich (1983) The identification of antigeno-
    toxic/anticarcinogenic agents in food. In Diet and Cancer: From
    Basic Research to Policy Implications, D.A. Roe, ed. Alan R. Liss,
    Inc., New York, pp. 141-154.
28. Schmid, W. (1976) The micronucleus test for cytogenetic analysis.
    In Chemical Mutagens: Principles and Methods for Their Detection,
    Vol. 4, A. Hollaender, ed. Plenum Press, New York, p. 31.

29. Shapiro, S.S., D.J. Mott, and L.J. Machlin (1984) Kinetic charac-
    teristics of beta-carotene uptake and depletion in rat tissue. J.
    Nutr. 114:1924-1933.
30. Sporn, M.B., A.B. Roberts, and D.S. Goodman (1984) The Reti-
    noids, Vol. 1 and 2, Academic Press, New York.
31. Stich, H.F., A.P. Hornby, and B.P. Dunn (1985) A pilot beta-caro-
    tene intervention trial with Inuits using smokeless tobacco. Int. J.
    Cancer 36:321-327.
32. Stich, H.F., A.P. Hornby, and B.P. Dunn (1986) Beta-carotene lev-
    els in exfoliated human mucosa cells of population groups at low and
    elevated risk for oral cancer. Int. J. Cancer 37:389-393.
33. Stich, H.F., A.P. Hornby, and B.P. Dunn (1986) Beta-carotene lev-
    els in exfoliated human mucosa cells following its oral administration.
    Cancer Lett. 30:133-141.
34. Stich, H.F., A.P. Hornby, B. Mathew, R. Sankaranarayanan, and
    M. Krishnan Nair (1988) Response of oral leukoplakias to the admin-
    istration of vitamin A. Cancer Lett. 40:93-101.
35. Stich, H.F., and M.P. Rosin (1983) Micronuclei in exfoliated human
    cells as an internal dosimeter for exposures to carcinogens. In
    Carcinogens and Mutagens in the Environment, Vol. II, Naturally
    Occurring Compounds: Endogenous Formation and Modulation, H.F.
    Stich, ed. CRC Press, Boca Raton, Florida, pp. 17-25.
36. Stich, H.F., and M.P. Rosin (1983) Quantitating the synergistic
    effect of smoking and alcohol consumption with the micronucleus test
    on human buccal mucosa cells. Int. J. Cancer 31:305-308.
37. Stich, H.F., and M.P. Rosin (1984) Micronuclei in exfoliated human
    cells as a tool for studies in cancer risk and cancer intervention.
    Cancer Lett. 22:241-253.
38. Stich, H.F., and M.P. Rosin (1985) Towards a more comprehensive
    evaluation of a genotoxic hazard in man. Mutation Res. 150:43-50.
39. Stich, H.F., M.P. Rosin, A.P. Hornby, B. Mathew, R. Sankarana-
    rayanan, and M. Krishnan Nair (1988) Remission of oral leukoplakias
    and micronuclei in tobacco/betel quid chewers treated with beta-
    carotene and with beta-carotene plus vitamin A. Int. J. Cancer
    42:195-199.
40. Stich, H.F., M.P. Rosin, A.P. Hornby, B. Mathew, R. Sankarana-
    rayanan, and M. Krishnan Nair (1989) Pilot intervention studies with
    carotenoids. In Carotenoids '87, N.I. Krinsky, M.M. Mathews-Roth,
    and R.F. Taylor, eds. Plenum Press, New York (in press).
41. Stich, H.F., M.P. Rosin, and M.O. Vallejera (1984) Reduction with
    vitamin A and beta-carotene administration of proportion of micro-
    nucleated buccal mucosal cells in Asian betel nut and tobacco chew-
    ers. Lancet 1:1204-1206.
42. Stich, H.F., R.H.C. San, and M.P. Rosin (1983) Adaptation of the
    DNA repair and micronucleus tests to human cell suspensions and
    exfoliated cells. Ann. N.Y. Acad. Sci. 407:93-105.
43. Stich, H.F., W. Stich, M.P. Rosin, and M.O. Vallejera (1984) Use of
    the micronucleus test to monitor the effect of vitamin A, beta-
    carotene and canthaxanthin on the buccal mucosa of betel nut/tobac-
    co chewers. Int. J. Cancer 34:745-750.
44. Swift, M., D. Morrell, E. Cromartie, A.R. Chamberlin, M.H.
    Skolnick, and D.T. Bishop (1986) The incidence and gene frequency
    of ataxia-telangiectasia in the United States. Am. J. Hum. Genet.
    39:573-583.
45. Swift, M., P.J. Reitnauer, D. Morrell, and C.L. Chase (1987)
    Breast and other cancers in families with ataxia-telangiectasia. New
    Engl. J. Med. 316:21-26.

46. Tates, A.D., I. Neuteboom, M. Hofker, and L. Den Enbelsen (1980)
     A micronucleus technique for detecting clastogenic effects of muta-
     gens/carcinogens (DEN, DMN) in hepatocytes of rat liver in vivo.
     Mutat. Res. 74:11.
47. Underwood, B.A. (1984) Vitamin A in animal and human nutrition.
     In The Retinoids, Vol. 1, M.B. Sporn, A.B. Roberts, and D.S.
     Goodman, eds. Academic Press, New York, pp. 281-392.
48. Wu, A.H., B.E. Henderson, M.C. Pike, and M.C. Uy (1985) Smok-
     ing and other risk factors for lung cancer in women. J. Natl.
     Cancer Inst. 74:747-751.
49. Wynder, E.L., I.J. Bross, and R.M. Feldman (1957) A study of the
     etiological factors in cancer of the mouth. Cancer 10:1300-1323.

PLANT ANTIMUTAGENS

Monroe E. Wall, Mansukh C. Wani, Thomas J. Hughes, and Harold Taylor

Research Triangle Institute
Box 12194
Research Triangle Park, North Carolina 27709

INTRODUCTION

The research group at Research Triangle Institute (RTI) has had many years of experience in the isolation and structural elucidation of antitumor agents.   Several of us attended the First International Conference on Antimutagenesis and Anticarcinogenesis Mechanisms (University of Kansas, 1985).   Stimulated by papers presented at this meeting and the subsequent book (7), we became interested in the question whether secondary plant metabolites have a role in antimutagenesis mechanisms. There was already considerable evidence that higher plants contain a variety of preformed secondary metabolites which represent a structurally diverse array of antimutagenic and mutagenic compounds (3,5).   Since we had on hand thousands of plant extracts from previous antitumor studies and had extensive experience in bioassay-directed isolation of naturally-occurring compounds, along with the availability of an active microbiology laboratory, we believed that we might be able to make a contribution to this interesting field.

GENERAL METHODS

Antimutagenic Screening

The procedure used for screening crude plant extracts for inhibition of the mutagenicity of various mutagens to Salmonella typhimurium has been described in detail in several of our papers (9-11) and is outlined in Chart 1.

The mutagen selected for general screening is 2-aminoanthracene (2AN).   Our methods are based on a procedure originally described by Birt et al. (1). This mutagen requires metabolic activation by the Ames S-9 liver preparation (4).   We prefer 2AN for screening to other mutagens which also require metabolic activation with S-9, such as 2-aminoacetylfluorene (AAF) or benzo(a)pyrene (BAP), because with 2AN,

Chart 1.   Screening procedures.

INHIBITION PROCEDURE

1. Salmonella typhimurium (T-98)

\+

2. Ames S-9 preparation

\+

3. 600 µg/plate of test substance

\+

4. 2-Aminoanthracene (2AN)
   (2.5 µg/plate)

5. Mix with top agar and incubate 48-72 hours.

TOXICITY PROCEDURE

1. Steps 1-3 as above

2. Omit 2AN in step 4 and replace by Histidine

3. Step 5 as above

positive controls give 2,500-3,000 colonies per culture plate, which is 2-4 times greater than we can achieve with AAF or BAP. However, all of these mutagens, as well as ethylmethanesulfonate (EMS), which does not require metabolic activation (2), are used for studying the mutagenic inhibition produced by the various pure natural products which we have isolated or in some cases procured.

The bioassay method has been described in considerable detail (9-11), but a few points may be emphasized. As shown in Chart 1, the mutagen is added last to the incubation mixture, a point quite properly emphasized by Mitscher et al. (5). This minimizes the possibility of reaction of mutagen with substrate. Crude organic solvent extracts are concentrated and a known concentration is prepared in dimethylsulfoxide (DMSO), an excellent solvent. Aliquots containing 600 µg are added to the incubation mixture (9). This high dose is necessary because the "active" plant constituent may be present in concentrations of <1%. Nevertheless, the method shows considerable selectivity, and active plants constitute only 3-4% of the total screened. Three plates are utilized for each assay. A crude extract is not considered inhibitory unless a duplicate determination conducted at another time gives similar results. We require at least a 30% reduction in the number of colonies found in the positive controls for an extract to be considered active in the inhibition of the mutagenic action of 2AN on S. typhimurium (T98) (9).

Toxicity determinations are essential, since the decrease in the number of viable cells may be due to a toxic rather than a true desmutagenic or antimutagenic effect. As shown in Chart 1, the determination is carried out in a manner similar to the determination of mutagenic inhibition, except that the mutagen is replaced by histidine. Various concentrations of suspected mutagen inhibitors are tested. In general, if a concentration of 600 μg/plate of crude extract does not significantly decrease the number of colonies found in the positive control, then the sample is earmarked for further purification and, hopefully, isolation of the pure, active compound. It should be realized that the initial finding that a crude extract is nontoxic may be revised when tests are made with more concentrated preparations or pure samples.

ACTIVE PLANTS

Edible Plants

Initially, we screened many edible plants, particularly members of the Brassicaceae (cabbage) family. This group had been studied by Wattenberg for carcinogen inhibitors (17). Our efforts using 2AN were all negative with a variety of edible plants, including many members of the cabbage family. Recently, a reexamination with extracts of collards, broccoli, cabbage and turnip greens has shown moderate activity in the AAF screen (Tab. 1) at high dose levels (600 μg/plate), and this is being further investigated. Negative results were obtained when testing these extracts with the 2AN and EMS screens.

Wild Plants

We have had much greater success in our assays with noncultivated plants collected in many countries. Table 2 gives a listing of plant extracts which show initial activity, as judged by the inhibition of the mutagenic activity of 2AN + S-9 enzyme toward S. typhimurium (T98).

Tab. 1.   Percent inhibition of various mutagens by chloroform extracts of Brassicaceae.

| | Microgram s Extract per Plate | | | | | | | | | | | |
| | 2AN[*] | | | AAF[*] | | | B(a)P[*] | | | EMS[*] | | |
	600	300	150	600	300	150	600	300	150	600	300	150
Cabbage	21	22	8	70	61	49	0	0	0	17	9	13
Collards	23	10	17	66	58	50	0	0	5	21	21	3
Broccoli	27	20	12	66	68	52	0	0	0·	40	24	11
Turnip Greens	17	18	0	48	55	48	0	0	0	56	17	0

[*]   2AN - 2-aminoanthracene
[*]   AAF - acetylaminofluorene
[*]   B(a)P - benzo[a]pyrene
[*]   EMS - ethyl methylsulfonate

Tab. 2.   Plants   active   in   2-aminoanthracene   (2AN)   antimutagenesis screening.

Family Genus Species	Plant Part	Source	Fraction	%Inhibition	Toxicity
Anacardiaceae					
Rhus undulata	tw,lf	S.Africa	CH_2Cl_2	45,83	No Tox
Annonaceae					
Artobotrys velutimus	tw,lf	Ghana	CH_2Cl_2	99,97	Toxic 300,600
Guatteria acutissima	tw	India	CH_2Cl_2	83,68	Toxic 600
Apiaceae					
Ferula sp.	st,sd,lf,fl	Turkey	EtOH	81,83	sl Tox 600
					No Tox 300
Cnidium monniere	sd	China	CH_2Cl_2	93,97,90	No Tox
Araceae					
Anthuricum lancea	lf	Ecuador	$CHCl_3$	98,94,99	Toxic 300,600
Wettenia angusta	st	Peru	CH_2Cl_2	44,42	No Tox
Aristolocheaceae					
Aristolochia debilis ws,sb,rt		China	CH_2Cl_2	69,89,99	No Tox
Asteraceae					
Lagascea mollis	rt,st,lf	India	CH_2Cl_2	96,60,59	No Tox
Carpesium abrolanoides st, lf		China	CH_2Cl_2	42,42	
Mikania cordata	rt,st,lf,fr	India	CH_2Cl_2	62,44,89,47	No Tox
Bignoniaceae					
Markhamia obtusifolia	tw,lf	Tanzania	CH_2Cl_2	80,78	No Tox
Oroxylum indicum	tw,lf	Thailand	CH_2Cl_2	78,99,89	No Tox
Boraginaceae					
Lithospermum erythrorhizin st		China	CH_2Cl_2	50,53	No Tox
Clusiaceae					
Vismia amazonia	tw	Peru	CH_2Cl_2	60,36	No Tox
Combretaceae					
Terminalia boivinii	lf	Tanzania	$CHCl_3$	85,87,88	No Tox
Euphorbiaceae					
Cleislanthus collinus	fr	India	$CHCl_3$	82,36,63	No Tox
Euphorbia corallata	pl	N.C.	H_2O, EtOAc	24,44	No Tox
Fabaceae					
Cassia sp.	rt	Thailand	CH_2Cl_2	50,58	No Tox
Indiofera elliptica	sd	Thailand	EtOH	95	Toxic 600
Psoralea corylifolia	sd	India	CH_2Cl_2	75,98,88	No Tox
Fagaceae					
Castanopsis hystrix	st,fr	India	EtOH	94	Toxic 600
Flacourtiaceae					
Casearia arborea	st	Costa Rica	CH_2Cl_2	43,46,87	No Tox
Lacistema aggregatum	ws-sb	Peru	CH_2Cl_2	44,53	No Tox
Guttiferae					
Garcinia mangostana	sd	India	CH_2Cl_2	95,40	No Tox
Juglandaceae					
Juglans nigra	fr	N.C.	H_2O,n.buoH	46,88	No Tox
Lamiaceae					
Mentha arvensis	lf	China	CH_2Cl_2	70,80	Toxic 600
Phlormis sp.	st,lf,fl	Turkey	CH_2Cl_2	54,36	No Tox
Salvia miltiorrhiza	rt	China	CH_2Cl_2	81,84,78	Toxic 600
Thymus sp.	st,lf,fl,rt	Turkey	CH_2Cl_2	30,44	No Tox

Tab. 2 (continued)

Family	Genus	Species	Plant Part	Source	Fraction	%Inhibition	Toxicity
Loganiaceae							
	Fagraea racemosa		sd	Thailand	EtOH	92	No Tox
Marcgraveaceae							
	Marcgravia trinititis		lf	Peru	CH_2Cl_2	72,47,79	No Tox
Melastomalaceae							
	Bellucia pentamera		ws,sb	Peru	CH_2Cl_2	44,54	No Tox
	Oxyspora paniculata		rt,st,lf	India	EtOH	95	No Tox
Monimiaceae							
	Siparuna decipiens		tw	Peru	EtOH	64,37,31,42	
Myrsinaceae							
	Maesa indica		sd	India	CH_2Cl_2	52,44	No Tox
	Maesa montana		sd	India	CH_2Cl_2	75,54	No Tox
	Maesa tabacifolia		ws,sb	W. Samoa	CH_2Cl_2	42,41	
	Rapanea allenii		tw,lf	Panama	CH_2Cl_2	86,52	No Tox
	Virola elongata		ws-sb	India	CH_2Cl_2	93,94	Toxic 600
Myrtaceae							
	Baeckea latragina		rt,st,lf,fl	W.Austr	CH_2Cl_2	45,73,66	No Tox
	Eugenia egensis		tw	Peru	$CHCl_3$	90,66	No Tox
	Syzygium gratum		sb	Thailand	CH_2Cl_2	27,75	No Tox
Olaceae							
	Heisteria spruceana		tw	Peru	CH_2Cl_2	47,51	No Tox
Papaveraceae							
	Papaver sp.		rt,st,lf,fl	Turkey	CH_2Cl_2	89,91,81	Toxic 600
Polygonaceae							
	Coccaloba densifrons		tw	Peru	EtOH	66,20,75,77	Toxic 600
Polytrichaceae							
	Polytrichia commune		moss	Maine	CH_2Cl_2	62,71	No tox
	Polytrichia juniperus		moss	P.A.	$CHCl_3$	61,65	No Tox
Proteaceae							
	Grevillia baxteri		rt	W.Austr.	CH_2Cl_2	64,98	No Tox
	Grevillia leucopterus		rt	W.Austr.	CH_2Cl_2	92,87,72	Toxic 600
	Hakea costata		rt	W.Austr.	CH_2Cl_2	70,95	No Tox
Rhamnaceae							
	Rhamnus prinoides		st,lf,fr	S.Africa	CH_2Cl_2	89,97,74	No Tox
Rhanunculaceae							
	Clematis buchaniana		st,lf,fr	India	EtOH	84,28,91	No Tox
Rubiaceae							
	Ixora ulei		rt,st,tw	Peru	CH_2Cl_2	78,33,45,65	No Tox
	Rubia cordifolia		st,ws,sb	China	CH_2Cl_2	78,83,88	No Tox
Rutaceae							
	Boronia inornata		rt,st,lf,fl	W.Austr.	CH_2Cl_2	52,92	No Tox
	Euodia ruticarpa		fr	China	CH_2Cl_2	45,49	No Tox
	Melicope sissiliflora		lf	Aust.	$CHCl_3$	91,99	Toxic 600
	Zanthoxylum leibmannianum		tw,lf,inf	Mexico	$CHCl_3$	40,41	No Tox
Sapindaceae							
	Dodonea viscosa		tw,lf	S.Africa	CH_2Cl_2	36,57	No Tox
Solanaceae							
	Brunfelsia grandiflora		tw	Peru	CH_2Cl_2	53,85	No Tox

Tab. 2 (continued)

Family Genus Species	Plant Part	Source	Fraction	%Inhibition	Toxicity
Thymelacaceae					
Daphne sp.	lf,fl	Turkey	CH_2Cl_2	59,89,52,43,48	No Tox
Daphne genkwa	fl	China	CH_2Cl_2	63,53	No Tox
Turneraceae					
Turnea acuta	lf	Peru	CH_2Cl_2	98,87	Toxic 600
Verbenaceae					
Vitex quinata	tw,lf	Thailand	CH_2Cl_2	49,59,47,56	No Tox
Vitex negundo	lf	Thailand	CH_2Cl_2	46,34	
Vitaceae					
Cayratia japonica	st,lf	China	CH_2Cl_2	78,33,45,65	No Tox
Xanthorrhoeaceae					
Xanthorrhoea preissii	st,lf	W.Austr.	CH_2Cl_2	81,84,88	Toxic 300,600

The solvents used for the crude extracts were largely 95% ethanol or organic solvents such as chloroform or methylene chloride, after partitioning the residue from ethanol extraction between organic solvent and water. The activity was almost invariably found in the organic solvent. A limited number of direct aqueous extractions were conducted, usually with negative results. Table 3 presents a summary of our screening data. Approximately 2,750 wild plants have been screened for mutagenicity inhibition of 2AN toward S. typhimurium, and 3.5% of all the species (samples) tested were nontoxic and active.

Marine samples have also been tested recently, including marine animals, corals and sponges, and algae. As shown in Tab. 3, algae seem to be the most promising group for investigation.

ISOLATION PROCEDURES

We have published in detail isolation and structure elucidation procedures (9-14). In brief, the methodology is based on our group's

Tab. 3. Summary of screening data.

TOTAL NUMBER OF WILD PLANTS	2747
NUMBER OF NON-TOXIC FAMILIES	58
NUMBER OF NON-TOXIC SPECIES	97 (3.5%)
MARINE SAMPLES	
ALGAE	
NUMBER SAMPLES TESTED	50
NUMBER NON-TOXIC ACTIVES	10 (20%)
CORAL AND SPONGES	
NUMBER SAMPLES TESTED	96
NON-TOXIC ACTIVES	3 (3.1%)
VARIOUS MARINE ANIMALS	
NUMBER TESTED	186
NON-TOXIC ACTIVES	0

previous experience with bioassay-supported isolation of antitumor agents (8), modified to operate on extracts from as little as 0.25-0.5 kg dry plant material.

Chromatography

The organic solvent extract ($CHCl_3$ or CH_2Cl_2) is chromatographed on silica gel and a number of fractions collected and assayed for 2AN inhibition at 0.6 mg/plate. Usually 6-10 fractions are collected, using gradient elution. We usually use CH_2Cl_2 with methanol gradients from 0.5 to 5% in CH_2Cl_2 as eluants, but many other solvent combinations can be utilized. Fractions which exhibit 2AN inhibition are further purified by possibly another chromatography on silica gel or Sephadex LH-20 (9-11). Frequently, we are able to achieve rapidly final purification by preparatory HPLC (9-11).

Table 4 exemplifies our chromatographic procedures. A crude CH_2Cl_2 extract from Psorolea corylifolia was chromatographed on silica gel, using a fraction collector. Fractions with similar thin-layer chromatography (TLC) patterns were combined and assayed for 2AN inhibition. Many active fractions were obtained, as shown in Tab. 4. All of the active fractions were further purified. Figure 1 shows the HPLC purification of one of these fractions, 232-255. Fractions with retention times from 24-36 min had similar TLC patterns, were combined and the pure active compound crystallized from the elution solvent as it was concentrated. From the melting point, mass spectral, and NMR properties, this compound was found to be the known compound neobavaisoflavone (1,6). In any event, one of the pitfalls in our procedure was demonstrated by the fact that although the crude extract was nontoxic, the pure compound was toxic at high dose levels.

Structure Determination

Purified material is crystallized and high resolution mass spectrometry analysis is immediately carried out on the compound so as to obtain the molecular formula. Then ultraviolet, infrared, and, in particular, high resolution proton nuclear magnetic spectroscopy are conducted.

Known Compounds

At this point a computer-based search of Chemical Abstracts is conducted, using the molecular formula as a basis for the search. In this manner, known compounds can be quickly identified; or, as frequently occurs, if a number of compounds with the same molecular formula are found, comparison of the spectroscopic properties of the known compound with the unknown can be made.

New Compounds

If the isolated compound is new, more sophisticated NMR spectroscopy is utilized, including ^{13}C-NMR, and various two-dimensional NMR procedures can be utilized (12-14). In most cases, the structure of the unknown compound can be readily obtained. If we are dealing with a complex structure involving a number of optical centers, and if the crystal form is appropriate, modern X-ray analysis quickly reveals the structure. Recently, in one week, we obtained in this manner the structure of a new bromine-containing compound from a marine alga (15).

Tab. 4.   Chromatographic procedures.

15 ml Fraction	Wt (mg)	Eluting Solvent	2AN Activity
1-10	35	1% MeOH + CH_2Cl_2	
11-27	1521		96%
28-46	279		
47-63	49		
64-77	62		85%
78-87	67		neg
88-110	107		
111-125	203		neg
126-133	328	2% MeOH + CH_2Cl_2	neg
134-142	577		97%
143-166	568		
167-173	92		91%
174-178	94		
179-186	216		98%
187-206	494		56%
207-218	381		
219-231	325		98%
232-255	345	4% MeOH + CH_2Cl_2	
256-282	200		neg
283-298	94		
299-330	129		neg
331-355	106	6% MeOH + CH_2Cl_2	
356-393	91		
394-415	55		
416-445	64	8% MeOH + CH_2Cl_2	
446-475	82		neg
S.M.			95%

11   12-27   28-46   47-63   64-77   88-87   88-110   111-125   126-133   134-142   143-166   SM   167-173   174-178   179-186   187-206   207-218   219-231   232-255   256-282

ANTIMUTAGENIC COMPOUNDS ISOLATED

Flavonoids

In our initial studies, certain flavonoids were initially isolated as a consequence of a bioassay-guided procedure, using the inhibition of 2AN mutagenicity.   Chart 2 lists a number of flavones and isoflavones and a few related compounds.   Those indicated by an asterisk were isolated at

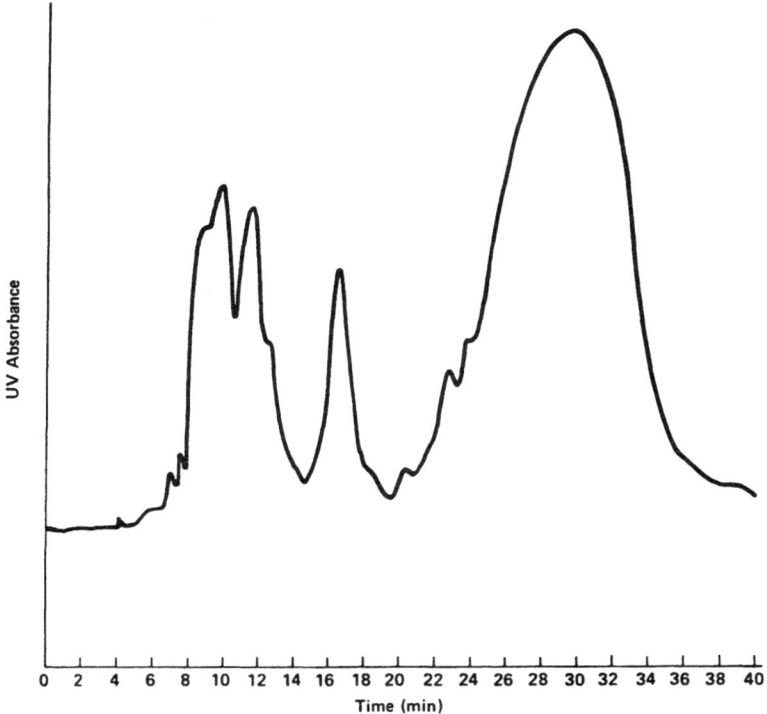

Fig. 1. HPLC purification.

RTI. Others were purchased, our idea being to test a variety of different structural types. Both inhibition and toxicity assays were conducted at high dose levels, 600 µg/plate. Until further studies are carried out at lower dose levels for both 2AN inhibition and toxicity, as well as inhibition of other mutagens such as AAF and EMS, relatively few conclusions can be made. The data found in Chart 2 and Tab. 5 and 6 permit some observations. Activity can be found in both flavones and isoflavones. Water-soluble flavonoids, i.e., those with sugars attached to the flavonoid skeleton, are in general of low activity or inactive. Table 5 shows differences in mutagen inhibition profiles of the various compounds. Apigenin was inactive in 2AN and AAF but quite active in BAP. Quercitin displayed moderate activity in 2AN inhibition but was inactive with other mutagens.

Table 6 presents mutagenic inhibition data for some highly active new compounds. Fremontin and fremontone are two new isoflavonoids recently isolated at our laboratory (13). They are identical except that the former has a 5'-prenyl substituent (cf. Chart 3), whereas the latter has both 3'-αα-dimethylallyl and 5'-prenyl substituents. Fremontin and fremontone are nontoxic at high dose levels. As shown in Tab. 6, they are both highly inhibitory to the mutagenic action of 2AN and EMS on S. typhimurium at dose levels as low as 37.5 µg/plate, but are much less active in AAF, thus showing interesting differences in their effects on various mutagens.

Chart 2. Inhibition of mutagenicity of 2-aminoanthracene (2AN) by various flavonoids.

Compounds	Structure	%Inhibition of 2AN Mutagenicity Dose 600 µg/plate

Flavonoids

Daidzein*

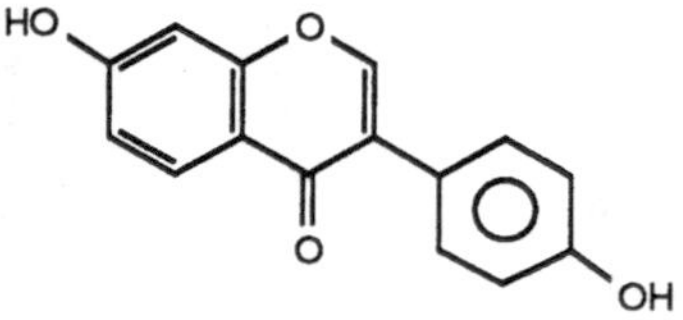

32% (N.T.)

Apigenin

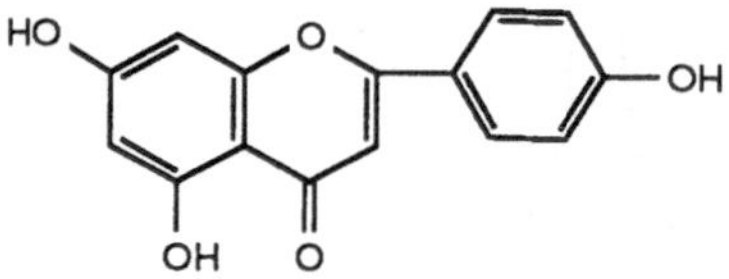

10% (N.T.)

Formononetin*

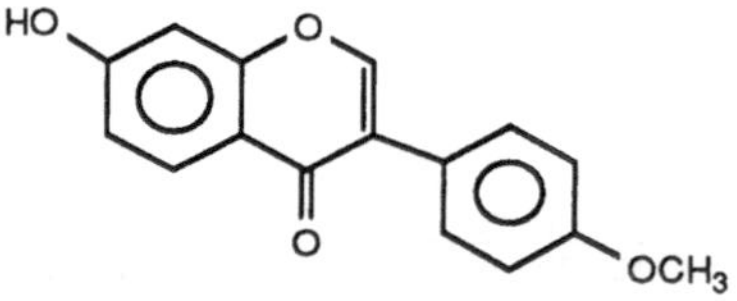

0% (N.T.)

Biochanin A*

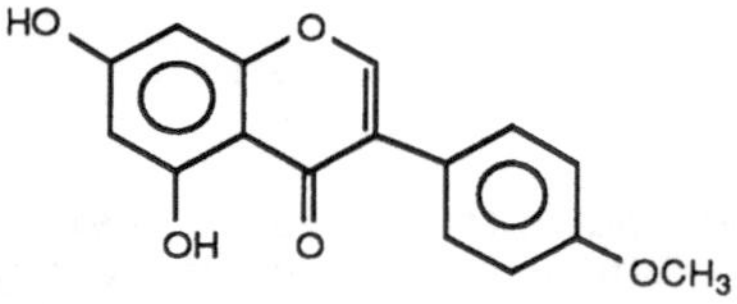

57% (N.T.)

Neobavaisoflavone*

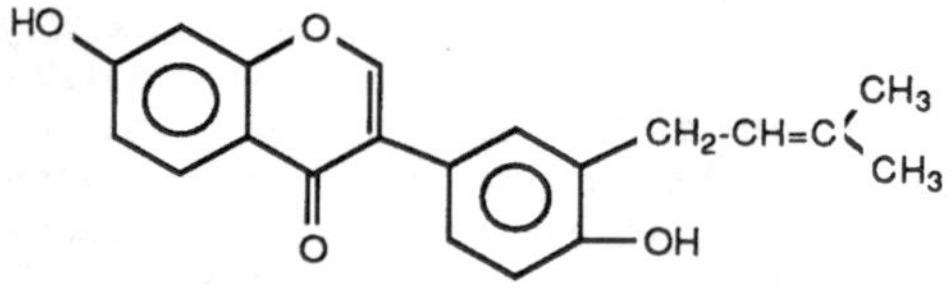

97% (T.)

Bavachin*

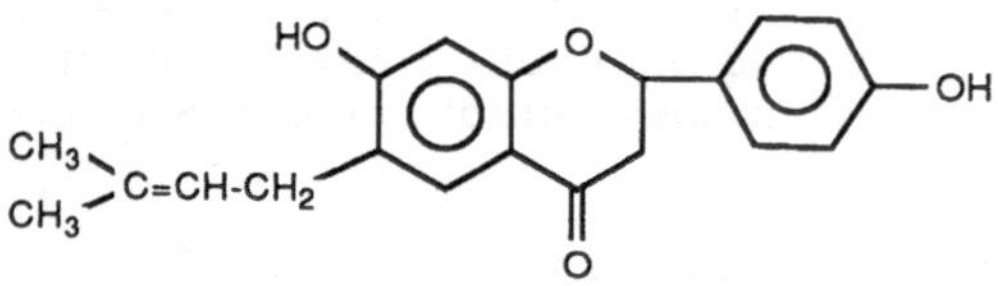

62% (T.)

Isobavachalcone*

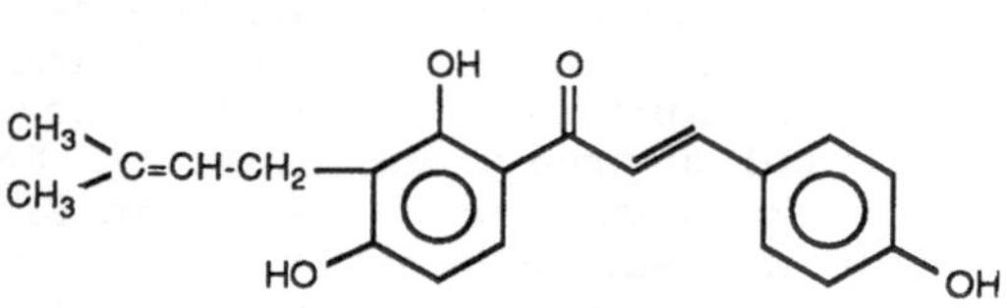

86% (T.)

* Isolated at RTI
N.T. = nontoxic
T. = toxic

Chart 2 (continued)

Compound	Structure	%Inhibition of 2AN Mutagenicity
		Dose 600 µg/plate

Galangin

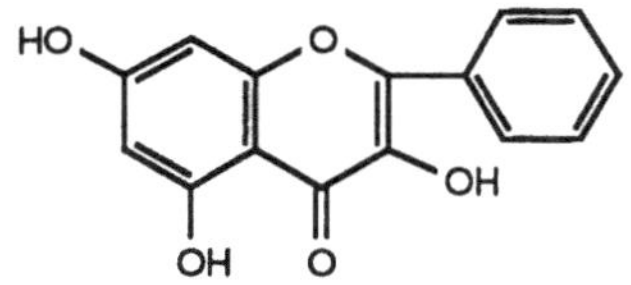

97% (N.T.)

Quercetin

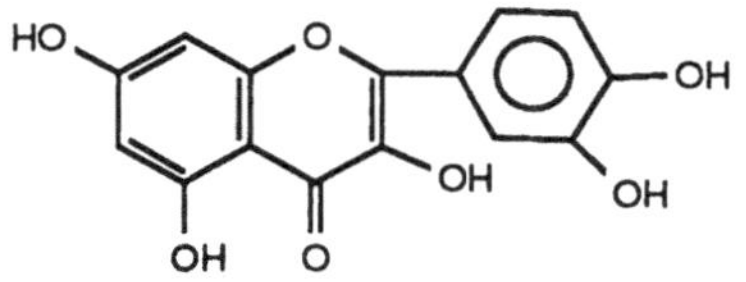

57% (N.T.)

Morin

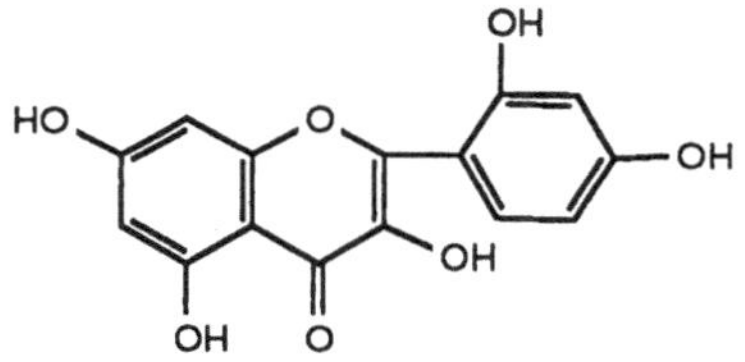

0% (N.T.)

Rhamnetin

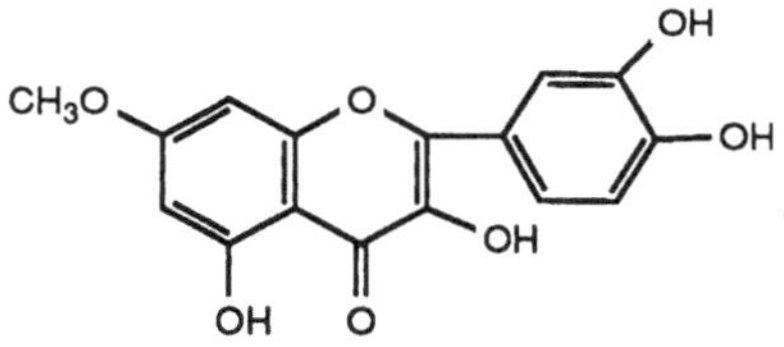

78% (N.T.)

Tangeretin

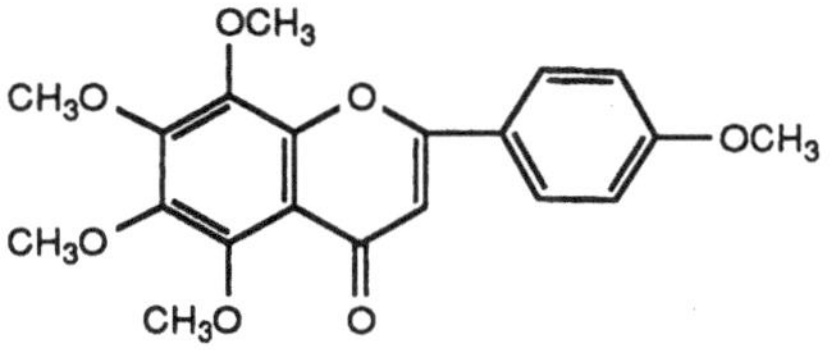

42% (N.T.)

Nobiletin

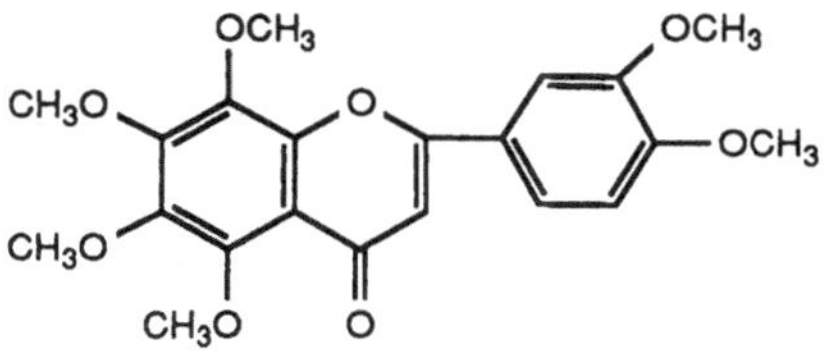

43% (N.T.)

Naringin

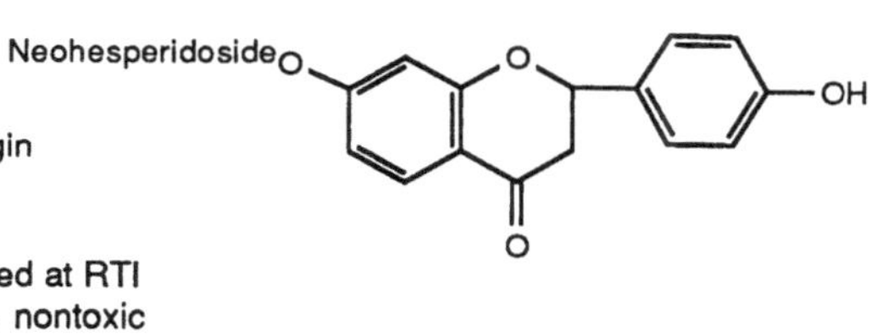

38% (N.T.)

* Isolated at RTI
N.T. = nontoxic
T. = toxic

Chart 2 (continued)

Compound	Structure	%Inhibition of 2AN Mutagenicity Dose 600 µg/plate

Hesperidin	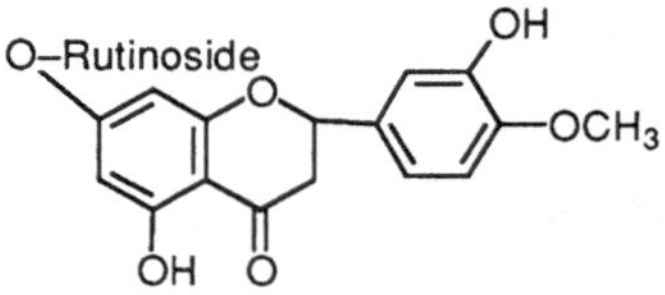	13% (N.T.)
Rutin	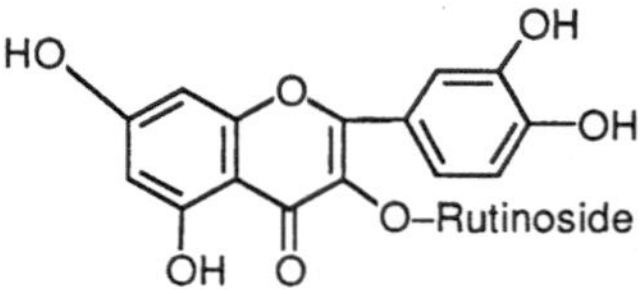	0% (N.T.)
Bakuchiol*	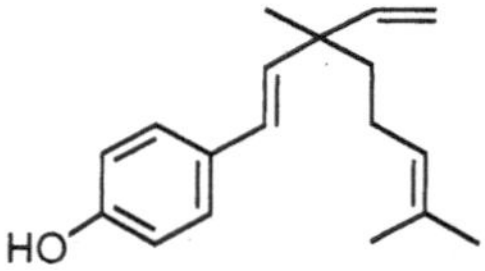	81% (T.)
Acacetin*	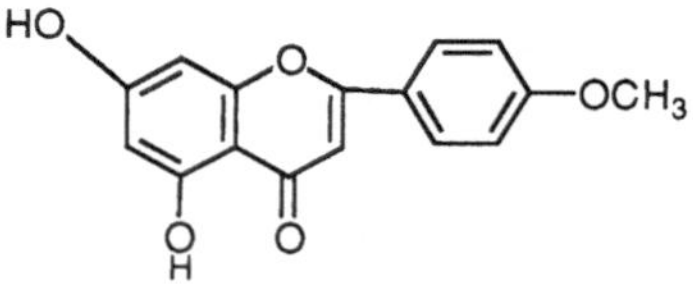	68% (N.T.)
Oroxylin A*	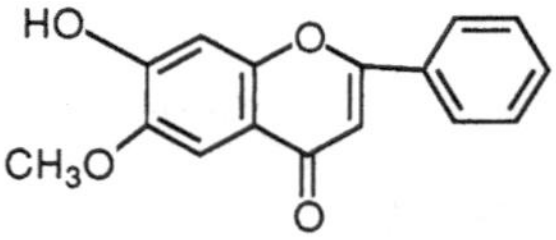	48% (N.T.)
Chrysin	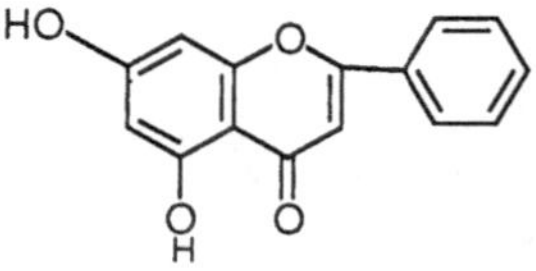	71% (T.)

* Isolated at RTI
N.T. = nontoxic
T. = toxic

Tab. 5. Percent inhibition of mutagenic activity by various flavonoids.

Micrograms Compound/Plate

	2AN[a,f]			AAF[b,f]			B[a]P[c,f]			2NF[d]			EMS[e]		
	600	300	150	600	300	150	600	300	150	600	300	150	600	300	150
Flavonoids															
Apigenin	11	0	0	0	14	0	75	64	67	88	70	60			
* Neobavaisoflavone	97	67	98	96	72	2	97	74	17	66	53	15			
* Bavachin	62	98	98		0	0		0	0		72	52			
Quercitin	57	55	51	0	0	0	0	0	0	0	0	0			
Rhamnetin	78	0	0		52	11		45	7		41	34			
* Acacetin	68	2	8	0	0	0	27	17	30				45	41	33
* Oroxylin A	57	56	64	64	65	61	53	81	77				19	23	49
Terpene phenol															
* Bakuchiol	81	84	91	0	0	0	43	28	0	92	79	75			

a 2AN = 2-aminoanthracene
b AAF = Acetylaminofluorene
c B[a]P = Benzo[a]pyrene
d 2NF = Nitrofluorene
e   = Ethyl methysulfonate
f   = Requires metabolic activation by S9
*   = Isolated at RTI

Tab. 6. Percent inhibition of various mutagens by various phenolic compounds.

Micrograms Compound/Plate

	2AN				AAF				EMS			
Compound	300	150	75	37.5	300	150	75	37.5	300	150	75	37.5
Fremontin	93	24	0	65	0	0	0	0	87	81	90	58
Fremontone	93	17	78	89	70	27	3	0	94	86	81	63
Intricatin	300	150	75	37.5	300	150	75	37.5	300	150	75	37.5
	62	84	92	70	0	0	0	0	45	27	0	0
Intricatol	300	150	75	37.5	300	150	75	37.5	300	150	75	37.5
	87	57	36	8	70	54	48	26	74	82	74	30
4,5-Dihydroxy-xanthone									68	70	41	
1,5-Dihydroxy-8-methoxyxanthone	72	60			52	41			96	95	14	
Barbatol	89	98	89		0	0	0		90	88	87	85
Isobarbatol	97	98	92									
2,3-Dihydroxy-5-ethoxy-acetophenone	38	0	0		49	12	6		99	84	82	52

Chart 3. Inhibition of mutagenicity by phenolic compounds.

Compounds	Structure	% Inhibition of 2AN Mutagenicity
4,6-Dihydroxyxanthone	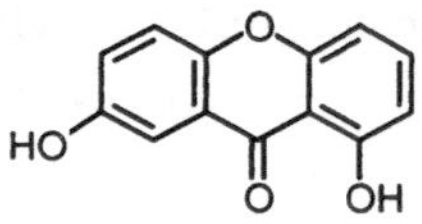	43% (NT.)
1,5-Dihydroxy - 8-methoxy Xanthone	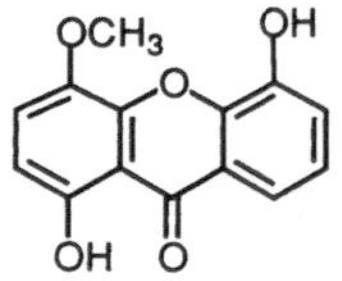	72% (NT.)
7,8-Dihydroxy-4'-methoxyl-Homoisoflavone (Intricatol)	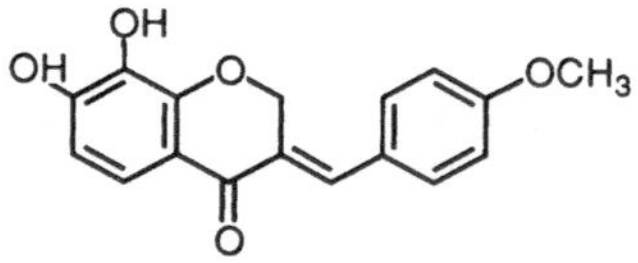	87% (NT.)
8-Hydroxy-7,4'-dimethoxy-Homoisoflavone (Intricatin)	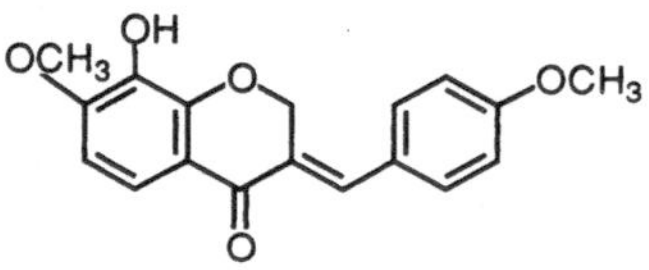	62% (NT.)
-2,3-Dihydroxy-ethoxy-acetophenone	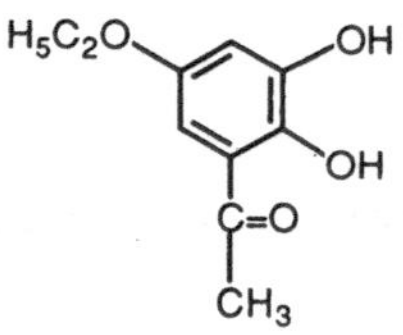	62% (N.T.)
Fremontin	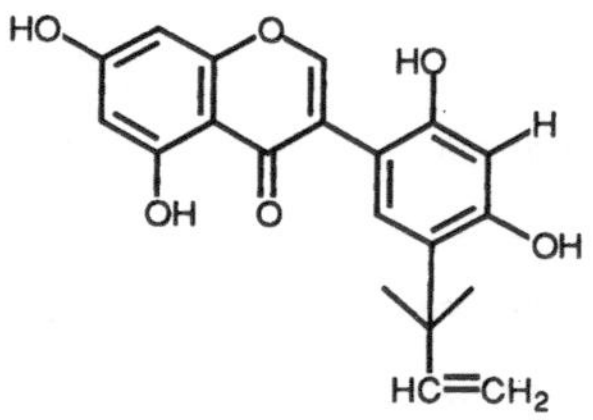	93% (NT.)
Fremontone	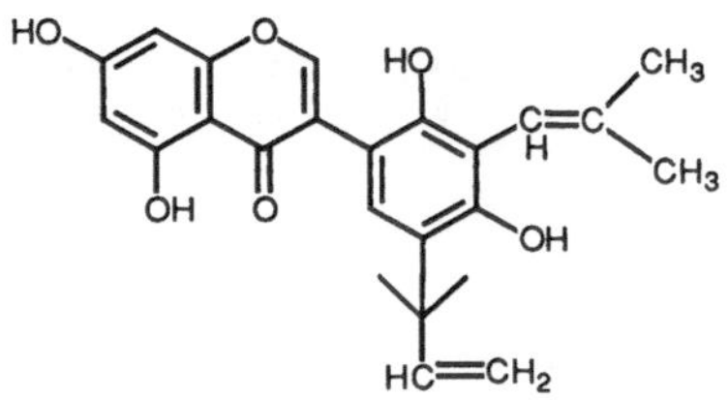	93% (NT.)
Barbatol	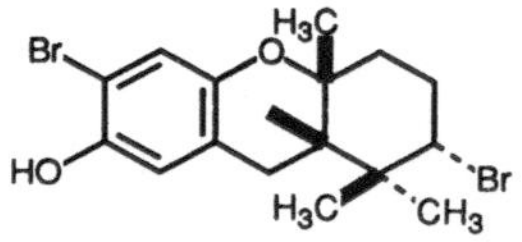	

All isolated at RTI.
N.T. = nontoxic

Intricatin and intricatol are two new homoisoflavonoids (14) which are nontoxic at high dose levels. Structurally, they differ only slightly (cf. Chart 3), intricatin being the 7-methoxy analog of intricatol. Both compounds are active in the inhibition of the mutagenic action of 2AN, but intricatol is much more active than intricatin in the inhibition of AAF and EMS. The differences in activity in this case seem due to the 7.8-dihydroxy-moiety present in intricatol.

PHENOLIC COMPOUNDS

Several highly active new phenolic compounds have been recently isolated in our laboratory, all of which are nontoxic at high dose levels (300-600 µg/plate). As shown in Tab. 6 and Chart 3, these include two new xanthones, a brominated phenolic compound, and an acetophenone derivative.

Xanthones

The mutagenic inhibitory effects of two new xanthones which differ slightly in structure (cf. Chart 3) are shown in Tab. 6. Both are nontoxic at high dose levels and show considerable activity in the inhibition of the mutagenic activity in 2AN and EMS toward S. typhimurium. A paper is in preparation (16).

Algae Compounds

Barbatol and isobarbatol are two new brominated compounds obtained from blue-green algae with novel structures (cf. Chart 3). Although nontoxic at high dose levels, both have extremely high mutagenic inhibitory activity in 2AN and, in the case of barbatol, in EMS also. Surprisingly, barbatol was inactive in AAF. Further testing is in progress.

Tab. 7.   Percent inhibition of mutagenic activity by various coumarins.

	Micrograms Compound/Plate														
	2AN[a]			AAF[b]			B[a]P[c]			2NF[d]			EMS[e]		
	600	300	150	600	300	150	600	300	150	600	300	150	600	300	150
Coumarins															
Coumarin	22	0	0	36	28	5	53	16	10	73	17	24			
Imperatorin*	81	45	41	58	42	30	92	92	94	58	43	22			
Osthol*	68	29	13	74	56	51	54	49	13	80	55	29			
8-methoxypsoralen	44	0	0	39	21	0	86	76	58	25	39	5			
5-methoxypsoralen	23	0	0	0	0	0	75	67	61	40	41	27			
Psoralen*	61	45	30	51	10	0	78	62	55				81	38	17

[a] 2AN   = 2-Aminoanthracene
[b] AAF   = Acetylaminofluorene
[c] B[a]P = Benzo[a]pyrene
[d] 2NF   = Nitrofluorene
[e] EMS   = Ethyl methylsulfonate
*   = Isolated at RTI

The structure of barbatol (Chart 3) was determined by X-ray analysis. The structure of isobarbatol is being studied (15).

Acetophenone Analog

2.3-Dihydroxy-5-ethoxy-acetophenone was obtained from a lichen. It shows strong activity in the inhibition of EMS mutagenicity but not in 2AF or AAF (16).

Chart 4. Structure and inhibition of 2-aminoanthracene (2AN) mutagenicity by various coumarins.

Compounds	Structure	%Inhibition of 2AN Mutagenicity
Coumarins		Dose 600 µg/plate
Coumarin	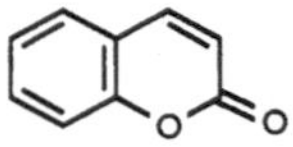	22% (N.T.)
Umbelliferone	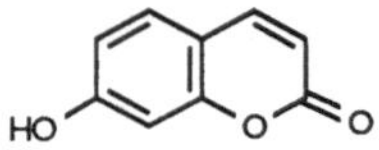	14% (N.T.)
Psoralen*	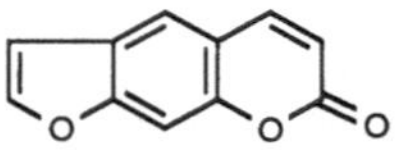	60% (N.T.)
Imperatorin*	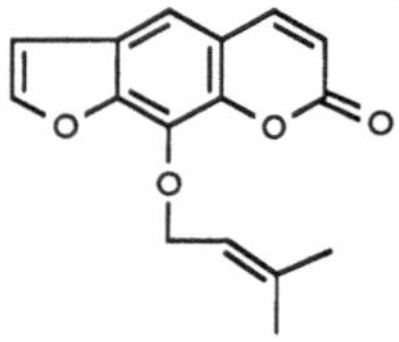	81% (N.T.)
Osthol*	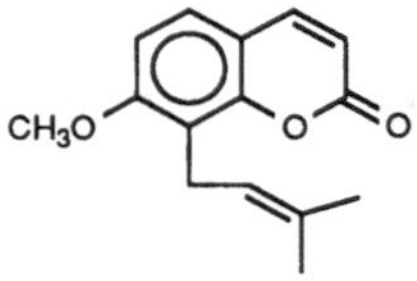	68% (N.T.)
8-methoxypsoralen	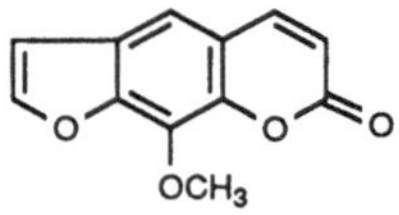	44% (N.T.)
5-methoxypsoralen	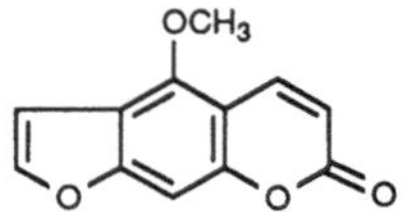	23% (N.T.)

* Isolated at RTI
N.T. = nontoxic

Coumarins

Coumarins are plant metabolites, almost as ubiquitous in nature as flavonoids. A few have been reported previously (9,11) (Chart 4 and Tab. 7). We have obtained the present structures and inhibition data with some of the compounds isolated in our laboratory and with several purchased for comparison purposes. All compounds we tested in this series were nontoxic at high dose levels (600 μg/plate). The coumarins seemed to range from moderate to poor activity in 2AN and AAF. Most of the compounds displayed better activity in inhibition of BAP mutagenicity.

REFERENCES

1. Bajwa, B.S., P.L. Khanna, and T.R. Seshadri (1974) Indian J. Chem. 12:15.
2. Birt, D.F., B. Walker, M.G. Tibbels, and E. Bresnick (1986) Antimutagenesis and anti-promotion by apigenin, robinetin and indole-3 carbinol. Carcinogenesis 7:959-963.
3. Ishii, R., H. Yoshikawa, H. Minakata, N.T. Komura, and Kada (1984) Specificity of bio-antimutagens in the plant kingdom. Agric. Biol. Chem. 48:2587-2591.
4. Maron, D.M., and B.N. Ames (1983) Revised methods for the Salmonella mutagenicity tests. Mutat. Res. 113:173-215.
5. Mitscher, L.A., S. Drake, S.R. Gollapudi, J.A. Harris, and D.M. Shankel (1986) Isolation and identification of higher plant agents active in antimutagenic assay systems: Glycyrrhiza glabra. In Antimutagenesis and Anticarcinogenesis Mechanisms, D.M. Shankel, P.E. Hartman, T. Kada, and A. Hollaender, eds. Plenum Press, New York, pp. 153-165.
6. Nakayama, M., S. Eguchi, S. Hayashi, M. Tsukayama, T. Horie, T. Yamada, and M. Masumura (1978) The synthesis of neobavaisoflavone and related compounds. Chem. Soc. Japan 51:2393-2400.
7. Shankel, D.M., P.E. Hartman, T. Kada, and A. Hollaender, eds. (1986) Antimutagenesis and Anticarcinogenesis Mechanisms, Plenum Press, New York.
8. Wall, M.E., M.C. Wani, and H.L. Taylor (1976) Isolation and chemical characterization of antitumor agents. Cancer Treat. Repts. 60:1011.
9. Wall, M.E., M.C. Wani, T.J. Hughes, and H. Taylor (1988) Plant antimutagenic agents. 1. General bioassay and isolation procedures. J. Nat. Prod. 51:866-873.
10. Wall, M.E., M.C. Wani, G. Manikumar, P. Abraham, H. Taylor, T.J. Hughes, J. Warner, and R. McGivney (1988) Plant antimutagenic agents. 2. Flavonoids. J. Nat. Prod. 51:1084-1091.
11. Wall, M.E., M.C. Wani, G. Manikumar, T.J. Hughes, H. Taylor, R. McGivney, and J. Warner (1988) Plant antimutagenic agents. 3. Coumarins. J. Nat. Prod. 51:1148-1152.
12. Wall, M.E., M.C. Wani, K. Gaetano, G. Manikumar, H. Taylor, and R. McGivney (1988) Plant antimutagenic agents. 4. Isolation and structure elucidation of maesol. J. Nat. Prod. 51:1226-1231.
13. Wall, M.E., G. Manikumar, K. Gaetano, M.C. Wani, H. Taylor, T.J. Hughes, J. Warner, and R. McGivney (1989) Plant antimutagenic agents. 5. Isolation and structure of two new isoflavonones, fremontin and fremontone, from Psorothamnus fremontii. J. Nat. Prod. (submitted for publication).

14. Wall, M.E., M.C. Wani, G. Manikumar, H. Taylor, and R. McGivney
     (1989) Plant antimutagen. 6. Intricatin and intricatol, new antimuta-
     genic homoisoflavonoids from Hoffmansseggia intricata. J. Nat.
     Prod. (submitted for publication).
15. Wall, M.E., M.C. Wani, K. Gaetano, H.L. Taylor, R. McGivney, and
     W. Gerwick (1989) Plant antimutagen. 7. Isolation, structure and
     antimutagenic properties of barbatol and related compounds. J. Nat.
     Prod. (submitted for publication).
16. Wall, M.E., M.C. Wani, G. Manikumar, H. Taylor, R. McGivney,
     T.J. Hughes, and J. Warner (1989) Plant antimutagenic agents. 8.
     Isolation, structure elucidation and antimutagenic properties of var-
     ious phenolic compounds. J. Nat. Prod. (submitted for publication).
17. Wattenburg, L.W. (1983) Inhibition of neoplasia by minor dietary
     constituents. Cancer Res. 43(Suppl.):2448-2453.

HUMAN BIOMONITORING IN EXPOSURE

TO ENVIRONMENTAL GENOTOXICANTS

M. Sorsa, M. Hayashi, H. Norppa, and H. Vainio

Institute of Occupational Health
Helsinki, Finland

INTRODUCTION

Humans have throughout their evolution been exposed to a multitude of genotoxicants via food and drink intake, natural dietary contaminants, and emissions of smoke. The present-day concern of health hazards with chemical exposures is, however, directed toward man-made chemicals, the numbers and production figures of which are constantly showing a growing trend. However, the "natural" sources of genotoxicant exposure have not been decreasing, especially the complex emissions of organic combustion, which have dramatically increased with fossil fuel energy production and transportation exhausts. Contaminants in the atmosphere are deposited in the soil and in waters, and may thus further contribute to exposures through water and food.

In real life, human exposure patterns are thus enormously complicated, with respect to both the multitude of single agents and the complex mixtures involved. The picture is further complicated by agents and exposures acting synergistically or with contradicting inhibitory effects, i.e., antigenotoxicity. In the evaluation of the biologically significant exposures, therefore, it may be more relevant to look for early effects directly in the exposed individuals or groups than to try to predict potential hazards of the complex exposure patterns from data with single compounds. Focusing on the applicable methodologies in human monitoring is getting increasing attention.

BIOMONITORING OF EXPOSURE TO GENOTOXICANTS

Human biomonitoring is an important tool in the evaluation of exposures to genotoxic chemicals and, thus, in the prevention of the adverse effects of chemicals. These methods are also useful in the assessment of health risks, although hazard assessments for most exposures are still largely in the research stage (see Tab. 1). Presently, there is a need for thorough multidisciplinary studies enabling the follow-up of the different biological parameters in conjunction with ambient monitoring and epidemiology.

Tab. 1.   Strategic scheme of methodologies available in assessment of
          exposure and adverse effects of genotoxicants (modified from
          Ref. 4).

EXTERNAL DOSE	Ambient monitoring
↓	
INTERNAL DOSE	Biological monitoring s. str. Mutagenicity in urine Thioethers in urine
↓	
BIOLOGICALLY EFFECTIVE DOSE	Adducts in proteins, DNA, RNA
↓	
EARLY BIOLOGICAL EFFECTS	UDS Gene mutations Chromosomal aberrations
IN SOMATIC CELLS	SCEs Micronuclei
IN GERMINAL CELLS	Gene mutations Structural and numerical chromosomal aberrations
↓	
CLINICAL RESPONSES	Epidemiology of end effects

Today, most experience on human biomonitoring still relates to cytogenetic surveillance of groups of people, mostly from "high-exposure" occupational situations.

Cytogenetic surveys of defined exposures can be accomplished, but
the difficulties in extrapolation from small populations to large populations
with very low level exposures pose substantial problems.   That is, the
elevated frequencies of chromosomal aberrations can signal a potential
hazard in the population, but extrapolation of the risk in quantitative
terms is extremely difficult.   Thus, chromosomal changes are an indicator
of cellular genetic damage in a population, but they cannot at present be
used to predict future health effects for a given individual.   The main
relevance of induced chromosomal damage is its indication that exposures
have been able to damage the genetic material in human cells.

Many potential confounders may occur in studies concerning human
populations; they may derive from technical factors, but also from factors
concerning the definition of exposure of the persons to be studied (see
Ref. 11).

ENVIRONMENTAL GENOTOXICANTS AND CYTOGENETIC DAMAGE

Environmental Genotoxicants

The definition of "environmental" compounds may vary--here it is
taken to cover environmentally derived man-made chemicals which expose
humans via food, drink, or inhalation.   Many, and, in fact, the most intense chemical exposures occur in the occupational environment, and since
exactly the same compounds expose humans outside or inside the work-
place, there should, purely on scientific grounds, be no difference in the

scope, even though legal and administrative functions in most countries are separate for the occupational and the general environment. The term "genotoxicant" is used here when referring to compounds that have been shown to be genotoxic in experimental systems or in humans, paying special attention to genotoxicity visualized as chromosomal damage--the endpoint from which most human data are derived.

Cytogenetic Endpoints and Environmental Carcinogens

Cytogenetically recognizable chromosomal damage includes structural aberrations (CAs), in which a gross change in the morphology of a chromosome has occurred; micronuclei (MN), which arise from acentric chromosome fragments or lagging whole chromosomes; and sister chromatid exchanges (SCEs), which represent symmetrical intrachromosomal exchanges between the two sister chromatids. The same endpoints, often in different target tissues, can be used both for experimental animals and for humans.

In the most recent evaluations of the International Agency for Research on Cancer (IARC) (10), 50 agents, processes and treatments were considered carcinogenic to humans (group 1), 37 agents probably carcinogenic to humans (group 2A), and 158 agents possibly carcinogenic to humans (group 2B). Many of these compounds can be considered environmental carcinogens in view of the criteria presented above. The preceding evaluation volume by the IARC (9) deals with genotoxicity and short-term test data, including data on genetic effects in humans. Cytogenetic data is available on only 44 of the environmental carcinogens; combining the information from human and animal studies reveals a shortage of data to a surprising degree (Tab. 2).

The data on micronuclei in human biomonitoring is almost completely lacking; for ethylene oxide and styrene/styrene oxide, single studies have shown increased levels of micronucleated lymphocytes in occupational exposures (7,8). For very few compounds--benzene, vinyl chloride, ethylene oxide, and styrene/styrene oxide--there is complete or almost complete concordance of results in experimental animals and in humans. The reasons for discrepancies are primarily the insensitivity of cytogenetic methodologies in vivo and, second, the obvious differences in the primary lesions of the three endpoints. A further message from the comparative data of Tab. 2 is the complete lack of human cytogenetic data on exposures in the general environment: normally such exposures are too low, too complex, and too confounded for cytogenetic studies. However, occupational studies have an indicator value to identify hazardous exposures of the general population.

An additional problem in cytogenetic surveys of populations exposed environmentally is in obtaining statistically significant results with regard to very low-level exposures, and then determining the meaning of such results for health effects in the population as a whole and in the individuals on which the observations were made. An increase of chromosomal aberrations can signal a potential problem in the population, but quantification of the abnormalities is extremely difficult. Thus, although chromosomal changes are an indicator of cellular genetic damage in a population, they cannot be used quantitatively to predict future health effects for a given individual.

Tab. 2.   Environmental carcinogens (10) and cytogenetic endpoints in humans and experimental animals (9).

Compound/Exposure	Humans			Animals		
	CA	MN	SCE	CA	MN	SCE
1' CARCINOGENIC TO HUMANS						
Arsenic & its compounds	?	..	?	(+)	(+)	..
Asbestos	..	..	?	(-)	(-)	..
Benzene	+	..	(-)	+	+	+
Chromium & its compounds (VI)	(+)	(-)	?	+	+	..
Nickel & its compounds	?	..	-	..	..	..
Vinyl chloride	+	..	?	(+)	+	+
2A PROBABLY CARCINOGENIC TO HUMANS						
Acrylonitrile	(-)	..	..	-	(-)	..
Benz[a]anthracene	..	..	..	?	(+)	(+)
Benzo[a]pyrene	..	..	..	+	..	+
Cadmium & its compounds	?	..	(-)	-	-	..
Dibenz[a,h]anthracene	..	..	..	(-)	(-)	(+)
Ethylene dibromide	-	..	-	(-)	(-)	(+)
Ethylene oxide	+	(+)	+	+	+	+
Formaldehyde	?	..	?	-	-	(-)
Polychlorinated biphenyls	..	..	..	-	?	..
Propylene oxide	..	..	..	..	(+)	..
Styrene oxide	..	..	..	?	-	-
2B POSSIBLY CARCINOGENIC TO HUMANS						
Acetaldehyde	..	..	..	(+)	..	+
Acrylamide	..	..	..	+	..	..
Amitrole	..	..	..	..	-	..
Auramine	..	..	..	..	(-)	..
Benzo[b]fluoranthene	..	..	..	(-)	..	(+)
1,3-Butadiene	..	..	..	..	(+)	(+)
Carbon tetrachloride	..	..	..	(-)	..	..
α-Chlorinated toluenes	..	..	..	..	-[1]	..
Chloroform	..	..		..	-	..
Chlorophenols	(+)[2],(+)[3]	..	(-)[2,3]	..	..	..
2,4-Diaminoanisole	..	..	..	..	-	..
Dichloroethane	..	..	..	(-)	(-)	..
Ethyl acrylate	..	..	..	..	(+)	..
Ethylene thiourea	..	..	..	(-)	-	(-)
Hexachlorocyclohexanes	(-)	..	..	-	..	..
Hydrazine	..	..	..	..	(-)	..
Lead and lead compounds,inorganic	?	..	?	?	?	(-)
4,4'-Methylene bis(2-methylaniline)	.. ..		..	..	(+)	..
4,4'-Methylenedianiline	..	..	..	..	..	+
Nitrofen (technical-grade)	..	..	..	(-)	(-)	..
Polybrominated biphenyls	..	..	..	-	(-)	..
Potassium bromate	..	..	..	..	+	..
Saccharin	..	..	..	-[4,5]	(-)[4,5]	(+)[5]
Styrene	+	+	?	?	(+)	+
2,3,7,8-Tetrachlorodibenzo-para-dioxin (TCDD)	?	..	..	-	(-)	-
Tetrachloroethylene	(-)	..	(-)	..	..	..
ortho-Toluidine	..	..	..	..	(-)	?

+: positive results   -: negative results
( ): single study only   ?: equivocal or inconclusive
1: Benzyl chloride
2: Pentachlorophenol
3: 2,4,5-Trichlorophenol
4: Saccharin, unspecified
5: Saccharin, sodium

The Relevance of Cytogenetic Damage

The main conceptual basis for the application of cytogenetic assays is that damage to the genetic material of cells represents initial events in a process that may eventually lead to manifestations of ill health. Thus, cytogenetic surveillance can serve as an early indicator, enabling the prevention of potential adverse effects.

Chromosomal abnormalities are characteristic features of malignant cells. An overwhelming amount of data exists concerning the association of chromosomal aberrations with cancer, and several correlations have been established between specific nonrandom chromosome aberrations and certain malignant or premalignant disorders. At the clinical level, chromosome analysis is of value in diagnosis, and malignant cells of most tumors show specific chromosomal defects (14). These abnormalities are usually represented by a translocation or a loss of a chromosome band. Several studies have indicated that chromosomal breakpoints in these aberrations tend to cluster identified locations of proto-oncogenes (5,6,14). Furthermore, genetic predisposition to cancer is known to be associated with inherited chromosomal instability conditions, such as Bloom's syndrome (BS), Fanconi's anemia (FA), and ataxia telangiectasia (AT), thus suggesting possible health significance of chromosomal breakage also at the individual level. No studies to confirm these possible associations have been published.

BIOMONITORING IN RISK ASSESSMENT

Epidemiologic studies on the final end-effects in humans are necessary for risk assessment approaches. However, the epidemiologic approach is limited for two main reasons: first, only relatively high risks can be detected, and, second, the observations on end-effects are the consequence of exposures that may have occurred several decades earlier. Improved epidemiology, ideally, needs direct and accurate estimates of individual exposures. Biomonitoring has become an essential part in the exposure assessment, its special objective being in defining biologically relevant doses.

A general strategy for this multidisciplinary approach to risk assessment of environmental genotoxicants is depicted in Fig. 1. Concerning the biological characteristics of genotoxicants, and their toxic activity to the genetic material, proper and relevant methods for genotoxicity assessment need to be used both at the experimental and at the human exposure assessment levels. "Molecular epidemiology" aims at combining the laboratory and epidemiologic tools in an effort which is more analytical and sensitive to allow individual exposure assessments (13).

SETTING OF PRIORITIES FOR STUDIES
ON ENVIRONMENTAL GENOTOXICANTS

The Chemical Abstracts Service listed a total of 8 million compounds in their registry by the end of 1986, and the rate of adding new synthetic or natural substances is about 400,000 per year (1).

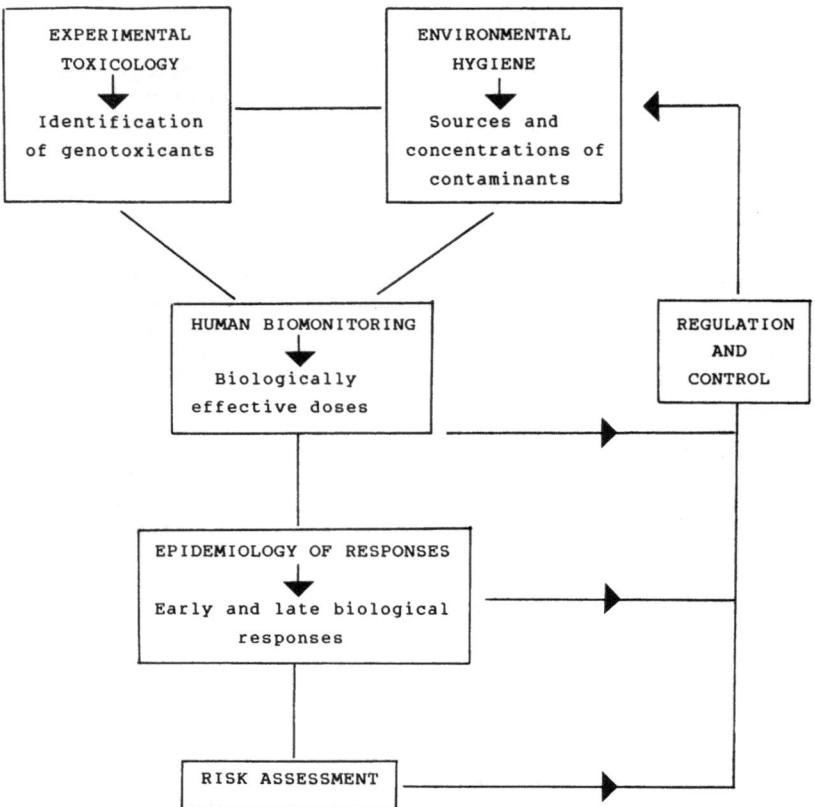

Fig. 1.   Disciplines and objectives in the general strategy for risk assessment in human exposure to environmental genotoxicants.

Quantitative Priorities

Generally, some 60,000 compounds are estimated to occur in industrialized societies; toxicological knowledge of them is, however, poor and is
estimated to be available only on some 12 percent of the compounds (3).
This unbalance between the increasing number of compounds and resources available for toxicological testing can only be handled by reasonable prioritization, on the basis of production quantities, exposure patterns, structure-activity relationships, etc.   A survey of quantitatively
important chemicals reveals that 99% of the total use of chemicals is based
on 1,000 top synthetic compounds, and for the 50 top chemicals the consumption figures per capita are surprisingly similar in all industrialized
countries (12).   These compounds, produced in million-ton quantities, fall
into the priority category for toxicological testing.

Testing for Genotoxicity

During the last years, application of short-term tests for mutagenicity in prediction of carcinogenic hazard of chemicals has become routine:

at present, approximately 13,000 single chemicals or complex mixtures have been tested in various mutagenicity assays (2).

The simple rationale behind using short-term mutagenicity tests in prediction of carcinogenicity is, first, in the universality of the genetic apparatus in living organisms, including the indicator test organisms and humans, and, second, in the empirical and theoretical relationship between carcinogenicity and mutagenicity. However, it must be realized that due to the mechanistic differences in the action of carcinogens, only the agents acting via a genotoxic mechanism can be expected to be identifiable with short-term tests. Short-term tests provide a possibility for monitoring complex ambient exposure and are thus of importance in following changes in occupational conditions. Short-term tests can also aid in identifying the special hazardous compounds involved and in establishing priorities for more definitive chemical analysis and monitoring, and for further testing in comparative systems, including whole animals, for mutagenesis and carcinogenesis. Experimental testing should, likewise, always precede human biomonitoring studies.

CONCLUSIONS

The use of genotoxicity testing is essential for the assessment of potential human toxicity so that hazards can be prevented. Laboratory tests are particularly necessary if a new product has been introduced or if exposure to an "old" product increases substantially. Situations where the quality of human data is low or difficult to obtain or interpret also require laboratory tests, and, principally, experimental testing should precede any human studies.

In human biomonitoring, emphasis should be given in validating the new "molecular epidemiology" endpoints in correlation with more traditional cytogenetic parameters. The sensitivity of the latter can be increased by automation of the methods, allowing more cells to be analyzed. These combined efforts should help us to learn about individual exposure assessment and biologically relevant doses, hopefully leading to quantitative assessment of human cancer risks.

REFERENCES

1. American Chemical Society (1987) Chemical Abstracts Index Guide, American Chemical Society, Washington, D.C.
2. Ashby, J. (1986) The prospects for a simplified and internationally harmonized approach to the detection of possible human carcinogens and mutagens. Mutagenesis 1:3-16.
3. Bingham, E. (1988) Extent of carcinogenicity testing of commercial chemicals. In Living in a Chemical World, C. Maltoni and I.J. Selikoff, eds. New York Academy of Sciences, New York, pp. 1038-1041.
4. Commission of the European Communities (1988) Indicators for assessing exposure and biological effects of genotoxic chemicals. EUR 11642, Luxemburg, pp. 191.
5. DeKlein, A. (1987) Oncogene activation by chromosomal rearrangements in chronic myelocytic leukemia. Mutat. Res. 186:161-172.
6. Heim, S., and F. Mitelman (1987) Nineteen of 26 cellular oncogenes precisely localized in the human genome map to one of the 83 bands

involved in primary cancer specific rearrangements. Hum. Genet. 75:70-72.

7. Högstedt, B., B. Gullberg, K. Hedner, A.-M. Konig, F. Mitelman, S. Skerfving, and B. Widegren (1981) Chromosome damage in bone marrow cells and peripheral blood lymphocytes in humans exposed to ethylene oxide. Hereditas 98:105-133.

8. Högstedt, B., B. Åkesson, K. Axell, B. Gullberg, F. Mitelman, R.W. Pero, S. Skerfving, and H. Welinder (1983) Increased frequency of lymphocyte micronuclei in workers producing reinforced polyester resin with low exposure to styrene. Scand. J. Work Environ. Health 49:271-276.

9. IARC (1987) IARC Monographs on the Evaluation of the Carcinogenic Risks to Humans, Supplement 6, Genetic and Related Effects: An Updating of Selected IARC Monographs from Volumes 1 to 42, IARC, Lyon, France.

10. IARC (1987) IARC Monographs on the Evaluation of the Carcinogenic Risks to Humans, Supplement 7, Overall Evaluations of Carcinogenicity: An Updating of IARC Monographs from Volumes 1 to 42, IARC, Lyon, France.

11. Sorsa, M., and J.W. Yager (1987) Cytogenetic surveillance of occupational exposures. In Cytogenetics, G. Obe and A. Basler, eds. Springer-Verlag, Berlin and Heidelberg, pp. 345-360.

12. Tossavainen, A. (1988) Selection of chemicals for carcinogenicity studies. Proceedings of the Fifth Finnish-Hungarian Joint Symposium, Institute of Occupational Health, Helsinki, pp. 165-171.

13. Vainio, H., and K. Hemminki (1988) Observational and experimental approaches in the evaluation and control of chemical carcinogens. Comments Toxicology 2:321-349.

14. Yunis, J.J. (1986) Chromosomal rearrangements, genes, and fragile sites in cancer: Clinical and biologic implications. In Important Advances in Oncology, V.T. DeVita, Jr., S. Hellman, and S.A. Rosenberg, eds. Lippincott Company, Philadelphia, pp. 93-128.

THE CONCEPT OF ACTIVITY PROFILES OF ANTIMUTAGENS*

Michael D. Waters,[1] Ann L. Brady,[2]
H. Frank Stack,[2] and Herman E. Brockman[3]

[1]U.S. Environmental Protection Agency
Research Triangle Park, North Carolina 27711

[2]Environmental Health Research and Testing, Inc.
Research Triangle Park, North Carolina 27709

[3]Department of Biological Sciences
Illinois State University
Normal, Illinois 61761

INTRODUCTION

There is considerable evidence that gene and chromosomal mutations are important factors in carcinogenesis (1,45). It follows, therefore, that the incidence of cancer can be reduced by decreasing the rate of mutation. The most obvious way to decrease the rate of mutation is to determine which environmental agents are mutagenic and then to reduce or prevent human exposure (1). Short-term genetic tests have been, and continue to be, useful in the detection of environmental mutagens and potential carcinogens (9).

The chemical induction of mutations involves a series of events including metabolic activation and/or detoxification, the formation of reactive electrophilic metabolites, the interaction of these metabolites with DNA, error-free and/or error-prone DNA repair, and altered cell selection. It is clear, therefore, that mechanisms such as detoxification and DNA repair exist which protect against the effects of many classes of naturally-occurring and synthetic mutagens. Furthermore, it has become clear that certain chemicals possess properties that directly or indirectly reduce or eliminate the mutagenic activity of other chemicals. In fact, the term "antimutagen" was used originally to describe those agents that

---

*This document has been reviewed in accordance with U.S. Environmental Protection Agency policy and approved for publication. Mention of trade names or commercial products does not constitute endorsement or recommendation for use.

reduce the frequency or rate of spontaneous or induced mutation independent of the mechanisms involved (39). This is the general meaning of "antimutagen" used in this paper. We also use "anticarcinogen" in the same general way to describe any agent that reduces the incidence of cancer.

Early research on the detection of antimutagens using mutation tests was reviewed by Clarke and Shankel (15). Research in the late 1970s (20,23,26,36,37,41,47,49,50-52) further suggested that the same kinds of short-term tests that had been useful in detecting mutagens might also be effective in identifying antimutagens. Thus, an alternative means to decrease the rate of mutation, and subsequently to decrease the incidence of cancer, may be to identify effective antimutagens and increase our exposure to them (2,46), especially through the diet (22). In conclusion, an appraisal of the potential carcinogenic hazard of a chemical for humans requires evaluation not only of its mutagenicity but also of the endogenous and exogenous factors that may modify its mutagenic activity.

Classifications

Kada and co-workers emphasized the importance of distinguishing between antimutagenic agents acting outside the cell (desmutagens) and those which function inside the cell (bioantimutagens). These authors further subdivided each of these two classes of antimutagens based on mechanisms of action (27).

From the point of view of carcinogenesis, one means of providing an organizational framework for the classification of "antimutagens" is to

Tab. 1.   Classification of antimutagens by mechanism.

---

1. **Extracellular**

   a.   Inhibitors of formation or uptake of mutagens

   b.   Inactivators of promutagens or mutagens

2. **Intracellular**

   a.   Blocking agents - prevent mutagens from reaching or reacting with target sites.

      i.   inhibit conversion to ultimate carcinogen
      ii. increase activity of detoxifying enzymes
      iii. direct reaction with electrophiles

   b.   Scavengers of radicals

   c.   Suppressing agents - prevent neoplastic expression of initiated cells

   d.   Agents affecting DNA repair

From: ICPEMC Publ. No. 12, C. Ramel et al., 1986, Mutation Res. 168: 47-65.

classify them according to the stage in the carcinogenic process in which they are effective (67). Such a classification may be useful in comparing the results of in vitro tests for antimutagenic activity with similar results for anticarcinogenic activity in the intact animal. Utilizing such a framework, Wattenberg (67) divided inhibitors of carcinogenesis into three major categories. The first category consists of compounds that prevent the formation of mutagens/carcinogens from precursor chemicals. Compounds that prevent carcinogens from reaching or reacting with critical cellular sites, i.e., "blocking agents," comprise the second category. The third category of carcinogenesis inhibitors are termed "suppressive agents," because they suppress the expression of neoplasia in cells previously exposed to carcinogenic doses of an agent. Thus, inhibitors of mutagenicity and carcinogenicity can be classified according to the stage during which they exert their protective effects. Building upon the classification schemes of Kada et al. (27) and Wattenberg (67), Ramel et al. (46) recognized two principal stages during which antimutagens/anticarcinogens can act: Stage 1, comprising agents acting extracellularly; and Stage 2, comprising agents acting intracellularly. Each of these stages is further subdivided as shown in Tab. 1. Generally, we have employed the classification scheme of Ramel et al. (46) in the classification of the antimutagens to be described in this paper (Tab. 1). Subsequent to presentation of this paper, De Flora and Ramel (16) published a refined classification of antimutagens/anticarcinogens based on mechanism of action.

The Concept of Activity Profiles of Antimutagens

In assemblying the data base for the present investigation, the literature was surveyed for the availability of antimutagenicity data. Publications were selected that presented original quantitative data for any of the genotôxicity assays that are in the scope of the genetic activity profiles (64). A table was prepared of the paired antimutagens and mutagens that were evaluated in each study. From this table, representative indirect- and direct-acting mutagens that had been tested in combination with the greatest range of antimutagens were selected. An effort was made to classify the antimutagens according to the scheme described by Ramel et al. (46). One mutagen and two antimutagens were selected as models for analysis to be presented here.

The literature on antimutagenesis was first reviewed by Clarke and Shankel (15). Later reviews (e.g., 2,3,16,24,46) demonstrate the extensiveness and complexity of this literature. The complexity is due in part to three simultaneous experimental variables: the mutagens (and/or spontaneous mutation) studied, the antimutagens studied, and the short-term tests used. We have addressed the question of how best to summarize and display the published results from these short-term tests for antimutagens tested in combination with a given mutagen or, conversely, for mutagens tested in combination with a given antimutagen. We have based the concept of "activity profiles for antimutagens" on the activity profiles used for the display of data on genetic and related effects (64).

The genetic activity profiles contained two of the three variables previously described, the mutagen and a series of short-term tests. The third variable, the antimutagen, presents a complex problem for graphics without using a three-dimensional profile. The plots are organized using two basic formats. A "single antimutagen" plot displays one antimutagen across the array of mutagens, and similarly a "single mutagen" plot dis-

plays one mutagen across the array of antimutagens. Since most muta-
gen/antimutagen combinations were not extensively studied across the full
range of genotoxicity tests, three formats were adopted for the organization of data. For the single mutagen, the profiles were arranged either
by the chemical class of the antimutagens or by the presumed mechanism
of action. For the single antimutagen, the profiles were organized by the
in vitro and in vivo groups, with the in vitro tests subdivided on the
basis of the presence or absence of S-9 activation.

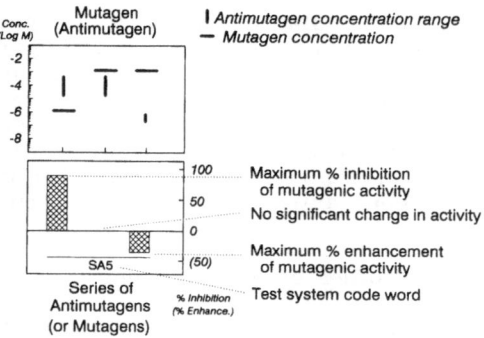

Fig. 1.   The genetic activity profiles are organized to show either the
           antimutagenic activity of various antimutagens in combination
           with a single mutagen or the activity of a single antimutagen
           with various mutagens. The upper bar graph displays the
           mutagen concentration and the range of antimutagen concentra-
           tions tested. The lower graph shows either the maximum per-
           cent inhibition, represented by a bar directed upward from the
           origin or the maximum percent enhancement of the genotoxic re-
           sponse, represented by a downward-directed bar. The absence
           of a bar indicates that no significant difference was detected.

A major modification in the antimutagenesis profile was implemented
to accommodate doses from both the mutagen and the antimutagen and to
represent the test response (either inhibition or enhancement). The
resultant profiles are actually two parallel sets of bar graphs (Fig. 1).
The upper graph displays the mutagen concentration and the range of
antimutagen concentrations tested. The lower graph shows either the
maximum percent inhibition, represented by a bar directed upward from
the origin, or the maximum percent enhancement of the genotoxic response, represented by a downward-directed bar. The absence of a bar
on the lower graph indicates that no significant difference in the response

Chart 1. Test code definitions.

Test	Definition
BFA	Body fluids from animals, microbial mutagenicity
BID	Binding (covalent) to DNA in vitro
BVD	Binding (covalent) to DNA, animal cells in vivo
CHL	Chromosomal aberrations, human lymphocytes in vitro
CIC	Chromosomal aberrations, Chinese hamster cells in vitro
ECW	Escherichia coli WP2 uvrA, reverse mutation
EC2	Escherichia coli WP2, reverse mutation
HMM	Host mediated assay, microbial cells in animal hosts
SAS	Salmonella typhimurium (other misc. strains), reverse mutation
SA0	Salmonella typhimurium TA100, reverse mutation
SA3	Salmonella typhimurium TA1530, reverse mutation
SA5	Salmonella typhimurium TA1535, reverse mutation
SA7	Salmonella typhimurium TA1537, reverse mutation
SA8	Salmonella typhimurium TA1538, reverse mutation
SA9	Salmonella typhimurium TA98, reverse mutation
SIC	Sister chromatid exchange, Chinese hamster cells in vitro
UIH	Unscheduled DNA synthesis, other human cells in vitro
URP	Unscheduled DNA synthesis, rat primary hepatocytes

was detected between the mutagen tested alone or in combination with the antimutagen. Codes used to represent the short-term tests have been reported previously (64), and the subset of tests represented in this paper are defined in the text as they are discussed and are shown in Chart 1.

ACTIVITY PROFILES OF ANTIMUTAGENS

Model Mutagen/Carcinogen

To provide a vehicle for the assessment of antimutagenesis, we selected N-methyl-N'-nitro-N-nitrosoguanidine (MNNG), a direct-acting mutagen that had been tested in combination with various antimutagens. The mutagenicity of MNNG was first reported by Mandell and Greenberg (34) using Escherichia coli. Sugimura et al. (59) reported that MNNG injected subcutaneously induced transplantable fibrosarcomas in rats. The carcinogenicity of MNNG was independently confirmed by Schoenthal (53) and Druckrey et al. (17). Sugimura and co-workers subsequently reported that MNNG added to the drinking water produced adenocarcinomas of the glandular stomach in rats (60,61).

The mechanism of action of MNNG is thought to involve its decomposition to short-lived, highly-reactive electrophiles, of which the alkonium ion is probably the ultimate mutagen (33). Electrophilic attack on nucleophilic sites of DNA bases leads to altered bases, some of which result in base-substitution mutations (31,38). It seems reasonable that an exogenous nucleophile could act as an antimutagen/anticarcinogen against MNNG by competing with DNA bases for the alkonium ion (20). To be effective,

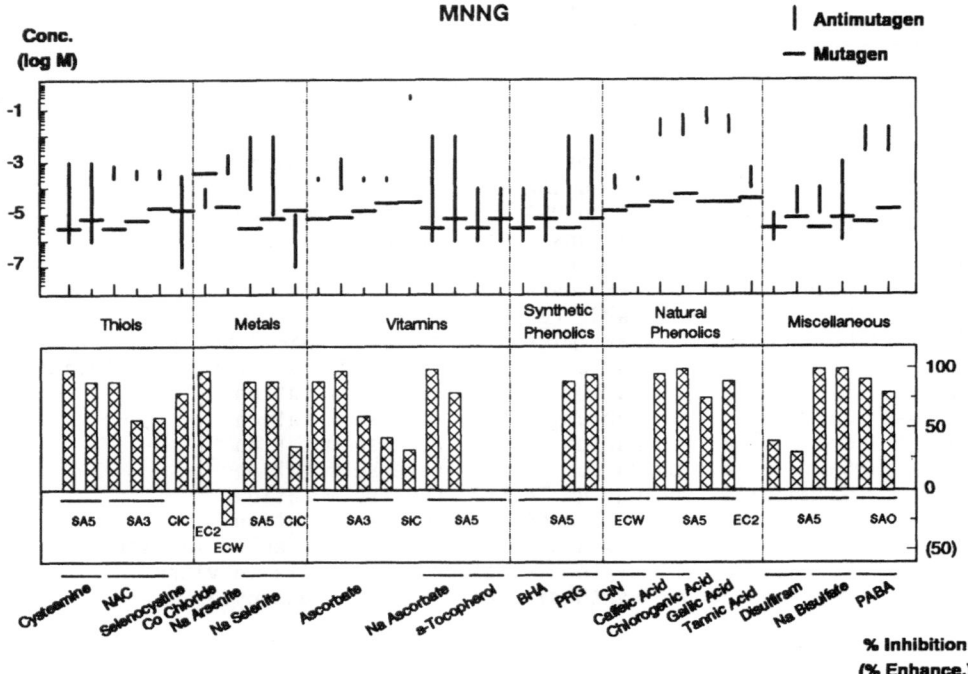

Fig. 2. Activity profile of compounds antimutagenic to N-methyl-N'-nitro-N-nitrosoguanidine (MNNG). The antimutagens are grouped by chemical classes.

the exogenous nucleophile would have to be more reactive than endogenous nucleophiles in the proximity of DNA. Alternatively, the antimutagen could be of comparable nucleophilicity but be present at a higher concentration than endogenous nucleophiles in the proximity of DNA (20).

Activity profiles of compounds antimutagenic to MNNG. Antimutagenic activity against MNNG has been shown for a wide variety of chemical agents, including both synthetic and natural compounds. In Fig. 2, these compounds have been grouped into the appropriate chemical classes. Figure 2 shows that the mutagenicity of MNNG is inhibited by several classes of chemicals, including thiols, metals, vitamins, synthetic phenolics, and others. The natural phenolic compounds cinnamaldehyde, tannic acid, and α-tocopherol (listed under vitamins) are inactive; however, certain other natural phenolic compounds (caffeic acid, chlorogenic acid, and gallic acid) effectively inhibit the mutagenicity of MNNG.

In Fig. 3, the overall activity profile of compounds antimutagenic to MNNG has been subdivided according to the mechanistic classifications of Ramel et al. (46). Figure 3a represents those compounds thought to act extracellularly with MNNG, while Fig. 3b represents those agents thought

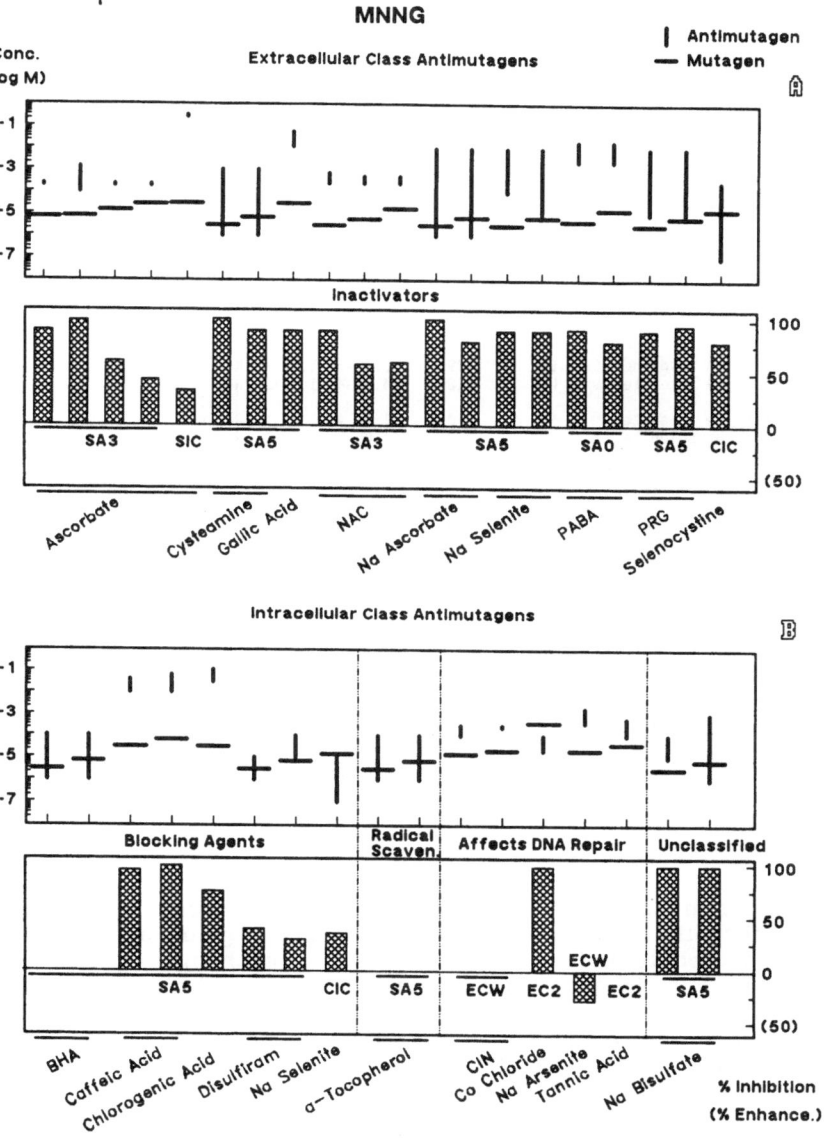

Fig. 3. Overall activity profile for agents antimutagenic to MNNG. Compounds presumed to inactivate MNNG extracellularly are shown in A, and compounds that may interact within the cell are displayed in B.

to act intracellularly. Table 2 is the detailed listing of the data upon which Fig. 2 and 3 are based, including compound names, test system code words, the percent response (inhibition or enhancement), doses, and short citations. As shown in Fig. 3a, the mutagenic activity of MNNG in S. typhimurium tester strains (SA3, SA0, SA5) is inhibited by N-acetyl-cysteine (NAC, a precursor of cysteine in vivo), para-aminobenzoic acid

Tab. 2. Antimutagenic effects of agents added with MNNG.

MNNG Dose (M)	Antimutagen	Test Code	% Inhib.[a] (% Enhance.)	LED(M) or LID(M)	HED(M) or HID(M)	Citation
	Thiols					
3.0E-6	Cysteamine	SA5	100	1.0E-6	1.0E-3	Rosin & Stich, 1979
7.0E-6	Cysteamine	SA5	90	1.0E-6	1.0E-3	Rosin & Stich, 1979
3.0E-6	NAC	SA3	90	2.5E-4	7.5E-4	Wilpart et al., 1985
6.0E-6	NAC	SA3	58	2.5E-4	5.0E-4	Wilpart et al., 1985
1.8E-5	NAC	SA3	60	2.5E-4	5.0E-4	Wilpart et al., 1985
1.5E-5	SEC	CIC	80	1.0E-7	3.0E-4	Whiting et al., 1981
	Metals					
4.0E-4	Co Chloride	EC2	99	2.0E-5	1.0E-4	Kada & Kanematsu, 1978
2.0E-5	Na arsenite	ECW	(28)	4.0E-4	1.9E-3	Nunoshiba & Nishioka, 1987
3.0E-6	Na selenite	SA5	90	1.0E-4	1.0E-2	Rosin & Stich, 1979
7.0E-6	Na selenite	SA5	90	1.0E-5	1.0E-2	Rosin & Stich, 1979
1.5E-5	Na selenite	CIC	36	1.0E-7	1.0E-5	Whiting et al., 1981
	Vitamins					
6.8E-6	Ascorbate	SA3	90	2.0E-4	2.0E-4	Guttenplan, 1978
8.0E-6	Ascorbate	SA3	99	1.0E-4	1.4E-3	Guttenplan, 1977
1.4E-5	Ascorbate	SA3	61	2.0E-4	2.0E-4	Guttenplan, 1978
2.7E-5	Ascorbate	SA3	43	2.0E-4	2.0E-4	Guttenplan, 1978
3.0E-6	Ascorbate	SIC	33	3.0E-1	3.0E-1	Galloway & Painter, 1979
3.0E-6	Na ascorbate	SA5	100	1.0E-6	1.0E-2	Rosin & Stich, 1979
7.0E-6	Na ascorbate	SA5	80	1.0E-6	1.0E-2	Rosin & Stich, 1979
3.0E-6	α-Tocopherol	SA5	0	1.0E-6	1.0E-4	Rosin & Stich, 1979
7.0E-6	α-Tocopherol	SA5	0	1.0E-6	1.0E-4	Rosin & Stich, 1979
	Synthetic phenolics					
3.0E-6	BHA	SA5	0	1.0E-6	1.0E-4	Rosin & Stich, 1979
7.0E-6	BHA	SA5	0	1.0E-6	1.0E-4	Rosin & Stich, 1979
3.0E-6	Propyl gallate	SA5	90	1.0E-5	1.0E-2	Rosin & Stich, 1979
7.0E-6	Propyl gallate	SA5	95	1.0E-5	1.0E-2	Rosin & Stich, 1979

(PABA, the sun screen), ascorbate, cysteamine, propyl gallate (PRG, a synthetic phenolic compound), sodium ascorbate, gallic acid, and sodium selenite. Ascorbate induces sister-chromatid exchanges (SCEs) in Chinese hamster cells in vitro; but depending upon relative concentrations, ascorbate also inhibits the induction of SCEs by MNNG (18), as shown in the SIC entry. Selenocystine (SEC) inhibits MNNG-induced chromosomal aberrations in Chinese hamster cells in vitro (CIC). All of these compounds are directly active in the extracellular environment.

In the intracellular environment (Fig. 3b), several classes of antimutagenic agents may be distinguished: blocking agents, radical scavengers, agents that affect DNA repair, and one agent (sodium bisulfate) that is unclassified. Among the blocking agents, the synthetic phenolic compound butylated hydroxyanisole (BHA) is inactive in S. typhimurium TA1535 (SA5), whereas the phenolics caffeic acid and chlorogenic acid cause almost complete inhibition, and disulfiram, a thiono-sulfur-containing compound, is moderately active in the same strain. Sodium selenite inhibits the induction of chromosomal aberrations by MNNG in Chinese hamster cells in vitro (CIC). The radical scavenger and natural phenolic

Tab. 2. (continued)

MNNG Dose (M)	Antimutagen	Test Code	% Inhib.[a] (% Enhance.)	LED(M) or LID(M)	HED(M) or HID(M)	Citation
	Natural phenolics					
1.4E-5	CIN	ECW	0	1.0E-4	3.0E-4	Ohta et al., 1983a
2.0E-5	CIN	ECW	0	2.6E-4	2.6E-4	Ohta et al., 1983b
3.0E-5	Caffeic acid	SA5	96	1.0E-2	4.0E-2	Chan et al., 1986
6.0E-5	Caffeic acid	SA5	100	1.0E-2	6.0E-2	Chan et al., 1986
3.0E-5	Chlorogenic acid	SA5	76	3.0E-2	1.1E-1	Chan et al., 1986
3.0E-5	Gallic acid	SA5	90	1.2E-2	5.5E-2	Stitch & Rosin, 1984
4.0E-5	Tannic acid	EC2	0	1.0E-4	6.0E-4	Shimoi et al., 1985
	Miscellaneous agents					
3.0E-6	Disulfiram	SA5	40	1.0E-6	1.0E-5	Rosin & Stich, 1979
7.0E-6	Disulfiram	SA5	30	1.0E-5	1.0E-4	Rosin & Stich, 1979
3.0E-6	Na bisulfate	SA5	100	1.0E-5	1.0E-4	Rosin & Stich, 1979
7.0E-6	Na bisulfate	SA5	100	1.0E-6	1.0E-3	Rosin & Stich, 1979
5.0E-6	PABA	SA0	91	2.5E-3	2.0E-2	Gichner et al., 1987
1.5E-5	PABA	SA0	80	2.5E-3	2.0E-2	Gichner et al., 1987

[a]A positive value represents the percentage inhibition of the effect induced by the mutagen in the short-term test system; a negative value ( ) is the percentage enhancement of the effect; and a value of zero indicates no significant difference between the effect observed when the mutagen was tested alone or in combination with the antimutagen. A value of zero is used to indicate no effect and is not a precise value; in general, the true value is less than 20% inhibition or enhancement.

compound α-tocopherol is inactive against MNNG in strain TA1535 (SA5). Agents thought to enhance error-free DNA repair, including cinnamaldehyde (CIN) and tannic acid, are inactive in E. coli strains (ECW, EC2) against MNNG, whereas cobaltous chloride and sodium arsenite are active in the same organism. Sodium bisulfite is very effective in inhibiting the mutagenic activity of MNNG in strain TA1535 (SA5). All in vitro tests for antimutagenesis against MNNG were carried out in the absence of an exogenous metabolic system (S-9).

Model Antimutagens

Activity profiles for two model antimutagens will now be presented that have been tested in combination with various mutagens. The antimutagens selected are butylated hydroxyanisole (BHA) and butylated hydroxytoluene (BHT), synthetic phenolics which are used as antioxidants in food. Antimutagenesis by these compounds occurs under a variety of experimental conditions and with a broad range of mutagens/carcinogens. BHA is preferable to BHT because it is considerably less toxic in vivo (65). The human consumption of phenolic antioxidants in the United States is estimated to be several milligrams a day (66). If the results of the animal experiments can be extrapolated to man, this amount of phenolic antioxidants could be important in inhibiting the effects of chronic exposure to low doses of carcinogens that occurs in human populations.

Blocking agents frequently induce increases in the activities of multiple enzyme systems that can detoxify genotoxic compounds; BHA causes such inductive effects. In vivo, BHA induces a marked increase in the activity of glutathione S-transferase (6), an enzyme that is of primary importance in the detoxification of electrophiles (13), and in the activity of UDP-glucuronyl transferase, an important enzyme for conjugation reactions (11). Furthermore, BHA enhances the activities of two other detoxifying enzymes: epoxide hydrolase and quinone reductase (7). Glutathione (GSH) synthesis is also increased by BHA (6).

Although BHA does not induce increased microsomal monooxygenase activity, this compound does modify this enzyme system (29). Thus, alteration of multiple enzyme systems could explain how BHA inhibits the activity of chemical carcinogens, especially that of benzo(a)pyrene (BAP) (5,11). These enzyme alterations seem to cause a decrease in epoxidation, possibly the most important activation process, while increasing the formation of 3-hydroxy-BAP, a product of detoxification (29,30).

Activity profiles of BHA and BHT. As shown in Fig. 4 and Tab. 3, negative results are obtained when BHA is tested in the absence of S-9 in

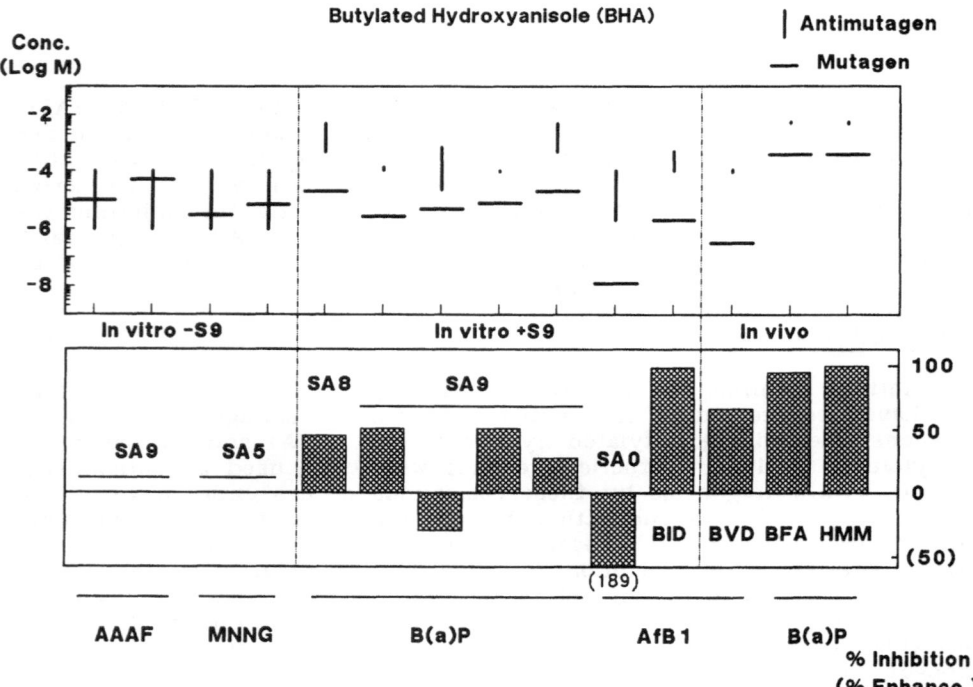

Fig. 4.   Antimutagenesis activity profile of butylated hydroxyanisole (BHA).

Tab. 3.   Antimutagenic effects of butylated hydroxyanisole (BHA).

Mutagen	Mutagen Dose (M)	Test Code	% Inhib.[a] (% Enhanc.)	LED(M) or LID(M)	HED(M) or HID(M)	Citation
			In vitro without S9			
AAAF	1.0E-5	SA9	0	1.0E-6	1.0E-4	Rosin & Stich, 1979
AAAF	5.0E-5	SA9	0	1.0E-6	1.0E-4	Rosin & Stich, 1979
MNNG	3.0E-6	SA5	0	1.0E-6	1.0E-4	Rosin & Stich, 1979
MNNG	7.0E-6	SA5	0	1.0E-6	1.0E-4	Rosin & Stich, 1979
			In vitro with S9			
B(a)P	2.0E-5	SA8	44	5.0E-4	5.0E-3	Mckee & Tometsko, 1979
B(a)P	2.7E-6	SA9	50	1.1E-4	1.4E-4	Calle & Sullivan, 1982
B(a)P	5.0E-6	SA9	(30)	2.2E-5	7.3E-4	Terwel & van der Hoeven, 1985
B(a)P	8.0E-6	SA9	50	1.0E-4	1.0E-4	Rahimtula et al., 1977
B(a)P	2.0E-5	SA9	27	5.0E-4	5.0E-3	McKee & Tometsko, 1979
AfB$_1$	1.2E-8	SA0	(189)	2.0E-6	1.0E-4	Shelef & Chin, 1980
AfB$_1$	2.0E-6	BID	98	1.0E-4	5.0E-4	Bhattacharya et al., 1984
			In vivo or host-mediated			
AfB$_1$	3.2E-7	BVD	65	1.0E-4	1.0E-4	Kensler et al., 1985
B(a)P	4.0E-4	BFA	94	5.3E-3	5.3E-3	Batzinger et al., 1978
B(a)P	4.0E-4	HMM	99	5.3E-3	5.3E-3	Batzinger et al., 1978

[a]See footnote to Table 2.

combination with N-acetoxy-2-acetylaminofluorene (AAAF) in TA98 (SA9) or with MNNG in TA1535 (SA5). BHA is marginally effective as an inhibitor of BAP mutagenesis in S. typhimurium strains TA1538 (SA8) and TA98 (SA9) in the presence of S-9. BHA is more effective against BAP in body fluid (BFA) and host-mediated (HMM) assays. BHA inhibits the binding of [³H]AFB$_1$ to DNA in vitro (BID) in the presence of metabolic activation and in vivo (BVD) but apparently enhances the mutagenicity of AFB$_1$ in TA100 (SA0) in the presence of S-9. As shown in Fig. 5a and Tab. 4, BHT is effective against 2-acetylaminofluorine (2AAF) and its proximate mutagenic metabolite, N-hydroxy-2-acetylaminofluorine (N-OH-2AAF) in primary rat hepatocytes in vitro (URP) without S-9. BHT also is antimutagenic to 2AAF in primary human hepatocytes in vitro (UIH) without S-9. These results agree with the inhibition of hepatocarcinogenic effects of 2AAF by BHT administered concurrently to rats (32,63). However, in these studies more bladder cancer was observed, presumably due to expulsion of activated metabolites from hepatocytes.

BHT also is antimutagenic to 7,12-dimethylbenz(a)anthracene (DMBA), malonaldehyde (MDA), and β-propiolactone (β-PRL) in various in vitro tests (CHL, SAS) without S-9. BHT gave negative results when tested in S. typhimurium strains (SA9, SA0) with aminacrine, 2,4-dinitro-

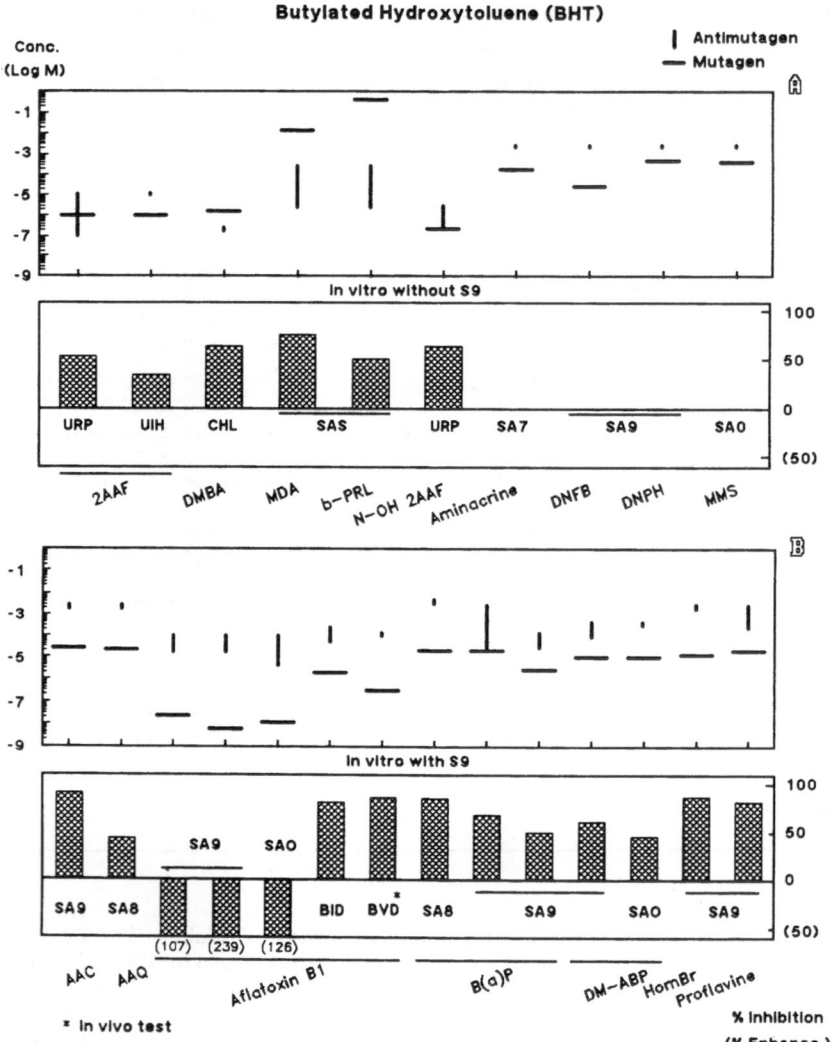

Fig. 5.   Antimutagenesis activity profile of butylated hydroxytoluene
          (BHT).

1-fluorobenzene (DNFB), 2,4-dinitrophenylhydrazine (DNPH), and methyl
methanesulfonate (MMS) in the absence of S-9.   As shown in Fig. 5b and
Tab. 4, BHT enhances the mutagenic activity of AFB$_1$ in TA98 (SA9) and
TA100 (SA0) in the presence of S-9, but inhibits the mutagenicity of BAP
in TA1538 (SA8) and TA98 (SA9) in the presence of S-9.   BHT inhibits
the binding of [^{3}H]AFB$_1$ to DNA in vitro (BID) and in vivo (BVD).   BHT
displays antimutagenic activity against 1-aminoanthracene (AAC), 2-amino-
anthraquinone (AAQ), 3,2'-dimethyl-4-aminobiphenyl (DM-ABP), homidium

Tab. 4. Antimutagenic effects of butylated hydroxytoluene (BHT).

Mutagen	Mutagen Dose (M)	Test Code	% Inhib.[a] (% Enhanc.)	LED(M) or LID(M)	HED(M) or HID(M)	Citation
In vitro without S9						
2AAF	1.0E-6	URP	53	1.1E-7	1.1E-5	Chipman & Davies, 1988
2AAF	1.1E-6	UIH	34	1.1E-5	1.1E-5	Chipman & Davies, 1988
DMBA	1.6E-6	CHL	64	2.1E-7	2.1E-7	Shamberger et al., 1973
MDA	1.5E-2	SAS	75	2.6E-6	2.6E-4	Shamberger et al., 1979
β-PRL	4.4E-1	SAS	50	2.6E-6	2.6E-4	Shamberger et al., 1979
N-OH 2AAF	2.5E-7	URP	64	3.2E-7	3.2E-6	Chipman & Davies, 1988
Aminacrine	2.0E-4	SA7	0	2.5E-3	2.5E-3	McKee & Tometsko, 1979
DNFB	2.7E-5	SA9	0	2.5E-3	2.5E-3	McKee & Tometsko, 1979
DNPH	5.0E-4	SA9	0	2.5E-3	2.5E-3	McKee & Tometsko, 1979
MMS	4.5E-4	SA0	0	2.5E-3	2.5E-3	McKee & Tometsko, 1979
In vitro with S9						
AAC	2.6E-5	SA9	89	2.5E-3	2.5E-3	McKee & Tometsko, 1979
AAQ	2.2E-5	SA8	42	2.5E-3	2.5E-3	McKee & Tometsko, 1979
AfB$_1$	2.4E-8	SA9	(107)	1.7E-5	1.0E-4	Shelef & Chin, 1980
AfB$_1$	6.0E-9	SA9	(239)	1.7E-5	1.0E-4	Shelef & Chin, 1980
AfB$_1$	1.2E-8	SA0	(126)	4.2E-6	1.0E-4	Shelef & Chin, 1980
AfB$_1$	2.0E-6	BID	80	5.0E-5	2.5E-4	Bhattacharya et al., 1984
AfB$_1$	3.2E-7	BVD[b]	85	1.0E-4	1.0E-4	Kensler et al., 1985
B(a)P	2.0E-5	SA8	84	2.9E-3	2.9E-3	McKee & Tometsko, 1979
B(a)P	2.0E-5	SA9	67	2.5E-5	2.5E-3	McKee & Tometsko, 1979
B(a)P	2.7E-6	SA9	50	2.7E-5	1.4E-4	Calle & Sullivan, 1982
DM-ABP	1.0E-5	SA9	60	9.0E-5	4.5E-4	Reddy et al., 1983
DM-ABP	1.0E-5	SA0	45	4.5E-4	4.5E-4	Reddy et al., 1983
HomBr	1.3E-5	SA9	86	2.5E-3	2.5E-3	McKee & Tometsko, 1979
Proflavine	2.0E-5	SA9	81	2.5E-4	2.5E-3	McKee & Tometsko, 1979

[a]See footnote to Table 2.
[b]*In vivo* test.

bromide (HomBr; also known as ethidium bromide), and proflavine (a disinfectant, 3,6-diaminoacridine) in <u>S</u>. <u>typhimurium</u> strains (SA9, SA8, SA0) in the presence of S-9.

CONCLUSIONS

More than 200 short-term tests have been used for evaluating the mutagenic effects of over 18,200 chemicals, and the Environmental Mutagen Information Center data base presently contains more than 60,000 references (64). This very large and diverse data base led Waters and coworkers to the concept of presenting the published data from short-term tests on the genetic effects of a given chemical in standardized formats: in tabular form as an activity profile listing and graphically as an activity profile plot (64). The activity profile of a given chemical permits rapid visualization of the quantitative results (as lowest effective dose and/or highest ineffective dose) in each of the short-term tests used for that chemical.

Although the literature on antimutagenicity is not nearly as extensive as that on mutagenicity, it is more complex in that the experimental variable of the antimutagens used must be considered in addition to the variables of the mutagens and short-term tests used. While considering the question of how to analyze the literature of results on antimutagens in short-term tests, it occurred to us that the concept of activity profile listings and plots already applied successfully to mutagenicity data might also be applied to antimutagenicity data. We have presented here our first attempt to develop the concept of activity profile listings and plots for antimutagens. The activity profiles have been organized in two general ways: for the antimutagens that have been tested in combination with a given mutagen (Tab. 2; Fig. 2 and 3) and for the mutagens that have been tested in combination with a given antimutagen (Tab. 3 and 4; Fig. 4 and 5). We think that the activity profile plots permit rapid visualization of considerable data and experimental parameters.

The short-term tests that have been used extensively to identify mutagens and potential carcinogens increasingly are being used to identify antimutagens and potential anticarcinogens. Just as the usefulness of short-term tests for identifying potential carcinogens continues to be controversial (9), it is difficult at the present time to answer unequivocally the question of whether the inhibition of carcinogen-induced mutation is a good indicator of anticarcinogenic properties. The available data on the inhibition of carcinogenesis and mutagenesis are still incomplete; ideally, one should have requisite information on both effects for a variety of carcinogens. Thus, to compare the data presented here on the activity profile of antimutagens vs MNNG-induced mutagenicity, a similar body of data on anticarcinogenicity against MNNG-induced carcinogenicity is required. Furthermore, the in vitro mutagenicity tests will not detect those compounds that act in the in vivo carcinogenicity test to change the activity of one or more enzyme systems not present in vitro. Rather, the in vitro tests will detect only those compounds that inhibit the metabolism of the carcinogen directly, react directly with the mutagenic species to inactivate them, or otherwise show an effect that is demonstrable in vitro.

Much additional research is needed in order to determine the degree of agreement among the results from short-term tests used in antimutagenicity studies and among the results from the laboratory animal assays used in anticarcinogenicity studies. Subsequently, the degree of concordance between the results from antimutagenicity and anticarcinogenicity studies needs to be ascertained. Finally, the utility of both kinds of studies for detecting human anticarcinogens as identified through epidemiological studies must be addressed.

ACKNOWLEDGEMENT

H.E.B. acknowledges the support of the U.S. Environmental Protection Agency Distinguished Visiting Scientist Program.

REFERENCES

1. Ames, B.N. (1979) Identifying environmental chemicals causing mutations and cancer. Science 204:587-593.
2. Ames, B.N. (1983) Dietary carcinogens and anticarcinogens. Science 221:1256-1264.

3. Ames, B.N. (1986) Overview: Food constituents as a source of mutagens, carcinogens and anticarcinogens. In Genetic Toxicology of the Diet, I. Knudsen, ed., Alan R. Liss, Inc., New York, pp. 3-32.

4. Batzinger, R.P., S.-Y.L. Ou, and E. Bueding (1978) Antimutagenic effects of 2(3)-tert-butyl-4-hydroxyanisole and of antimicrobial agents. Cancer Res. 38:4478-4485.

5. Benson, A.M., R.P. Batzinger, S.-Y.L. Ou, E. Bueding, Y.-N. Cha, and P. Talalay (1978) Elevation of hepatic glutathione S-transferase activities and protection against mutagenic metabolites of benzo(a)pyrene by dietary antioxidants. Cancer Res. 38:4486-4495.

6. Benson, A., Y. Cha, E. Bueding, H. Heine, and P. Talay (1979) Elevation of extrahepatic glutathione S-transferase and epoxide hydratase activities by 2(3)tert.-butyl-4-hydroxyanisole. Cancer Res. 39:2971-2977.

7. Benson, A., M. Hunkeler, and P. Talalay (1980) Increases of NAD-(P)H:quinone reductase by dietary antioxidants: Possible role in protection against carcinogenicity and toxicity. Proc. Natl. Acad. Sci., USA 77:5216-5220.

8. Bhattacharya, R.K., P.F. Firozi, and V.S. Aboobaker (1984) Factors modulating the formation of DNA adduct by aflatoxin B_1 in vitro. Carcinogenesis 5:1359-1362.

9. Brockman, H.E., and D.M. DeMarini (1988) Utility of short-term tests for genetic toxicity in the aftermath of the NTP's analysis of 73 chemicals. Environ. Molec. Mutag. 11:421-435.

10. Calle, L.M., and P.D. Sullivan (1982) Screening of antioxidants and other compounds for antimutagenic properties towards benzo(a)pyrene-induced mutagenicity in strain TA98 of Salmonella typhimurium. Mutat. Res. 101:99-114.

11. Cha, Y., and E. Bueding (1979) Effects of 2(3)-tert.-butyl-4-hydroxyanisole administration on the activities of several hepatic microsomal and cytoplasmic enzymes in mice. Biochem. Pharmacol. 28:1917-1921.

12. Chan, R.I.M., R.H.C. San, and H.F. Stich (1986) Mechanisms of inhibition of N-methyl-N-nitro-N-nitrosoguanidine induced mutagen by phenolic compounds. Cancer Lett. 31:27-34.

13. Chasseaud, L. (1979) The role of glutathione and glutathione-S-transferases in the metabolism of chemical carcinogens and other electrophilic agents. Adv. Cancer Res. 29:175-274.

14. Chipman, J.K., and J.E. Davies (1988) Reduction of 2-acetylamino-fluorene-induced unscheduled DNA synthesis in human and rat hepatocytes by butylated hydroxytoluene. Mutat. Res. 207:193-198.

15. Clarke, C.H., and D.M. Shankel (1975) Antimutagenesis in microbial systems. Bacteriol. Rev. 39:33-53.

16. De Flora, S., and C. Ramel (1988) Mechanisms of inhibitors of mutagenesis and carcinogenesis. Classification and overview. Mutat. Res. 202:285-306.

17. Druckrey, H., R. Preussmann, S. Ivankovic, B.T. So, and C.H. Schmidt (1966) Bucheler J. Zur Erzeugung subcutaner Sarkome an Ratten. Carcinogene Wirkung von Hydrazodicarbonsaure-bis-(methyl-nitrosamid), N-nitroso-N-n-butylharnstoff, N-methyl-N-nitroso-nitrosoguanidin und N-nitroso-imidazolidon. Z. Krebsforsch. 68:87-102.

18. Galloway, S.M., and R.B. Painter (1979) Vitamin C is positive in the DNA-synthesis inhibition and sister-chromatid exchange tests. Mutat. Res. 60:321-327.

19. Gichner, T., J. Veleminsky, I.A. Rapoport, and S.V. Vasilieva (1987) Antimutagenic effect of p-aminobenzoic acid on the mutagenic-

ity of N-methyl-N'-nitro-N-nitrosoguanidine in <u>Salmonella</u> <u>typhi-</u>
<u>murium</u>. <u>Mutat. Res.</u> 192:95-98.

20. Guttenplan, J.B. (1977) Inhibition by L-ascorbate of bacterial mutagenesis induced by two N-nitroso compounds. <u>Nature</u> 268:368-370.

21. Guttenplan, J.B. (1978) Mechanisms of inhibition by ascorbate of
microbial mutagenesis induced by N-nitroso compounds. <u>Cancer Res.</u>
38:2018-2022.

22. Hayatsu, H., S. Arimoto, and T. Negishi (1988) Dietary inhibitors
of mutagenesis and carcinogenesis. <u>Mutat. Res.</u> 202:429-446.

23. Jacobs, M.M., T.S. Matney, and C.A. Griffin (1977) Inhibitory effects of selenium on the mutagenicity of 2-acetylaminofluorene (AAF)
and AAF derivatives. <u>Cancer Lett.</u> 2:319-322.

24. Kada, T. (1984) Desmutagens and bio-antimutagens: Their action
mechanisms and possible role in the modulation of dose relationships.
In <u>Problems of Threshold in Chemical Mutagenesis</u>, Y. Tazima, S.
Kondo, and Y. Kuroda, eds. Environmental Mutagen Society of
Japan, Mishima, pp. 73-82.

25. Kada, T., and N. Kanematsu (1978) Reduction of N-methyl-N'-nitro-
N-nitrosoguanidine induced mutations by cobalt chloride in <u>Esche-</u>
<u>richia</u> <u>coli</u>. <u>Proc. Japan Acad.</u> 54:234-237.

26. Kada, T., K. Morita, and T. Inoue (1978) Antimutagenic action of
vegetable factor(s) on the mutagenic principle of tryptophan pyrolysate. <u>Mutat. Res.</u> 53:351-353.

27. Kada, T., T. Inoue, and M. Namiki (1981) Environmental mutagenesis and plant biology. In <u>Environmental Desmutagens and Antimuta-</u>
<u>gens</u>, E.J. Klekowski, Jr., ed. Praeger, New York, pp. 134-151.

28. Kensler, T.W., P.A. Egner, M.A. Trush, E. Bueding, and J.D.
Groopman (1985) Modification of aflatoxin B$_1$ binding to DNA in vivo
in rats fed phenolic antioxidants, ethoxyquin and a dithiothine.
<u>Carcinogenesis</u> 6:759-763.

29. Lam, L.K., and L.W. Wattenberg (1977) Effects of butylated hydroxyanisole on the metabolism of benzo(a)pyrene by mouse liver microsomes. J. Natl. Cancer Inst. 58:413-417.

30. Lam, L.K.T., A.V. Fladmoe, J.B. Hochalter, and L.W. Wattenberg
(1980) Short-time interval effects of butylated hydroxyanisole on the
metabolism of benzo(a)pyrene. <u>Cancer Res.</u> 40:2824-2828.

31. Loveless, A. (1969) Possible relevance of O-6 alkylation of deoxyguanosine to the mutagenicity and carcinogenicity of nitrosamines and
nitrosamides. Nature 223:206-207.

32. Maeura, Y., J.H. Weisburger, and G.M. Williams (1984) Dose-
dependent reduction of N-2-fluorenylacetamide-induced liver cancer
and enhancement of bladder cancer in rats by butylated hydroxytoluene. Cancer Res. 44:1604-1610.

33. Magee, P.N., R. Montesano, and R. Preussmann (1976) N nitroso
compounds and related carcinogens. In Chemical Carcinogens, American Chemical Society Monographs, Vol. 173, Charles E. Searle, ed.
American Chemical Society, Washington, D.C., pp. 491-625.

34. Mandell, J.D., and J. Greenberg (1960) A new chemical mutagen for
bacteria, 1-methyl-3-nitro-1-nitrosoguanidine. <u>Biochem. Biophys.</u>
Res. Commun. 3:575-577.

35. McKee, R.H., and A.M. Tometsko (1979) Inhibition of promutagen
activation by the antioxidants butylated hydroxyanisole and butylated
hydroxytoluene. J. Natl. Cancer Inst. 63:473-477.

36. Moriya, M., K. Kato, and Y. Shirasu (1978) Effects of cysteine and
a liver metabolic activation system on the activities of mutagenic
pesticides. <u>Mutat. Res.</u> 57:259-263.

37. Nagao, M., T. Yahagi, T. Kawachi, T. Sugimura, T. Kosuge, K. Tsuji, K. Wakabayashi, S. Mizusaki, and T. Matsumoto (1977) Co-mutagenic action of norharman and harman. Proc. Japan Acad. 53:95-98.
38. Nicoll, J.W., P.F. Swann, and A.E. Pegg (1975) Effect of dimethyl-nitrosamine on persistence of methylated guanines in rat liver and kidney DNA. Nature 254:261-262.
39. Novick, A., and L. Szilard (1952) Anti-mutagens. Nature 170:926-927.
40. Nunoshiba, T., and H. Nishioka (1987) Sodium arsenite inhibits spontaneous and induced mutations in Escherichia coli. Mutat. Res. 184:99-105.
41. Oesch, F., P. Bentley, and H.R. Glatt (1976) Prevention of benzo(a)pyrene-induced mutagenicity by homogenous epoxide hydratase. Int. J. Cancer 18:448-452.
42. Ohta, T., K. Watanabe, M. Moriya, Y. Shirasu, and T. Kada (1983a) Antimutagenic effects of cinnamaldehyde on chemical mutagenesis in Escherichia coli. Mutat. Res. 107:219-227.
43. Ohta, T., K. Watanabe, M. Moriya, Y. Shirasu, and T. Kada (1983b) Analysis of the antimutagenic effect of cinnamaldehyde on chemically induced mutagenesis in Escherichia coli. Molec. Gen. Genet. 192:309-315.
44. Rahimtula, A.D., P.K. Zachariah, and P.J. O'Brien (1977) The effects of antioxidants on the metabolism and mutagenicity of benzo(a)pyrene in vitro. Biochem. J. 164:473-475.
45. Ramel, C. (1986) Deployment of short-term assays for the detection of carcinogens: Genetic and molecular considerations. Mutat. Res. 168:327-342.
46. Ramel, C., U.K. Alekperov, B.N. Ames, T. Kada, and L.W. Wattenberg (1986) Inhibitors of mutagenesis and their relevance to carcinogenesis. Mutat. Res. 168:47-65.
47. Rannug, U., A. Sundvall, and C. Ramel (1978) The mutagenic effect of 1,2-dichloroethane on Salmonella typhimurium. 1. Activation through conjugation with glutathione in vitro. Chem.-Biol. Interact. 20:1-16.
48. Reddy, B.S., D. Hanson, L. Mathews, and C. Sharma (1983) Effect of micronutrients, antioxidants and related compounds on the mutagenicity of 3,2-dimethyl-4-aminobiphenyl, a colon and breast carcinogen. Food Chem. Toxicol. 21:129-132.
49. Rosin, M.P., and H.F. Stich (1978a) Inhibitory effect of reducing agents on N-acetoxy- and N-hydroxy-2-acetylaminofluorene-induced mutagenesis. Cancer Res. 38:1307-1310.
50. Rosin, M.P. and H.F. Stich (1978b) The inhibitory effect of cysteine on the mutagenic activities of several carcinogens. Mutat. Res. 54:73-81.
51. Rosin, M.P., and H.F. Stich (1979) Assessment of the use of the Salmonella mutagenesis assay to determine the influence of antioxidants on carcinogen-induced mutagenesis. Int. J. Cancer 23:722-727.
52. Rosin, M.P., and H.F. Stich (1980) Enhancing and inhibiting effects of propyl gallate on carcinogen-induced mutagenesis. J. Environ. Pathol. Toxicol. 4:159-167.
53. Schoenthal, R. (1966) Carcinogenic activity of N-methyl-N-nitroso-N'-nitrosoguanidine. Nature 209:726-727.
54. Shamberger, R.J., F.F. Baughman, S.L. Kalchert, C.E. Willis, and G.C. Hoffman (1973) Carcinogen-induced chromosomal breakage decreased by antioxidants. Proc. Natl. Acad. Sci., USA 70:1461-1463.

55. Shamberger, R.J., C.L. Corlett, K.D. Beaman, and B.L. Kasten (1979) Antioxidants reduce the mutagenic effect of malonaldehyde and β-propiolactone. Mutat. Res. 66:349-355.

56. Shelef, L.A., and B. Chin (1980) Effect of phenolic antioxidants on the mutagenicity of aflatoxin B_1. Appl. Environ. Microbiol. 40:1039-1043.

57. Shimoi, K., Y. Nakamura, I. Tomita, and T. Kada (1985) Bioantimutagenic effects of tannic acid on UV and chemically induced mutagenesis in Escherichia coli B/r. Mutat. Res. 149:17-23.

58. Stich, H.F., and M.P. Rosin (1984) Naturally occurring phenolics as antimutagenic and anticarcinogenic agents. In Nutritional and Toxicological Aspects of Food Safety, M. Friedman, ed. Plenum Press, New York, pp. 1-29.

59. Sugimura, T., M. Nagao, and Y. Okada (1966) Carcinogenic action of N-methyl-N'-nitro-N-nitrosoguanidine. Nature 210:962-963.

60. Sugimura, T., and S. Fujimura (1967) Tumor production in glandular stomach of rat by N-methyl-N'-nitro-N-nitrosoguanidine. Nature 216:943-944.

61. Sugimura, T., S. Fujimura, and T. Baba (1970) Tumor production in the glandular stomach and alimentary tract of the rat by N-methyl-N'-nitro-N-nitrosoguanidine. Cancer Res. 30:455-465.

62. Terwel, L., and J.C.M. van der Hoeven (1985) Antimutagenic activity of some naturally occurring compounds towards cigarette-smoke condensate and benzo(a)pyrene in the Salmonella/microsome assay. Mutat. Res. 152:1-4.

63. Ulland, B.M., J.H. Weisburg, R.S. Yamamoto, and E.K. Weisburger (1973) Antioxidants and carcinogenesis: Butylated hydroxytoluene, but not diphenyl-p-phenylenediamine, inhibits cancer induction by N-2-fluorenylacetamide in rats. Food Cosmet. Toxicol. 11:199-207.

64. Waters, M.D., H.F. Stack, A.L. Brady, P.H.M. Lohman, L. Haroun, and H. Vainio (1988) Use of computerized data listings and activity profiles of genetic and related effects in the review of 195 compounds. Mutat. Res. 205:295-312.

65. Wattenberg, L.W. (1978a) Inhibition of chemical carcinogenesis. J. Natl. Cancer Inst. 60:11-18.

66. Wattenberg, L.W. (1978b) Inhibitors of chemical carcinogenesis. Adv. Cancer Res. 26:197-226.

67. Wattenberg, L.W. (1983) Inhibition of neoplasia by minor dietary constituents. Cancer Res. 43(Suppl.):2448s-2453s.

68. Wattenberg, L.W. (1985) Chemoprevention of cancer. Cancer Res. 45:1-8.

69. Whiting, R.F., L. Wei, and H.F. Stich (1981) Mutagenic and antimutagenic activities of selenium compounds in mammalian cells. In Selenium in Biology and Medicine, J.E. Spallholz, J.L. Martin, and H.E. Ganther, eds. AVI Publishing Company, Inc., Westport, Connecticut, pp. 325-330.

70. Wilpart, M., P. Mainguet, D. Geeroms, and M. Roberfroid (1985) Desmutagenic effects of N-acetylcysteine on direct and indirect mutagens. Mutat. Res. 142:169-177.

PREVENTION OF FORMATION OF IMPORTANT MUTAGENS/CARCINOGENS IN THE HUMAN FOOD CHAIN

J.H. Weisburger and R.C. Jones

American Health Foundation
Valhalla, New York   10595

ABSTRACT

Etiological factors for gastric cancer, among others, involve consumption of smoked, salted, and pickled fish of certain types.   Their chemical nature is not yet fully established but probably involves diazo phenols, and their formation can be prevented either by omitting the salting and pickling process, or by using vitamins C and E on the food prior to salting, pickling, or smoking.   Both preventive approaches would limit the formation of mutagenic and carcinogenic diazo phenols.   Sugimura and associates discovered new types of mutagens as heterocyclic amines that are formed during frying or broiling of meats and fish.   In rats, these amines induce cancer specifically in organs such as breast, colon, or pancreas, associated with Western-type nutrition where promotional elements such as dietary fat play an enhancing role.   Thus, inhibition of the formation of these new carcinogens during cooking would remove the genotoxic components from the diet.   Mixing 10% soy protein with ground meat prior to frying prevents the formation of these mutagens presumably by affording a lower surface temperature.   More effective is the addition of tryptophan, proline, or mixtures thereof, which specifically blocks the formation of these mutagens/carcinogens, probably by competing for reactive intermediary aldehydes, so that these cannot interact with the normal essential target, creatinine.   Thus, we have available practical, yet science-based, mechanistically understood procedures to prevent the formation of carcinogens associated with important types of cancer prevalent in many countries.

INTRODUCTION

Early in the development of the relationship between mutagens and carcinogens, there were many conflicting findings.   Renowned investigators in the field of mutagenesis found that a number of the then-known classic carcinogens were not at all active in the test systems they used (4,6).   The lack of biochemical activation in the assays was the underlying reason for the negative outcome.   It was Sugimura (see Ref. 50) who in 1967 reported on the carcinogenicity of $\underline{N}$-methyl-$\underline{N}$'-nitro-$\underline{N}$-nitrosoguanidine (MNNG), discovered to be mutagenic in prokaryotic systems just a few years earlier by Mandell (34). This discovery provided the basis for a renewed interest in the association between mutagens and carcinogens. Shortly thereafter, Malling (33) found that the liver carcinogen dimethylnitrosamine yielded mutagens in Neurospora, if the organism was injected directly into the vascular system of the liver, but not elsewhere. This study demonstrated the importance of metabolism and the transfer of a reactive metabolite to the detector organism (see Ref. 49).

Based on these findings, I organized a workshop in 1971, chaired by the late Dr. Alexander Hollaender, to provide the basis for a careful examination of systems that could be used for rapid bioassays in connection with the development of the NCI bioassay system. These tests were carried out with different classes of known carcinogens and analogous noncarcinogens, and led to the demonstration that there were validated test systems available, provided there was provision for metabolic activation (42). A few years later, Ames (35) developed the now widely used test system involving Salmonella typhimurium and liver S-9 fractions. Shortly thereafter, Williams utilized the intrinsic metabolic competence of hepatocytes and produced a test system involving unscheduled DNA synthesis, and, using liver cell lines, a test for mutagenesis at the HGPRT (hypoxanthine-guanine-phosphoribosyl transferase) locus (65). The combination of the Ames system and the Williams approach formed a useful battery in which known human carcinogens yielded positive responses (66). Thus, at first approximation it would seem that chemicals that are positive in both the Ames and Williams tests might be human carcinogenic risks.

About ten years ago, Sugimura and his associates (see Ref. 50) made another major discovery, namely, that during cooking leading to browning of meat or fish, powerful Ames-positive chemicals are produced. Their isolation and identification as a new class of heterocyclic amines led to tests of these chemicals in the Williams test system, where they were also uniformly active (69). Thus, based on past experience, it would seem that these chemicals should be not only carcinogenic in animal models but also are probable human cancer risks. The carcinogenicity of this class of compounds was indeed demonstrated in mice and rats (51,53). The target organs included breast, colon, and pancreas, which are important sites of cancer in the Western world (62). Thus, it became important to find a means of lowering or preventing the formation of these new carcinogens during cooking.

At the same time, methods were developed to block the formation of carcinogens for cancer of the stomach, prevalent in the Far East, Japan and China, and cancer of the esophagus in parts of China, stemming from the tradition of salting and pickling certain foods, especially fish (8,25,56,59,70) and certain vegetables. Here also, means have been developed to lower the production of these types of carcinogens (27,37,62). These elements will be briefly reviewed.

INHIBITION OF NUTRITIONAL CARCINOGENESIS:
BREAST, COLON, AND PANCREAS

Extraction of the mutagenic components from fried meat led to the identification of a totally new class of mutagens, heterocyclic amines, typified by 2-amino-3-methylimidazo[4,5-f]quinoline (IQ) and related quinoxalines (MeIQx; 4,8-diMeIQx, etc.) (10,12,26,51,57).

These chemicals are genotoxic, being highly active in the Ames test and in the DNA repair test of Williams, and indeed in almost all tests measuring genotoxicity (32,54,64,69). Bioassays revealed them to be toxic and carcinogenic for several specific target organs, including the liver, urinary bladder, pancreas, and intestinal tract, with emphasis on colon, and mammary gland (see Ref. 2, 51, and 53). In the area of nutrition and cancer, there is evidence that the amount and type of dietary fat can exert powerful promoting effects on cancers in the colon, mammary gland, and pancreas (45). On the other hand, fat appears to have little enhancing effect on carcinogenesis in the liver and urinary bladder, perhaps accounting for the fact that neoplasms in these organs are at low incidence in the West.

The mutagens formed during cooking are easily detected because of high specific mutagenic activity, but the actual amount present and thence consumed per day by meat-eating populations is relatively small (9,51). However, such foods are consumed daily from childhood onwards. Thus, the occurrence of colon, breast, and perhaps pancreas cancer (in the latter case carcinogens from tobacco smoke may also be involved) may stem from the daily intake of small amounts of such genotoxic carcinogens, together with continuing promotion derived from the customary intake of fats from a high-fat diet of 35-45% of daily calories (61). The mechanisms of the effect of fat are reasonably well established, and therefore recommendations have been made to limit the daily fat intake to 20-25% of calories, in order to reduce the promoting potential (45).

Another approach to cancer prevention in these important target organs is to study the possibility of avoiding the formation of these mutagens and carcinogens during cooking. One such method, applicable to ground meat, is to add about 10% or more of soy protein, which has been shown to inhibit the formation of mutagens (58). Another advantage of soy protein/meat mixtures is that total fat is reduced by dilution with good-quality protein in the final product. Antioxidants also lowered the formation of mutagens during cooking.

We discovered the addition of L-tryptophan (L-trp) and related indoles to be a more specific means of lowering the formation of mutagens. This method is based on the following background information (20,21). The laboratory of Jägerstad (19) first demonstrated that the formation of mutagens during cooking requires creatinine as the essential component. Jägerstad and others (36,39,43,44) have demonstrated that the higher the creatinine level in model systems or in fried foods, the greater the formation of these mutagens. The overall mechanism of Maillard reactions in forming IQ-type mutagens, including the determination of the reactive intermediates that interact with creatinine, is as yet obscure. We have observed, however, that the addition of L-trp to in vitro reflux systems such as that of Jägerstad (18) yielded a dose-related inhibition of the formation of mutagens (Tab. 1). Mixtures of L-trp and L-proline are also effective inhibitors (Tab. 2).

The role of tryptophan may be to compete with creatinine for the reactive intermediates. For example, the usual system leading to this class of mutagens involves heating a solution of a carbohydrate, such as glucose or fructose, an amino acid such as glycine or alanine, and creatinine. Depending on conditions and reactants, such mixtures typically produce carcinogens such as 2-amino-3,8-dimethyl[4,5-f]quinoxaline, or the 3,4,8-trimethyl analog (19,26).

Tab. 1.   Inhibition of the formation of IQ-type mutagens by L-tryptophan
          in liquid-reflux models (from Ref. 20).

mM L-Trp added	TA98 + S9 Avg. rev. col./pl. per 0.1 ml reflux sample		% Inhibition	
	Study 1	Study 2	Study 1	Study 2
Complete model[a]				
0 (Control)	545 ± 77	1170 ± 58	Control	
1.75	460 ± 70 (p>0.05)	–	20	–
3.5	235 ± 17 (p<0.05)	–	70	–
17	230 ± 5 (p<0.05)	–	70	–
70	–	250 ± 8 (p<0.05)	–	90
105	–	140 ± 16 (p<0.05)	–	100
Vehicle control	50	70		
IQ[a] (5 ng/pl)	960[a]	1130[a]		
(Positive control)				

[a]2-Amino-3-methylimidazo[4,5-f]quinoline.

Heating a solution of hydroxyamino acid threonine leads to intermediates that react with creatinine and yield two mutagenic products that
have been characterized.   The major product (2-amino-5-ethylidene-1-
methylimadazol-4-one) can also be formed directly by the reaction between
acetaldehyde and creatinine, thus (1) proving by synthesis the structure
of the adduct formed from threonine, and (2) demonstrating that aldehydes can react with creatinine (22).   Threonine and creatinine in the
presence of L-trp do not form significant amounts of the mutagens noted
above; rather, the intermediate reacts with L-trp to yield 2-carboxy-8-
(1'-hydroxyethyl)pyrrolo(2,3-b)indole, a product of the reaction of acetaldehyde with a cyclization complex derived from L-tryptophan (R.C.
Jones and J.H. Weisburger, unpubl. results).   This finding directs attention to the likelihood that reactive aldehydes are the key intermediates
in the Maillard reactions during frying or broiling of meat or fish leading
to IQ or IQx carcinogens (23) (Fig. 1).

Indeed, it has been discovered that the formation of powerful mutagens and carcinogens during frying or broiling of fish or meat can
be lowered, most likely through similar mechanisms, by the addition of
relatively small amounts of L-trp and other indoles (Tab. 3).   We have
formulated practical means of performing this by adding L-trp to meat
sauce applied to meat prior to cooking (20,21).

Tab. 2. Inhibition of the formation of IQ-type mutagens by L-proline and L-tryptophan in liquid-reflux models (from Ref. 22).

Ingredients (mM)	TA98 + S9 Avg. rev. col./plate			% Inhibition		
	Exp. 1	Exp. 2	Exp. 3	Exp. 1	Exp. 2	Exp. 3
Control (gluc + gly + cr)	1590	1665	590			
Pro (7)	1460			8		
Pro (35)	1330			16		
Trp (7)	1015			35		
Pro (7) + Trp (7)	820			50		
Pro (35) + Trp (7)	460			75		
Pro (35) + Trp (14)		245			90	
Pro (35) + Trp (35)			160			80
Pro (70)		720			60	
Trp (70)			160			80
Pro (105)		740			55	
Pro (140)		1045			40	
DEG neg. control	60	45	60			
5 ng IQ-plate (positive control)	915	950	830			

INHIBITION OF NUTRITIONAL CARCINOGENESIS: STOMACH AND ESOPHAGUS

The traditional intake of salted, pickled, or smoked foods in the Far East is associated with a finding of cancer in the stomach, esophagus, and liver. Other specific chronic diseases, such as hypertension and stroke, are also prevalent in these geographic areas. Time trends for these diseases exhibited a pronounced decrease in incidence in the last 50-60 years in some areas of the world (7,16,25). For example, the decline of gastric cancer and hypertension and stroke in the United States has been the subject of detailed research. At the same time, pronounced changes in nutritional traditions were occurring. Prior to the advent, first, of commercial refrigeration and then, later, of household refrigeration, foods were preserved for long-term storage by salting and pickling with as much as 7-10% NaCl, or brine solutions, or mixtures of salt with saltpeter, with amounts of up to 5,000 ppm of saltpeter and a few percent of salt (1). These traditions have changed. Lower levels of nitrite, replacing nitrate--around 300 ppm and, more recently, as low as 50-80 ppm--are now being used.

Gastric cancer stems from the reaction of nitrite with certain substrates. This concept is supported by the discovery of Sugimura (50) that MNNG in drinking water is a reliable means of inducing glandular stomach cancer in many species. Such compounds can readily form from nitrite and appropriate reactants. Among others, Mirvish (37) has contributed sound data on the kinetics of formation of nitrosamines and nitrosamides as a function of such variables as concentration of nitrite, concentration of amine, substrate, pH, temperature, and other conditions.

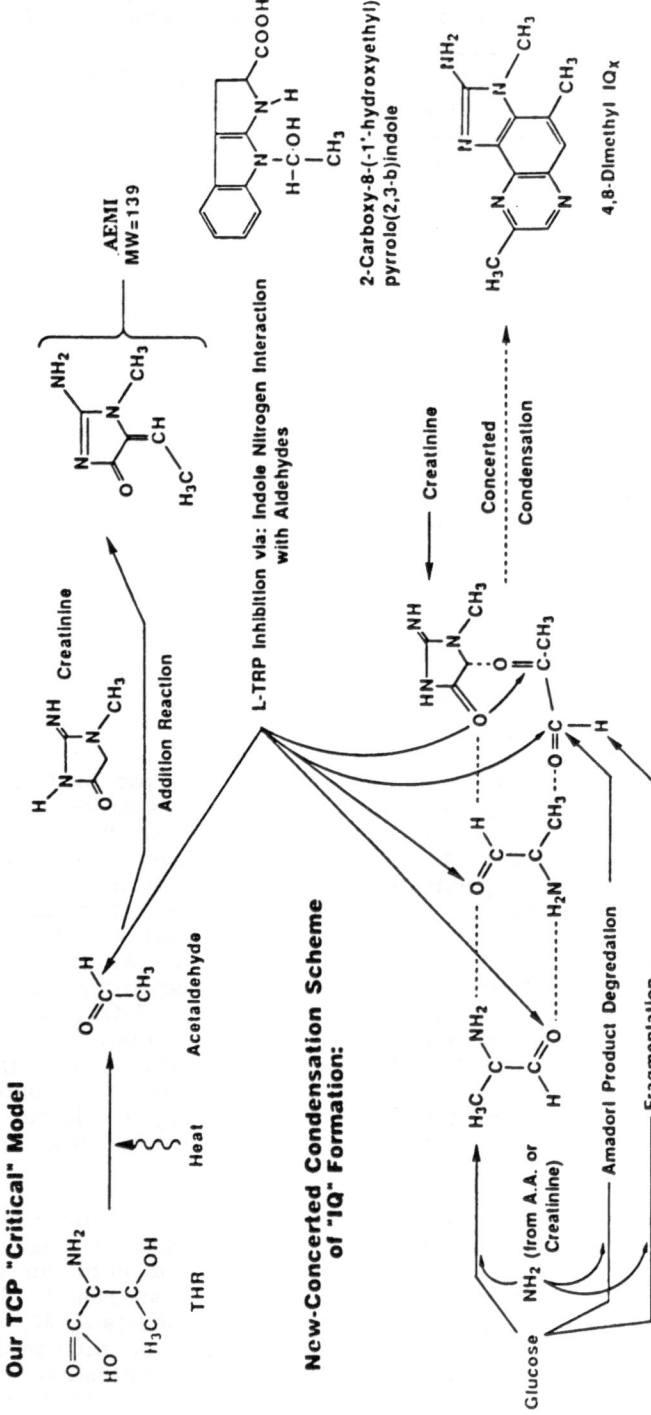

Fig. 1. Possible mechanism of inhibition of formation of IQ-like and IQ-type mutagens by competition between L-tryptophan, through the indole nitrogen, and creatinine (from Ref. 23).

Tab. 3.   Anticarcinogenesis and underlying mechanisms in specific nutritionally-linked cancers.

Site	Carcinogen	Inhibitors	Mechanisms
Esophagus	Tobacco, pickled, salted foods	Vegetables, fruits	?
Stomach	Pickled, salted foods; nitrate, nitrite	Vegetable, fruits, vitamins C and E	Less atrophic gastritis; destroy nitrite
Breast, colon, pancreas	Heterocyclic amines in fried, broiled meat and fish	Soy protein (for ground meat only)	Lowers cooking temperature through water retention
		Antioxidants	Trap intermediary reactive products during cooking
		L-Tryptophan, L-proline	Compete with creatinine for reactive products during cooking

The nitrite required might come from certain procedures of food preservation, such as salting and pickling, a hypothesis we favor on the basis of a number of lines of evidence. Nitrite might also derive from the presence of high geochemical nitrate levels, leading to foods containing high nitrate levels, or the formation of nitrite in the oral cavity or stomach, thence leading to endogenous nitrosation mainly in the stomach (8,38). In addition, such nitrosation reactions likely can occur during the exposure of certain foods such as fish or meats to wood smoke.

To account for the high risk of gastric cancer in parts of Latin America, the groups of Correa (8,38) and Tannenbaum (68) have studied the nitrosation of fava beans and have identified a product stemming from this reaction that might be the carcinogen. Mirvish (37) has subjected fish to nitrosation and has developed an indication for the production of a nitroso derivative. The group of Weisburger (62) has reacted a number of different fish, particularly those customarily used in northern Japan-- still a high-risk region for gastric cancer--with nitrite and has noted the rapid formation of a direct-acting mutagen (without requiring S-9 fractions) in the Ames assay. An extract of this mutagen, when administered to Wistar strain rats in a small pilot experiment, yielded cancer of the glandular stomach and associated preneoplastic lesions.

In the meantime, the group of Sugimura (41) found an association between the development of mutagens from different kinds of soy sauce upon nitrosation and the presence of tyramine. They have demonstrated that the product of the reaction of nitrite with tyramine is a novel form of reactive, yet fairly stable, diazonium salt. Weisburger's group has preliminary evidence that the product stemming from the reaction of an extract of Sanma fish with nitrite also yields this kind of reactive product. A Hawaiian fish has yielded a mutagen with similar properties (17). Also, nitrosamines were identified in nitrite-treated foods typical of Kashmir, a region with an appreciable incidence of cancer of the stomach and esophagus (46).

Of great relevance is the observation by Mirvish (37) that nitrosation of substrates could be blocked almost completely, depending on conditions, by vitamin C. This finding was extended to vitamin E (40).
These observations have been utilized in practice to lower the formation
and thence the presence of specific nitrosamines in foods such as bacon
and other meat products consumed by the public. Also, Haenszel (11)
and collaborators (see Ref. 13) and Hirayama (15) have noted that yellow-
green vegetables may have an inhibiting effect in gastric cancer and, by
extension in mechanistic terms, even in head and neck cancers. Yellow-
green vegetables are sources of vitamins C and E (5,37,60). We have not
found a protective effect of β-carotene, given at a high level of 2% in the
diet, in the MNNG-induced gastric cancer model in rats (24). Thus, the
protective effect reliably found with yellow-green vegetables may be due
to complex inhibiting factors, including small amounts of micronutrients
like β-carotene (71).

At the Ohito meeting, a number of papers described the inhibiting
effects of green tea and extracts therefrom in several in vitro and in vivo
systems of carcinogenesis and mutagenesis. An important contribution
was that of Furihata and Matsushima (see these Proceedings) that calcium
chloride depressed MNNG-induced gastric carcinogenesis, a finding analogous to the lower cell cycling in the colon of high-risk individuals given
calcium salts, reported by Lipkin and Newmark (30).

DISCUSSION

Lowering individual fat intake is likely effective at any time in reducing the risk of cancer development because the underlying mechanism
involves promotion, which is a highly dose-dependent and reversible
phenomenon (45). For example, the Western intake of fat of 40% of calories yields a concentration of about 12 mg/g of bile acids in stool, with
demonstrated promoting potential for colon cancer. In contrast, the traditional low-fat intake in Japan (15-20% of calories) yields a bile acid
concentration of 4 mg/g. Likewise, Western populations switching from a
high to a low fat level lower the effective concentration of bile acids
within days (14). Thus, the recommendation for lower risk is likely to
be effective almost immediately upon changing lifestyle. This approach
has been proposed for patients in addition to their having the usual medical management, including surgical removal of a primary colon or breast
cancer, to delay or avoid future recurrence (67).

On the other hand, with genotoxic carcinogens, knowledge of the
underlying mechanisms suggests that exposure to such agents leads to a
permanent increase in risk. For example, gastric cancer stems from exposure to genotoxic agents without much influence of promoters (8,16,37,
61). This neoplastic disease was prevalent in traditional Japan, and
migrants from high-risk Japan to much lower-risk U.S. have maintained
their risk for disease (11). Likewise, Japanese used to have a lower risk
for colon cancer, albeit they consumed fried fish containing IQ-type carcinogens, but in the presence of a low-fat diet the disease was not expressed. However, when individuals migrate from Japan and acquire
Western high-fat dietary traditions, they acquire the high risk for colon
cancer in the same generation (11,48). As the Japanese, particularly in
urban areas, begin to eat Western style, which involves a higher fat intake, their incidence of distal, but not proximal, colon cancer increases
(52,63). The cell duplication rate of intestinal mucosa is a principal

Tab. 4. Long-term goals for chronic disease prevention.

Action	Benefit
Control smoking--less harmful cigarette	Coronary heart disease; cancers of the lung, kidney, bladder, pancreas
Lower total fat intake	Coronary heart disease; cancers of the colon, breast, prostate, ovary, endometrium
Lower salt Na^+ intake, balance $K^+ + Ca^{2+}/Na^+$ ratio	Hypertension; stroke; cardiovascular disease
Increase natural fiber, Ca^{2+} (?)	Colorectal cancer
Avoid pickled, smoked, highly salted foods	Cancers of the stomach, esophagus, nasopharyngx, liver (?); hypertension; stroke
Avoid mycotoxins, senecio alkaloids, bracken fern	Cancers of the liver, stomach, bladder (?)
Increase and balance micronutrients, vitamins, minerals	Cardiovascular disease; several types of cancer
Lower intake of fried foods	Cancers of colon, breast, pancreas (?)
Practice sexual hygiene	Cancers of the cervix, penis
Have regular moderate exercise	Coronary heart disease; cancers of the colon, breast
Maintain proper weight, avoid obesity	Coronary heart disease; cancers of the endometrium, kidneys

factor to lock in abnormal expression of the genotoxic carcinogen-DNA interaction. In the colon, the cell turnover is fairly constant and is a function of familial factors and intestinal pH, calcium ions, or bile acid concentration, in turn, controlled by dietary fat (28,29,47).

In contrast to the rapid acquisition of risk for colon cancer, migrants from Japan continue at a relatively low risk for breast cancer (31,55). A possible explanation for this difference between colon and breast cancer risk in migrants may stem from the fact that during sexual maturation, mammary tissue displays a burst of high-level DNA synthesis and cell duplication, conditions favoring the transformation of normal to early neoplastic cells in the presence of a carcinogen. Thus, the group aged 10-15 at the time of the single high-radiation exposure from the atomic bomb in Hiroshima led to a four times higher breast cancer rate than in individuals who were older with a fully developed breast (3). Perhaps, the food-derived genotoxic carcinogens affecting mammary tissue yield a relatively lower exposure in Japan than in the U.S., and the required metabolism to a reactive carcinogen may also be fat-dependent.

In summary, the combined results presented suggest that avoidance early in life of the nutritionally-linked genotoxic carcinogens affecting breast, intestinal tract, and pancreas, on the one hand, and the distinct classes of carcinogens for stomach and esophagus may be a significant means to decrease the risk of later cancer development in these organs. The approach proposed--the addition of small amounts of L-tryptophan and L-proline prior to cooking of meats and fish--in lowering the formation of putative carcinogens for breast, colon, and pancreas is effective and can be readily and practically implemented. In the Orient especially, lowering the use of salted, pickled fish and vegetables and increasing the intake of yellow-green vegetables and fruits and low-fat milk products as a source of calcium are designed to significantly lower the risk for cancer of the stomach and esophagus. Thus, we have in hand relatively simple lifestyle adjustments and procedures to effectively decrease factors associated with major human neoplasms in the world (Tab. 4).

ACKNOWLEDGEMENTS

This paper is respectfully dedicated to the memory of the late Dr. Tsuneo Kada, Genetics Institute, Mishima, an active, distinguished researcher with broad interests and important contributions, who also displayed superb musical talents. Research in our laboratory is supported by USPHS grants CA-24217, CA-42381, and CA-45720 from the National Cancer Institute. We are indebted to Mrs. Clara Horn for expert editorial assistance.

REFERENCES

1. Binkerd, E.F., and O.E. Kolari (1975) The history and use of nitrate and nitrite in the curing of meat. Food Cosmet. Toxicol. 13:655-661.
2. Bird, R.P., and W.R. Bruce (1984) Damaging effect of dietary components to colon epithelial cells in vivo: Effect of mutagenic heterocyclic amines. J. Natl. Cancer Inst. 73:237-240.
3. Boyce, J.D., and R.N. Hoover (1981) Radiogenic breast cancer: Age effects and implications for models of human carcinogenesis. In Cancer: Achievements, Challenges, and Prospects for the 1980s, Vol. 1, J.H. Burchenal and H.F. Oettgen, eds. Grune and Stratton, New York, pp. 209-221.
4. Boyland, E. (1954) Mutagens. Pharmacol. Rev. 6:345-364.
5. Bright-See, E. (1983) Vitamins C and E (or fruits and vegetables) and the prevention of human cancer. In Nutritional Factors in the Induction and Maintenance of Malignancy, C.E. Butterworth, Jr., and M.L. Hutchinson, eds. Academic Press, Inc., New York, pp. 217-223.
6. Burdette, W.J. (1955) The significance of mutation in relation to the origin of tumors: A review. Cancer Res. 15:201-226.
7. Coggon, D., and E.D. Acheson (1984) The geography of cancer of the stomach. Br. Med. Bull. 40:335-341.
8. Correa, P. (1983) The gastric precancerous process. Cancer Surv. 2:437-450.
9. Felton, J.S., M.G. Knize, N.H. Shen, B.D. Andresen, L.F. Bjeldanes, and F.T. Hatch (1986) Identification of the mutagens in cooked beef. Environ. Health Perspect. 67:17-24.
10. Felton, J.S., M.G. Knize, N.H. Shen, P.R. Lewis, B.D. Andresen, J. Happe, and F.T. Hatch (1986) The isolation and identification of a new mutagen from fried ground beef: 2-amino-1-methyl-6-phenyl-imidazo[4,5-b]pyridine (PhIP). Carcinogenesis 7:1081-1086.
11. Haenszel, W. (1975) Migrant studies. In Persons at High Risk of Cancer: An Approach to Cancer Etiology and Control, J.F. Fraumeni, Jr., ed. Academic Press, Inc., New York, pp. 361-371.
12. Hatch, F.T., S. Nishimura, W.D. Powrie, and L.N. Kolonel, eds. (1986) Formation of mutagens during cooking and heat processing of food. Environ. Health Perspect. 67:3-157.

13. Henderson, B.E., ed. (1982) Third Symposium on Epidemiology and Cancer Registries in the Pacific Basin. NCI Monogr. 62, National Cancer Institute, Bethesda, Maryland.
14. Hill, M.J. (1971) The effect of some factors on the faecal concentration of acid steroids, neutral steroids and urobilins. J. Pathol. 104:239-245.
15. Hirayama, T. (1979) Diet and cancer. Nutr. Cancer 1:67-79.
16. Howson, C.P., T. Hiyama, and E.L. Wynder (1986) The decline in gastric cancer: Epidemiology of an unplanned triumph. Epidemiol. Rev. 8:1-27.
17. Ichinotsubo, D.Y., and H.F. Mower (1982) Mutagens in dried/salted Hawaiian fish. J. Agric. Food Chem. 30:937-939.
18. Jägerstad, M., S. Grivas, K. Olsson, A.L. Reutersward, C. Negishi, and S. Sato (1986) Formation of food mutagens via Maillard reactions. In Genetic Toxicology of the Diet, I. Knudsen, ed. Alan R. Liss, Inc., New York, pp. 155-168.
19. Jägerstad, M., A.L. Reutersward, R. Olsson, S. Grivas, T. Nyhammar, K. Olsson, and A. Dahlqvist (1983) Creatin(in)e and Maillard reaction products as precursors of mutagenic compounds: Effects of various amino acids. Food Chem. 12:255-264.
20. Jones, R.C., and J.H. Weisburger (1988) L-Tryptophan inhibits formation of mutagens during cooking of meat and in laboratory models. Mutat. Res. 206:343-349.
21. Jones, R.C., and J.H. Weisburger (1989) Characterization of amino-alkylimidazol-4-one mutagens from liquid-reflux models. Mutat. Res. 222:43-51.
22. Jones, R.C., and J.H. Weisburger (1988) Inhibition of aminoimidazoquinoxaline-type and aminoimidazol-4-one type mutagen formation in liquid-reflux models by the amino acids L-proline and/or L-tryptophan. Environ. Molec. Mutag. 11:509-514.
23. Jones, R.C., and J.H. Weisburger (1988) Inhibition of aminoimidazoquinoxaline-type and aminoimidazol-4-one type mutagen formation in liquid reflux models by L-tryptophan and other selected indoles. Jpn. J. Cancer Res. (Gann) 79:222-230.
24. Jones, R.C., S. Sugie, J. Braley, and J.H. Weisburger (1989) Dietary beta-carotene in rat models of gastrointestinal cancer. J. Nutr. 119:503-514.
25. Joossens, J.V., M.J. Hill, and J. Geboers, eds. (1985) Diet and Human Carcinogenesis, Excerpta Medica, Amsterdam.
26. Knudsen, I., ed. (1986) Genetic Toxicology of the Diet, Alan R. Liss, Inc., New York.
27. Licht, W.R., S.R. Tannenbaum, and W.M. Deen (1988) Use of ascorbic acid to inhibit nitrosation: Kinetic and mass transfer considerations for an in vitro system. Carcinogenesis 9:365-372.
28. Lipkin, M. (1988) Biomarkers of increased susceptibility to gastrointestinal cancer: New application to studies of cancer prevention in human subjects. Cancer Res. 48:235-245.
29. Lipkin, M., E. Friedman, S.J. Winawer, and H. Newmark (1989) Colonic epithelial cell proliferation in responders and nonresponders to supplemental dietary calcium. Cancer Res. 49:248-254.
30. Lipkin, M., and H. Newmark (1985) Effect of added dietary calcium on colonic epithelial-cell proliferation in subjects at high risk for familial colonic cancer. N. Engl. J. Med. 313:1381-1384.
31. Locke, F.B., and H. King (1980) Cancer mortality risk among Japanese in the United States. J. Natl. Cancer Inst. 65:1149-1156.

32. Loury, D.J., and J.L. Byard (1985) Genotoxicity of the cooked-food mutagens IQ and MeIQ in primary cultures of rat, hamster, and guinea pig hepatocytes. Environ. Mutag. 7:245-254.
33. Malling, H.V. (1966) Mutagenicity of two potent carcinogens dimethylnitrosamine and diethylnitrosamine in Neurospora crassa. Mutat. Res. 3:537-540.
34. Mandell, J.D., and J. Greenberg (1960) A new chemical mutagen for bacteria, 1-methyl-3-nitro-1-nitrosoguanidine. Biochem. Biophys. Res. Commun. 3:575-577.
35. Maron, D.M., and B.N. Ames (1983) Revised methods for the Salmonella mutagenicity test. Mutat. Res. 113:173-215.
36. Matsushima, T. (1982) Mechanisms of conversion of food components to mutagens and carcinogens. In Molecular Interrelations of Nutrition and Cancer, M.S. Arnott, J. van Eys, and Y.-M. Wang, eds. Raven Press, New York, pp. 35-42.
37. Mirvish, S.S. (1983) The etiology of gastric cancer. Intragastric nitrosamide formation and other theories. J. Natl. Cancer Inst. 71:631-647.
38. Montes, G., C. Cuello, P. Correa, W. Haenszel, G. Zarama, and G. Gordillo (1984) Mutagenic activity of nitrosated foods in an area with a high risk for stomach cancer. Nutr. Cancer 6:171-175.
39. Nes, I.F. (1987) Formation of mutagens in an amino acid-glucose model system and the effect of creatine. Food Chem. 24:137-146.
40. Newmark, H.L., and W.J. Mergens (1981) Block nitrosamine formation using ascorbic acid and alpha-tocopherol. Banbury Rep. 7:285-290.
41. Ochiai, M., K. Wakabayashi, M. Nagao, and T. Sugimura (1984) Tyramine is a major mutagen precursor in soy sauce, being convertible to a mutagen by nitrite. Gann 75:1-3.
42. Poirier, L.A., and E.K. Weisburger (1979) Selection of carcinogens and related compounds tested for mutagenic activity. J. Natl. Cancer Inst. 62:833-840.
43. Reutersward, A.L., K. Skog, and M. Jägerstad (1987) Effects of creatine and creatinine content on the mutagenic activity of meat extracts, bouillons and gravies from different sources. Fd. Chem. Toxicol. 25:747-754.
44. Reutersward, A.L., K. Skog, and M. Jägerstad (1987) Mutagenicity of pan-fried bovine tissues in relation to their content of creatine, creatinine, monosaccharides and free amino acids. Food Chem. Toxicol. 25:755-762.
45. Rose, D.P., ed. (1987) Proceedings: Workshop on new developments on dietary fat and fiber in carcinogenesis. Prev. Med. 16:449-595.
46. Siddiqi, M., A.R. Tricker, and R. Preussmann (1988) Formation of N-nitroso compounds under simulated gastric conditions from Kashmir foodstuffs. Cancer Lett. 39:259-265.
47. Sorenson, A.W., M.L. Slattery, and M.H. Ford (1988) Calcium and colon cancer: A review. Nutr. Cancer 11:135-145.
48. Stemmermann, G.N., A.M.Y. Nomura, and L.N. Kolonel (1987) Cancer among Japanese-Americans in Hawaii. Gann Monogr. Cancer Res. 33:99-108.
49. Stoltz, D.R., L.A. Poirier, C.C. Irving, H.F. Stich, J.H. Weisburger, and H.C. Grice (1974) Evaluation of short-term tests for carcinogenicity. Toxicol. Appl. Pharmacol. 29:157-180.
50. Sugimura, T. (1979) Topics of studies in 25 years under Dr. Waro Nakahara. GANN Monogr. Cancer Res. 24:3-50.
51. Sugimura, T. (1988) New environmental carcinogens in daily life. Trends Pharmacol. Sci. 9:205-209.

52. Tajima, K., K. Hirose, N. Nakagawa, T. Kuroshishi, and S. Tominaga (1985) Urban-rural difference in the trend of colo-rectal cancer mortality with special reference to the subsites of colon cancer in Japan. Jpn. J. Cancer Res. (Gann) 76:717-728.

53. Tanaka, T., W.S. Barnes, J.H. Weisburger, and G.M. Williams (1985) Multipotential carcinogenicity of the fried food mutagen 2-amino-3-methylimidazo[4,5-f]quinoline (IQ) in rats. Jpn. J. Cancer Res. (Gann) 76:570-576.

54. Terada, M., M. Nagao, M. Nakayasu, H. Sakamoto, F. Nakasato, and T. Sugimura (1986) Mutagenic activities of heterocyclic amines in Chinese hamster lung cells in culture. Environ. Health Perspect. 67:117-119.

55. Tominaga, S. (1985) Cancer incidence in Japanese in Japan, Hawaii, and Western United States. Natl. Cancer Inst. Monogr. 69:83-92.

56. Ubukata, T., A. Oshima, K. Morinaga, T. Hiyama, S. Kamiyama, A. Shimada, and J.-P. Kim (1987) Cancer patterns among Koreans in Japan, Koreans in Korea and Japanese in Japan in relation to life style factors. Jpn. J. Cancer Res. (Gann) 78:437-446.

57. Vuolo, L.L., and G.J. Schuessler (1985) Review: Putative mutagens and carcinogens in foods. Environ. Mutag. 7:577-598.

58. Wang, Y.Y., L.L. Vuolo, N.E. Spingarn, and J.H. Weisburger (1982) Formation of mutagens in cooked foods. V. The mutagen reducing effect of soy protein concentrates and antioxidants during frying of beef. Cancer Lett. 16:179-189.

59. Watanabe, S., S. Tsugane, and Y. Ohno (1988) Prediction of the gastric cancer mortality in 2000 in Japan. Jpn. J. Cancer Res. (Gann) 79:439-444.

60. Weisburger, J.H. (1985) Causes of gastric and esophageal cancer. Possible approach to prevention by vitamin C. Intl. J. Vitamin Nutr. Res. Suppl. 27:381-402.

61. Weisburger, J.H. (1986) Application of the mechanisms of nutritional carcinogenesis to the prevention of cancer. In Diet, Nutrition and Cancer, Y. Hayashi, M. Nagao, T. Sugimura, S. Takayama, L. Tomatis, L.W. Wattenberg, and G.N. Wogan, eds. Japan Science Societies Press, Tokyo, pp. 11-26.

62. Weisburger, J.H., and C.L. Horn (1984) Human and laboratory studies on the causes and prevention of gastrointestinal cancer. Scand. J. Gastroenterol. 19(Suppl. 104):15-26.

63. Weisburger, J.H., and E.L. Wynder (1987) Etiology of colo-rectal cancer with emphasis on mechanism of action and prevention. In Important Advances in Oncology 1987, V.T. DeVita, Jr., S. Hellman, and S.A. Rosenberg, eds. J.B. Lippincott Co., Philadelphia, pp. 197-219.

64. Wild, D., E. Gocke, D. Harnasch, G. Kaiser, and M.-T. King (1985) Differential mutagenic activity of IQ (2-amino-3-methylimidazo[4,5-f]quinoline) in Salmonella typhimurium strains in vitro and in vivo, in Drosophila, and in mice. Mutat. Res. 156:93-102.

65. Williams, G.M. (1979) Review of in vitro test systems using DNA damage and repair for screening of chemical carcinogens. J. Assn. Official Analytical Chemists 62:857-863.

66. Williams, G.M., and J.H. Weisburger (1988) Application of a cellular test battery in the decision point approach to carcinogen identification. Mutat. Res. 205:79-90.

67. Wynder, E.L., D.P. Rose, and L.A. Cohen (1986) Diet and breast cancer in causation and therapy. Cancer 58:1804-1813.

68. Yang, D., S.R. Tannenbaum, G. Bochi, and G.C.M. Lee (1984) Chloro-6-methoxyindole is the precursor of a potent mutagen (4-

chloro-6-methoxy-2-hydroxy-1-nitroso-indolin-3-one-oxime) that forms during nitrosation of the fava bean (Vicia faba). Carcinogenesis 5:1219-1224.

69. Yoshimi, N., S. Sugie, H. Iwata, H. Mori, and G.M. Williams (1988) Species and sex differences in genotoxicity of heterocyclic aminopy-rolysis and cooking products in the hepatocyte primary culture/DNA repair test using rat, mouse, and hamster hepatocytes. Environ. Molec. Mutag. 12:53-64.

70. You, W.-C., W.J. Blot, Y.-S. Chang, A.G. Ershow, Z.-T. Yang, Q. An, B. Henderson, G.-W. Xu, J.F. Fraumeni, Jr., and T.-G. Wang (1988) Diet and high risk of stomach cancer in Shadong, China. Cancer Res. 48:3518-3523.

71. Ziegler, R.G. (1989) A review of epidemiologic evidence that carote-noids reduce the risk of cancer. J. Nutr. 119:116-122.

FREE RADICAL REACTIONS WITH DNA AND ITS NUCLEOTIDES

Ronald P. Mason, Klaus Stolze and W. D. Flitter

Laboratory of Molecular Biophysics
National Institute of Environmental Health Sciences
Research Triangle Park, North Carolina 27709

INTRODUCTION

The interaction of free radicals with DNA has been a major area of interest and speculation, but previous electron spin resonance investigations of this area have been very limited. Many free radicals, like other chemicals will interact physically with DNA before they react chemically.

$$R' + DNA \rightarrow [R^{\bullet} \cdots DNA] \rightarrow R\text{-}DNA^{\bullet}$$

One example of a physical interaction with DNA is intercalation. We have begun ESR studies using redox active compounds such as chlorpromazine to study the intercalation of free radicals between the bases of DNA (1,2). The cation radi- cal of chlorpromazine was obtained by oxidation with horseradish peroxidase and hydrogen peroxide as described by Piette et al. (3). The chlorpromazine cation radical was made in the presence of calf thymus DNA at 2 mg/ml, which was flowed at 1 ml/min through a folded capillary of 0.25 mm (Fig. 1). The folded capillary can be rotated by 90° within the ESR cavity so that measurements can be made with the flow either perpendicular or parallel (Fig. 1) to the magnetic field of the ESR spectrometer. The flow orients the DNA helix within the capillary. When the flow is perpendicular to the magnetic field, a singlet with a shoulder due to unresolved hyperfine structure is observed (Fig. 2). When the flow is parallel to the magnetic field, the magnetic field is perpendicular to the DNA bases and the intercalated chlorpromazine cation radical gives the largest hyperfine interaction of the unpaired electron with the ring nitrogen, which results in three lines.

Chlorpromazine is one member of a class of phenothiazines that are oxidized to cation radicals by horseradish peroxidase, including phenothiazine itself (2). In the case of phenothiazine, four lines are found when the intercalated free radical is flowed parallel to the magnetic field. This type of spectrum is due to the interaction of the unpaired electron with one nitrogen atom and one proton. This proton is exchanged with deuterium in buffer made with D_2O, resulting in a three line spectrum very similar to that of the chlorpromazine cation radical (1).

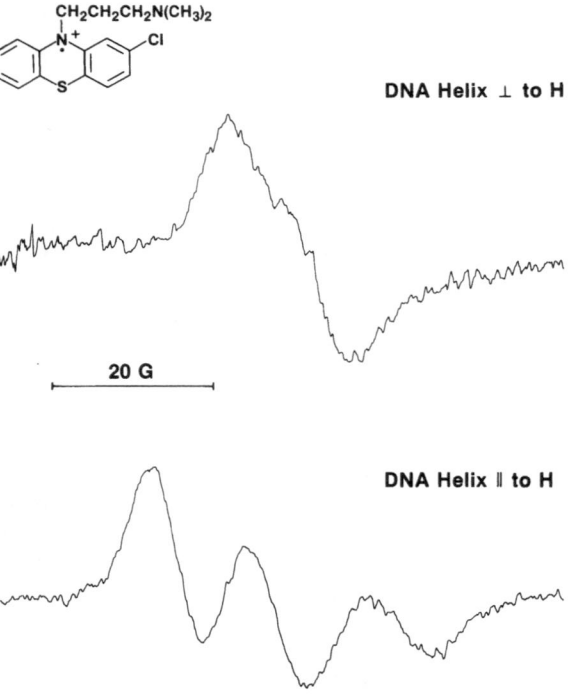

Fig. 1.   ESR spectra of the chlorpromazine cation radical intercalated
into calf thymus DNA. Reaction conditions: chlorpromazine, 0.5
mM; hydrogen peroxide, 0.25 mM; horseradish peroxidase, 28
units/ml; DNA, 2 mg/ml; final pH 4.6.   (A) Flow parallel to
the magnetic field; flow rate 3.5 ml/min. room temperature.
(B) Flow perpendicular to the magnetic field; flow rate 3.5
ml/min, room temperature.

When chlorpromazine is oxidized by horseradish peroxidase to its
cation free radical, covalent binding of a metabolite of chlorpromazine to
DNA occurs in addition to the intercalation (4,5).   The cation radical is
proposed to be the electrophile, but a cation radical-derived species may
actually be responsible for this covalent binding.

When a free radical reacts chemically with DNA, at least three types
of reactions can be described. First, free radical chemistry may not be
directly involved.

$$R^{\cdot} + DNA \rightarrow R^{\cdot}\text{-DNA}$$

For example, mitomycin C forms a semiquinone which can become covalently bound to the DNA as shown by Tomaz et al. (6), this process involves
even-electron reactions of the mitomycin C and DNA, and the semiquinone
of mitomycin C, although present, plays no direct role.

Reactions of the hydroxyl radical with DNA are central to radiation
damage caused by ionizing radiation; the second and third type of reactions of free radicals with DNA are illustrated by the hydroxyl radical.

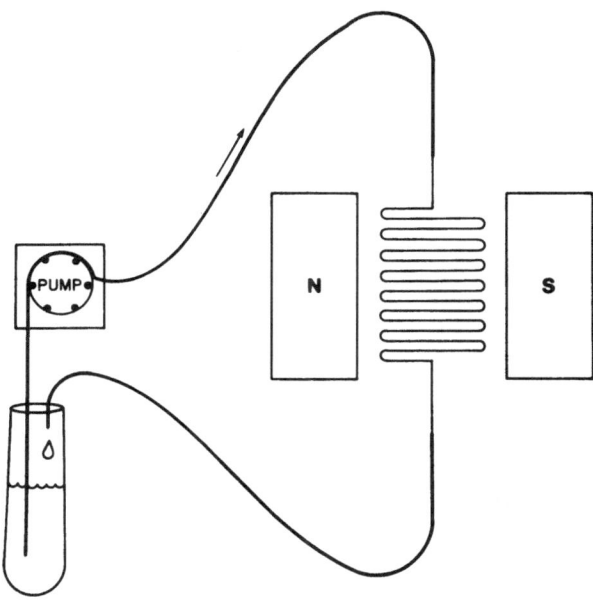

Fig. 2.   Schematic of the folded capillary, i.d. 0.25 mm, o.d. 1 mm. made of pyrex glass showing configuration with flow parallel to the magnetic field.

$$\cdot OH + DNA \rightarrow R\text{-}DNA\cdot$$

The hydroxyl radical is known to add across the double bonds of the DNA bases or to abstract hydrogen atoms from the DNA bases as well as the deoxyribose, forming DNA-centered free radicals of different types (7). The hydroxyl radical reacts at a near diffusion-limited rate, which means that this free radical reacts without forming a physical complex.

An oxygen-derived radical, presumably the hydroxyl radical, is responsible for the mutagenicity of paraquat. This mutagenicity is superoxide dismutase inhibitable (8). Even oxygen can be a mutagen in bacteria with low superoxide dismutase levels (9). In line with the hydroxyl radical being a mutagen in chemical systems as well as in radiation damage is the mutagenicity of ferrous iron (10). Damage to DNA by hydrogen peroxide is also thought to occur through the Fenton reaction (11).

Presumably, the hydroxyl radical generated in vivo would come from the metal ion-dependent breakdown of H_2O_2 according to the general equation

$$M^{n+} + H_2O_2 \rightarrow M^{(n+1)+} + \cdot OH + OH^- \tag{1}$$

in which M^{n+} is an unidentified endogenous metal ion. In chemical systems this metal ion could be titanium III, copper I, iron II or cobalt II. Realistically only the iron II and copper I reactions could occur in vivo, and these would probably occur only under abnormal physiological conditions. Considerable attention has been focused on the iron-dependent decomposition of H_2O_2, the Fenton reaction.

$$Fe^{2+} + H_2O_2 \rightarrow Fe^{3+} + {}^{\bullet}OH + OH^{-} \tag{2}$$

Several authors have recently suggested that at least two separate oxidizing species are produced by the action of hydrogen peroxide on iron chelates. Youngman and Elstner have proposed that a crypto' hydroxyl radical could be formed in some in vitro systems (12,13). Koppenol has proposed that ferrous iron and hydrogen peroxide produce a ferryl iron or ferrous hydrogen peroxide complex (14,15). Winterbourn proposes that when ferrous iron is chelated with EDTA, the predominant species formed is the hydroxyl radical, while with nonchelated iron, a tetravalent iron complex is formed which is capable of oxidizing ribose (16). Iron can bind to a wide range of ligands including nucleotides and DNA. If these ligands are capable of supporting 'OH production, it would seem likely that the ${}^{\bullet}OH$ formed in this way would attack the ligand binding the metal ion. Indeed this has brought forward the concept of site-specific damage (17).

The reaction of the hydroxyl radical, generated by a Fenton system, with pyrimidine deoxyribonucleotides was investigated using the ESR technique of spin trapping. The spin trap t-nitrosobutane was employed to trap secondary carbon-centered free radicals formed by the reaction of the hydroxyl radical with these nucleotides. In the spin-trapping technique the short-lived radicals (R.) are generated in a system containing a spin trap (18). t-Nitrosobutane is a commonly used spin trap for carbon-centered free radicals, and the resultant radical adducts are relatively stable nitroxide radicals which can be conveniently identified by ESR spectroscopy. The hydroxyl radical adduct of t-nitrosobutane is very unstable, so we do not detect the hydroxyl radical with this spin trap (19).

With ionizing radiation, hydroxyl radical attack at the C(6) position of the thymidine C(5)C(6) double bond leads to a carbon-centered free radical at the C(5) position (Fig. 3). Hydroxyl radical attack at the other end of the double bond leads to a carbon-centered free radical at the C(6) position, which is spin trapped to form a nitroxide (Fig. 4). Attack of the hydroxyl radical at either position would lead to the formation of thymine glycol. The ESR spectra that are detected result from

Structure 1.

two different radical species, two thymidine-derived radical adducts (20). Both free radicals are entirely dependent upon the presence of thymidine. Identical spectra of reduced intensity were obtained in the absence of the iron chelator EDTA. The free radical formation is dependent on added iron. The radical adduct is of course dependent on spin trap. A low concentration of the radical adducts is formed in the absence of added hydrogen peroxide. In this case some hydrogen peroxide is presumably produced via autoxidation of the ferrous iron (20).

The hyperfine coupling constants from C(5)- and C(6)-centered radical adducts formed in a Fenton system are in good agreement with those reports where irradiation was the source of the hydroxyl radical (21,22). This result suggests that the primary reactive species generated by the Fenton system is indeed the hydroxyl radical. We could find no unusual radicals as might be expected from the involvement of a ferryl iron species or site specific reactions of the hydroxyl radical.

2-Deoxycytidine-5-monophosphate also forms C(5)- and C(6)-centered free radicals in our Fenton system (20). The radical formation is dependent on all reaction components. 2-Deoxyuridine 5-monophosphate once again forms C(5)- and C(6)-centered free radicals, with spectra very similar to those of 2-deoxycytidine.

Gutteridge has demonstrated that TBA-reactive material can be produced from 2-deoxyribose on exposure to ferrous iron (23). Since this work by Gutteridge, many investigators have used the 2-deoxyribose/TBA assay as a measure of hydroxyl radical formation. Recently, Cheeseman et al. (24) have shown that this TBA-reactive material is malondialdehyde, and presumably the 2-deoxyribose moiety of the nucleotides is yielding malondialdehyde in our system as well. The results of our study with the TBA assay show that the formation of malondialdehyde is almost entirely dependent on the presence of the nucleotide and ferrous ion. Interestingly, there is still a considerable amount of malondialdehyde formed when EDTA is removed from the reaction system, which is in agreement with the spin-trapping studies. The removal of the spin trap from the reaction system, which is present at one-tenth the concentration of the nucleotides, has little overall effect. The removal of added hydrogen peroxide produces only a small decrease in the amount of malondialdehyde. Apparently sufficient hydrogen peroxide can be produced by the autoxidation of the ferrous ion.

Structure 2.

If the TBA assay is repeated in the presence of the hydroxyl radical scavenger formate, and assays are performed in the presence and absence of both EDTA and t-nitrosobutane. some interesting observations can be made. If we consider the complete system, addition of sodium formate at a final concentration of 100 mM produces a 71% inhibition in the formation of malondialdehyde material. This result suggests that formate is acting as a hydroxyl radical scavenger.   In the system with no EDTA, there is a decrease in the amount of malondialdehyde produced, but even in this system, formate can still produce a degree of inhibition (47%).   This result suggests that not all of the damage inflicted on the nucleotide is produced by a site-specific generation of the hydroxyl radical, which would not be inhibited by formate, and that some "free" ˙OH must still be produced.   This would infer that a species capable of binding iron is still present.   The only candidate for this role is the spin trap t-nitroso-butane.

When both EDTA and t-nitrosobutane are removed from the reaction system, the protective effect of formate is greatly decreased. and only 23% inhibition is found.   This clearly suggests that in this system the hydroxyl radical is produced by a site-specific mechanism.   That is, the iron necessary for the Fenton reaction is bound to the nucleotide.   The results of the TBA assays suggest that the spin trap t-nitrosobutane can serve to bind iron in such a way that it can still undergo the necessary redox chemistry to produce the hydroxyl radical, free in solution.

In summary, our investigations demonstrate that free radical metabolites of redox active compounds, such as the phenothiazine cation radical intercalate between DNA bases (1,2), and that a Fenton system produces the same secondary carbon-centered free radicals from pyrimidine nucleotides as high energy ionizing radiation (20), thus implicating the hydroxyl radical as the damaging species.   This study also shows that hydroxyl radical attack on the nucleotide produces TBA-reactive material, presumably malondialdehyde, and that there is qualitative agreement between the formation of malondialdehyde from the deoxyribose group and pyrimidine-derived free radicals.

REFERENCES

1.   Stolze, K., and R.P. Mason (1987) A new flat cell for flow-orientation ESR experiments. J. Mag. Res. 73:287-292.
2.   Stolze, K., and R.P. Mason (1989) ESR spectroscopy of flow-oriented cation radicals of phenothiazine derivatives and phenoxathiin intercalated in DNA (submitted for publication).
3.   Piette, L.H., G. Bulow, and I. Yamazaki (1964) Electron paramagnetic-resonance studies of the chlorpromazine free radical formed during enzymatic oxidation by peroxidase-hydrogen peroxide. Biochim. Biophys. Acta. 88:120-129.
4.   de Mol, N.J., and R.W. Busker (1984) Irreversible binding of the chlorpromazine radical cation and of photoactivated chlorpromazine to biological macromolecules. Chem.-Biol. Interact. 52:79-92.
5.   de Mol, N.J., A.B.C. Becht, J. Koenen, and G. Lodder (1986) Irreversible binding with biological macromolecules and effects in bacterial mutagenicity tests of the radical cation of promethazine and photoactivated promethazine. Comparison with chlorpromazine. Chem.-Biol. Interact. 57:73-83.

6. Tomasz, M., R. Lipmann, D. Chowdary, J. Pawlak, G.L. Verdine, and K. Nakanishi (1987) Isolation and structure of a covalent cross-link adduct between mitomycin C and DNA. Science 235:1204-1208.

7. Pryor, W.A. (1988) Why is the hydroxyl radical the only radical that commonly adds to DNA? Hypothesis: It has a rare combination of high thermochemical reactivity, and a mode of production that can occur near DNA. Free Radical Biol. Med. 4:219-223.

8. Moody, C.S., and H.M. Hassan (1982) Mutagenicity of oxygen free radicals. Proc. Natl. Acad. Sci., USA 79:2855-2859.

9. Farr, S.B., R. D'Ari, and D. Touati (1986) Oxygen-dependent mutagenesis in Escherichia coli lacking superoxide dismutase. Proc. Natl. Acad. Sci., USA 83:8268-8272.

10. Loeb, L.A., E.A. James, A.M. Walterdorph, and S.J. Klebanoff (1988) Mutagenesis by the autoxidation of iron with isolated DNA. Proc. Natl. Acad. Sci., USA 85:3918-3922

11. Imlay, J.A., and S. Linn (1988) DNA damage and oxygen radical toxicity. Science 240:1302-1309.

12. Youngman, R.J., and E.F. Elstner (1981) Oxygen species in paraquat toxicity: The crypto-OH radical. FEBS Lett. 129:265-268.

13. Youngman, R.J. (1984) Oxygen activation: Is the hydroxyl radical always biologically relevant? Trends Biochem. Sci. 9:280-283.

14. Koppenol, W.H. (1985) The reaction of ferrous EDTA with hydrogen peroxide: Evidence against hydroxyl radical formation. J. Free Rad. Biol. Med. 1:281-285.

15. Rush, J.D., and W.H. Koppenol (1986) Oxidizing intermediates in the reaction of ferrous EDTA with hydrogen peroxide. J. Biol. Chem. 261:6730-6733.

16. Winterbourn, C.C. (1987) The ability of scavengers to distinguish OH· production in the iron-catalyzed Haber-Weiss reaction: Comparison of four assays for OH·. Free Rad. Biol. Med. 3:33-39.

17. Salmuni, A.M. Chevion, and G. Czapski (1981) Unusual Copper-induced Sensitization of the biological damage due to superoxide radicals. J. Biol. Chem. 256:12632-12635.

18. Mason, R.P. (1984) Spin Labeling in Pharmacology, J.L. Holtzman, ed. Academic Press, Orlando, pp. 87-129.

19. Sargent, F.P., and E.M. Gardy (1976) Spin trapping of radicals formed during radiolyis of aqueous solutions. Direct electron spin resonance observations. Can. J. Chem. 54:275-279.

20. Flitter. W.D., and R.P. Mason (1989) The spin trapping of pyrimidine nucleotide free radicals in a Fenton system. Biochem. J. (in press).

21. Joshi, A., S. Rustgi. and P. Riesz (1976) E.S.R. of spin-trapped radicals in gamma-irradiated aqueous solutions of nucleic acids and their constituents. Int. J. Radiat. Biol. 30:152-170.

22. Kawabara. M., A. Mingishi, A. Ito, and T. Ito (1986) Photochem. Photobiol. 44:265-272.

23. Gutterdige. J.M.C. (1981) Thiobaroituric acid-reactivity following iron-dependent free-radical damage to amino acids and carbohydrates. FEBS Lett. 128:343-346.

24. Cheesmman, K.H., A. Beavis, and H. Esterbauer (1988) Hydroxyl-radical-induced iron-catalysed degradation of 2-deoxyribose. Biochem. J. 252:649-653.

MECHANISMS OF INACTIVATION OF OXYGEN RADICALS
BY DIETARY ANTIOXIDANTS AND THEIR MODELS

Michael G. Simic[1] and Slobodan V. Jovanovic[2]

[1]National Institute of Standards and Technology
Gaithersburg, Maryland 20899

[2]Boris Kidric Institute of Nuclear Sciences
Beograd 11001, Yugoslavia

INTRODUCTION

Numerous anticarcinogens (38) and antimutagens (15,39) have antioxidant properties, i.e., the ability to inhibit propagation of peroxidation processes by peroxy radicals (31), ROO·. Peroxy products (H_2O_2, ROOH) and various oxygen radicals (·OH, ·O_2^-, ROO·, RO·, ArO·) (35), which are sometimes called active oxygen species, appear to play an important role in carcinogenesis in general and promotion in particular (7). Consequently, the suggestion that dietary antioxidants (2) are critical in cancer prevention is gaining recognition and interest (8,10,21,27).

Although there are many other modes of anticarcinogenic activities (9) at various stages of carcinogenesis (initiation, promotion, progression, and metastasis), this brief overview deals only with the kinetics and mechanisms of inhibition of generation and inactivation of oxy radicals. An important point to remember is that antioxidants that appear to be active as anticarcinogens via nonfree radical processes (e.g., activation of enzymes) may nevertheless be involved in a nonapparent inactivation or inhibition of oxy radicals (30). This possibility reflects the intrinsic difficulties of detecting and monitoring free radical processes in living systems.

OXYGEN (OXY) RADICALS

There are five major classes of oxy radicals: hydroxyl (·OH); peroxyls (ROO·); superoxide (·O_2^-), a special class of peroxyls; alkoxyls (RO·); and aroxyls (ArO·). Current terminology is not uniform, and both the "oxy" and "oxyl" endings are encountered in the literature. The "yl" ending (IUPAC convention) designates a free radical, and "oxyl," an oxygen free radical. In this paper, for example, hydroxy radical or simply hydroxyl will be used to designate the free radical $\cdot$OH. In all cases R represents an aliphatic group, and Ar an aromatic group.

Oxy radicals are most conveniently studied by time-resolved techniques, such as pulse radiolysis (10) and laser photolysis (5). Using these two techniques, free radicals can be detected and monitored directly. Because of their low concentrations and short lifetimes, they can be followed in vivo only by indirect measurements using specific markers (spin-labels, stable products) (14).

Generation of oxy radicals in biosystems has been demonstrated by numerous workers (35), many of them conducting pioneering research that led to the exponential growth of investigations in this area. Proof of direct causality of biological consequences by specific free radicals, however, is still missing. Experimental approaches designed to provide the information needed to resolve that question are confounded and complicated by the large number of different radicals present in biosystems.

The following reactions have been studied as potential sources of free radicals in vivo. Hydrogen peroxide, a well-known cellular metabolite, generates hydroxy radicals by the Haber-Weiss reaction (13):

$$H_2O_2 + Fe(II) \rightarrow \cdot OH + OH^- + Fe(III) \qquad (1)$$

Being one of the most reactive radicals, $\cdot$OH rapidly reacts with biomolecules by abstracting hydrogen or by adding to unsaturated bonds or aromatic rings, e.g.,

$$RH + \cdot OH \rightarrow R\cdot + H_2O \qquad (2)$$

$$k \sim 10^9 \ M^{-1} \ s^{-1}$$

The resulting radicals, $R\cdot$, react rapidly with oxygen:

$$R\cdot + O_2 \rightarrow ROO\cdot \qquad (3)$$

$$k \sim 10^9 \ M^{-1} \ s^{-1}$$

Consequently, hydrogen peroxide generates a variety of peroxy radicals that may abstract hydrogen from weaker C-H bonds, e.g., bisallylic,

$$ROO\cdot + H_2L \rightarrow ROOH + HL\cdot \qquad (4)$$

or that may accept an electron from good electron donors (17,31-33)

$$ROO\cdot + D^-(or\ DH) \rightarrow ROO^- + \cdot D(DH^+) \qquad (5)$$

which is followed rapidly by:

$$ROO^- + H^+ \rightleftharpoons ROOH \qquad (6)$$

$$pK_a = 12.8$$

In a reaction similar to reaction (1), hydroperoxides react with ferrous ion:

$$ROOH + Fe(II) \rightarrow RO\cdot + OH^- + Fe(III) \qquad (7)$$

Alkoxy radicals, which are similar to $\cdot OH$, are very reactive with biomolecules, although the rate constants (22) are somewhat lower than for $\cdot OH$ radicals (6).

Because of their extremely high reactivities (6), $\cdot OH$ radicals yield a variety of $R\cdot$, $ROO\cdot$, and $RO\cdot$ radicals. For this reason, despite starting with only one type of radical, it is very difficult to assign specific biological consequences to each particular radical.

ANTIOXIDANTS

Reactions (1) to (4) may lead to multiple peroxidations via chain reactions (14). Such reactions take place in micelles, membranes, and lipids. In cytosol, peroxy radicals may damage biomolecules:

$$ROO\cdot + Bio \rightarrow Bio' + Products \tag{8}$$

Either in lipids [reaction (4)] or in cytosol [reaction (8)], damage may be inhibited by preventing the formation of the initial damaging species [e.g., by preventing reactions (1) and (7)] or by eliminating the peroxy radical [reaction (5)]. Antioxienzymes (catalase, glutathione, and peroxidase) and metal-complexing agents (Desferal) act as inhibitors of initiation, whereas antioxidants act as eliminators of peroxy radicals (35). For example, uric acid eliminates peroxy radicals (34):

$$UH_2^- + ROO\cdot \rightarrow \cdot UH_2 + ROO^- \tag{9}$$

$$k \sim 3 \times 10^6 \ M^{-1} \ s^{-1}$$

and may prevent biological damage (3). A similar reaction has been observed for ascorbate (23):

$$AH^- + CH_3OO\cdot \rightarrow \cdot AH + CH_3OO^- \tag{10}$$

$$k \sim 2.2 \times 10^6 \ M^{-1} \ s^{-1}$$

Another important class of antioxidants is the phenolic antioxidants (ArOH), i.e., hydroxylated aromatic rings. The aromatic group, Ar, may be a simple benzene ring, condensed nuclei (naphthalene), or heterocyclic (indole). In addition to the hydroxy group, the rings may have other substituents that define the antioxidant properties. Phenolic antioxidants are well represented in foods and are important dietary antioxidants. Phenolic antioxidants are good electron donors, as already pointed out, and are therefore ideal agents for inactivation of peroxy radicals (19):

$$ArOH + ROO\cdot \rightarrow ArOH^+ + ROO^- \tag{11}$$

followed by rapid deprotonation

$$ArOH^+ \rightarrow ArO\cdot + H^+ \tag{12}$$

Previously it was believed that phenolic antioxidants act as H-atom donors, but two facts provide evidence against this hypothesis (17,31-33). First, peroxy radicals are strongly electrophilic (17) and, second, the bond strength of ArO-H is too high to promote rapid H-atom transfer.

The overall scheme for generation of linoleic acid radicals and prevention of chain autoxidation (peroxidation) is shown in Fig. 1.   The initiating agent, X, could be a free radical or a metal complex (4).

The classic phenolic antioxidant vitamin E (ChrOH) is one of the best phenolic antioxidants due to its high reactivity with peroxy radicals (24,29,31):

$$ChrOH + ROO\cdot \rightarrow ChrOH^+ + ROO^- \qquad (13)$$

$$k \sim 6 \times 10^6 \ M^{-1} \ s^{-1}$$

followed by

$$ChrOH^+ \rightarrow ChrO\cdot + H^+ \qquad (14)$$

Because vitamin E is membrane bound, its value in protecting the cell as a whole is limited.   Its major role in protecting the integrity of membranes, however, is crucial for the survival of cells exposed to oxidants both under normal conditions and under conditions of oxidative stress (28).

INACTIVATION OF OXY RADICALS BY ANTIOXIDANTS

There are five classes of oxy radicals, and they all may react with a suitable antioxidant.   Whether an antioxidant can inhibit the damaging effects of oxy radicals in vivo depends on the type of radical and the experimental conditions.   The reactivities of oxy radicals with a few selected substrates are shown in Tab. 1.

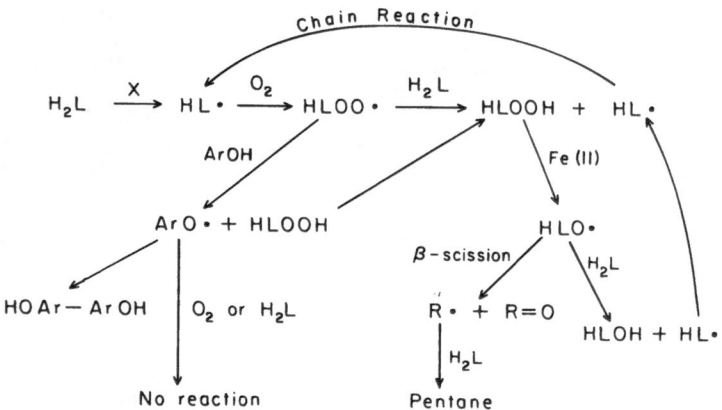

Fig. 1.   Peroxidation (autoxidation) mechanisms for linoleic acid and inhibition by antioxidants.

Tab. 1. Reaction rate constants for oxy radical with bio-related substrates in solutions at room temperature (from Ref. 6, 11, 24, and 30).

Substrate	$k(R\cdot + S)$, M^{-1} s^{-1}				
	HROO·	·O_2H	·O_2^-	HRO·	·OH
Stearic acid	10^{-3}–10^{-4}	low	low	2.3×10^6	$\sim10^9$
Oleic acid	0.1–1	low	low	3.3×10^6	$\sim10^{10}$
Linoleic acid	~60	1.2×10^3	low	8.8×10^6	$\sim10^{10}$
Linolenic acid	~120	1.7×10^3	low	1.3×10^7	$\sim10^{10}$
Arachidonic acid	~180	3.0×10^3	low	2.0×10^7	$\sim10^{10}$
Aldehyde	2.7×10^3	50	n.m.[a]	n.m.	$\sim10^9$
RSH	$<10^6$	1.8×10^5	<15	n.m.	$\sim10^{10}$
BHT	10^4	2.4×10^3	n.m.	4×10^7	$\sim10^{10}$
BHA	2.6×10^6	n.m.	n.m.	n.m.	$\sim10^{10}$
QH_2	1.2×10^5	10^4–10^5	n.m.	n.m.	$\sim10^{10}$
α-Tocopherol	5.7×10^6	2.0×10^5	6	n.m.	$\sim10^{10}$
Ascorbic acid (AH_2)	n.m.	1.6×10^4	---	n.m.	$\sim10^{10}$
Ascorbate (AH^-)	2.2×10^6	---	5.0×10^4	n.m.	$\sim10^{10}$
Thymine	n.o.[b]	n.o.	n.o.	n.m.	7×10^9
Guanine	n.o.	n.o.	n.o.	n.m.	9×10^9

[a] n.m., not measured.
[b] n.o., not observed.

Hydroxy Radical, ·OH

Because of their high reactivity with biomolecules (6), ·OH radicals cannot be scavenged easily by antioxidants despite maximum reaction rate constants, k(·OH + antioxidant) $\sim10^{10}$ M^{-1} s^{-1}. For example, a 2% concentration of antioxidant would scavenge only a few percent of ·OH radicals in tissue.

Peroxy Radicals, ROO·

Peroxy radicals are only moderately reactive with antioxidants, as discussed previously. Because of their much lower reactivity with biomolecules, peroxy radicals can be inactivated efficiently, even in vivo, by suitable antioxidants.

Superoxide, ·O_2^-, and Hydroperoxyl, ·O_2H

Superoxide radical is much less reactive than hydroperoxyl (11). Hydroperoxyl, ·O_2H, exhibits reactivities similar to alkyl-peroxyls, ROO·. Although the ·O_2^- form predominates in biosystems, hydroperoxyl may also participate in the reactions because

$$\cdot O_2^- + H^+ \rightleftharpoons \cdot O_2H \qquad (15)$$
$$pK_a = 4.75 \pm 0.1$$

For example, at pH = 6.75, which is on average a biological pH, $\cdot O_2H$ will be 1% of the concentration of the $\cdot O_2^-$ form. At lower pH and in cellular regions with a lower dielectric constant than that of water, the fraction of $\cdot O_2H$ will increase, especially in membranes.

Alkoxyls, RO·

Alkoxyl reactivities are similar to ·OH reactivities, although some-what lower (2-3 orders of magnitude). In addition to H abstraction, ad-dition to unsaturated bonds, and oxidation reactions, alkoxyls undergo a specific unimolecular reaction known as β-scission (Fig. 1):

$$\underset{\displaystyle k \sim 10^7 \ s^{-1}}{-\overset{\displaystyle \overset{\textstyle O\cdot}{|}}{\underset{|}{C}}-\overset{|}{\underset{|}{C}}- \quad \rightarrow \quad -\overset{|}{\underset{|}{C}}\cdot \ + \ O{=}C{<}} \tag{16}$$

The rate of β-scission is an order of magnitude slower in nonpolar media (26). Reaction (16) greatly reduces the alkoxyls' lifetime and conse-quently their effective range of damage. If generated in the immediate vicinity of the biomolecule, e.g., a BioFe(II) complex, an alkoxyl could inflict substantial damage to the biomolecule, Bio.

Aroxy Radicals, ArO·

If the redox potential of ArO· is higher than the redox potential of another antioxidant, then ArO· can be repaired by that antioxidant. For example, ascorbate (E_7 = 0.28 V) (36) readily repairs vitamin E radicals (24) (E_7 = 0.48 V) (19) generated in reactions (13) and (14),

$$ChrO\cdot + AH^- \rightarrow ChrO^- + \cdot AH \tag{17}$$

followed in neutral solutions by

$$ChrO^- + H^+ \rightarrow ChrOH \tag{18}$$

$$\cdot AH \rightleftharpoons \cdot A^- + H^+ \tag{19}$$

$$pK_a = 4.2$$

The ascorbate one-electron redox potential [$E_7(\cdot A^-/AH^-)$ = 0.28 V], not only is lower than that for α-tocopherol [$E_7(ChrO\cdot/ChrOH)$ = 0.48 V], but also is one of the lowest redox potentials for any antioxidant.

DIETARY ANTIOXIDANTS

Dietary antioxidants may be divided into three groups: water solu-ble (ascorbate), lipid soluble (β-carotene), and interfacial (vitamin E and bilirubin). The prosthetic hydroxy group of vitamin E is at the water-lipid interface of the membrane, whereas bilirubin complexes with proteins (e.g., on the surface of albumin). The classic water-soluble antioxi-dant ascorbate is an important dietary component with a well-understood biochemical role as a vitamin. It participates as an electron donor in numerous biochemical reactions. Other physiological antioxidants may have multiple (serotonin) or exclusive (vitamin E) roles.

The role of an antioxidant is to inactivate an oxy radical and prevent or minimize the damage to biomolecules. Because oxy radicals have high electron affinity, the most efficient way to neutralize them is by donating an electron, as indicated in reaction (11). This can be demonstrated by the following example involving phenol, p-methoxyphenol, and p-hydroquinone.

Phenol is a poor electron donor (E_7 = 1.0 V) and is not a good antioxidant. Another hydroxy or a methoxy group in the para position greatly enhances the electron-donating properties of phenol, as is evident by the decreased redox potentials in substituted derivatives. Consequently, both p-methoxyphenol (E_7 = 0.74 V) and p-hydroquinone (E_7 = 0.46 V) react with peroxy radicals via electron transfer:

$$p\text{-}CH_3OC_6H_4OH + ROO\cdot \rightarrow p\text{-}CH_3OC_6H_4OH^+ + ROO^- \qquad (20)$$

$$p\text{-}CH_3OC_6H_4OH^+ + H_2O \rightarrow p\text{-}CH_3OC_6H_4O\cdot + H_3O^+ \qquad (21)$$

and similarly,

$$HOC_6H_4OH + ROO\cdot \rightarrow HOC_6H_4OH^+ + ROO^- \qquad (22)$$

$$HOC_6H_4OH^+ + H_2O \rightarrow HOC_6H_4O\cdot + H_3O^+ \qquad (23)$$

In aqueous media, however, the semiquinone intermediates undergo another significant reaction

$$HOC_6H_4O\cdot \rightleftharpoons {}^-OC_6H_4O\cdot + H^+ \qquad (24)$$

$$pK_a = 4.0$$

which leads to a major difference between the properties of p-hydroxymethoxy and hydroquinone intermediates, namely,

$$CH_3OC_6H_4O\cdot + O_2 \rightarrow \text{not observed (Ref. 16)} \qquad (25)$$

$$k < 10^{-2} \ M^{-1} \ s^{-1}$$

whereas

$${}^-OC_6H_4O\cdot + O_2 \rightarrow \text{slow reaction} \qquad (26)$$

Reaction (26) becomes more prominent with semiquinone radicals with lower redox potential, e.g., the radical of ubiquinone. Reaction (25), i.e., the unreactivity of the antioxidant radical intermediate with oxygen, is one of the basic features of true antioxidants.

The mechanisms of other nonphenolic antioxidants are not completely understood. Bilirubin is a good electron donor (1,37) that may act in a way similar to phenolic antioxidants. The mechanisms associated with the antioxidant properties of β-carotene are not understood and therefore will not be discussed, despite the importance of β-carotene as a major dietary antioxidant.

ANTIOXIDANTS AS ANTICARCINOGENS

Compounds that exhibit anticarcinogenic and antimutagenic activity have diverse natures and act in numerous ways. They may inhibit metabolic formation of mutagens, deactivate mutagens, modulate cell replication, activate cytosolic enzymes (GSH S-transferase and DT diaphorase), inhibit oncogene expression, inhibit activation of protein kinase C, inactivate oxy radicals, etc. Their extracellular and intracellular activities have been classified recently by DeFlora and Ramel (9).

We propose, however, a novel working hypothesis for the anticarcinogenic properties of antioxidants, which is based on the existing, simple multistep model of carcinogenesis (25):

$$\text{Initiation} \rightarrow \text{Promotion} \rightarrow \text{Progression} \rightarrow \text{Cancer} \qquad (27)$$

In our proposed model, each phase of carcinogenesis consists of a much larger number of steps, some of them repetitive, which is consistent with the long-term development of a tumor. The chain of events might not be linear, i.e., it may be branched. To simplify the concept, however, we shall use a linear model (Scheme 1) and indicate only a few of the multitude of steps:

```
        ·R              ·R                       ·R
 Cell → → → Mutants → → → Precancerous cell → → → Cancer cell
     └─┴──────── Antioxidants ──────────┴─┘              (Scheme 1)
```

Some of the steps may be driven by free radicals (·R), but each step within a phase does not necessarily involve free-radical reactions. The model would require, however, that at least one of the steps be a free-radical process, whether in the initiation, promotion, or progression phase. An antioxidant may have an inhibitory effect on some of the non-free-radical steps and a definite inhibitory effect on a free-radical step. This would explain why BHT, for example, inhibits promotion by enhancing GSH S-transferase (9) but BHT may also inactivate oxy radicals (30), an expectation based on its antioxidant properties. Which of these inhibitory processes prevails would depend on the nature of the antioxidant, carcinogen, and promoter.

The concept presented in Scheme 1 is based on the observation that an inhibitory anticarcinogenic effect is achieved by antioxidants with very diverse structures. The structures may vary, from a simple methoxyphenol, such as BHA, to a long-chain conjugated molecule, such as β-carotene (2); long dimeric molecules, such as curcumin (15); polynuclear flavonoids, such as quercetin (20); bulky catechine derivatives, such as EGCG (39); and many others.

PROTO-ANTIOXIDANTS

A dietary component, though anticarcinogenic, may not be an efficient antioxidant. Some dietary components are anticarcinogens but do not have antioxidant properties in their native states; they may, however, be converted metabolically into efficient antioxidants. We propose

that indole derivatives from cruciferous vegetables (e.g., indole-3-car-
binol, 3,3'-diindolylmethane) may in fact be converted into efficient antioxidants by enzymatic hydroxylation in the C5 position. Such a process
normally occurs with tryptophan as a substrate,

$$\text{R-Ind} \xrightarrow{\text{Hydroxylase}} \text{5-R-Ind-OH} \qquad (28)$$

Although Trp is not an antioxidant, 5-Trp-OH is (Ref. 18):

$$\text{ROO} \cdot + \text{5-R-Ind-OH} \rightarrow \text{ROO}^- + \text{5-R-Ind-OH}^+ \qquad (29)$$

followed by

$$\text{5-R-Ind-OH}^+ + H_2O \rightarrow \text{5-R-Ind-O} \cdot + H_3O^+ \qquad (30)$$

Decarboxylation of 5-Trp-OH generates serotonin, another excellent antioxidant of the 5-Ind-OH family.

Therefore, we conclude that, in addition to dietary anticarcinogenic
antioxidants, there may exist a whole new class of dietary proto-antioxidants that can be metabolically converted into efficient anticarcinogenic
antioxidants capable of inactivating superoxide and peroxy radicals. The
search for such molecules should be one of the highest priorities in our
fight against cancer.

ACKNOWLEDGEMENTS

We wish to thank E.P.L. Hunter for useful comments, K.A. Taylor
for editorial assistance, and G. Wiersma for technical execution of the
paper.

REFERENCES

1. Al-Sheikhly, M., and M.G. Simic (1987) Inhibition of autoxidation by
   vitamin E and bilirubin. In Anticarcinogenesis and Radiation
   Protection, P. Cerutti, O.F. Nygaard, and M.G. Simic, eds. Plenum
   Press, New York, pp. 47-50.
2. Ames, B.N. (1983) Dietary carcinogens and anticarcinogens. Science
   221:1256-1264.
3. Ames, B.N., R. Cathcart, E. Schwiers, and P. Hochstein (1981)
   Uric acid provides an antioxidant defense in humans against oxidant
   and radical-caused aging and cancer. Proc. Natl. Acad. Sci., USA
   78:6858-6862.
4. Aust, S.D. (1988) Iron redox reactions and lipid peroxidation. In
   Oxygen Radicals in Biology and Medicine, M.G. Simic, K.A. Taylor,
   J.F. Ward, and C. von Sonntag, eds. Plenum Press, New York,
   p. 137.
5. Bensasson, R.V., E.J. Land, and T.G. Truscott, eds. (1983) Flash
   Photolysis and Pulse Radiolysis, Pergamon Press, New York.
6. Buxton, G.V., C.L. Greenstock, W.P. Helman, and A.B. Ross
   (1988) Critical review of rate constants for reactions of hydrated
   electrons, hydrogen atoms and hydroxyl radicals. Phys. Chem. Ref.
   Data 17:513-886.
7. Cerutti, P.A. (1985) Prooxidant stress and tumor promotion. Science 227:375-381.

8. Cerutti, P.A., A.F. Nygaard, and M.G. Simic, eds. (1987) Anticarcinogenesis and Radiation Protection, Plenum Press, New York.
9. DeFlora, S., and C. Ramel (1988) Mechanisms of inhibitors of mutagenesis and carcinogenesis: Classification and overview. Mutat. Res. 202:279-284.
10. Farhataziz, and M.A.J. Rodgers, eds. (1987) Radiation Chemistry, VCH, New York.
11. Farhataziz, and A.B. Ross (1977) Selected specific rates of reactions of transients from water in aqueous solution. III. Hydroxyl radical and perhydroxyl radical and their radical ions. National Standard Reference Data Series, NSRDS-NBS 59, National Institute of Standards and Technology, Gaithersburg, Maryland.
12. Grice, H.C., ed. (1986) Food antioxidants: International perspectives. Food Chem. Toxicol. 24:997-1255.
13. Haber, F., and J. Weiss (1932) Uber die katalyse des hydroperoxydes. Naturwiss. 20:948-950.
14. Halliwell, B., and J.M.C. Gutteridge, eds. (1985) Free Radicals in Biology and Medicine, Clarendon Press, Oxford.
15. Huang, M.-T., R.C. Smart, C.-Q. Wong, and A.H. Conney (1988) Inhibitory effect of curcumin, chlorogenic acid, caffeic acid, and ferulic acid on tumor promotion in mouse skin by TPA. Cancer Res. 48:5941-5946.
16. Hunter, E.P.L., M.F. Desrosiers, and M.G. Simic (1989) The effect of oxygen antioxidants and superoxide radicals on tyrosine phenoxyl radical dimerization. Inhibition of tyrosine phenoxy radical dimerization by antioxidants and superoxide radical. Free Radical Biol. Med. 6:581.
17. Hunter, E., and M.G. Simic (1983) Kinetics of peroxy radical reactions with antioxidants. In Oxy Radicals and Their Scavenger Systems, G. Cohen and R. Greenwald, eds. Elsevier Biomedical, New York, pp. 32-37.
18. Jovanovic, S.V., and M.G. Simic (1985) Tryptophan metabolites as antioxidants. Life Chemistry Reports 3:124-130.
19. Jovanovic, S.V., and M.G. Simic (1988) Redox properties of oxy and antioxidant radicals. In Oxygen Radicals in Biology and Medicine, M.G. Simic, K.A. Taylor, J.F. Ward, and C. von Sonntag, eds. Plenum Press, New York, pp. 115-122.
20. Kato, R., T. Nakadate, S. Yamamoto, and T. Sugimura (1983) Inhibition of 12-O-tetradecanoylphorbol-13-acetate-induced tumor promotion and ornithine decarboxylase activity by quercetin: Possible involvement of lipoxygenase inhibition. Carcinogenesis 4:1301.
21. Kuroda, Y., D.M. Shankel, and M.D. Waters, eds. (1989) Mechanisms of Antimutagenesis and Anticarcinogenesis, Plenum Press, New York.
22. Neta, P., M. Dizdaroglu, and M.G. Simic (1983) Radiolytic studies of the cumyloxyl radical in aqueous solutions. Israel J. Chem. 24:25-28.
23. Packer, J.E.. J.S. Mahood, R.L. Willson, and R. Wolfenden (1981) Reactions of the trichloromethylperoxy free radical ($Cl_3COO\cdot$). Int. J. Radiat. Biol. 39:135-141.
24. Packer, J.E., T.F. Slater, and R.L. Willson (1979) Direct observation of a free radical interaction between vitamin E and vitamin C. Nature (London) 278:737-738.
25. Rotstein, J.B., and T.J. Slaga (1988) Anticarcinogenesis mechanisms as evaluated in the multistage mouse skin model. Mutat. Res. 202:421-427.

26. Scaiano, J.C. (1988) Kinetic studies of alkoxyl radicals. In Oxygen Radicals in Biology and Medicine, M.G. Simic, K.A. Taylor, J.F. Ward, and C. von Sonntag, eds. Plenum Press, New York, p. 59.

27. Shankel, D.M., P.E. Hartman, T. Kada, and A. Hollaender, eds. (1986) Antimutagenesis and Anticarcinogenesis Mechanisms, Plenum Press, New York.

28. Sies, H., ed. (1985) Oxidative Stress, Academic Press, London.

29. Simic, M.G. (1980) The kinetics of peroxy radical reactions with α-tocopherol. In Autoxidation in Food and Biochemical Systems, M.G. Simic and M. Karel, eds. Plenum Press, New York, p. 17.

30. Simic, M.G. (1989) Mechanisms of inhibition of free-radical processes in mutagenesis and carcinogenesis. Mutat. Res. 202:377-386.

31. Simic, M., and E. Hunter (1983) Interaction of free radicals and antioxidants. In Radioprotectors and Anticarcinogens, O. Nygaard and M. Simic, eds. Academic Press, New York, p. 449.

32. Simic, M.G., and E. Hunter (1986) Reaction mechanisms of peroxyl and C-centered radicals with sulfhydryls. J. Free Radicals Biol. Med. 2:227-230.

33. Simic, M.G., E.P.L. Hunter, and S.V. Jovanovic (1987) Electron vs. H-atom transfer in chemical repair. In Anticarcinogenesis and Radiation Protection, P. Cerutti, O.F. Nygaard, and M.G. Simic, eds. Plenum Press, New York, pp. 17-24.

34. Simic, M.G., and S.V. Jovanovic (1989) Antioxidation mechanisms of uric acid. J. Am. Chem. Soc. (in press).

35. Simic, M.G., K.A. Taylor, J.F. Ward, and C. von Sonntag, eds. (1988) Oxygen Radicals in Biology and Medicine, Plenum Press, New York.

36. Steenken, S., and P. Neta (1982) One-electron redox potentials of phenols. J. Phys. Chem. 86:3661-3667.

37. Stocker, R., Y. Yamamoto, A.F. McDonagh, A.N. Glaser, and B.N. Ames (1987) Bilirubin is an antioxidant of possible physiological importance. Nature 235:1043-1046.

38. Wattenberg, L.W. (1978) Inhibitors of chemical carcinogens. Adv. Cancer Res. 26:197-226.

39. Yoshizawa, S.. T. Horiuchi, H. Fujiki, T. Yoshida, T. Okada, and T. Sugimura (1987) Antitumor promoting activity of epigallocatechin gallate, the main constituent of "tannin" in green tea. Phytotherapy Res. 1:44-47.

ROLE OF DIETARY ANTIOXIDANTS IN PROTECTION

AGAINST OXIDATIVE DAMAGE

Toshihiko Osawa, Mitsuo Namiki, and Shunro Kawakishi

Department of Food Science and Technology
Nagoya University
Chikusa, Nagoya 464-01, Japan

INTRODUCTION

Recently, much attention has been focused on studies which suggest the involvement of active oxygens and free radicals in a variety of pathological events, cancer, and even the aging process (21,30). Oxygen is indispensable for aerobic organisms including, of course, human beings; however, it is believed that oxygen also may be responsible for undesired phenomena (4). In particular, oxygen species such as hydrogen peroxide, superoxide anion radical and singlet oxygen, and other radicals, are proposed as agents attacking polyunsaturated fatty acid in cell membranes, giving rise to lipid peroxidation (3). Several reports have suggested that lipid peroxidation may result in destabilization and disintegration of cell membranes, leading to liver injury and other diseases, and finally, to aging and susceptibility to cancer. However, normal cell membranes do not undergo lipid peroxidation so severely in vivo because of the extremely efficient protective mechanisms against damage caused by active oxygens and free radicals. Such systems include enzymatic inactivation by, for example, superoxide dismutase, glutathione-peroxidase and catalase, as well as nonenzymatic protection of polyunsaturated fatty acid by physiological and biological antioxidants such as vitamin E, vitamin C, β-carotene, and uric acid (2). More recently, bilirubin (35) and carnocine (18) have been reported as being biologically significant antioxidants. In addition, several antioxidants have been reported to play an important role in the prevention of carcinogenesis related to active oxygen radicals, and in some cases, to extend the life span of animals (8).

On the other hand, there are also some indications that not only endogenous antioxidants but also dietary antioxidants may be effective in protection from peroxidative damage in living systems. However, quite a few dietary antioxidants other than α-tocopherol have been evaluated using in vivo and in vitro lipid peroxidation systems (25). For this reason, an intensive search has begun for novel antioxidants from natural resources, in particular, plant foods, because dietary antioxidants other than α-tocopherol may also play an important role in protecting the cell against damage caused by active oxygens and free radicals, and also may have the potential of inhibiting mutagenicity induced during lipid peroxidation (17).

NATURAL ANTIOXIDANTS

Natural antioxidants commonly include an aromatic ring as part of the molecular structure. These may be associated with a variety of cyclic ring structures and may have one or more hydroxy groups to act as an electron donor. Antioxidative activity is dependent on whether antioxidants provide a labile hydrogen (hydrogen donor) or act as an electron donor; however, antioxidants are proposed to be poor hydrogen donors (34). Natural antioxidants may also function as peroxide decomposers, as quenchers of singlet oxygen, and as an inhibitor of lipoxygenase. In some cases, antioxidants may function as synergists.

Antioxidant compounds have been found in numerous plant materials (10) such as oil seeds, crops, vegetables, fruits, leaves and leaf wax, bark, and roots. Spices, herbs, and crude drugs are also important sources of natural antioxidants, although the chemical properties and physiological role of active principles have not fully been understood.

Typical molecules of natural antioxidants are derivatives or isomers of plant phenols. Many different types of natural antioxidants have been examined for their use as food antioxidants, and some of them were found to provide enhanced stability of lipids in foods (29). However, most of these dietary antioxidants have not been examined for effectiveness in protection from peroxidative damage in living systems.

TOCOPHEROLS

An intensive search has been made for novel antioxidants from natural resources, including those used as foods (5). Tocopherols have been known for many years to have antioxidant properties and are widely distributed in plant foods such as vegetable oils, crops, beans, fruits, and nuts. There are four kinds of tocopherol derivatives, and the differences are caused by the substitution of a methyl group on the chroman ring structure (Fig. 1). Many reports which have been published on biological antioxidant activity have focused on tocopherols readily available from wheat germ oil and on those taken from deodorization traps in soybean oil processing.

The distribution of tocopherols in vegetable oils is very wide. The content of α-tocopherol is very high in cottonseed, sunflower, rice bran, and safflower oils; however, corn oil contains a large amount of γ-tocopherol, and soybean oil contains γ-tocopherol and some δ-tocopherol (Tab. 1). Sesame oil contains mainly γ-tocopherol; however, the amount of γ-tocopherol is not sufficient for strong antioxidative activity of sesame oil. When tocopherols are used in food systems, δ-tocopherol is the most effective antioxidant, and β- or δ-tocopherol has medium antioxidant activity (31). However, α-tocopherol shows the strongest biological activity in estimating the vitamin E requirement for humans and animals (22).

From this background, we have been involved in isolation and identification of antioxidants from sesame seeds, because these seeds have

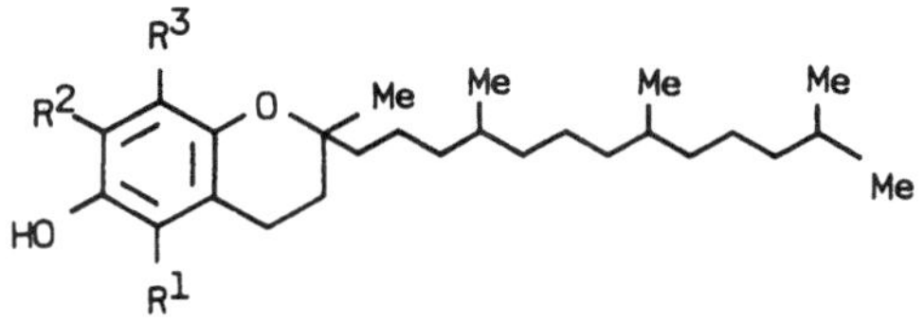

Compound	R^1	R^2	R^3
α-Tocopherol	Me	Me	Me
β-Tocopherol	Me	H	Me
γ-Tocopherol	H	Me	Me
δ-Tocopherol	H	H	Me

Fig. 1. Structure of natural tocopherols.

been used traditionally in Japan, China, and other Eastern Asian countries from the viewpoint of wholesomeness. Sesame oils, especially roast-
ed sesame oils, are widely used in Chinese and Japanese dishes and have
been evaluated as being highly antioxidative. However, correlation between the chemical constituents, their wholesomeness, and their antioxidative nature has not been clearly interpreted yet, and this prompted us
to investigate the antioxidative components in sesame seeds.

LIGNAN-TYPE ANTIOXIDANTS IN SESAME SEEDS

Sesamol and γ-tocopherol were reported as being present in sesame
seeds as lipid-soluble antioxidants, especially in sesame oils (19). However, our data suggest that other types of antioxidants than sesamol and
γ-tocopherol may be present in sesame seeds and may play an important
role for stability of sesame seed oil from oxidative deterioration (13).
Therefore, we have made a large-scale isolation and identification of antioxidative components in sesame seeds, and finally resulted in the isolation
and identification of two types of antioxidants--lipid-soluble and water-
soluble antioxidants (Fig. 2) (14).

Tab. 1. Tocopherol contents (mg/100 g) of vegetable oils.

	α-T	β-T	γ-T	δ-T	Total
Soybean	8.87	1.68	77.09	27.27	114.91
Cottonseed	45.49	0.51	43.25	0.48	87.73
Corn	21.47	1.08	90.58	3.71	116.84
Rapeseed	19.10	0.30	43.17	1.24	63.81
Sunflower	62.71	1.80	2.24	0.58	67.33
Rice bran	39.52	2.30	5.21	0.61	47.64
Safflower	46.31	1.17	2.18	0.39	50.04
Palm	13.80	0.44	1.45	0.20	15.90
Sesame	0.36	-	43.66	0.66	44.68
Peanut	7.00	0.40	11.80	0.70	19.90
Olive	9.80	0.13	0.97	-	10.90
Camelia	10.93	0.10	0.15	-	11.18
Mustard	5.85	-	33.80	0.85	40.50

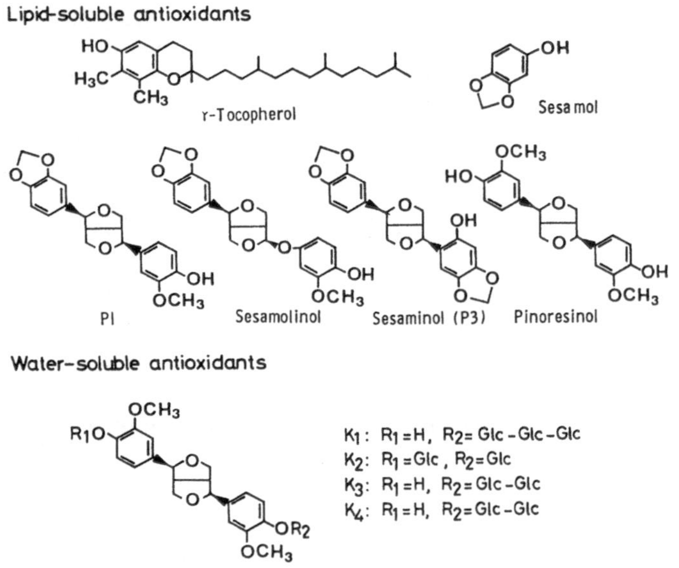

Fig. 2.   Natural antioxidants isolated from sesame seeds.

In order to determine their antioxidative activity, we have evaluated several in vitro lipid peroxidation systems, in particular, the rabbit erythrocyte ghost system (28).   Like many other biological membranes, red blood cell membranes are prone to lipid peroxidation because of their high polyunsaturated lipid content (7).   Although glutathione and ascorbic acid can reduce t-butoxyradical to t-butanol normally, the phospholipid of biological membrane in blood is also prone to autoxidation, as measured by a positive thiobarbituric acid (TBA) test.

As shown in Fig. 3, two novel lignan-type antioxidants, sesamolinol and sesaminol, are found to be present only in sesame seeds (27).   All lipid-soluble antioxidants have substantial antioxidant activity compared to α-tocopherol; however, we decided to focus mainly on sesaminol, because the content of sesaminol has been increased dramatically during the refining process of sesame seed oil production (15).   By HPLC analysis, four brands of commercially available sesame seed oils out of six contain a large amount of sesaminol (16).   Sesaminol is a novel antioxidative lignan, and its stereochemical structure has been determined by instrumental analyses, including X-ray crystallographic analysis (24).

Usually, unroasted sesame oil is extracted by an expeller and refined by alkaline treatment, water washing, bleaching with acid clay, and a deodorizing process.   These refining processes are almost the same for other vegetable oils such as corn oil, soybean oil, safflower oil, and sunflower oil.   By quantification using HPLC, the content of sesaminol was found to be dramatically increased during the bleaching process (15).

These results suggested to us that sesaminol is chemically transformed during the bleaching process.   By chemical analysis, we confirmed that sesaminol has been produced by intermolecular transformation from sesamolin (16); there exist four stereoisomers, and all of these isomers

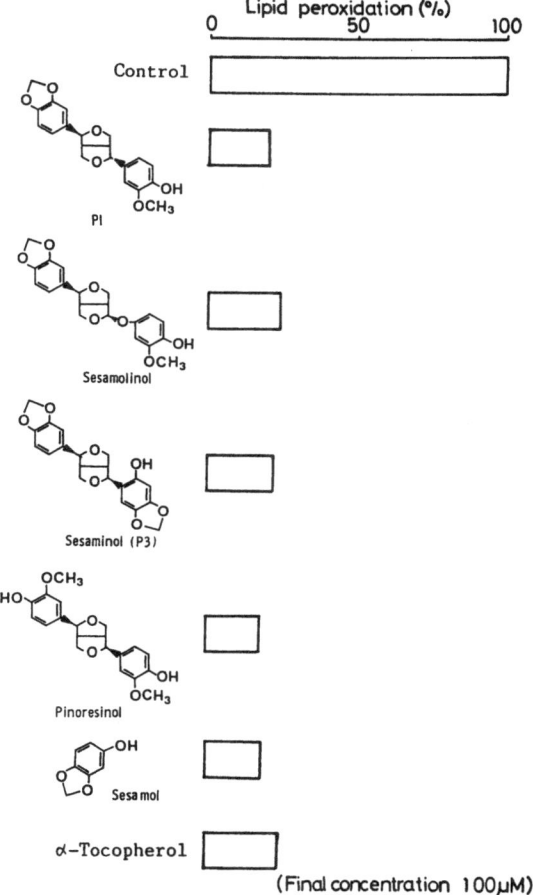

Fig. 3.   Antioxidative activity of lipid-soluble antioxidants isolated from sesame seeds by the rabbit erythrocyte ghost system.

have quite strong antioxidative activity.   When the amounts of sesaminol were quantified by HPLC in commercially-available sesame oils, the total amounts of sesaminol isomers were about four times that of γ-tocopherol, as shown in Tab. 2.   Therefore, it is concluded that sesaminol is the main antioxidative component present in sesame oil.   The protective role of sesaminol against oxidative damage has also been reported, using normal human diploid fibroblast induced by t-butylhydroperoxide (32).

CATECHIN-TYPE ANTIOXIDANTS IN TEA LEAVES

The custom of drinking tea, especially green tea, is very important from the viewpoint of dietary antioxidants for Japanese.   Green tea is rich in strong antioxidants, and six different tea catechins are detected in the HPLC trace of crude extracts (T. Osawa et al., ms. in prep.). The concentration of epigallocatechin gallate is the highest, and quite a

Tab. 2.   Amount of sesaminol and tocopherol in different commercial ses-
           ame oils (mg/100 g oil).

Commercial sesame oil	Sesaminol	Epi-sesaminol	Total sesaminol	γ-Toco-pherol
A	61.2	81.6	142.8	25.5
B	58.2	76.0	134.2	29.3
C	52.2	69.6	121.8	25.2
D	17.9	23.9	41.8	23.5
E	52.2	69.6	121.8	n.d.
F	6.6	8.8	15.8	n.d.

Amount of antioxidants was analyzed by HPLC.

large amount of epicatechin gallate is also present in the crude catechin
fraction of green tea, as shown in Fig. 4.   Although the antioxidative
activity of tea catechins has been examined using many different types of
in vitro systems, the relative antioxidant activity, determined by using
the same rabbit erythrocyte ghost system, has been used for comparison.
The result showed that two water-soluble antioxidants, epicatechin gallate
and epigallocatechin gallate, showed strong protection of erythrocyte membranes from oxidative damage induced by t-butylhydroperoxide (Fig. 5).
Bioantimutagenicity of tea catechins has been examined, and it was reported that the pyrogalloyl moiety is essential for bioantimutagenicity

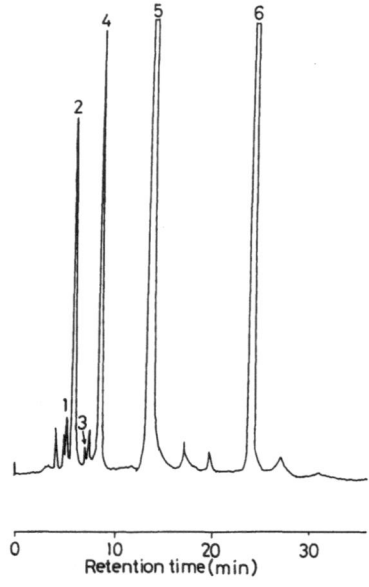

Fig. 4.   HPLC chromatogram of crude tea catechin mixture.   1: (+) gal-
           locatechin (peak area: 0.35%); 2: (-) epigallocatechin (4.50%);
           3: (+) catechin (0.16%); 4: (-) epicatechin (6.02%); 5: (-) epi-
           gallocatechin gallate (62.49%); and 6: (-) epicatechin gallate
           (21.12%).

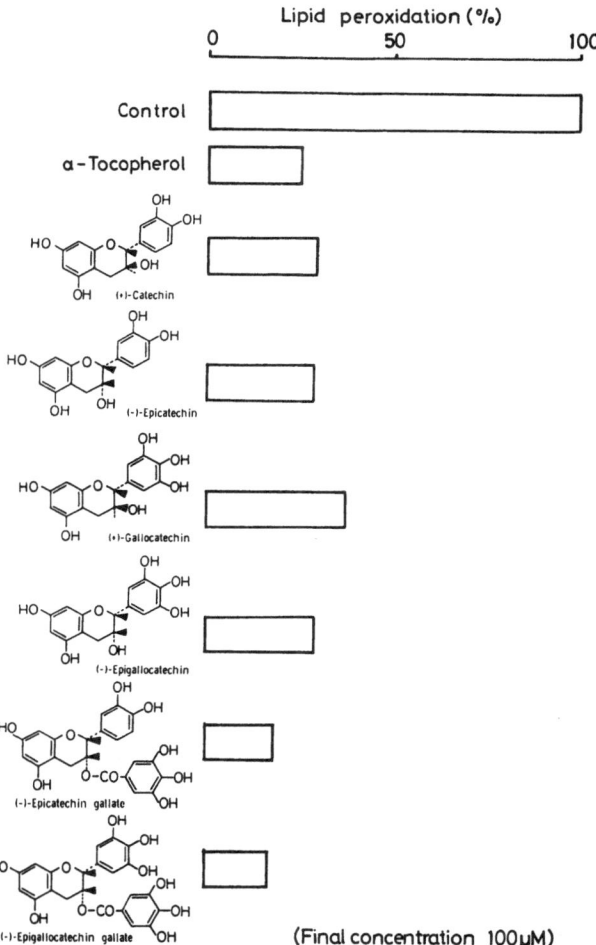

Fig. 5.   Antioxidative activity of tea catechins determined by the rabbit erythrocyte ghost system.

(33).   However, our data suggest that the gallic acid moiety enhanced the antioxidative activity, which is correlated strongly with the antipromoter activity of tea catechins (H. Fujiki, pers. comm.).

TANNIN-TYPE ANTIOXIDANTS IN HERBS AND CRUDE DRUGS

Herbs and crude drugs prepared from plant materials are traditionally used, and their pharmacological effects have been extensively studied; however, only a few reports are available on the antioxidative components in herbs and crude drugs. From this background, we have started our study on screening of 32 different types of herbs and crude drugs obtained from a domestic drugstore in Taipei. As shown in Tab. 3, 22 species of herbs and crude drugs showed quite strong antioxidative activity (36). Of course, vitamin E seems to be one of the most important

Tab. 3.   Antioxidative activity of organic extracts from herbs/crude
          drugs.

No.	Crude drug	Antioxidative activity		
		Ether extr.	EtOAc extr.	MeOH extr.
15	*Dryopteris crassirhizoma* Naikai	+	+	+
18	*Tussilago farfara* L.	−	−	+
19	*Siegesbeckia orientalis* L.	−	+ +	+ + + +
20	*Verbena officinalis* L.	−	+	+ + + +
28	*Punica granatum* L.	−	+	+ + + +
34	*Scutellaria barbata* D. Don	−	−	+ +
36	*Terminalia chebula* Retzus	−	+	+ +
39	*Asiasarum heterotropoides* F. Maekawa	+	−	+
41	*Sparganium stoloniferum* Buch. Hamil.	−	−	+
85	*Gynura japonica* (Thunb.) Juel	+ + +	+ + + +	+ +
88	*Perilla frutescens* (L.)	+	+ +	+ +
95	*Polygonum avicalare* L.	−	−	+ +
96	*Hepericium japonicum* Thunb.	−	−	+ + +
105	*Gleditsia sinensis* Lam.	−	−	+ + +
122	*Osbeckia chinensis* L.	+ +	+ + + +	+ + + +
144	*Euryale ferox* Salisb.	+ + +	+ + + +	+ + +
147	*Artemisia capillaris* Thunb.	+ +	+ + +	+ + +
159	*Uncaria kawakamii* Hayata	−	+ + + +	+
160	*Mentha arvensis* L.	+	−	+
172	*Mosla formosana* Maxim.	+ +	+ + +	+ + +
193	*Crocus sativa* L.	−	−	+ + +
194	*Syzygium aromaticum* (L.) Merr. et Perry	−	−	+

0.2 mg of the samples, α-tocopherol and BHA were analyzed by the thiocyanate method. The incubation period
(days) when absorbance increased to 0.3 at O.D. 500 nm were control, 25; α-tocopherol, 41; BHA, 52.
−, below 32; +, 33∼38; + +, 39∼44; + + +, 45∼50; + + + +, ≧51.

antioxidative components and may be responsible for antioxidative activity
in the extracts of herbs and crude drugs.   The amounts of tocopherol
derivatives have been quantified using HPLC, and it was concluded that
seven species of herbs and crude drugs, except Osbeckia chinensis, have
tocopherols (Tab. 4).   Therefore, we decided to isolate and identify the
active. principles in the extracts of O. chinensis.

Tab. 4.   Quantitative determination of tocopherols in herbs/crude drugs.

No.	Crude drug	α (μg/g)	β (μg/g)	γ (μg/g)	δ (μg/g)
19	*Siegesbeckia orientalis* L.	5.17	ND	ND	ND
85	*Gynura japonica* (Thunb.) Juel	8.85	ND	ND	ND
88	*Perilla frutescens* (L.)	2.25	ND	3.42	ND
122	*Osbeckia chinensis* L.	ND	ND	ND	ND
144	*Euryale ferox* Salisb.	416.04	181.08	34.06	47.89
147	*Artemisia capillaris* Thunb.	7.72	ND	8.16	ND
159	*Uncaria kawakamii* Hayata	1.76	ND	2.19	ND
172	*Mosla formosana* Maxim.	5.21	ND	14.64	ND

These results are the average of three determinations.
ND = none detected.

The whole plant of O. chinensis has long been commonly used as an anodyne, an anti-inflammation agent, and an antipyretic in Taiwan, Japan, and China. After large-scale purification, many different types of antioxidative components have been isolated and identified (Fig. 6) (37,38). Most of the antioxidative activity of the extracts of O. chinensis is assumed to be due to tannin-type antioxidants (39). Relative

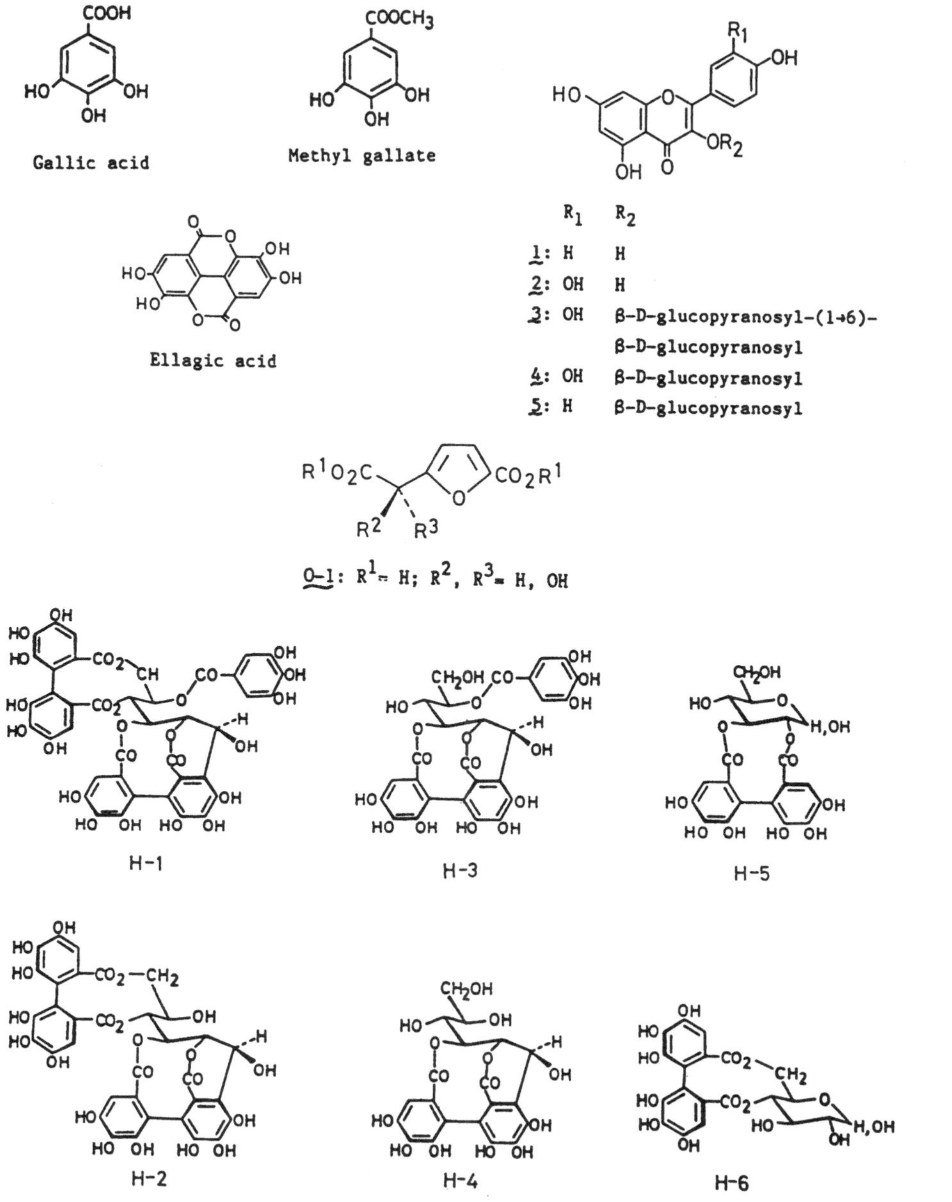

Fig. 6. Structures of antioxidative substances isolated from Osbeckia chinensis L.

antioxidative activity of the isolated tannins has been examined using the rabbit erythrocyte system, and it is shown that all of the isolated tannins have quite strong antioxidative activity (Fig. 7). These isolated tannins contain the ellagic acid moiety in their structures, and ellagic acid itself has almost the same antioxidant activity. These results suggest that the ellagic acid moiety must be responsible for the antioxidative activity of the isolated tannins.

LIPID PEROXIDATION AND MUTAGENICITY

There are many indications of participation of lipid peroxidation in carcinogenesis, although there is no definitive evidence. However, the peroxidative breakdown of the membrane polyunsaturated fatty acids is known to be accompanied by the formation of a complex mixture of many

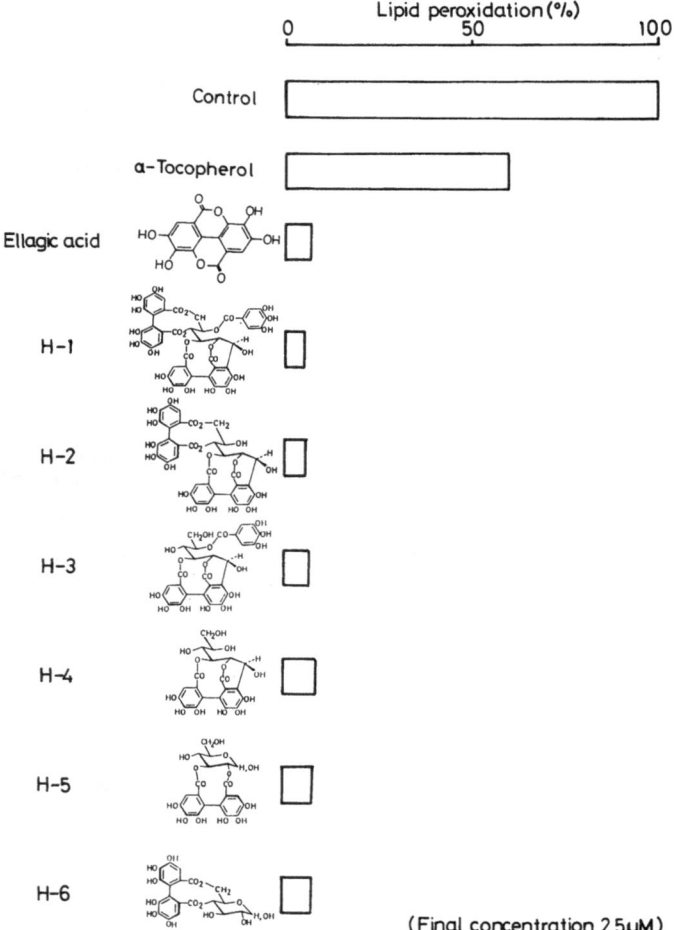

Fig. 7.   Antioxidative activity of isolated tannins from <u>Osbeckia</u> <u>chinensis</u> L. by the rabbit erythrocyte ghost system.

different types of secondary products. One of the main degradation products of lipid peroxidation, malondialdehyde (MDA), has been reported to be mutagenic in Escherichia coli (40) and Salmonella typhimurium (23). Other aldehydic products, among which was 4-hydroxynonenal, were found to cause DNA fragmentation and also to be mutagenic in V79 Chinese hamster cells (6). Recently, Akasaka and Yonei (1) detected mutation induction in E. coli incubated in a reaction mixture of NADPH-dependent lipid peroxidation of rat liver microsome, although identification of the mutagenic products has not been carried out.

In order to obtain some idea that dietary antioxidants can protect from mutagenicity induced during lipid peroxidation, we examined whether mutagenicity can be induced during lipid peroxidation of the erythrocyte membranes when initiated by addition of t-butylhydroperoxide. E. coli WP2s (uvrA⁻, trp⁻), an excision repair-deficient strain, was used for the mutagenicity assay (26), and dose-response mutagenicity was detected parallel to the formation of TBA-reactive substances, as shown in Fig. 8. A lipid-soluble aldehydic fraction (9) has been extracted from the oxidized erythrocyte ghost using ethyl acetate (12), but it exhibited no mutagenicity. Moreover, the amount of MDA formed by oxidation in the erythrocyte ghost system was not sufficiently high to cause mutagenicity (T. Osawa et al., ms. in prep.). At the present stage, it is not clear what kinds of secondary products are formed during lipid peroxidation; however, it is suggested that types of mutagens other than MDA and the known lipid-soluble aldehydes must be formed (11).

In order to determine the mechanism of mutagenicity, a wild type of bacterial strain, E. coli WP2 (trpE⁻) (33), and a umc mutant, E. coli ZA51 (uvrE⁻, umuC⁻, trpE⁻), were used for comparison. However, mutagenicity has been detected only in the excision repair-deficient strain E. coli WP2s, and it was suggested that degradation products or free radicals formed during lipid peroxidation of the erythrocyte membrane ghost caused DNA damages and induced mutations, although the active species were not identified (20).

Correlation between antioxidative activity and inhibitory effect on mutagenicity has also been investigated. As shown in Fig. 9, superoxide dismutase, catalase, and mannitol were not effective; therefore, active oxygens such as superoxide anion radical, hydroxy radical, and hydrogen peroxide are assumed not to induce both lipid peroxidation and

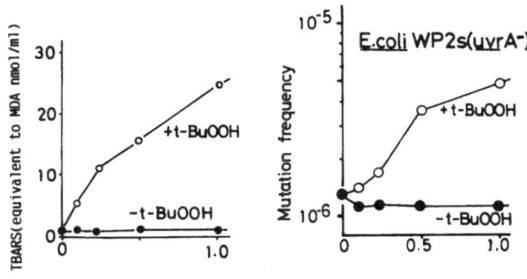

Concentration of erythrocyte ghosts (mg protein/ml)

Fig. 8.   Lipid peroxidation and mutagenicity induced by a rabbit erythrocyte ghost initiated by addition of t-butylhydroperoxide.

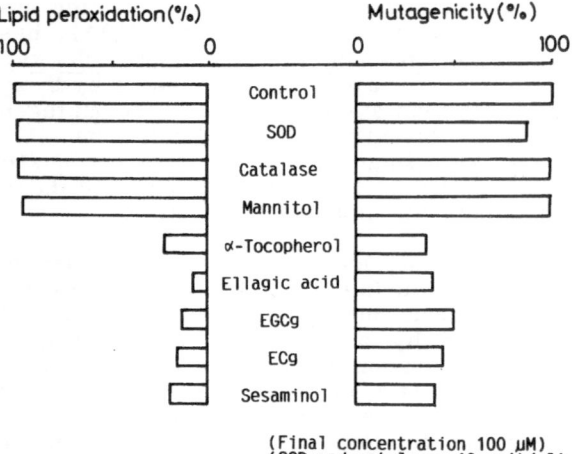

(Final concentration 100 μM)
(SOD and catalase, 10 unit/ml)

Fig. 9.   Correlation between antioxidative activity and inhibitory effects
          of mutagenicity of a rabbit erythrocyte ghost induced by t-
          butylhydroperoxide using E. coli WP2s.

mutagenicity.   Moreover, it was shown that both lipid-soluble antioxidants, such as α-tocopherol and sesaminol, and water-soluble antioxidants, such as ellagic acid, epigallocatechin gallate, and epicatechin
gallate, all inhibited lipid peroxidation and mutagenicity effectively.

      These antioxidants were added into the systems together with bac-
teria after completion of lipid peroxidation.   In this case, lipid-soluble
antioxidants such as α-tocopherol and sesaminol had no inhibitory effect
on mutagenicity; however, water-soluble antioxidants had quite strong inhibitory effects on mutagenicity (Tab. 5).

      In summary, we are suggesting two mechanisms for inhibition of
mutagenicity by lipid-soluble and water-soluble antioxidants (Fig. 10).
When both lipid- and water-soluble antioxidants are added at stage 1,
they can inhibit the free-radical chain reaction of erythrocyte membrane
lipids; therefore, DNA-damaging activity has been inhibited.   On the
other hand, when added at stage 2, water-soluble antioxidants only inhibited mutagenicity, although a detailed examination is under way.

Tab. 5.   Inhibitory effects of mutagenicity induced by an oxidized rabbit
          erythrocyte ghost by selected dietary natural antioxidants using
          E. coli WP2s (uvr⁻).

Antioxidants	Mutation Frequency $(\times 10^{-6})$
α -Tocopherol	2.72
Sesaminol	3.01
Ellagic acid	1.29
Epigallocatechin gallate	1.05
Epicatechin gallate	1.12
Control	3.12

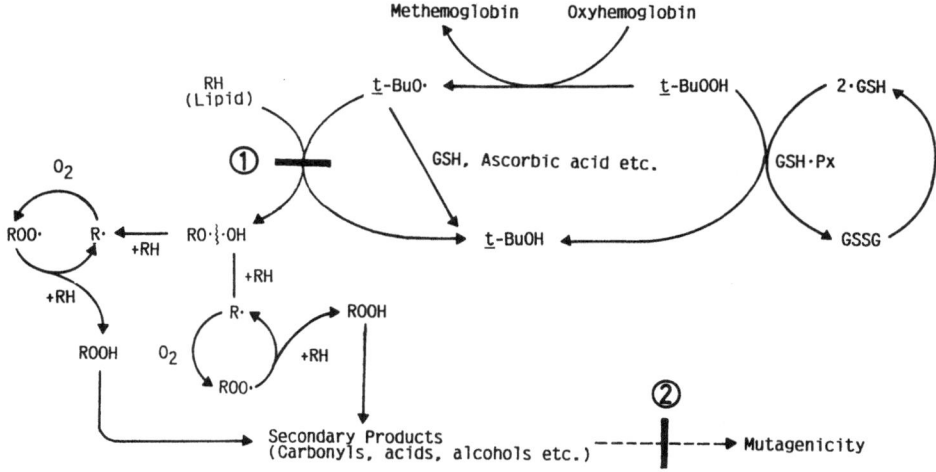

Fig. 10. Proposed mechanisms of inhibition of natural antioxidants on mutagenicity induced by erythrocyte membrane lipid peroxidation initiated by addition of t-butylhydroperoxide.

ACKNOWLEDGEMENTS

The authors wish to thank Ms. Harue Kumon for excellent technical assistance and Drs. Y. Fukuda, K. Shimoi, and T. Ohta for their cooperation. This work was supported by a Grant-in-Aid for Cancer Research from the Ministry of Health and Welfare, and by a Scientific Research Grant from the Ministry of Education.

REFERENCES

1. Akasaka, S., and S. Yonei (1985) Mutation induction in Escherichia coli incubated in the reaction mixture of NADPH-dependent lipid peroxidation of rat-liver microsomes. Mutat. Res. 149:321-326.
2. Ames, B.N., R. Cathcart, E. Schwiers, and P. Hochstein (1981) Uric acid provides an antioxidant defense in humans against oxidant- and radical-caused aging cancer: A hypothesis. Proc. Natl. Acad. Sci., USA 78:6858-6862.
3. Aust, S.D., and B.A. Svingen (1982) The role of ion in enzymatic lipid peroxidation. In Free Radicals in Biology, Vol. V, W.A. Pryor, ed. Academic Press, Inc., New York.
4. Balentino, D.B. (1982) Pathology of Oxygen Toxicity, Academic Press, Inc., New York.
5. Bauernfeind, J. (1980) Tocopherols in foods. In Vitamin E: A Comprehensive Treatise, Marcel Dekker, Inc., New York, pp. 99-167.
6. Cajelli, E., A. Ferraris, and G. Brambilla (1987) Mutagenicity of 4-hydroxynonenal in V79 Chinese hamster cells. Mutat. Res. 190:169-171.
7. Chiu, D., B. Lubin, and S.B. Shohet (1982) Peroxidative reactions in red cell biology. In Free Radicals in Biology, Vol. V, W.A. Pryor, ed. Academic Press, Inc., New York, pp. 115-160.

8. Cutler, R.G. (1984) Antioxidants, aging, and longevity. In Free Radicals in Biology, Vol. VI, W.A. Pryor, ed. Academic Press, Inc., New York, pp. 371-428.

9. Dianzani, M.U. (1982) Biochemical effects of saturated and unsaturated aldehydes. In Free Radicals, Lipid Peroxidation, and Cancer, D.C.H. McBrien and T.F. Slater, eds. Academic Press, Inc., New York, pp. 129-171.

10. Dugan, L.R. (1980) Natural antioxidants. In Autoxidation in Food and Biological Systems, M.G. Simic and H. Karel, eds. Plenum Press, New York, pp. 261-282.

11. Esterbauer, H. (1982) Aldehydic products of lipid peroxidation. In Free Radicals, Lipid Peroxidation, and Cancer, D.C.H. Mcbrien and T.F. Slater, eds. Academic Press, Inc., New York, pp. 101-128.

12. Esterbauer, H., K.H. Cheeseman, M.U. Dianzani, G. Poli, and T.F. Slater (1982) Separation and characterization of the aldehydic products of lipid peroxidation stimulated by ADP-Fe^{2+} in rat liver microsomes. Biochem. J. 208:129-140.

13. Fukuda, Y., T. Osawa, and M. Namiki (1981) Antioxidants in sesame seed. Nippon Shokuhin Kogyo Gakkaishi 28:461-464.

14. Fukuda, Y., T. Osawa, M. Namiki, and T. Ozaki (1985) Studies on antioxidative substances in sesame seed. Agric. Biol. Chem. 49:301-306.

15. Fukuda, Y., M. Nagata, T. Osawa, and M. Namiki (1986) Contribution of lignan analogues to antioxidative activity of refined unroasted sesame seed oil. J. Am. Oil Chem. Soc. 63:1027-1031.

16. Fukuda, Y., M. Isobe, M. Nagata, T. Osawa, and M. Namiki (1986) Acidic transformation of sesamolin, the sesame oil constituent into an antioxidant bisepoxylignan, sesaminol. Heterocycles 24:923-926.

17. Hochstein, P., and A.H. Atallah (1988) The nature of oxidants and antioxidant systems in the inhibition of mutation and cancer. Mutat. Res. 202:363-375.

18. Kohen, R., Y. Yamamoto, K.C. Cundy, and B.N. Ames (1988) Antioxidant activity of carnosine, homocarnosine, and anserine present in muscle and brain. Proc. Natl. Acad. Sci., USA 85:3175-3179.

19. Lyon, C.K. (1972) Sesame: Current knowledge of composition and use. J. Am. Oil Chem. Soc. 49:245-249.

20. Marnett, L.J., H.K. Hurd, M.C. Hollstein, D.E. Levin, H. Esterbauer, and B.N. Ames (1985) Naturally occurring carbonyl compounds are mutagens in Salmonella tester strain TA104. Mutat. Res. 148:25-34.

21. McBrien, D.C.H., and T.F. Slater, eds. (1982) Free Radicals, Lipid Peroxidation and Cancer, Academic Press, Inc., New York.

22. McCay, P.B., and M.M. King (1980) Vitamin E: Its role as a biological free radical scavenger and its relationship to the microsomal mixed-function oxidase system. In Vitamin E: A Comprehensive Treatise, L.J. Mcchlin, ed. Marcel Dekker, Inc., New York, pp. 289-317.

23. Mukai, F.H., and B.D. Goldstein (1976) Mutagenicity of malonaldehyde, a decomposition product of peroxidized polyunsaturated fatty acids. Science 191:868-869.

24. Nagata, M., T. Osawa, M. Namiki, Y. Fukuda, and T. Ozaki (1987) Stereochemical structures of antioxidative bisepoxylignans, sesaminol and its isomers, transformed from sesamolin. Agric. Biol. Chem. 51:1285-1289.

25. Namiki, M., and T. Osawa (1986) Antioxidants/antimutagens in foods. In Antimutagenesis and Anticarcinogenesis Mechanisms, D.M.

Shankel, P.E. Hartman, T. Kada, and A. Hollaender, eds. Plenum Press, New York, pp. 131-142.

26. Ohta, T., K. Watanabe, M. Moriya, Y. Shirasu, and T. Kada (1983) Anti-mutagenic effects of coumarin and umbelliferone on mutagenesis induced by 4-nitroquinoline 1-oxide or UV irradiation in E. coli. Mutat. Res. 117:135-138.

27. Osawa, T., M. Nagata, M. Namiki, and Y. Fukuda (1985) Sesamolinol, a novel antioxidant isolated from sesame seeds. Agric. Biol. Chem. 49:3351-3352.

28. Osawa, T., A. Ide, J.-De Su, and M. Namiki (1987) Inhibition of lipid peroxidation by ellagic acid. J. Agric. Food Chem. 35:808-812.

29. Porter, W.L. (1980) Recent trend in food applications of antioxidants. In Autoxidation in Food and Biological Systems, M.G. Simic and M. Karel, eds. Plenum Press, New York, pp. 295-365.

30. Pryor, W.A. (1986) Cancer and free radicals. In Antimutagenesis and Anticarcinogenesis Mechanisms, D.M. Shankel, P.E. Hartman, T. Kada, and A. Hollaender, eds. Plenum Press, New York, pp. 45-59.

31. Pziezak, J.D. (1986) Preservatives: Antioxidants. The ultimate answer to oxidation. Food Tech. 40(9):94-102.

32. Shima, A. (1988) Effect of food components on cellular aging. In Food Functionalities, M. Fujimaki, ed. Gakkai Press Center, Tokyo, pp. 227-231.

33. Shimoi, K., Y. Nakamura, I. Tomita, Y. Hara, and T. Kada (1986) The pyrogallol related compounds reduce UV-induced mutations in Escherichia coli, B/r WP2. Mutat. Res. 173:239-244.

34. Simic, M.G. (1988) Mechanisms of inhibition of free-radical processes in mutagenesis and carcinogenesis. Mutat. Res. 202:377-386.

35. Stocker, R., Y. Yamamoto, A.F. McDonagh, A.N. Glazer, and B.N. Ames (1987) Bilirubin is an antioxidant of possible physiological importance. Science 235:1043-1046.

36. Su, J.-De, T. Osawa, and M. Namiki (1986) Screening for antioxidative activity of crude drugs. Agric. Biol. Chem. 50:199-203.

37. Su, J.-De, T. Osawa, S. Kawakishi, and M. Namiki (1987) Antioxidative flavonoids isolated from Osbeckia chinensis L. Agric. Biol. Chem. 51:2801-2803.

38. Su, J.-De, T. Osawa, S. Kawakishi, and M. Namiki (1987) A novel antioxidative synergist isolated from Osbeckia chinensis L. Agric. Biol. Chem. 51:3449-3450.

39. Su, J.-De, T. Osawa, S. Kawakishi, and M. Namiki (1988) Tannin antioxidants from Osbeckia chinensis. Phytochemistry 27:1315-1319.

40. Yonei, S., and H. Furui (1981) Lethal and mutagenicity effects of malonaldehyde, a decomposition product of peroxidized lipids, on Escherichia coli with different DNA repair capacities. Mutat. Res. 88:23-32.

INHIBITION OF CARCINOGENESIS BY NATURALLY-OCCURRING

AND SYNTHETIC COMPOUNDS

Lee W. Wattenberg

Laboratory of Medicine and Pathology
University of Minnesota
Minneapolis, Minnesota 55455

INTRODUCTION

Cancer can be prevented by administration of a variety of chemical compounds. Some of these are naturally-occurring. Others are purely synthetic. Compounds belonging to over 25 different classes of chemicals have preventive capacities (30). There are two implications of the existence of a wide chemical diversity of inhibitors. One is that it enhances the likelihood that a variety of approaches can be made to cancer prevention by chemopreventive compounds. The second is that the occurrence of many naturally-occurring preventive agents in food indicates that the composition of the diet can be an important factor in determining the response of individuals to carcinogenic agents to which they are exposed.

The mechanisms of action of many of the inhibitors of carcinogenesis are poorly understood, making it difficult to organize them into a precise pattern. One way of providing an organizational framework is to classify inhibitors according to the time in the carcinogenic process at which they are effective. Utilizing this framework, inhibitors of carcinogenesis can be divided into three categories. The first consists of compounds that prevent the formation of carcinogens from precursor substances. In the second are compounds that inhibit carcinogenesis by preventing carcinogenic agents from reaching or reacting with critical target sites in the tissues. These inhibitors are called "blocking agents," which is descriptive of their mechanism of action. They exert a barrier function. The third category of inhibitors acts subsequent to exposures to carcinogenic agents. These inhibitors are termed "suppressing agents," since they act by suppressing the expression of neoplasia in cells previously exposed to doses of carcinogens that otherwise would cause cancer (30). During the course of studies of various inhibitors it has become apparent that some inhibitors act at more than one timepoint in the carcinogenic process. The implication of these observations will be discussed subsequently.

The classic example of inhibition of carcinogenesis by preventing the formation of carcinogens from precursor substances entails the use of ascorbic acid to prevent formation of nitroso carcinogens from the reactions of nitrite with appropriate amines or amides.   This prevention can occur in vivo, particularly in the alimentary tract (17).   It also can be used as a means of preventing formation of nitroso carcinogens in food.   Prevention of carcinogen formation, along with procedures for inactivating or destroying carcinogens in foods and beverages by chemical means prior to consumption, constitutes an extension of chemoprevention strategies into the area of environmental health.   Possible examples of such strategies would include those preventing the formation of imidazoquinoxalines in meat products, those preventing the formation of urethane in alcoholic beverages, and destruction of hydrazines in mushrooms (15,16,18).   An additional area of interest pertains to a group of suspect carcinogens, the fecapentaenes that are formed by the microbial flora of the large intestine of the human (1,14).   The fecapentaenes are highly mutagenic in the Ames test and also produce a wide variety of other genotoxic reactions. If the fecapentaenes prove to be full or partial carcinogens in the human, strategies to prevent their synthesis or to inactivate them could become a component of an overall program for preventing cancer of the large bowel.

BLOCKING AGENTS

Mechanisms of Action

A major portion of this presentation will focus on the inhibitory effects of blocking agents.   Blocking agents prevent carcinogens from reaching or reacting with critical target sites by several mechanisms. One of these is to inhibit activation reactions of carcinogens requiring metabolic activation.   A second entails induction of increases in the activities of enzyme systems that can detoxify carcinogens.   A third mechanism is the trapping of reactive carcinogenic species.   Each of these will be discussed in turn.

Blocking Agents That Act by Inhibiting Carcinogen Activation

Several groups of chemopreventive agents prevent carcinogenesis by inhibiting activation reactions of carcinogens.   Such activation reactions are required to produce the reactive chemical species that attack critical target sites.   An early example of a chemopreventive agent that acts by this mechanism is disulfiram.   Disulfiram is a potent inhibitor of 1,2-dimethylhydrazine (DMH)-induced carcinogenesis of the large bowel in rodents.   DMH undergoes a series of metabolic reactions that result in the formation of the ultimate carcinogenic species, a highly reactive electrophile.   Disulfiram has been found to block this sequence of activation reactions at two steps.   The elucidation of these enzyme inhibitions has been the result of extensive investigations by Fiala et al. (12).   In experimental animals, administration of disulfiram prior to and during the course of administration of DMH prevents large bowel carcinogenesis (25). Disulfiram is a compound with substantial toxicity, so that its use as a chemopreventive agent in the human is not practical.

More recently, several groups of naturally-occurring compounds also have been found to prevent carcinogenesis by inhibiting activation reactions.   Benzyl isothiocyanate is one such compound.   Benzyl isothiocyanate occurs in cruciferous vegetables.   It is formed by the hydrolysis of its glucosinolate, glucotropaeolin.   The glucosinolate is the storage form of the compound in plant material.   When the plant is processed or damaged, the glucosinolate undergoes hydrolysis, with resultant formation of the free aromatic isothiocyanate (11). In early experiments, benzyl isothiocyanate was found to inhibit 7,12-dimethylbenz(a)anthracene (DMBA)-induced mammary tumor formation in rats when given by oral intubation 2 hr prior to the carcinogen (26).

In more recent work, the effects of benzyl isothiocyanate on nitrosamine metabolism in rat liver microsomes and cultured esophagus was studied by Chung et al. (4,5). These investigators found that administration of benzyl isothiocyanate 2 hr prior to sacrifice markedly inhibited activation of N-nitrosodimethylamine (NDMA) and the tobacco-specific carcinogen, 4-(methylnitrosamino)-1-(3-pyridyl)-1-butanone (NNK). Subsequently, the inhibitory effects of benzyl isothiocyanate administered by gavage 15 min prior to oral carcinogen challenge with diethylnitrosamine (NDEA) or benzo(a)pyrene (BAP) were investigated. The results of these experiments are shown in Tab. 1. Benzyl isothiocyanate almost completely inhibited NDEA-induced forestomach tumor formation, but did not produce a statistically significant reduction of pulmonary tumors. With BAP as the carcinogen, benzyl isothiocyanate inhibited carcinogenesis in both the forestomach and lungs, the latter being greater (31). A closely related aromatic isothiocyanate, phenethyl isothiocyanate, has been shown to inhibit NNK-induced neoplasia of the lung in both rats and mice (Morse et al., this Volume).

Organosulfur compounds found in Allium species, including garlic, onions, leeks, and shallots (2,8-10), are a second group of naturally-occurring compounds that also have been found to inhibit carcinogenesis by preventing carcinogen activation. In one series of experiments, they inhibited NDEA-induced carcinogenesis of the forestomach and, to a lesser extent, the lungs in female A/J mice (32). The most potent of the naturally-occurring organosulfur compounds in this experimental model was diallyl disulfide (Tab. 2). In other studies, Wargovich has shown that diallyl sulfide inhibits N-nitrosomethylbenzylamine-induced esophageal cancer in rats when the organosulfur compound was administered orally 3 hr prior to the nitrosamine (24). In related work, Brady et al. have

Tab. 1. Effects of benzyl isothiocyanate on carcinogen-induced neoplasia in female A/J mice.

Exp. No.	Carcinogen	Number of mice per group[a]	Wt gain from 8 to 35 weeks of age (g)	Pre-treatment 15 min prior to carcinogen	Pulmonary ademomas No. of tumors/mouse[b]	Forestomach tumors All neoplasms % of mice with tumors	No. of tumors tumors/mouse[c]	Carcinomas % of mice with carcinomas
1	20 mg/kg DEN p.o. weekly for 8 weeks	12	5	H$_2$O	18.7 ± 4.5[d]	100	>30	25
		12	4	Cottonseed oil	12.7 ± 5.1	100	>30	25
		13	6	Benzyl isothiocyanate 1 mg	12.9 ± 4.9	50[e]	0.5[e]	0[f]
2	2.0 mg BP p.o. every other week for three administrations	14	6	Cottonseed oil	15.5 ± 5.3[d]	93	3.3 ± 3.0[d]	21
		15	6	Benzyl isothiocyanate 1.0 mg	7.4 ± 6.7[g]	87	2.4 ± 1.5	7
		15	7	Benzyl isothiocyanate 2.5 mg	3.9 ± 3.1[g]	73	1.1 ± 1.0[h]	0

[a] Female A/J mice were randomized by weight at 8 weeks of age and subjected to the procedures indicated. The experiment was terminated when the mice were 35 weeks of age.
[b] Pulmonary adenomas at the termination of the experiment. Numbers of tumors in the entire group/number of mice at risk.
[c] Gastric tumors at the termination of the experiment. Number of gastric tumors in the entire group/number of mice at risk.
[d] Mean $\pm$ SD.
[e] Cottonseed oil versus benzyl isothiocyanate P < 0.0001.
[f] Cottonseed oil versus benzyl isothiocyanate P = 0.07.
[g] Cottonseed oil versus benzyl isothiocyanate P < 0.001.
[h] Cottonseed oil versus benzyl isothiocyanate P < 0.05.

Tab. 2.   Effects of organosulfur compounds and D-limonene on NDEA-induced neoplasia in female A/J mice. Cottonseed oil (0.2 ml) or test compound in 0.2 ml cottonseed oil was administered by oral intubation once a week for 8 wk to female A/J mice at the time intervals designated prior to NDEA (20 mg/kg body weight). The experimental treatments were started when the mice were 9 wk old; the experiment was terminated 26 wk after the first dose of NDEA.

Pretreatment	Time interval prior to NDEA	Number of mice at risk	Forestomach tumors				Pulmonary adenomas	Wt. gain from 8 to 35 weeks of age (g)
			% of mice with papillomas	% of mice with >30 papillomas[a]	Median number of papillomas	% of mice with carcinomas	Number of tumors per mouse[b]	
None		14	100	100	>30	29	15.9 ± 1.0[c]	5.9
Cottonseed oil		13	100	100	>30	38	16.6 ± 0.8	7.0
Diallyl disulfide (0.02 mmol)		14	100	0[d]	7.0[d]	7	13.6 ± 1.1[e]	5.6
Diallyl sulfide (0.02 mmol)	15 min	14	100	100	>30	21	12.0 ± 0.9[f]	5.6
Allyl mercaptan (0.01 mmol)		12	92	0[d]	3.5[d]	0[e]	14.2 ± 1.6	6.8
Allyl methyl disulfide (0.02 mmol)		15	100	73	>30	27	12.5 ± 1.0[g]	7.3
Cottonseed oil		15	100	73	>30	27	10.4 ± 1.4	5.3
Diallyl disulfide (0.02 mmol)	1 hr	12	58[e]	0[h]	1.0[h]	0[i]	10.8 ± 1.0	4.9
D-Limonene (0.2 mmol)		15	67[e]	0[h]	1.0[h]	0[j]	6.5 ± 0.6[e]	4.2

[a]The number of papillomas per forestomach beyond 30 was not counted. Further counting is not accurate because of fusion of lesions.
[b]Number of tumors in the entire group/number of mice at risk.
[c]Mean ± SE.
[d]$p<0.001$ vs either vehicle or absolute control.
[e]$p<0.05$ vs vehicle control.
[f]$p<0.01$ vs either vehicle or absolute control.
[g]$p<0.05$ vs either vehicle or absolute control.
[h]$p<0.001$ vs vehicle control.
[i]$p=0.05$ vs vehicle controls combined from all four experiments.
[j]$p<0.05$ vs vehicle controls combined from all four experiments.

demonstrated that diallyl sulfide inhibits the microsomal metabolism of nitrosamines when administered 3 hr prior to sacrifice of rats (3). Inhibition of DMH-induced neoplasia of the large bowel in mice has also been shown to be brought about by diallyl sulfide when this compound was administered orally 3 hr prior to the carcinogen (23).

A third group of naturally-occurring compounds that inhibit carcinogen activation and carcinogenesis when given shortly before carcinogen exposure are the monoterpenes. Two such compounds have been investigated, D-limonene and D-carvone. D-Limonene is a major constituent of citrus fruit oils (19). For example, orange oil contains over 90% D-limonene. D-Carvone is a major constituent of caraway seed oil. This oil contains approximately 50% D-carvone (20). D-Limonene, D-carvone, orange oil, and caraway seed oil all have been found to inhibit activation of NDMA (L.W. Wattenberg and J.L. Leong, unpubl. results). Carcinogenesis experiments have been carried out in which D-limonene, D-carvone, or caraway seed oil was administered p.o. to female A/J mice 1 hr prior to NDEA (32). All three substances had a profound inhibitory effect on forestomach tumor formation and also inhibited the occurrence of pulmonary adenomas in these animals. The results of a study with D-limonene are shown in Tab. 2.

In summary, data are now available showing that three widely-occurring groups of naturally-occurring compounds, i.e., aromatic isothiocyanates, organosulfur compounds found in Allium species, and monoterpenes, have the capacity to inhibit carcinogen activation and, correspondingly, to inhibit carcinogenesis when administered shortly before carcinogen challenge.

Blocking Agents That Act by Increasing Carcinogen Detoxification

Blocking agents can prevent the occurrence of neoplasia by increasing the detoxification of carcinogens. Two general categories of blocking agents acting by this mechanism have been identified. They are designated Type A and Type B inhibitors (30). Type A inhibitors induce an increase in Phase 2 enzymes, i.e., conjugating enzymes and some related protective systems. A prominent feature of these enzyme inductions is a marked increase in glutathione S-transferase activity. UDP-Glucuronosyl transferase activity is also enhanced, as are epoxide hydrolase and NAD(P)H-quinone reductase activities. Another feature of Type A inhibitors is an alteration in microsomal metabolism in which there is little change in activity. However, a pronounced alteration in metabolite pattern occurs, as demonstrated with BAP as a substrate. In addition to enzymatic changes, an increase in the levels of glutathione in tissues is found. Type B inhibitors enhance both Phase I and Phase 2 enzyme activities. Reviews of the effects of Type A and Type B inhibitors have been published previously (30).

Since Type A inhibitors have been found to be highly effective, a continuing effort has been made to identify compounds belonging to this category of chemopreventive agents. Work of this nature has focused recently on the organosulfur compounds found in Allium species. In initial studies, the capacity of allyl methyl trisulfide to induce increased glutathione S-transferase activity in rodent tissues was investigated. Allyl methyl trisulfide induces increased activity of this enzyme system when given 4 and 2 da prior to assay. Using this administration schedule, allyl methyl trisulfide (AMT) then was found to inhibit BAP-induced neoplasia of the forestomach of female A/J mice (22). Experimental protocols in which the test compound is given 96 and 48 hr prior to carcinogen challenge are commonly used for studying the inhibitory effects of agents that act by inducing increased activity of detoxification systems. In subsequent work, additional organosulfur compounds were studied under these same conditions (Tab. 3).

Ultimately, the experiments have included four allyl group-containing compounds with one, two, or three linearly-connected sulfur atoms: AMT, allyl methyl disulfide (AMD), diallyl trisulfide (DAT), and diallyl sulfide (DAS). Four corresponding saturated compounds in which propyl groups have been substituted for the allyl groups also were studied. All four allylic compounds inhibited BAP-induced neoplasia of the forestomach. Their saturated analogs were almost without inhibitory activity, indicating the importance of the allyl groups. DAT, which contains two allyl groups, is more potent than AMT, which contains only one allyl group, thus providing further evidence for the role of allyl groups in the inhibitory effects observed. DAS and AMD, but not DAT or AMT, inhibited pulmonary adenoma formation (21). The fact that in the lung the monosulfide and disulfide inhibit, but the trisulfide does not inhibit, indicates that the number of sulfur atoms in the molecule can control the organ sites at which protection against carcinogenesis will occur. All four allylic compounds induce increased glutathione S-transferase activity in the forestomach, but vary in their capacity to induce glutathione S-transferase activity in the lung, liver, and small bowel. Their saturated

Tab. 3.   Effects of organosulfur compounds on NDEA-induced neoplasia in female A/J mice. Cottonseed oil (0.2 ml) or test compound in 0.2 ml cottonseed oil was administered by oral intubation once a week for 8 wk to female A/J mice at the time intervals designated prior to NDEA (20 mg/kg body weight). The experimental treatments were started when the mice were 9 wk old; the experiment was terminated 26 wk after the first dose of NDEA.

Pretreatment	Time interval prior to NDEA	Number of mice at risk	Forestomach tumors				Pulmonary adenomas	
			% of mice with papillomas	% of mice with >30 papillomas[a]	Median number of papillomas	% of mice with carcinomas	Number of tumors per mouse[b]	Wt. gain from 8 to 35 weeks of age (g)
None		13	100	100	>30	31	14.2 ± 1.3[c]	7.6
Cottonseed oil	96 hr	15	100	100	>30	20	14.1 ± 1.3	7.3
Diallyl disulfide (0.02 mmol)	and	13	100	0[d]	3.0[d]	0[e]	9.9 ± 1.2[f]	7.0
Allyl mercaptan (0.01 mmol)	48 hr	14	100	57[f]	>30	0[e]	9.6 ± 1.3[f]	7.1
Allyl methyl disulfide (0.02 mmol)		15	100	60[f]	>30	0[g]	9.3 ± 0.8[h]	6.9

[a]The number of papillomas per forestomach beyond 30 was not counted. Further counting is not accurate because of fusion of lesions.
[b]Number of tumors in the entire group/number of mice at risk.
[c]Mean ± SE.
[d]p<0.001 vs either vehicle or absolute control.
[e]p=0.05 vs vehicle controls combined from all four experiments.
[f]p=<0.05 vs either vehicle or absolute control.
[g]p<0.05 vs vehicle controls combined from all four experiments.
[h]p<0.001 vs either vehicle or absolute control.

analogs produced little or no induction (21). Like the organosulfur compounds from Allium species, aromatic isothiocyanates from cruciferous vegetables (in particular, benzyl isothiocyanate) induce increased glutathione S-transferase activity and have other properties of Type A inhibitors.

SUPPRESSING AGENTS

Suppressing agents have been found in cruciferous vegetables and also in orange oil (29). In the initial work with cruciferous vegetables, dehydrated powders prepared from Chieftain Savoy cabbage were shown to inhibit DMBA-induced mammary tumor formation. In further investigations, it was found that frozen, thawed segments of cabbage leaf added directly to cages of rats produced a similar level of reduction of neoplasia (33). The amounts of cabbage leaf employed were of the order of magnitude consumed by humans. More recently, similar studies have been done with frozen-thawed material from the floret portion of broccoli. In these experiments in which the suppressive capacities of frozen-thawed cabbage and broccoli have been investigated, the cabbage (Chieftain Savoy) or broccoli (Premium Crop) was given beginning 1 wk after the DMBA and was continued for the duration of the experiments (18 wk after DMBA). The rats were maintained on a semipurified diet. The vegetable additions were given 6 days a week. The results are presented in Tab. 4. In Experiment 1 the effects of the cabbage are shown. Fewer animals receiving cabbage had mammary tumors, as compared to those not receiving the supplement (100% vs 56%), and the average number of tumors per rat was diminished markedly (2.5 vs 0.8). Similar findings were obtained with broccoli additions.

Tab. 4.   Suppressive effects of cabbage and broccoli on 7,12-dimethyl-
benz(a)anthracene (DMBA)-induced mammary carcinogenesis in
female Sprague-Dawley rats.

Experiment[a]	Additions to the diet	Amount Gm/day[b]	No. of rats at risk	Final Weight (g)	Mammary tumors[c]	
					% of rats with tumors	No. of tumors per rat[d]
1	None	-	16	204	100	2.5 ± 0.3^e
	Cabbage (Chieftain Savoy)	1	16	210	56[f]	0.8 ± 0.2^g
2	None	-	24	261	79	2.0 ± 0.3
	Broccoli (Premium Crop)	0.25	24	270	62	1.5 ± 0.3
	Broccoli (Premium Crop)	1	24	265	42[h]	1.2 ± 0.3^h

[a] Experiments 1 and 2 were carried out several years apart using similar, but not identical, conditions. The Sprague Dawley rats used were from different colonies. In Experiment 1 a fixed dose of DMBA of 12 mg/rat was given. In Experiment 2 the dose of DMBA employed was 6 mg/100 Gms body weight. The average body weight when the rats were 7 weeks of age was approximately 170 gms, so that the dose of DMBA received was approximately 10 mg.
[b] The indicated amounts of frozen-thawed vegetable were added daily (x6days/wk) to each cage beginning 7 days after DMBA administration. The vegetable supplements were generally consumed within minutes. Rats were fed a semisynthetic diet (32).
[c] Mammary tumors at termination of the experiment, i.e. 18 weeks after administration of DMBA.
[d] Number of tumors in the entire group divided by the number of rats at risk.
[e] Mean ± S.E.
[f] p<0.01
[g] p<0.001
[h] p<0.05

Possibly related to the suppressive effects of cruciferous vegetables has been the finding that benzyl isothiocyanate, a constituent of these plants, will inhibit DMBA-induced mammary carcinogenesis when administered beginning 1 wk after challenge by DMBA (Tab. 5) (28). In other work, suppressive effects have been produced by administration of orange oil or its principal constituent, D-limonene (7,29). The observations that suppressing agents occur in commonly consumed dietary constituents indicate that further searches should be made for this category of chemopreventive agent. The current test systems for identifying and then further studying suppressing agents are cumbersome. The single most effective method for these investigations entails full carcinogenesis protocols. To

Tab. 5.   Suppressive effects of benzyl isothiocyanate on DMBA-induced mammary carcinogenesis in female Sprague-Dawley rats.

Additions to the diet[a]	Concentration (mmol/g)	No of rats at risk	Wt. gain (g)[c]	Mammary Neoplasms[b]	
				% of rats with tumors	No. of tumors per rat
None	-	27	53	100	3.4 ± 0.3^d
Benzyl isothiocyanate	0.017	16	48	63[e]	1.4 ± 0.4^e

[a] Female Sprague Dawley rats were given 12 mg DMBA in 1.0 ml olive oil p.o. when 7 weeks of age. One week later they were fed a control diet (Purina rat chow) or a diet containing benzyl isothiocyanate. These diets were continued for the duration of the experiment.
[b] Mammary neoplasms when rats were 23 weeks of age.
[c] Weight gain from 7 to 23 weeks of age.
[d] Mean ± S. E.
[e] p<0.001

date, no reliable short-term assay has been developed. The burdensome
methodology has made efforts at finding suppressing agents more difficult, but, obviously, this is an important endeavor.

INHIBITORS WITH BOTH BLOCKING AND SUPPRESSING ACTIVITY

During the course of studies of inhibitors of carcinogenesis, it has
become apparent that some substances have both blocking and suppressing effects. The first such compound that we have studied is sodium
cyanate. This compound has both blocking and suppressing effects
against DMBA-induced mammary carcinogenesis in the rat (27,28). In
later work, cruciferous vegetables were shown to have these characteristics, as did one of their constituents, i.e., benzyl isothiocyanate
(Tab. 6). In the studies with cabbage, consumption of a semi-purified
diet to which dehydrated powders prepared from Chieftain Savoy cabbages
had been added, produced blocking effects against DMBA-induced mammary carcinogenesis in the rat. In addition, it was found that consumption of the cabbage powders beginning 1 wk after administration of DMBA
resulted in suppression of mammary tumor formation. In subsequent
work, benzyl isothiocyanate was shown to have a blocking effect against
DMBA-induced mammary carcinogenesis and also to suppress formation of
mammary tumors under similar conditions to those used with crude plant
materials (28).

In other work, citrus fruit oils have been found to have both blocking and suppressing capacities. The most extensively studied of these
oils has been orange oil. This oil has a blocking effect against DMBA-
induced mammary neoplasia and also exerts a suppressive effect in this
experimental model (29). In experiments with constituents of orange oil,
D-limonene has been shown to have both blocking and suppressing effects
against DMBA-induced mammary tumor formation (7). In addition to inhibition of neoplasia, cruciferous vegetable powders, benzyl isothiocyanate, orange oil, and D-limonene all have another property in common,
i.e., they all induce increased glutathione S-transferase activity in mouse
tissues. Comparable studies have not yet been reported for the rat.

Tab. 6.   Inhibitors with both blocking and suppressing activities.

Inhibitor	Blocking effect		Suppressing effect	
	Tissue	Carcinogen	Tissue	Carcinogen
Cabbage	Breast[a]	DMBA (35)[b]	Breast	DMBA (29,33)
Benzyl isothiocyanate	Breast	DMBA (26)	Breast	DMBA (28)
Orange oil	Breast	DMBA (31,35)	Breast	DMBA (29)
D-Limonene	Breast	DMBA (7)	Breast	DMBA (7)
Sodium cyanate	Breast	DMBA (27)	Breast	DMBA (28)

[a]Female Sprague-Dawley rats.
[b]Literature references in parentheses.

The finding of inhibitors having both blocking and suppressing activity would appear to be important. However, the mechanism or mechanisms which result in these dual inhibitory actions is uncertain. Two distinctive possibilities merit consideration, although others also may exist. One is that a coordinated pleiotropic protective system against potentially toxic compounds exists that entails both a detoxification component and a suppressive component. The latter would be a "failsafe" means of protection in the event that the former was not completely effective. On exposure to toxic compounds, the detoxification system constitutes the first line of defense. Phase 2 enzymes, as well as glutathione, are components of this detoxification (blocking) defense. Since blocking effects might not always be completely successful, a second line of defense would be required that would entail preventing the manifestations of any damage that occurred, i.e., "suppression."

Whereas a coordinated protective mechanism entailing increased activity of detoxification systems and suppression of manifestations of toxicity would appear to be a reasonable possibility, the inclusion into such a coordinated mechanism of a decrease in cytochrome P450 enzyme activities might appear incongruous. However, studies by Mannering et al. offer some insight as to how such enzyme inhibitions might fit into an overall protective system. These investigators have shown that some compounds that reduce cytochrome P450 activity increase interferon levels in tissues. Such findings suggest that a decrease in activity of one mechanism that is involved in protective events can be accompanied by an enhanced activity of a different protective system (6,13). The inclusion of DNA repair systems as a component of suppressive mechanisms also is a consideration. Additional ones also may exist for preventing manifestations of damage to other cellular constituents. If a coordinated defense mechanism with both blocking and suppressing components exists, it is possible that it is triggered by a receptor mechanism. This certainly would be a point for future investigation.

A second possibility that could account for a compound having both blocking and suppressing activity would be that the chemical is highly reactive and simply has multiple effects due to this attribute. If such is the case, then the blocking and suppressing effects would be triggered separately. Aromatic isothiocyanates are highly reactive compounds, and their capacity to act as both blocking and suppressing agents might be accounted for on the basis of this high chemical reactivity. In contrast, D-limonene is not a highly reactive molecule. However, it is metabolized to epoxides which do have high reactivity. At the present time, the paucity of data and the characteristics of the inhibitors make it difficult to evaluate the two major mechanisms of action considered. The identification of a receptor activating both blocking and suppressing action, if such occurs, would be a major achievement. But even if such a unified defense system does not exist, a knowledge of the characteristics of inhibitors having both blocking and suppressing effects might enhance our information base for obtaining chemopreventive agents with maximum protective capacities.

SUMMARY

A continuing study of chemopreventive agents has focused on three categories of naturally-occurring compounds that inhibit carcinogen activation and are effective in preventing carcinogen-induced neoplasia when administered at short time intervals prior to carcinogen challenge. The three are: aromatic isothiocyanates found in cruciferous vegetables, monoterpenes from citrus fruits and caraway seed oils, and organosulfur compounds occurring in Allium species. The short time-interval effects could be significant in terms of their impact on responses of humans to carcinogen exposures.

The capacity of sodium cyanate, cruciferous vegetables, orange oil, benzyl isothiocyanate, and D-limonene to act as both blocking and suppressing agents has been discussed. Two possible mechanisms for this multiphase activity were presented. The first is that these inhibitory substances activate a complex integrated defense mechanism against toxic compounds which entails both blocking and suppressing components. The blocking component is the initial line of defense, and the suppressing component constitutes a "fail-safe" backup to assure that if any of the toxic material attacks cellular constituents, its effects will be nullified. The second possible mechanism considered is that the inhibitors, because of high reactivity, have multiple biological effects that are separate and not part of a single, coordinated response. Inhibitors that have both blocking and suppressing effects could be particularly useful as chemo-preventive agents.

REFERENCES

1. Baptista, J., W.R. Bruce, I. Gupta, J.J. Krepinsky, R.L. Van Tassell, and T.D. Wilkins (1984) On distribution of different feca-pentaenes, the fecal mutagens, in the human population. Cancer Lett. 22:299-303.
2. Block, E. (1985) The chemistry of garlic and onions. Scient. Am. 252:114-119.
3. Brady, J.F., D. Li, H. Ishizaki, and C.S. Yang (1988) Effect of di-allyl sulfide on rat liver microsomal nitrosamine metabolism and other monooxygenase activities. Cancer Res. 48:5937-5940.
4. Chung, F.L., A. Juchatz, and S.S. Hecht (1984) Effects of dietary compounds on α-hydroxylation of N-nitrosopyrrolidine and N'-nitroso-nornicotine in rat target tissues. Cancer Res. 44:2924-2928.
5. Chung, F.L., M. Wang, and S.S. Hecht (1985) Effects of dietary indoles and isothiocyanates on N-nitrosodimethylamine and 4-(methyl-nitrosamine)-1-(3-pyridyl)-1-butanone-α-hydroxylation and DNA methylation in rat liver. Cancer Res. 45:539-543.
6. Deloria, L., V. Abbott, N. Gooderham, and G.J. Mannering (1985) Induction of xanthine oxidase and depression of cytochrome P-450 by interferon inducers: Genetic differences in the responses of mice. Biochem. Biophys. Res. Commun. 131:109-114.
7. Elson, C.E., T.H. Maltzman, J.L. Boston, and M.A. Tabber (1988) Anticarcinogenic activity of d-limonene during initiation and promotion/progression stages of DMBA-induced mammary carcinogenesis. Carcinogenesis 9:331-332.
8. Fenwick, G.R., and A.B. Hanley (1985) The genus Allium. CRC Crit. Rev. Food Sci. Nutrit. 22:199-271.
9. Fenwick, G.R., and A.B. Hanley (1985) The genus Allium. Part 2. CRC Crit. Rev. Food Sci. Nutrit. 22:273-377.
10. Fenwick, G.R., and A.B. Hanley (1985) The genus Allium. Part 3. CRC Crit. Rev. Food Sci. Nutrit. 23:1-73.

11. Fenwick, G.R., R.K. Heaney, and W.J. Mullin (1983) Glucosinolates and their breakdown products in food and food plants. CRC Crit. Rev. Food Sci. Nutr. 18:123-201.

12. Fiala, E.S., G. Bobotas, C. Kulakis, L.W. Wattenberg, and J. Weisburger (1977) The effects of disulfiram and related compounds on the in vivo metabolism of the colon carcinogen 1,2-dimethylhydrazine. Biochem. Pharmacol. 26:1763-1768.

13. Gooderham, N.J., and G.J. Mannering (1986) Depression of cytochrome P-450 and alterations of protein metabolism in mice treated with the interferon inducer polyriboinosinic acid, polyribocytidylic acid. Arch. Biochem. Biophys. 250:418-425.

14. Gupta, I., J. Baptista, W.R. Bruce, C.T. Che, R. Furrer, J.S. Gingerich, A.A. Grey, L. Marai, P. Yates, and J.J. Krepinsky (1983) Structures of fecapentaenes, the mutagens of bacterial origin isolated from human feces. Biochemistry 22:241-245.

15. Jagerstad, M., A.L. Reutersward, S. Griuas, K. Olson, C. Neigishi, and S. Sato (1985) Effects of meat composition and cooking conditions on the formation of mutagenic imidazoquinoxalines. In Diet, Nutrition and Cancer, Y. Hayashi, M. Nagao, T. Sugimura, S. Takayama, L. Tomatis, L. Wattenberg, and G. Wogan, eds. Japan Science Societies Press, Tokyo, pp. 87-96.

16. Miller, E.C., and J.A. Miller (1979) Naturally occurring chemical carcinogens that may be present in foods. Int. Rev. Biochem. 27:123-165.

17. Mirvish, S.S. (1981) Inhibition of the formation of carcinogenic N-nitroso compounds by ascorbic acid and other compounds. Cancer Achievements, Challenges and Prospects for the 1980's, J.H. Burchenal and H.F. Oettgen, eds. Grune and Stratton, New York, pp. 557-588.

18. Ough, C.S. (1976) Ethyl carbamate in fermented beverages and foods. 1. Naturally-occurring ethyl carbamate. J. Agric. Food Chem. 24:323-331.

19. Shaw, P.E. (1979) Review of quantitative analyses of citrus essential oils. Agric. Food Chem. 27:246-257.

20. Solzin, U.J. (1977) The analysis of essential oils and extracts (oleoresins) from seasonings--A critical review. CRC Crit. Rev. Food Sci. Nutrit. 9:345-373.

21. Sparnins, V.L., G. Barany, and L.W. Wattenberg (1988) Effects of organo-sulfur compounds from garlic and onions on benzo(a)pyrene-induced neoplasia and glutathione S-transferase activity. Carcinogenesis 9:131-134.

22. Sparnins, V.L., A.W. Mott, G. Barany, and L.W. Wattenberg (1986) Effects of allyl methyl trisulfide on glutathione S-transferase activity and BP-induced neoplasia in the mouse. Nutrit. Cancer 8:211-215.

23. Wargovich, M.J. (1987) Diallyl sulfides, a flavor component of garlic (Allium sativum) inhibits dimethylhydrazine-induced colon cancer. Carcinogenesis 8:487-489.

24. Wargovich, M.J., C. Woods, V.W.S. Eng, L.C. Stephens, and K. Gray (1988) Chemoprevention of N-nitrosomethylbenzylamine-induced esophageal cancer in rats by the naturally-occurring thioether, diallyl sulfide. Cancer Res. 48:6872-6875.

25. Wattenberg, L.W. (1975) Inhibition of dimethylhydrazine-induced neoplasia of the large intestine by disulfiram. J. Natl. Cancer Inst. 54:1005-1006.

26. Wattenberg, L.W. (1977) Inhibition of carcinogenic effects of polycyclic hydrocarbons by benzyl isothiocyanate and related compounds. J. Natl. Cancer Inst. 58:395-398.

27. Wattenberg, L.W. (1980) Inhibition of polycyclic aromatic hydrocar-
     bon-induced neoplasia by sodium cyanate. Cancer Res. 40:232-234.
28. Wattenberg, L.W. (1981) Inhibition of carcinogen-induced neoplasia
     by sodium cyanate, tert-butylisocyanate and benzyl isothiocyanate
     administered subsequent to carcinogen exposure. Cancer Res.
     41:2991-2994.
29. Wattenberg, L.W. (1983) Inhibition of neoplasia by minor dietary
     constituents. Cancer Res. 43:2448S-2453S.
30. Wattenberg, L.W. (1985) Chemoprevention of cancer. Cancer Res.
     45:1-8.
31. Wattenberg, L.W. (1987) Inhibitory effects of benzyl isothiocyanate
     administered shortly before diethylnitrosamine or benzo(a)pyrene on
     pulmonary forestomach neoplasia in A/J mice. Carcinogenesis 8:1971-
     1973.
32. Wattenberg, L.W., V.L. Sparnins, and G. Barany (1989) Inhibition
     of N-nitrosodiethylamine carcinogenesis by naturally-occurring
     organosulfur compounds and monoterpenes. Cancer Res. 49:2689-
     2692.
33. Wattenberg, L.W., H.W. Schafer, L. Waters, and D.W. Davis (1989)
     Inhibition of mammary tumor formation by broccoli and cabbage.
     Proc. Am. Assoc. Cancer Res. 30:181.
34. Wattenberg, L.W., A.B. Hanley, G. Barany, V.L. Sparnins, L.K.
     Lam, and G.R. Fenwick (1985) Inhibition of carcinogenesis by some
     minor dietary constituents. In Diet, Nutrition and Cancer, Y.
     Hayashi et al., eds. Japan Science Societies Press, Tokyo, pp. 193-
     203.
35. Wattenberg, L.W., unpublished data.

NEWLY RECOGNIZED ANTICARCINOGENIC FATTY ACIDS

Michael W. Pariza and Yeong L. Ha

Food Research Institute
Department of Food Microbiology and Toxicology
University of Wisconsin, Madison
Madison, Wisconsin 63706

ABSTRACT

Evidence leading to the recognition of the anticarcinogenic activity of the conjugated dienoic derivatives of linoleic acid (CLA) is reviewed. New data indicate that CLA has potent antioxidant activity. Because the c-9,t-11 CLA isomer is esterified in phospholipid, it may represent a heretofore unrecognized in situ defense mechanism against membrane attack by oxygen radicals.

INTRODUCTION

In the quest to conquer cancer, the search for safe and effective anticarcinogens--substances that inhibit cancer or reduce its severity--is an important endeavor. Many potential anticarcinogens have already been identified in various foods, most notably those of plant origin. This paper will review the data on a newly recognized cancer inhibitor isolated first from an unexpected food source: beef fat.

DISCOVERY OF ANTICARCINOGENIC ACTIVITY OF CONJUGATED LINOLEIC ACID

We were the first to report that extracts of fried ground beef contained a mutagenic inhibitory activity (7). The observation was notable in that the inhibitory activity appeared to act at the stage of metabolic activation, and it seemed to be selective. Specifically, mutagenesis mediated by normal rat liver S-9 mix was inhibited, whereas when S-9 mix from rat liver treated with Aroclor 1254 was used no inhibition was seen. Hence, it appeared that the inhibition was not due to a substance that was merely interfering with the assay.

The mutagenic inhibitory activity was partially purified and tested against rat liver-mediated mutagenesis by 2-amino-3-methylimidazo[4,5-f]-quinoline (IQ) and 2-aminofluorene (8). A differential inhibitory effect was observed depending on the mutagen and the rat liver S-9 mix that was utilized. For IQ, inhibition was evident when liver S-9 mix from normal, phenobarbital-treated, or Aroclor-treated rats was employed, although in keeping with previous findings the Aroclor S-9 mix was more "resistant" to the inhibitory effects.

By contrast, with 2-aminofluorene, mutagenesis mediated by normal rat liver S-9 mix was inhibited, whereas that mediated by liver S-9 mix from Aroclor-treated rats was enhanced. There was no effect when liver S-9 mix from phenobarbital-treated rats was used. Based on these findings we began referring to the inhibitory activity as a mutagenesis modulating activity (8).

The effect of the mutagenesis modulator activity on carcinogenesis in animals was then investigated (6). The first model selected was the well-studied two-stage mouse epidermal carcinogenesis system, and the carcinogen used was 7,12-dimethylbenz[a]anthracene (DMBA). We found that a partially purified extract containing mutagenesis modulator activity reduced both the number of papillomas per mouse and the number of mice with papillomas. An anticarcinogenic principal was then purified from the grilled ground beef extracts and identified as a mixture of linoleic acid isomers, containing a conjugated double bond system (3). Because these fatty acid isomers are all derivatives of linoleic acid, we suggested referring to them collectively with the acronym CLA (for conjugated linoleic acid).

The observation that these newly recognized anticarcinogens are related chemically to the essential fatty acid, linoleic acid, is of interest. Linoleic acid is the only fatty acid that has been proved to enhance carcinogenesis in experimental animals (5).

Recently we found that CLA is effective in inhibiting carcinogen-induced forestomach neoplasia in mice (M.W. Pariza and Y.L. Ha, unpubl. data), using methods described in Ref. 1. In these experiments CLA was given by stomach tube 4 and 2 da prior to administration of a carcinogen, benzo(a)pyrene. This cycle was then repeated each week for 4 wk. The mice were sacrificed 20 wk after the last such treatment and the number of forestomach neoplasms determined. CLA treatment reduced forestomach neoplasia by 46-67% relative to controls (animals given linoleic acid or olive oil).

CONJUGATED LINOLEIC ACID SOURCES

We have found (M.W. Pariza and Y.L. Ha, ms. sub. for publ.) that following administration of CLA by stomach tube to mice, the material is incorporated into phospholipid. For example, in mice fed standard laboratory diets, only very low levels of CLA are present in the membranes of the cells in the forestomach. However, following administration of CLA by gastric tube using the protocol described earlier, the level in the forestomach cell membranes increased 300-fold. Moreover, one isomer (c-9,t-11-octadecadienoic acid) represented over 95% of the incorporated material, even though a mixture of nine isomers had been administered. These findings point out the importance of investigating dietary sources of CLA.

In addition to grilled beef, CLA is also present in dairy products, particularly cheeses (4). The CLA content varied considerably from product to product, but one cheese spread contained twice the CLA concentration of grilled hamburger. CLA may be present in other cooked foods as well. In the U.S. CLA consumption may be at least one gram per person per day (3,4).

The diet, however, is not the only source of CLA. There is strong evidence (2) that CLA is produced endogenously via carbon-centered free radical oxidation of linoleic acid. Further, certain serum proteins may mediate this reaction (2). The CLA content of human blood appears to be modulated by oxidative stress. For example, it is elevated in the blood of alcoholics and decreases during periods of abstinence (2).

We propose that the modulation of CLA production in the body by oxidative stress is more than just a consequence of such stress. Our working hypothesis is that the generation of CLA is in fact the key to a feedback mechanism that helps protect cells from the damaging effects of oxygen radicals.

CONJUGATED LINOLEIC ACID AS AN ANTIOXIDANT

In studies on the mechanism of action of CLA we have found that it is an effective antioxidant (M.W. Pariza and Y.L. Ha, ms. sub. for publ.). In a water/ethanol system involving the incubation of linoleic acid for 14 da under air (9), CLA reduced the oxidation of the oxidizable substrate (linoleic acid) by 86%. By comparison, alpha-tocopherol reduced oxidation by only 63%, whereas butylated hydroxytoluene (BHT) reduced oxidation by 92%. Hence, under these test conditions (where the ratio of linoleic acid to test antioxidant was 1,000 to 1) CLA was a more potent antioxidant than alpha-tocopherol, and almost as effective as BHT.

These findings lead to the conclusion that the in vivo formation and action of CLA may represent a feedback loop which serves to protect the cell from oxidative damage. Dietary CLA may also be important in this regard.

ACKNOWLEDGEMENTS

Supported in part by grants from the National Cancer Institute, DHHS; the Wisconsin Milk Marketing Board; and gift funds administered by the University of Wisconsin-Madison Food Research Institute.

REFERENCES

1. Benjamin, H., J. Storkson, P.G. Tallas, and M.W. Pariza (1988) Reduction of benzo[a]pyrene-induced forestomach neoplasms in mice given nitrite and dietary soy sauce. Food Chem. Toxicol. 26:671-678.
2. Dormandy, T.L., and D.G. Wickens (1987) The experimental and clinical pathology of diene conjugation. Chem. Phys. Lipids 45:353-364.
3. Ha, Y.L., N.K. Grimm, and M.W. Pariza (1987) Anticarcinogens from fried ground beef: Heat-altered derivatives of linoleic acid. Carcinogenesis 8:1881-1887.

4. Ha, Y.L., N.K. Grimm, and M.W. Pariza (1989) Newly recognized anticarcinogenic fatty acids: Identification and quantification in natural and processed cheeses. J. Agric. Food Chem. 37:75-81.
5. Pariza, M.W. (1988) Dietary fat and cancer risk: Evidence and research needs. Annu. Rev. Nutr. 8:167-183.
6. Pariza, M.W., and W.A. Hargraves (1985) A beef-derived mutagenesis modulator inhibits initiation of mouse epidermal tumors by 7,12-dimethylbenz[a]anthracene. Carcinogenesis 6:591-593.
7. Pariza, M.W., S.H. Ashoor, F.S. Chu, and D.B. Lund (1979) Effects of time and temperature on mutagen formation in pan fried hamburger. Cancer Lett. 7:63-69.
8. Pariza, M.W., L.J. Loretz, J.M. Storkson, and N.C. Holland (1983) Mutagens and modulator on mutagenesis in fried ground beef. Cancer Res. 43(Suppl.):2444s-2446s.
9. Ramarathnam, N., T. Osawa, M. Namiki, and S. Kawakishi (1988) Chemical studies on novel rice hull antioxidants. 1. Isolation, fractionation, and partial characterization. J. Agric. Food Chem. 36:732-737.

CALORIC RESTRICTION IN EXPERIMENTAL CARCINOGENESIS

David Kritchevsky

The Wistar Institute of Anatomy and Biology
Philadelphia, Pennsylvania 19104

The inhibiting influence of energy restriction on tumor growth has been known for 80 years. The early studies involved underfeeding of the test animals and were flawed in the sense that all macro- and micronutrients were reduced proportionately, which opened the results to the speculation that tumorigenesis might have been influenced by lack of a necessary growth factor rather than by calorie restriction per se. Current studies involve the use of semipurified diets which are designed specially to be replete in minerals and vitamins and to exert restriction at the expense of specified macronutrients. It is noteworthy that the older studies gave results similar to those obtained in the recent, more "sophisticated" experiments. There are available several reviews of the effects of caloric (energy) restriction on tumorigenesis (1,18,19,43).

Moreschi (25) was the first investigator to report that underfeeding inhibited the growth of transplanted sarcomas in mice (Tab. 1). A few years later, Rous (28,29) showed that underfeeding inhibits the growth of spontaneous tumors as well as the proliferation of tumors from remnants remaining after excision of the bulk of the tumor. Tumors recurred in 83% of the mice fed freely and in only 41% of those which were underfed.

Dietary restriction was shown to reduce spontaneous mammary tumors (34,40), pulmonary tumors (22), and lymphoid leukemia (33) in mice. Bischoff et al. (4) found that a 50% reduction in calories reduced the growth of mouse sarcoma 180; 33% caloric reduction had a slightly inhibiting effect; and 20% reduction had none. The first systematic investigation into the influence of diet and underfeeding on tumorigenesis was begun by Tannenbaum in the 1940s. In his first study (36), Tannenbaum found that underfeeding inhibited the growth of chemically-induced or spontaneous tumors in mice. The age at which underfeeding was instituted had a slight effect on outcome, but it was underfeeding that principally inhibited tumor growth (37).

In his later studies, Tannenbaum (37) turned to caloric restriction rather than underfeeding. The diet consisted of fox chow, skim milk powder, and cornstarch; calories were restricted by reducing the amount

Tab. 1. Effect of underfeeding on growth of sarcoma 7 in mice (after Moreschi [25]).

Group	Number	Food Intake (g/day)	Weight Change (g)	Tumor Weight (g)
1	13	1.0	-4.2	1.29 ± 0.24
2	7	1.5	-1.9	3.61 ± 0.54
3	8	2.0	-2.4	5.18 ± 0.46
4	8	Ad lib	+1.8	7.59 ± 0.81

of cornstarch. In these experiments, induced epithelial tumors or sarcomas and spontaneous lung or mammary tumors were reduced in DBA, ABC, Swiss, and C57 Black mice (Tab. 2).

Our interest in this area of research was stimulated by the report of Lavik and Baumann (23) that a diet high in fat but low in calories was about half as tumor-promoting as one low in fat but high in calories (Tab. 3). This finding, along with those of Tannenbaum (38) and Boutwell et al. (5), suggested that calories per se might be more important to tumor growth than dietary fat.

In our first study (13,20), we examined the effects of 40% caloric restriction on the development of 7,12-dimethylbenz(a)anthracene (DMBA)-induced mammary tumors in female Sprague-Dawley rats and of 1,2-dimethylhydrazine (DMH)-induced colon tumors in male F-344 rats. Table 4 describes the diets used. The control diet provided 358 kcal/100 g, with 0.011 g of fat per kcal; the calorie-restricted diet as constituted provided 363 kcal/100 g. But when fed at 60% of the control level, the calorie-

Tab. 2. Influence of caloric restriction on tumor formation (after Tannenbaum [37]).

Mouse Strain	Reduction of Tumorigenicity (%)*			
	Epithelial Induced	Sarcoma Induced	Mammary Spontaneous	Lung Spontaneous
ABC	58	--	--	47
Swiss	37	55	--	81
DBA	66	--	79	—
C57 Black	--	36	—	--

* Compared to control incidence

Tab. 3. Effect of fat and calories on methylcholanthrene-induced skin tumors in mice (after Lavik and Baumann [23]).

Regimen		Tumor Incidence
Calories	Fat	(%)
High	High	66
High	Low	54
Low	High	28
Low	Low	0

MCA applied twice weekly for 10 weeks. Study terminated at 6 months.

restricted diet provided 218 kcal, with 0.036 g of fat per kcal. The results of studies with saturated (coconut oil or butter oil) and unsaturated (corn oil) fat demonstrate the inhibitory effects of energy on both mammary and colon tumors (Tab. 5) and also confirm the observation of Carroll and Khor (6) on the different co-carcinogenic effects of saturated and unsaturated fats.

Tab. 4. Composition of diets used in caloric restriction study (after Klurfeld et al. [13]).

Ingredient	Ad Libitum	Restricted	
		As Formulated	As Fed
Sucrose	59.0	25.0	15.0
Casein	21.6	36.2	21.7
DL-Methionine	0.3	0.5	0.3
Corn Oil	4.1	13.1	7.9
Cellulose	10.1	16.8	10.1
Mineral Mix	3.8	6.4	3.8
Vitamin Mix	1.0	1.7	1.0
Choline Dihydrogen Citrate	0.2	0.3	0.2
Kcal/100 g	358	363	217

Tab. 5.   Influence of 40% caloric restriction on DMBA-induced mammary
tumors in female Sprague-Dawley rats and on DMH-induced
colon tumors in male F344 rats (after Ref. 13 and 20, and
unpubl. data).

Regimen	% Fat	Incidence (%)	Multiplicity*
Mammary Tumors			
Ad Lib	4.0 a	38	2.8 ± 0.5
Restricted	7.9	0	$\overline{0}$
Ad Lib	4.0 b	80	4.0 ± 0.5
Restricted	7.9	20	1.0 ± 0.0
Colon Tumors			
Ad Lib	4.0 c	85	1.8 ± 0.3
Restricted	7.9	35	1.0 ± 0.0
Ad Lib	4.0 b	100	3.5 ± 0.4
Restricted	7.9	53	2.1 ± 0.6

*   Tumors per tumor-bearing rat

a)   Coconut oil plus 1% corn oil
b)   Corn oil
c)   Butter oil plus 1% corn oil

We next addressed the question of the level of caloric restriction at
which a significant reduction in DMBA-induced tumorigenesis might be
seen.   We fed diets in which calories were restricted by 10, 20, 30, and
40% and in which the fat (corn oil) content was 5%, similar to that in the
control diet.   Thus, the test groups were consuming fewer calories but
slightly higher levels of corn oil.   The control diet provided 0.0136 g of
fat per kcal, and the diets restricted by 10, 20, 30, and 40% provided
0.0155, 0.0177, 0.020, and 0.024 g of fat/kcal, respectively.   The tumor
data are summarized in Tab. 6 (15).   It is evident that tumor multiplicity
and tumor burden are decreased with decreasing energy intake.   The insulin level in the control rats was 122 ± 16 μU/ml compared to 42 ± 5 and
41 ± 8 μU/ml in rats whose caloric intake was reduced by 30 or 40%.
Analysis of body composition showed that, while weight gain was reduced
proportionately to the degree of caloric restriction, body fat was reduced
by 16, 43, 63, and 72% in rats whose caloric intake was reduced by 10,
20, 30, or 40%, respectively.   These findings suggested a possible relationship between body fat and tumorigenicity.

In a subsequent study (16), we investigated the effects of caloric
restriction on DMBA-induced tumors in rats fed high levels of fat.   Rats
were fed ad libitum on diets containing 5, 15, or 20% corn oil.   Two other
groups of rats were pair-fed to the last two ad libitum groups on diets
containing 25% fewer calories and 20 or 26.7% corn oil, respectively.

Tab. 6. Effect of stepwise reduction of calories on DMBA-induced mammary tumors in female Sprague-Dawley rats (after Klurfeld et
al. [15]).

Group[a]	Incidence (%)	Multiplicity	Tumor Weight (g)	Tumor Burden (g)
Ad Lib	60	4.7 ± 1.3	2.0 ± 0.8	10.1 ± 3.3
10% Restricted	60	3.0 ± .8	1.8 ± 0.5	5.4 ± 3.0
20% Restricted	40*	2.8 ± 0.7	1.9 ± 0.7	4.7 ± 1.9*
30% Restricted	35*	1.3 ± 0.3	0.7 ± 0.6	0.9 ± 0.8*
40% Restricted	5*	1.0	—	—
	$p < 0.005$	$p > 0.05$	$p > 0.05$	$p < 0.05$

* Significantly different from ad lib value
a Twenty rats/group

Thus, the paired groups ingested identical amounts of corn oil. The
necropsy data are given in Tab. 7. In the ad libitum-fed groups, increasing the level of dietary fat led to increased tumor incidence, multiplicity, and tumor burden. Rats whose calories were restricted by 25%
but whose fat intake was three or four times that of the group fed 5% fat
ad libitum had lower tumor incidence, weight, and multiplicity. Their insulin levels (100 ± 12 or 117 ± 13 μU/ml) were significantly lower than
those of rats freely fed 5, 15, or 20% fat. Insulin levels (μU/ml) of the
ad libitum-fed rats were 143 ± 16, 164 ± 15, and 158 ± 11, respectively.
We did not determine the carcass composition of the rats used in this experiment, but the weight of the retroperitoneal fat pads of the calorie-
restricted rats fed 20 or 26.7% fat were 70 and 62% lower than those of
their freely fed counterparts.

The last study also suggested a possible connection between body fat
and tumorigenicity; so we next examined the influence of caloric restriction on DMBA-induced mammary tumors in a strain of genetically obese female rats (LA/N-cp). Obese rats and their phenotypically lean littermates
were fed ad libitum, and one group of obese rats was subjected to 40%
caloric restriction (17). As Tab. 8 shows, decreasing energy intake in
the obese rats reduced tumor incidence by 73%. Insulin levels were also
reduced by caloric restriction, but the level of carcass lipid was not.

This finding casts doubt on a possible direct connection between
body fat level per se and susceptibility to tumorigenesis, and suggests
that lean body mass may also be correlated with the risk of tumor development. Recent epidemiologic studies (2,35) indicate that this may also
be the case for man. Several recent studies (10,39,46) have not found a
significant independent effect of dietary fat on occurrence of breast
cancer in women. DeWaard and Trichopoulos (7) and Kato et al. (11)
suggest that reduced incidence of breast cancer might be achieved by
minimizing body mass and fatness and decreasing the rate of growth.

Tab. 7. Effect of 25% caloric restriction on DMBA-induced mammary tumors in female Sprague-Dawley rats (after Klurfeld et al. [16]).

Group[a]	Incidence (%)	Multiplicity	Tumor Weight (g)	Tumor Burden (g)
Ad Lib				
5% Corn Oil	65	1.9 ± 0.3	2.0 ± 0.7	4.2 ± 1.9
15% Corn Oil	85	3.0 ± 0.6	2.3 ± 0.7	6.6 ± 2.7
20% Corn Oil	80	4.1 ± 0.6	2.9 ± 0.5	11.8 ± 3.2
Restricted				
20% Corn Oil	60	1.9 ± 0.4	0.8 ± 0.2	1.5 ± 0.5
26.7% Corn Oil	30	1.5 ± 0.3	1.4 ± 0.1	2.3 ± 1.6
	$p < 0.005$	$p < 0.0001$	$p < 0.0001$	$p < 0.0001$

a Twenty rats/group

We also investigated the influence of duration or time of institution of caloric restriction on DMBA-induced tumorigenicity (21). In addition to groups of rats fed ad libitum or subjected to 25% energy restriction for the entire four-month period of the experiment, other groups were restricted for the first one or two months, the second and third month, or the third and fourth months. The data are presented in Tab. 9. Within the framework of this study, the duration of caloric restriction is important, but so is the experimental period during which the restriction is

Tab. 8. Necropsy data: DMBA-induced mammary tumors in female LA/N-cp (corpulent) rats (after Klurfeld et al. [17]).

| Regimen | Group | | |
	Obese Ad Lib	Obese 40% Restricted	Lean Ad Lib
Final Weight, g	538 ± 9	364 ± 8	210 ± 4
Tumor Incidence (%)	100	27	21
Multiplicity	4.1 ± 0.6	1.5 ± 0.4	1.0 ± 0.0
Tumor Weight (g)	5.3 ± 1.4	6.4 ± 3.2	1.7 ± 0.4

Tab. 9. Effect of variable (25%) caloric restriction on DMBA-induced mammary tumors in female Sprague-Dawley rats (after Kritchevsky et al. [21]).

Regimen* (Group)	Palpable Tumor Incidence (%)	Weight Gain (g)	Total Caloric Intake (kcal)	Feed Efficiency $(\times 10^2)$
(A) A-A-A-A	50	156	7508	2.08
(B) R-R-R-R	20	76	5624	1.35
(C) R-A-A-A	60	152	7401	2.05
(D) R-R-A-A	40	126	6746	1.87
(E) A-R-R-A	45	126	6691	1.88
(F) A-A-R-R	30	99	6958	1.42
C^a		r=0.96	r=0.83	r=0.94

* A: ad libitum; R: restricted. Each letter = 1 month.
a Correlation coefficient with tumor incidence

imposed. Group C was restricted for only the first month, and the rats approached the weight of the ad libitum-fed controls within a few weeks. The feed efficiencies of groups A and C were identical. Groups D, E, and F were all restricted calorically for two of the four months of the study. Groups D and E were on ad libitum ration for the last two months of the study. They showed similar tumor incidence and feed efficiency, which is reflected in the observation that they never approached the

Tab. 10. Effects of caloric restriction (CR) and exercise on DMH-induced carcinogenesis in male F344 rats (after Klurfeld et al. [14]).

Regimen*	Incidence (%)	Multiplicity	Body Weight (% of Control)	Body Fat (g)
Sedentary - AL	75	2.1 ± 0.4	—	21.8
Exercised - AL	36	1.3 ± 0.2	82	21.6
Sedentary - 25% CR	35	1.3 ± 0.2	75	13.0
Exercised - 25% CR	29	1.1 ± 0.1	63	8.3
Sedentary - 40% CR	21	1.2 ± 0.2	60	8.0

* AL: ad libitum; CR: calorie restricted

weight of the ad libitum-fed controls. Similarly, tumor incidence, while rising during the period of ad libitum feeding, did not reach the level of the controls. Group F was calorically restricted for the last two months of the study. The body weight of the rats plateaued at this point, as did tumor incidence. These findings suggest that caloric restriction later in the experimental period is most effective as an inhibitor of tumor growth.

Tannenbaum (37) has shown that underfeeding effectively reduced tumor incidence in mice even when imposed late in their life span. Weindruch et al. (41,42) restricted calories in one-year-old mice and found a decreased incidence of hepatomas and lymphomas as well as an increased life span.

Another means of increasing energy flux is exercise. An early study showed that transplanted tumors grew larger in sedentary mice than in mice placed in an activity wheel for 16 hours each day (32). We examined the effects of exercise on DMH-induced colon tumors in rats, with results shown in Tab. 10 (14). The rats were exercised on a treadmill (24 M/min x 60 min x 5 da/wk). Exercise reduced tumor incidence in ad libitum-fed rats by 52% and reduced body weight by 18%, but had no effect on body fat. Sedentary rats subjected to 25% caloric restriction exhibited the same tumor incidence as did the ad libitum-fed exercised rats, but weighed 9% less. Significantly, their body fat was 40% below that of the sedentary controls. The combination of 25% caloric restriction and exercise had a slight effect on tumor incidence, but reduced body fat by 36% compared to that of sedentary, calorie-restricted rats. These findings show that exercise reduces tumor incidence without affecting body fat, whereas caloric restriction reduces both tumor incidence and body fat. Apparently, a combination of these modalities must be used to reduce body fat effectively.

The mechanism(s) by which caloric restriction inhibits carcinogenesis remains moot. In a study of enzyme profiles of mammary tumors obtained from DMBA-treated rats, we found that the specific activities of hexokinase, pyruvate kinase, lactate dehydrogenase, glucose-6-phosphate dehydrogenase, malic enzyme, and fructose-1,6-bisphosphatase were elevated to varying degrees in large, palpable and small, nonpalpable tumors (less than 100 mg) from calorie-restricted rats compared to activities of these enzymes in tumors of control (ad libitum-fed) rats. Phosphofructokinase activity was increased in palpable tumors from calorie-restricted rats but reduced in nonpalpable tumors. The findings suggest compensatory changes in tumor enzyme profiles in response to the altered nutritional status of the host (30).

In the studies described above, we found that dietary fat had little effect on fasting serum insulin levels, whereas caloric restriction had a significant insulin-lowering effect. Further study (31) has shown that levels of serum insulin-like growth factor (IGF-I) decline with degree of caloric restriction in the early stages of caloric restriction but then revert to normal levels. Insulin levels decline shortly after institution of 40% caloric restriction, and we have seen significantly reduced insulin levels after termination of studies in which calories were restricted by 25% or more (15-17). The roles of insulin and IGF-I over the entire period of tumor promotion remain to be elucidated.

Results of many studies designed to test effects of specific nutrients in carcinogenesis can be explained as due to caloric inefficiency. White and Belkin (44) reported that transplanted tumors grew poorly in mice fed a low-nitrogen diet but over the course of the study (3 weeks), control mice gained 4 g, whereas those fed the low-nitrogen diet lost 4 g. A number of diets restricted in specific amino acids reduced tumorigenicity, but all inhibited growth of the mice (45). In a comparison of the effects of lactose, sucrose, and starch on DMBA-induced tumors, we found that rats fed lactose gained significantly less weight and had significantly fewer tumors than the rats fed sucrose or starch, in which weight gain and tumor incidence were virtually the same (12).

Doll and Peto (8) estimated that nutritional factors might be responsible for 35% of cancer deaths but also said, "It must be emphasized that the figure (35% of cancer related to diet) chosen is highly speculative and chiefly refers to dietary factors which are not yet reliably identified." More recently, Peto (26) has suggested that at the present time diet may be related to only 1% of cancer deaths.

The correlation of fat intake in developed countries with incidence of certain tumors may be, in reality, a reflection of increased caloric intake and decreased caloric expenditure rather than a connection with fat per se. Berg (3) suggested that hormone-dependent tumors are tumors of affluence. Garfinkel (9), in his review of overweight and cancer risk, concluded, "The regularity with which obesity is found as a risk factor in a number of studies implies but does not prove that high caloric intake may be a risk factor for a number of cancer sites." In a recent study of diet and colon cancer in Utah (24), the authors found a strong relationship between risk and caloric intake and stated, "Total energy intake must be evaluated before attempting to assign a causal role to any food or nutrient that may be postulated to play a role in colon cancer." It is interesting that in 1945 Potter (27) suggested exercise and reduction of caloric intake as possible means of cancer prevention.

The work on caloric restriction and cancer is in its third cycle. The first cycle was between 1909 and 1920; the second in the 1940s, principally the work of Tannenbaum; and the third began about 5-6 years ago. Every time, there has been a refinement in technique and an increased base of information. Perhaps the third cycle will be the charmed one, in which we will be able to elucidate mechanisms and explain the effects of caloric restriction on tumor growth.

ACKNOWLEDGEMENTS

This work was supported, in part, by a Research Career Award (HL00734) from the National Institutes of Health, and by funds from the Commonwealth of Pennsylvania.

REFERENCES

1. Albanes, D. (1987) Total calories, body weight and tumor incidence in mice. Cancer Res. 47:1987-1992.
2. Albanes, D., D.Y. Jones, A. Schatzkin, M.S. Micozzi, and P.R. Taylor (1988) Adult stature and risk of cancer. Cancer Res. 48:1658-1662.

3. Berg, J.W. (1975) Can nutrition explain the pattern of international epidemiology of hormone-dependent cancers? Cancer Res. 35:3345-3350.

4. Bischoff, F., M.L. Long, and L.C. Maxwell (1935) Influence of caloric intake upon the growth of sarcoma 180. Am. J. Cancer 24:549-553.

5. Boutwell, R.K., M.K. Brush, and H.P. Rusch (1949) The stimulating effect of dietary fat on carcinogenesis. Cancer Res. 9:741-746.

6. Carroll, K.K., and H.T. Khor (1971) Effect of level and type of dietary fat on incidence of mammary tumors induced in female Sprague-Dawley rats by 7,12-dimethylbenz(a)anthracene. Lipids 6:415-420.

7. DeWaard, F., and D. Tricchopoulos (1988) A unifying concept of the aetiology of breast cancer. Int. J. Cancer 41:666-669.

8. Doll, R., and R. Peto (1981) Avoidable risks of cancer in the United States. J. Natl. Cancer Inst. 66:1191-1308.

9. Garfinkel, L. (1985) Overweight and cancer. Ann. Intern. Med. 103:1034-1036.

10. Jones, D.Y., A. Schatzkin, S.B. Green, G. Block, L.A. Brinton, R.G. Ziegler, R. Hoover, and P.R. Taylor (1987) Dietary fat and breast cancer in the National Health and Nutrition Examination Survey. I. Epidemiological follow up study. J. Natl. Cancer Inst. 79:465-471.

11. Kato, I., S. Tominaga, and T. Suzuki (1988) Factors related to late menopause and early menarche as risk factors for breast cancer. Jpn. J. Cancer Res. (Gann) 79:165-172.

12. Klurfeld, D.M., M.M. Weber, and D. Kritchevsky (1984) Comparison of dietary carbohydrates for promotion of DMBA-induced mammary tumorigenesis in rats. Carcinogenesis 5:423-425.

13. Klurfeld, D.M., M.M. Weber, and D. Kritchevsky (1987) Inhibition of chemically-induced mammary and colon tumor promotion by caloric restriction in rats fed increased dietary fat. Cancer Res. 47:2759-2762.

14. Klurfeld, D.M., C.B.. Welch, E. Einhorn, and D. Kritchevsky (1988) Inhibition of colon tumor promotion by caloric restriction or exercise in rats. FASEB J. 2:A433.

15. Klurfeld, D.M., C.B. Welch, M.J. Davis, and D. Kritchevsky (1989) Determination of degree of energy restriction necessary to reduce DMBA-induced mammary tumorigenesis in rats during the promotion phase. J. Nutrit. 119:286-291.

16. Klurfeld, D.M., C.B. Welch, L.M. Lloyd, and D. Kritchevsky (1989) Inhibition of DMBA-induced mammary tumorigenesis by caloric restriction in rats fed high fat diets. Int. J. Cancer 42:922-925.

17. Klurfeld, D.M., L.M. Lloyd, C.L. Buck, M.J. Davis, O.L. Tulp, and D. Kritchevsky (1987) Inhibition of mammary tumorigenesis by caloric restriction in LA/N-cp (corpulent rats). Fed. Proc. 46:436.

18. Kritchevsky, D., and D.M. Klurfeld (1986) Influence of caloric intake on experimental carcinogenesis: A review. Adv. Exper. Mod. Biol. 206:55-68.

19. Kritchevsky, D., and D.M. Klurfeld (1987) Caloric effects in experimental mammary tumorigenesis. Am. J. Clin. Nutr. 45:236-242.

20. Kritchevsky, D., M.M. Weber, and D.M. Klurfeld (1984) Dietary fat versus caloric content in initiation and promotion of 7,12-dimethylbenz(a)anthracene-induced mammary tumorigenesis in rats. Cancer Res. 44:3174-3177.

21. Kritchevsky, D., C.B. Welch, and D.M. Klurfeld (1989) Response of

mammary tumors to caloric restriction for different time periods during the promotion phase. Nutrition and Cancer 12:259-269.

22. Larsen, C.D., and W.E. Heston (1945) Effects of cystine and caloric restriction on the incidence of spontaneous pulmonary tumors in strain A mice. J. Natl. Cancer Inst. 6:31-40.

23. Lavik, P.S., and C.A. Baumann (1943) Further studies on tumor promoting action of fat. Cancer Res. 3:749-756.

24. Lyon, J.L., A.W. Mahoney, D.W. West, J.W. Gardner, K.R. Smith, A.W. Sorenson, and W. Stanish (1987) Energy intake: Its relationship to colon cancer. J. Natl. Cancer Inst. 78:853-861.

25. Moreschi, C. (1909) Beziehungen zwischen ernahrung und tumorwachstum. Z. Immunitatsforsch. 2:651-675.

26. Peto, R. (1985) The preventability of cancer. In Cancer Risks and Prevention, M.P. Vessey and M. Gray, eds. Oxford University Press, Oxford, pp. 1-14.

27. Potter, V.R. (1945) The role of nutrition in cancer prevention. Science 101:105-109.

28. Rous, P. (1914) The influence of diet on transplanted and spontaneous mouse tumors. J. Exp. Med. 20:433-451.

29. Rous, P. (1915) The influence of dieting upon the course of cancer. Johns Hopkins Hosp. Bull. 26:146-148.

30. Ruggeri, B.A., D.M. Klurfeld, and D. Kritchevsky (1987) Biochemical alterations in 7,12-dimethylbenz(a)anthracene-induced mammary tumors from rats subjected to caloric restriction. Biochim. Biophys. Acta 929:239-246.

31. Ruggeri, B.A., D.M. Klurfeld, D. Kritchevsky, and R.W. Furlanetto (1989) Calorie restriction and DMBA-induced mammary tumor growth in rats: Alterations in circulating insulin, insulin-like growth factors I and II and epidermal growth factor. Cancer Res. (in press).

32. Rusch, H.P., and B.E. Kline (1944) The effect of exercise on the growth of a mouse tumor. Cancer Res. 4:116-118.

33. Saxton, J.A., Jr., M.C. Boon, and J. Furth (1944) Observations on the inhibition of development of spontaneous leukemia in mice by underfeeding. Cancer Res. 4:401-409.

34. Sivertsen, I., and W.H. Hastings (1938) A preliminary report on the influence of food and function on the incidence of mammary gland tumors in "A" stock albino mice. Minnesota Med. 21:873-875.

35. Swanson, C.A., D.Y. Jones, A. Schatzkin, L.A. Brinton, and R.G. Ziegler (1988) Breast cancer risk assessed by anthropometry in the NHANES-I epidemiological follow up study. Cancer Res. 48:5363-5367.

36. Tannenbaum, A. (1940) The initiation and growth of tumors. Introduction. I. Effects of underfeeding. Am. J. Cancer 38:335-350.

37. Tannenbaum, A. (1942) The genesis and growth of tumors. II. Effects of caloric restriction per se. Cancer Res. 2:460-467.

38. Tannenbaum, A. (1945) The dependence of tumor formation and the degree of caloric restriction. Cancer Res. 5:609-615.

39. Thind, I.S. (1988) Diet and cancer. An international study. Int. J. Epidemiol. 15:160-163.

40. Visscher, M.B., Z.B. Ball, R.H. Barnes, and I. Sivertsen (1942) The influence of caloric restriction upon the incidence of spontaneous mammary carcinoma in mice. Surgery 11:48-55.

41. Weindruch, R., and R.L. Walford (1982) Dietary restriction in mice beginning at 1 year of age: Effect of life span and spontaneous cancer incidence. Science 215:1415-1418.

42. Weindruch, R., S.R.S. Gottesman, and R.L. Walford (1982) Modifi-
     cation of age related immune decline in mice dietarily restricted from
     or after mid-adulthood. Proc. Natl. Acad. Sci., USA 79:898-902.
43. White, F.R. (1961) The relationship between underfeeding and tumor
     formation, transplantation, and growth in rats and mice. Cancer
     Res. 21:281-290.
44. White, F.R., and M. Belkin (1944) Source of tumor proteins. I. Ef-
     fect of a low nitrogen diet on the establishment and growth of a
     transplanted tumor. J. Natl. Cancer Inst. 5:261-263.
45. White, J., F.R. White, and G.B. Mider (1946) Effects of diets defi-
     cient in certain amino acids on the induction of leukemia in dba mice.
     J. Natl. Cancer Inst. 7:199-202.
46. Willett, W.C, M.J. Stampfer, G.A. Colditz, B.A. Rosner, C.H.
     Hennekens, and F.E. Speizer (1987) Dietary fat and the risk of
     breast cancer. N. Engl. J. Med. 316:22-28.

CARCINOGENICITY AND MODIFICATION OF CARCINOGENIC RESPONSE BY ANTIOXIDANTS

Nobuyuki Ito, Masao Hirose, Akihiro Hagiwara, and Satoru Takahashi

First Department of Pathology
Nagoya City University Medical School
1 Kawasumi, Mizuho-cho, Mizuho-ku
Nagoya 467, Japan

INTRODUCTION

Synthetic or naturally occurring antioxidants such as butylated hydroxyanisole (BHA), butylated hydroxytoluene (BHT), propyl gallate sodium L-ascorbate, and α-tocopherol have been widely used as food additives in various processed foods to prevent auto-oxidation of fatty acids. In addition, many naturally occurring antioxidants are present at appreciable levels in plants. In the light of studies which showed that antioxidants lack mutagenic activity and indeed even inhibit mutagenesis induced by carcinogens (13), they have been considered safe for use as food additives. In fact, since they have further been observed to inhibit chemical carcinogenesis in various organs when administered to rats concurrently with carcinogens (5,6,17), they have been considered as anticarcinogenic agents. However, BHA was recently found to be carcinogenic in the rat forestomach (9), and when antioxidants were given to rats after carcinogen exposure, they enhanced carcinogenesis in some organs while exerting an inhibitory influence in others (5-8). Therefore, antioxidants have both hazardous and nonhazardous effects in rodents, and by analogy also possibly in man.

This paper reports recent studies on the carcinogenic and modifying effects of synthetic and naturally-occurring antioxidants.

DOSE-RESPONSE STUDY OF BUTYLATED HYDROXYANISOLE IN RATS

Groups of 50 male F344 rats, initially 6 wk old (Charles River Japan, Inc., Kanagawa), were given an Oriental MF powdered diet (Oriental Yeast Co., Tokyo) containing BHA (>98% 3-tert-BHA and <2% 2-tert-BHA, Wako Pure Chemical Industries, Osaka) at levels of 0, 0.125, 0.25, 0.5, 1.0, or 2.0% for 104 wk. There was a significant decrease in final body weight compared with controls in all groups, except in those given 0.125% BHA in the diet. Table 1 shows the resultant incidences of proliferative

Tab. 1. Proliferative and neoplastic lesions of the forestomach epithelium
         in F344 rats given diet containing BHA.

BHA in diet (%)	No. of rats[a]	No. of rats with lesions (%)		
		Hyperplasia	Papilloma	Squamous cell carcinoma
0	50	0 (0)	0 (0)	0 (0)
0.125	50	1 (2)	0 (0)	0 (0)
0.25	50	7 (14)**	0 (0)	0 (0)
0.5	50	16 (32)***	0 (0)	0 (0)
1	50	44 (88)***	10 (20)**	0 (0)
2	50	50 (100)***	50 (100)***	11 (22)***

[a] Survived more than 50 weeks.

** $P < 0.01$, *** $P < 0.001$ vs control group

and neoplastic forestomach lesions. No squamous-cell carcinomas were induced at the 1.0% BHA level, although papillomas of the forestomach developed in 20% and 100% of the rats given a diet containing 1.0% and 2.0%
BHA, respectively. The incidence of epithelial hyperplasia of the forestomach also increased with the dose of BHA, to 100% at the highest dose
(10). These results thus demonstrated that the carcinogenic dose of BHA
in rats is 2% in diet, and that a clear dose dependency exists regarding
the development of hyperplasia and papillomas.

CARCINOGENICITY OF BUTYLATED HYDROXYANISOLE IN DIFFERENT SPECIES

Male F344 rats and male Syrian golden hamsters (Kagawa Tsuda Animal Farm, Kagawa) were given an Oriental MF pelleted diet containing
2.0% and 1.0% BHA, and male B6C3F$_1$ mice (Charles River Japan, Inc.,
Kanagawa) were treated with 1.0% and 0.5% BHA for up to 104 wk. The
results are shown in Tab. 2. In rats, papillomas were induced in 75.5%
and 91.5% of animals treated with 1% and 2% BHA, respectively, and carcinomas were again observed in the 2% group. Similar results were
gained with hamsters, where the 1% level was also sufficient to induce
malignant lesions. Although there were significant increases in the incidences of papillomas in mice given 0.5% and 1% BHA, the numbers of
squamous cell carcinomas did not significantly differ from those observed
in controls (15).

These results suggest that BHA is carcinogenic to rat and hamster
forestomach epithelium, and that the respective carcinogenic doses are 2%
and 1%. The data further indicate a possible weak carcinogenicity of BHA
for mouse forestomach epithelium.

Tab. 2.   Proliferative and neoplastic lesions of the forestomach epithelium in rats, hamsters, and mice given different levels of BHA.

Species	% BHA in diet	No. of animals[a]	No. of animals with lesions (%)		
			Hyperplasia	Papilloma	Squamous cell carcinoma
Rat	0	92	1 (1.1)	0 (-)	0 (-)
	1	94	92 (97.9)***	71 (75.5)***	0 (-)
	2	94	93 (98.9)***	86 (91.5)***	13 (13.8)***
Hamster	0	52	9 (17.3)	0 (-)	0 (-)
	1	55	53 (96.4)***	54 (98.2)***	4 (7.3)*
	2	40	40 (100)***	38 (95.0)***	4 (10.0)*
Mouse	0	39	0 (-)	0 (-)	0 (-)
	0.5	37	10 (27.0)***	5 (13.5)*	1 (2.7)
	1	43	35 (81.4)***	5 (14.3)*	2 (4.7)

a   Survived more than 48 weeks for rats, 64 weeks for hamsters and 88 weeks for mice.

*** $P < 0.001$ vs control group

* $P < 0.05$ vs control group

TUMORIGENICITY OF CAFFEIC ACID AND SESAMOL

Previously, we demonstrated that out of 8 naturally-occurring phenolic antioxidants, continuous oral treatment with 2% caffeic acid or 2% sesamol in the diet for 4 wk induced pronounced hyperplasia with epithelial damage in 6-wk-old male F344 rats (2). These findings prompted us to examine the long-term effects of these chemicals.

Groups of 6-wk-old male F344 rats were given Oriental MF powdered basal diet containing 2% BHA, 2% caffeic acid (purity >98%, Tokyo Chemical Industries, Tokyo), 2% sesamol (purity >98%, Tokyo Chemical Industries, Tokyo), 2% gallic acid (purity >98%, Tokyo Chemical Industries, Tokyo) or 0.5% BHT (purity >98%, Wako Pure Chemical Industries, Osaka) for 60 wk and then killed for histological examination.

Histological findings in the forestomach are summarized in Tab. 3. Hyperplasia was found in all rats treated with BHA, caffeic acid, or

Tab. 3.   Histological findings in the forestomach of rats treated with various antioxidants for 60 wk.

		No. of rats with lesions (%)		
Chemicals	No. of rats	Hyperplasia	Papilloma	Squamous cell carcinoma
BHA	19	19 (100)***	3 (15.8)	0
Caffeic acid	21	21 (100)***	6 (28.6)**	0
Sesamol	22	22 (100)***	4 (18.2)	1 (4.5)
Gallic acid	20	8 (40)	0	0
BHT	10	1 (10)	0	0
Control	19	8 (42.1)	0	0

***P<0.001, **P<0.002 vs control group

sesamol.   Papillomas were induced in 15.8%, 28.6% (p<0.02), and 18.2% of rats treated with BHA, caffeic acid, and sesamol, respectively.   In addition, one squamous cell carcinoma was found in a rat treated with sesamol.   No hyperplasias or tumors were observed in the glandular stomach.

Since caffeic acid and sesamol are as effective as carcinogenic BHA in inducing papillomas in forestomach epithelium, they might be expected to demonstrate complete carcinogenicity in the long term.   An experiment to clarify this point is now in progress.

REVERSIBILITY OF ANTIOXIDANT-INDUCED
FORESTOMACH HYPERPLASIA

The reversibility of BHA-induced forestomach lesions in rats had been studied by several investigators (12,14).   In our laboratory, the reversibility of forestomach lesions in rats treated with the genotoxic agents 2-(2-furyl)-3-(5-nitro-2-furyl)-acrylamide (AF2), 8-nitroquinoline (8NQ), or N-methyl-N'-nitro-N-nitrosoguanidine (MNNG), or with nongenotoxic BHA or caffeic acid were compared.   Groups of 20 male F344 rats aged 6 wk at the commencement were treated with 0.2% AF2 (Ueno Pharmaceutical Co., Osaka), 0.1% 8NQ (Nakarai Chem. Ltd., Kyoto), 2% BHA, or 2% caffeic acid in the diet, or a single intragastric dose of 20 mg/kg b.w. MNNG (Tokyo Chemical Industries, Tokyo) once a week for 24 wk.   Ten animals per group were sacrificed at the end of week 24, and another ten animals each were maintained on a basal diet alone for a further 24 wk and then sacrificed.   The results are summarized in Tab. 4.   The incidences of hyperplasia were low in rats treated with 8NQ and AF2 for 24 wk, but increased to 100% after a further 24 wk on basal diet alone. Squamous cell carcinomas were found in only 25% of rats at the cessation of MNNG treatment, but a 100% incidence was evident 24 wk later.   In

Tab. 4. Reversibility experiment: Histological findings in the forestomach epithelium.

Chemicals	Treatment(wk)		No. of	No. of rats with (%)		
	Chemicals	Basal diet	rats	Hyperplasia	Papilloma	Carcinoma
8NQ	24	0	10	2 (20)	0	0
	24	24	10	10 (100)	1 (10)	0
AF2	24	0	10	3 (30)	0	0
	24	24	8	8 (100)	0	0
MNNG	24	0	8	8 (100)	8 (100)	2 (25)
	24	24	6	6 (100)	6 (100)	6 (100)
BHA	24	0	9	7 (78)	0	0
	24	24	10	0	0	0
Caffeic acid	24	0	9	9 (100)	1 (11)	0
	24	24	10	0	0	0

contrast, whereas severe hyperplasia was observed in 78% and 100% of
animals treated with BHA and caffeic acid for 24 wk, respectively, it disappeared within 24 wk of cessation of chemical administration, although
mild hyperplasia was still present in these animals.

These results clearly showed that whereas genotoxic carcinogen-
induced forestomach lesions continue to proliferate and develop after
removal of the inducing stimulus, the nongenotoxic BHA- and caffeic acid-
induced lesions are reversible. Therefore it is likely that different
mechanisms underlie genotoxic and nongenotoxic carcinogenesis in forestomach epithelium, and that a continued proliferation stimulus is necessary for induction of squamous cell carcinomas by antioxidants.

MODIFICATION OF CARCINOGENESIS BY ANTIOXIDANTS

Modification by Simultaneous Treatment With Antioxidants

The modifying effects of concurrent treatment with high- or low-dose
BHA on wide-organ spectrum carcinogenesis have been studied in rats
(11).

Groups of 20 rats were treated with 2% or 0.04% BHA for 24 wk. After 2 wk of antioxidant administration the animals were given subcutaneous injections of 50 mg/kg body weight 3,2'-dimethyl-4-aminobiphenyl (DMAB, Matsugaki Pharmaceutical Co., Osaka) once a week, intragastric administrations of 200 mg/kg body weight 2,2'-dihydroxy-di-n-propylnitrosamine (DHPN, Nakarai Chemical Co., Osaka) once every 2 wk, or intraperitoneal injections of 15 mg/kg body weight N-methylnitrosourea (MNU, Katayama Chemical Co., Osaka) once every 2 wk for 22 wk. Further groups of 20 rats were treated with DMAB, DHPN, MNU, or 2% or 0.04% BHA alone. Surviving animals were killed at week 24 and all organs were examined histopathologically. The results are summarized in Tab. 5.

DMAB produced liver hyperplastic foci in 100% of the animals, but the BHA treatment dose-dependently decreased the incidences to 60% in the low dose and 25% in the high dose groups. DMAB did not induce tumors in the forestomach and urinary bladder, but concurrent treatment of 2% BHA clearly enhanced the incidence of forestomach papillomas (40%, p<0.01) and urinary bladder carcinomas (80%, p<0.001). Treatment with DHPN alone induced alveolar hyperplasia in 95% of rats, whereas concurrent treatment with 2% BHA exerted a significant inhibitory effect (15%, p<0.001). On the other hand, development of papillomas in the forestomach (p<0.05) was observed in animals receiving both DHPN and 2% BHA, but neither of these chemicals alone. Similarly, whereas simultaneous treatment with MNU and 2% BHA induced forestomach carcinomas in 75% of the animals (p<0.001), no malignant lesions were observed after MNU itself.

These results suggest that concurrent treatment with BHA can modify carcinogenesis irrespective of whether the carcinogens require metabolic activation. Thus, in addition to changes in metabolic pathways, modification may be caused by increased DNA synthesis in the target organ by BHA, or possibly by changes in DNA repair systems. It is very important that BHA was observed to potently modify carcinogenesis even at 1/50 of the carcinogenic dose. Therefore it is very possible that phenolic antioxidants may play a role in human carcinogenesis.

Modification by Antioxidants After Carcinogen Exposure

Modification of carcinogenesis by antioxidants after carcinogen exposure, namely, during the second stage of two-step carcinogenesis, has been demonstrated for many synthetic and naturally occurring antioxidants (5-8). Generally, different modifying effects (promotion and inhibition) have been observed in various target organs (Tab. 6). For example, BHA promotes forestomach and urinary bladder carcinogenesis but inhibits carcinogenesis in the liver, lung, and mammary gland of rats. On the other hand, BHT enhances carcinogenesis in the rat esophagus, urinary bladder, and thyroid gland, whereas it inhibits carcinogenesis in the lung and mammary gland. Recently, some phenolic antioxidants which are structurally similar to BHA were examined for their modifying effects on rat forestomach and glandular stomach carcinogenesis initiated with MNNG (17).

Groups of 20 rats were given an intragastric dose of 150 mg/kg body weight MNNG, and, starting 1 wk later, received a diet supplemented with 0.8% catechol (CC, purity >98%), 1.0% 2-tert-butyl-4-methylphenol (TBMP, purity >99%), 1.5% p-tert-butylphenol (PTBP, purity >95%), 1.5%

Tab. 5. Neoplastic and preneoplastic lesions in rats treated concurrently with BHA and carcinogens.

Treatment	No. of rats	Forestomach		Urinary bladder		Liver	Lung
		Papilloma	Carcinoma	Papilloma	Carcinoma	Hyperplastic focus	Alveolar hyperplasia
DMAB+BHA (2%)	15	6 (40)**	0	8 (53)***	12 (80)***	5 (25)***	0
DMBA+BHA (0.04%)	20	0	0	0	0	12 (60)**	0
DMBA	20	0	0	0	0	20 (100)	0
DHPN+BHA (2%)	20	6 (30)*	2 (10)	0	0	0	3 (15)***
DHPN+BHA (0.04%)	20	0	0	0	0	0	16 (80)
DHPN	20	0	0	0	0	0	19 (95)
MNU+BHA (2%)	20	6 (30)	15 (75)***	0	0	0	0
MNU+BHA (0.04%)	20	6 (30)	1 (5)	0	0	0	0
MNU	20	3 (15)	0	0	0	0	0
BHA (2%)	20	0	0	0	0	0	0
(0.4%)	20	0	0	0	0	0	0

*** $p < 0.001$, ** $p < 0.01$, * $p < 0.05$ as compared to the corresponding control group

Tab. 6. Modifying effects of antioxidants on carcinogenesis in rats after exposure to carcinogens.

| Target organ (carcinogen) | Antioxidants | | | | | |
| | Synthetic | | | Naturally occurring | | |
	BHA	BHT	EQ	SA	α-TP	CAT
Esophagus (DBN, MNAN)	→	↑	→	→	NE	↑
Forestomach (MNNG, MNU)	↑	→	→	↑	NE	↑
Glandular stomach (MNNG)	→	→	↑	→	→	↑
Colon (DMH)	→	→	→	↑	→	↓
Liver (EHEN)	↓	→	↓	→	→	NE
Lung (DHPN, MNAN)	↓	↓	↓	→	↓	↓
Kidney (EHEN)	→	→	↑	→	NE	NE
Urinary bladder (BBN, MNU, DBN)	↑	↑	→	↑	→	→
Mammary gland (DMBA)	↓	↓	↓	→	→	NE
Thyroid (MNU, DHPN)	→	↑	↑	→	→	→

↑, enhancement; →, no effect; ↓, inhibition; NE, not examined

methylhydroquinone (MHQ, purity >98%), 1.5% 4-methoxyphenol (4MP, purity >99%), or basal diet alone for 51 wk. These antioxidants had already been shown to induce hyperplasia in the forestomach epithelium or increase the labeling index of glandular stomach epithelial cells in hamsters (1). Further groups of 10-15 rats were maintained as controls without prior MNNG treatment. The results are summarized in Tab. 7 and 8.

The incidences of forestomach squamous cell carcinomas in MNNG-treated animals were significantly elevated by administration of diets containing CC (p<0.001), TBMP (p<0.001), or PTBP (p<0.01). Treatment with CC, TBMP, or 4MP alone, furthermore, induced forestomach hyperplasia at incidences of 86.7, 40, 93.3, and 100%, respectively. In the pyloric region of the glandular stomach, the development of adenomatous hyperplasias and adenocarcinomas after MNNG treatment was significantly enhanced by CC treatment (p<0.001). Moreover, treatment with CC alone induced a 100% yield of adenomatous hyperplasias and adenocarcinomas in 20% of the animals.

Of these phenolic antioxidants, catechol is of prime interest because it is widely distributed in our environment, being present in cigarette smoke, coffee, plants, hair dyes, and photographic developers. The modifying effects of catechol were also examined in rats treated with methyl-n-amylnitrosamine (MNAN) (18). Groups of 15 rats were given

Tab. 7.   Forestomach histological changes in rats treated with MNNG followed by antioxidants.

	No. of rats	No. of rats with lesions (%)		
Treatment	rats	Hyperplasia	Papilloma	Squamous cell carcinoma
MNNG → CC	19	19 (100)	18 (94.7)	19 (100)***
MNNG → TBMP	18	18 (100)	18 (100)	17 (94.4)***
MNNG → PTBP	20	20 (100)	19 (95)	15 (75)**
MNNG → MHQ	20	20 (100)	13 (65)	6 (30)
MNNG → 4MP	18	18 (100)	16 (88.9)	0
MNNG → BD	19	19 (100)	13 (68.4)	5 (26.3)
CC	15	13 (86.7)***	1 (6.7)	0
TBMP	15	6 (40)*	0	0
PTBP	15	14 (93.3)***	1 (6.7)	0
MHQ	15	0	0	0
4MP	15	15 (100)***	1 (6.7)	0
BD	10	0	0	0

*** $P<0.001$, ** $P<0.01$, * $P<0.05$ as compared to the corresponding control group

three intraperitoneal injections of 25 mg/kg b.w. MNAN within the initial 2-wk period and, commencing one week thereafter, were administered 0.8% catechol in powdered basal diet or the basal diet alone for 49 wk. Additional groups of 10-15 rats were similarly treated without prior carcinogen exposure.

Histological examination after sacrifice at week 52 revealed incidences of tongue papillomas and esophageal squamous cell carcinomas (57.1% and 64.3%, respectively) in the group given MNAN followed by catechol which were significantly higher than those in the carcinogen only controls (9.1% and 0%, respectively). The incidence of alveolar hyperplasia in the lungs of the animals given MNAN followed by catechol (0%) was, in contrast, significantly reduced as compared to the control value (54.5%).

These results thus indicated the possibility that the environmental contaminant catechol may play a role in the development of human upper gastrointestinal cancer, and that this antioxidant may also exert modifying effects in other organs.

Tab. 8. Glandular stomach histological changes in rats treated with
         MNNG followed by antioxidants.

Treatment	No. of rats	No. of rats with lesions at pyloric region(%)	
		Adenomatous hyperplasia	Adenocarcinoma
MNNG → CC	19	19 (100)***	18 (94.7)***
MNNG → TBMP	18	0	2 (11.1)
MNNG → PTBP	20	0	0
MNNG → MHQ	20	0	1 (5)
MNNG → 4MP	18	0	0
MNNG → BD	19	1 (5.3)	0
CC	15	15 (100)***	3 (20)
TBMP	15	0	0
PTBP	15	0	0
MHQ	15	0	0
4MP	15	0	0
BD	10	0	0

*** P<0.001, as compared to the corresponding control group

CONCLUSIONS

Since the carcinogenicity of BHA, a synthetic phenolic antioxidant,
was first demonstrated for rat and hamster forestomach epithelium, several naturally-occurring phenolic antioxidants such as caffeic acid,
sesamol, and catechol have been found to induce tumors in the glandular
stomach and/or forestomach of rats.

In addition to this complete carcinogenic activity, BHA and other
synthetic and naturally occurring antioxidants modify carcinogenesis in
various organs. Until recently, BHA has been considered a chemopreventive agent, because it often inhibits carcinogenesis when given to animals
simultaneously with carcinogens. However, the experiments reported
above demonstrated that BHA can strongly enhance DMAB- or MNU-induced forestomach and urinary bladder carcinogenesis when administered
concurrently. Furthermore, antioxidants exert independent modifying effects on second-stage carcinogenesis in different organs within the same
animal. For example, BHA enhances forestomach and urinary bladder
carcinogenesis in rats previously receiving MNNG, BBN, MNU, or DBN,

while, in contrast, inhibiting carcinogenesis in the liver, lung, and mammary gland of rats given N-ethyl-N-hydroxyethylnitrosamine, DHPN, or 7,12-dimethylbenz(a)anthracene. BHT, which is structurally similar to BHA, enhances lesion development in the esophagus and thyroid gland, whereas it reduces lung and mammary gland tumor yield. Both naturally-occurring and synthetic antioxidants modify the second stage of carcinogenesis. For example, catechol strongly enhances tongue, esophagus, forestomach, and glandular stomach carcinogenesis in rats initiated with MNAN, while inhibiting lung lesion development.

The mechanism(s) underlying antioxidant-induced carcinogenesis remains obscure. However, the finding that genotoxic 8NQ, AF2, or MNNG-induced forestomach changes progressively developed after withdrawal of carcinogens, whereas nongenotoxic BHA- or caffeic acid-induced forestomach hyperplasia regressed after cessation of antioxidant administration, indicated that genotoxic compounds cause irreversible changes, presumably involving forestomach DNA, while with nongenotoxic chemicals, continuous long-term proliferative stimuli are required for induction of tumors. In support of this conclusion, BHA and its metabolites were found not to bind to DNA (4) or to form DNA adducts in rat forestomach epithelium (16).

The results available indicate that antioxidants have not only beneficial but also harmful effects in small rodents, and possibly, by analogy, in man. It is important that efforts are made to find new antioxidants which have strong antioxidant activity and only chemopreventive effects on carcinogenesis. Rapid detection systems for carcinogens and/or modifiers of multiple-organ carcinogenesis are necessary for this purpose.

ACKNOWLEDGEMENTS

This work was supported by grants-in-aid for cancer research from the Ministry of Education, Science and Culture and from the Ministry of Health and Welfare, Japan, a grant from the Society for Promotion of Pathology, Nagoya, and a grant-in-aid from the Experimental Pathological Research Association, Nagoya.

REFERENCES

1. Hirose, M., T. Inoue, M. Asamoto, Y. Tagawa, and N. Ito (1986) Comparison of the effects of 13 phenolic compounds in induction of proliferative lesions of the forestomach and increase in the labelling indices of the glandular stomach and urinary bladder epithelium of Syrian golden hamsters. Carcinogenesis 7:1285-1289.
2. Hirose, M., A. Masuda, K. Imaida, M. Kagawa, H. Tsuda, and N. Ito (1987) Induction of forestomach lesions in rats by oral administration of naturally occurring antioxidants for 4 weeks. Jpn. J. Cancer Res. (Gann) 78:317-321.
3. Hirose, M., S. Fukushima, Y. Kurata, H. Tsuda, M. Tatematsu, and N. Ito (1988) Modification of N-methyl-N'-nitro-N-nitrosoguanidine-induced forestomach and glandular stomach carcinogenesis by phenolic antioxidants in rats. Cancer Res. 48:5310-5315.
4. Hirose, M., M. Asamoto, A. Hagiwara, N. Ito, H. Kaneko, K. Saito, Y. Takamatsu, A. Yoshitake, and J. Miyamoto (1987) Metabolism of 2- and 3-tert-butyl-4-hydroxyanisole (2- and 3-BHA) in the rat.

II: Metabolism in forestomach and covalent binding to tissue macro-molecules. Toxicology 45:13-24.

5. Ito, N., and M. Hirose (1987) The role of antioxidants in chemical carcinogenesis. Jpn. J. Cancer Res. (Gann) 78:1011-1026.

6. Ito, N., and M. Hirose (1989) Antioxidants--Carcinogenic and chemopreventive properties. Adv. Cancer Res. (in press).

7. Ito, N., S. Fukushima, and H. Tsuda (1985) Carcinogenicity and modification of the carcinogenic response by BHA, BHT, and other antioxidants. CRC Crit. Rev. Toxicol. 15:109-150.

8. Ito, N., S. Fukushima, and M. Hirose (1987) Modification of the carcinogenic response by antioxidants. In Toxicological Aspects of Foods, K. Miller, ed. Elsevier Applied Science, London and New York, pp. 253-293.

9. Ito, N., S. Fukushima, A. Hagiwara, M. Shibata, and T. Ogiso (1983) Carcinogenicity of butylated hydroxyanisole in F344 rats. J. Natl. Cancer Inst. 70:343-352.

10. Ito, N., S. Fukushima, S. Tamano, M. Hirose, and A. Hagiwara (1986) Dose-response in butylated hydroxyanisole induction of forestomach carcinogenesis in F344 rats. J. Natl. Cancer Inst. 77:1261-1265.

11. Ito, N., M. Hirose, M. Shibata, T. Shirai, and H. Tanaka (1989) Modifying effects of simultaneous treatment with different doses of butylated hydroxyanisole (BHA) on rat tumor development induced by the wide-spectrum carcinogens 3,2'-dimethyl-4-aminobiphenyl, 2,2'-dihydroxy-di-n-propylnitrosamine or N-methylnitrosourea (submitted for publication).

12. Iverson, F., E. Lok, E. Nera, K. Karpinski, and D.B. Clayson (1985) A 13 week feeding study of butylated hydroxyanisole: The subsquent regression of the induced lesion in male Fischer 344 rat forestomach epithelium. Toxicology 35:1-11.

13. Kahl, R. (1984) Synthetic antioxidants: Biochemical action and interference with radiation, toxic compounds, chemical mutagens and chemical carcinogens. Toxicology 33:185-228.

14. Masui, T., M. Asamoto, M. Hirose, S. Fukushima, and N. Ito (1987) Regression of simple hyperplasia and papillomas and persistence of basal cell hyperplasia in the forestomach of F344 rats treated with butylated hydroxyanisole. Cancer Res. 47:5171-5174.

15. Masui, T., M. Hirose, K. Imaida, S. Fukushima, S. Tamano, and N. Ito (1986) Sequential changes of the forestomach of F344 rats, Syrian golden hamsters, and B6C3F$_1$ mice treated with butylated hydroxyanisole. Jpn. J. Cancer Res. (Gann) 77:1083-1090.

16. Saito, K., A. Yoshitake, J. Miyamoto, M. Hirose, and N. Ito (1989) No evidence of DNA adduct formation in forestomach of rats treated with 3-BHA and its metabolites as assessed by an enzymatic ^{32}P-postlabeling method (submitted for publication).

17. Wattenberg, L.W. (1986) Chemoprevention of cancer. Cancer Res. 45:1-8.

18. Yamaguchi, S., M. Hirose, S. Fukushima, R. Hasegawa, and N. Ito (1989) Modification by catechol and resorcinol of upper digestive tract and lung carcinogenesis in rats treated with methyl-n-amylnitrosamine (submitted for publication).

INHIBITION OF TUMOR PROMOTION BY DL-α-DIFLUOROMETHYL-ORNITHINE, A SPECIFIC IRREVERSIBLE INHIBITOR OF ORNITHINE DECARBOXYLASE

Ajit K. Verma

Department of Human Oncology
University of Wisconsin Clinical Cancer Center
Madison, Wisconsin 53792

INTRODUCTION

Although a chemoprevention program can be targeted for intervention at any one of the steps (initiation, promotion, and progression) involved in the induction of cancer, interference with the promotion stage of carcinogenesis appears to be most appropriate and practical. The rationale behind this comes from the facts that promotion of tumor formation requires a repeated and prolonged exposure to a promoter and that tumor promotion is reversible, at least in the early stages. In contrast, initiation can be accomplished by a single exposure to a sufficiently small dose of a carcinogen, and this step is rapid and irreversible (3,36,41).

In light of the facts that humans are constantly exposed to environmental carcinogens and that even the human diet contains nitrites and nitrates which are converted in the gut to potent carcinogen nitrosamines (48), then one may believe that the initiation step of carcinogenesis is inevitable. Furthermore, progression, a critical and late step of tumorigenesis, which involves conversion of a preneoplastic cell to a cancer cell within a benign tumor, has not been extensively investigated (11). Thus, cancer prevention strategies should be based on knowledge of the mechanism of the promotion step of oncogenesis.

Available data indicate that the induction of ornithine decarboxylase (ODC) activity and the resultant accumulation of putrescine are essential components of the mechanism of tumor promotion (42). These findings have led to the synthesis of DL-α-difluoromethylornithine (DFMO), an enzyme-activated irreversible inhibitor of ODC (5,20) (Fig. 1). The evidence, indicating that agents which either inhibit the induction of ODC activity or inactivate ODC may be useful antitumor-promoting agents, is presented in this chapter.

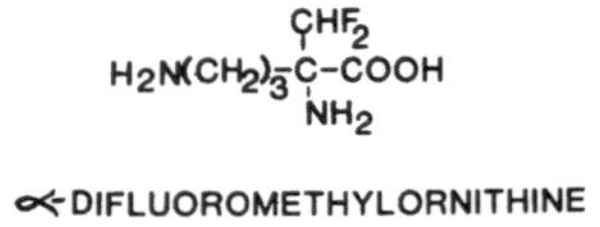

$$\begin{array}{c} CHF_2 \\ | \\ H_2N(CH_2)_3C-COOH \\ | \\ NH_2 \end{array}$$

α-DIFLUOROMETHYLORNITHINE

Fig. 1. Structure of DL-α-difluoromethylornithine (DFMO).

EXPERIMENTS AND RESULTS

Ornithine decarboxylase, which decarboxylates ornithine to form putrescine, is the key enzyme in mammalian polyamine biosynthesis (27-29) (Fig. 2). Polyamines play important roles in normal cell proliferation and differentiation as well as in malignant transformation (27-29). ODC activity in quiescent cells is extremely low and is increased within a few hours in response to many different stimuli, including hormones, growth factors, and tumor promoters (28). The induction of ODC activity and subsequent accumulation of putrescine, in general, constitute part of a cascade of biochemical events that accompany cell proliferation and differentiation (27-29). Evidence implicating the role of ODC in neoplastic growth is the finding that higher than normal ODC activity is observed in established tumors. Basal ODC levels are increased in normal-appearing colon mucosa of patients with familial polyposis, an autosomal dominant disorder, with high incidence of colon polyps and colon carcinoma (17). The role of ODC induction in mouse skin tumor promotion by 12-O-tetra-decanoylphorbol-13-acetate (TPA) has been extensively investigated, and evidence indicates that ODC induction plays an important role in mouse skin tumor promotion by TPA (11).

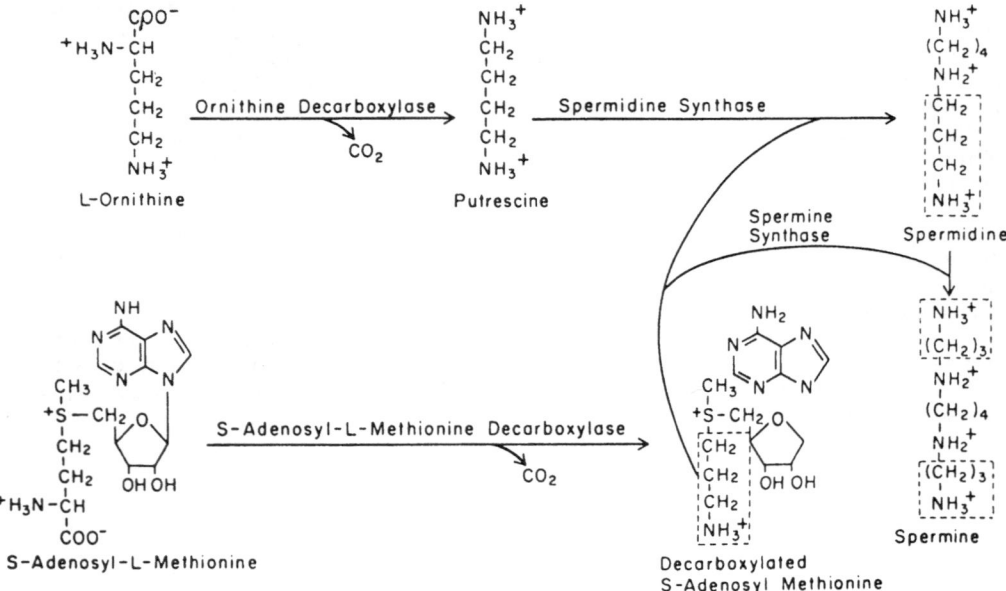

Fig. 2. Polyamine biosynthesis.

There is no single mechanism which may explain the rapid turnover (half-life, 15 min) of ODC activity. ODC activity has been shown to be regulated at the levels of synthesis (7) and degradation (35), and also by post-transcriptional modifications (34) and an interaction with macromolecules (10). Our results (12) indicate that TPA-induced ODC activity is regulated at the transcriptional level. An initial signal in ODC induction by TPA is not clear. We have suggested that TPA-increased accumulation of epidermal prostaglandins is required, but not sufficient, for ODC induction by TPA (43). Others have suggested the role of a lipoxygenase product(s) in ODC induction (22). Also, generation of free radicals appears to be involved in ODC induction by TPA (15). We have also shown that activation of protein kinase C may be an initial step in ODC induction by TPA (12,45).

Based on the knowledge of the mechanism of ODC induction by TPA, various agents were selected and tested for their ability to inhibit mouse skin tumor promotion by TPA (Tab. 1). One of the mechanisms of inhibition of tumor promotion by these agents (Tab. 1) is to inhibit the induction of ODC activity by TPA. However, these pharmacologically-active drugs are by no means specific inhibitors and cannot be used to conclusively define the role of ODC in tumor promotion. Availability of α-difluoromethylornithine, a suicide inhibitor of ODC (5,20), has helped to determine the role of ODC in induction of cancer in experimental animals.

DFMO

DFMO (Fig. 1) is a specific irreversible inhibitor of ODC. DFMO is enzymatically decarboxylated, generating an intermediate carbanionic species which, with the loss of fluorine, alkylates a nucleophilic residue at or near the active site, thereby covalently binding the inhibitor to the enzyme (20). DFMO treatment inhibits ODC activity and concomitantly depletes putrescine and spermidine levels, but it does not affect the

Tab. 1. List of agents which inhibit ODC induction and skin tumor promotion by TPA.

Agents	Reference
Indomethacin	41
Nordihydroguaiaretic acid	23
Copper (II) 3,5-diisopropylsalicylic acid	6
Quercetin	13
Palmitoylcarnitine	46,22
Certain retinoids	47
Cyclic AMP	32,33
3-Isobutyl-1-methylxanthine	32,33
Putrescine	50

spermine level in intact animals and cells in culture (5,20). Also, DFMO-treated animals have elevated S-adenosyl-L-methionine decarboxylase (EC 4.1.1.50) activity and increased levels of decarboxylated S-adenosylmethionine (30,31).

DFMO is a water-soluble compound and can be administered intraperitoneally or in drinking water. The plasma half-life of DFMO in rat, rabbit, dog, monkey (8), and healthy man (19) has been quantified, and it varies with species, from 83 min to 353 min. Treatment of animals with high doses of DFMO for a long period of time results in several side effects such as suppression of early embryogenesis and arrest of embryonic development (9), enhanced ovulation in rats (4), weight loss, diarrhea, thrombocytopenia (18), and impairment of the development of brain in preweanling rats (37). However, such toxic side effects of DFMO were not observed in mice kept for 32 wk on 0.25% DFMO in the drinking water (44). Toxicity associated with high doses of DFMO in humans is thrombocytopenia and/or reversible ototoxicity (1,2,21).

DFMO Inhibits the Induction of Cancer in Various Animal Models

The results indicating that DFMO in the drinking water inhibits the induction of skin, breast, colon, intestinal, and bladder cancer in experimental animals are summarized in this section.

The inhibition of mouse skin carcinogenesis by DFMO. DFMO, when given in the drinking water (1% w/v), was an effective inhibitor of skin tumor promotion by TPA (38,50). DFMO (1% w/v) treatment inhibited the number of papillomas by 56% (38). In a separate experiment, DFMO at 0.25% concentration inhibited the number of papillomas per mouse 50-80% and the carcinoma incidence by about 90% (44). DFMO in the drinking water (1% w/v) failed to treat the established skin carcinomas but led to a marked regression of the established skin papillomas, the benign skin lesions (Fig. 3). The inhibition of mouse skin tumor promotion caused by the DFMO treatment was reversible; as soon as DFMO treatment was discontinued, skin papillomas grew back (Tab. 2 and 3).

DFMO inhibits the induction of cancer in organs other than skin. The concept that DFMO can be a useful chemopreventive agent is substantiated by the finding that DFMO in the drinking water inhibits carcinogenesis in various organs in organ-specific models (42). A few

Tab. 2.   Effect of oral DFMO on promotion of mouse skin tumor formation.

DFMO administered at weeks of promotion	At 14 weeks of promotion papillomas/mouse
None	14.26 ± 1.91
0-14	7.20 ± 2.71
0-11	16.14 ± 1.84
11-14	8.95 ± 0.40

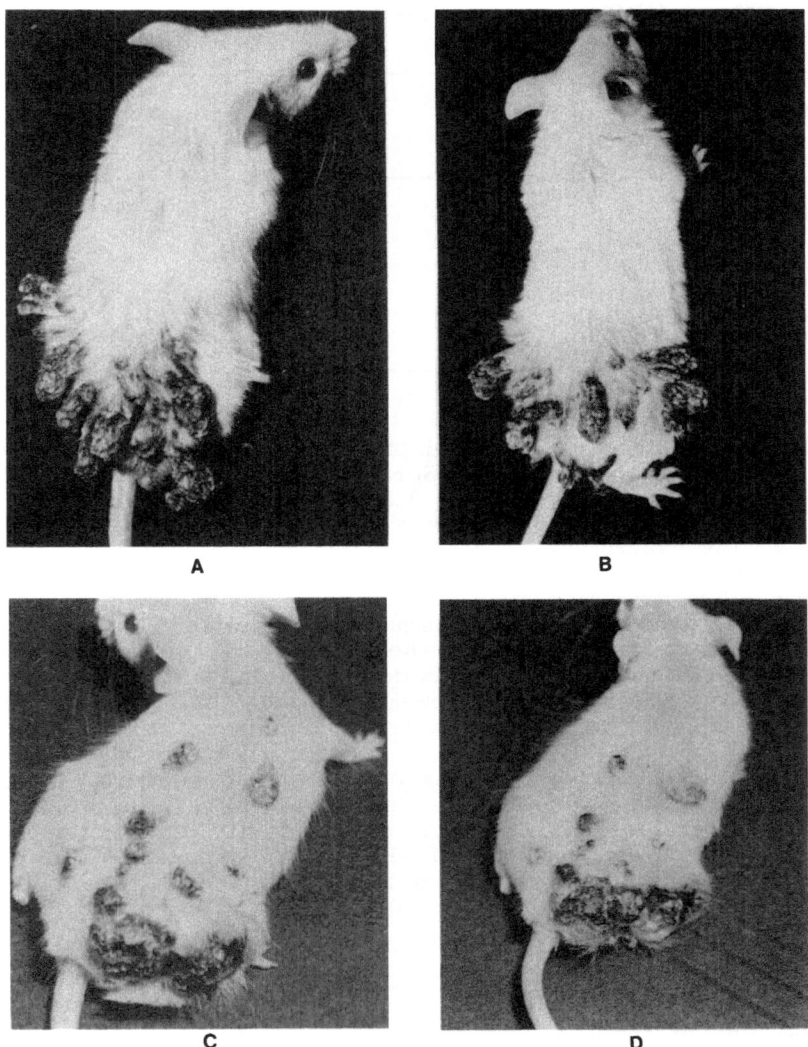

Fig. 3.   Effect of DFMO in the drinking water (1%) on established skin
           papillomas and carcinomas.   Representative mice were photo-
           graphed at the end of the experiment described in Tab. 3.
           (A) Mice on tap water for 7 wk; (B) mice on 1% DFMO for 7
           wk.   Mice with both papillomas and a carcinoma were kept for
           7 wk on (C) tap water, (D) 1% DFMO.

examples will be cited.   DFMO inhibited the formation of 1,2-dimethyl-
hydrazine (DMH)-induced colon cancer in mice (14).   In this study, mice
received 1% DFMO in the drinking water throughout the experiment and
were given injections of DMH (20 mg/kg) weekly for 28 wk.   DFMO reduced the colonic putrescine level and mucosal hyperplasia and also decreased the incidence of DMH-induced colon cancer in mice.   DFMO treatment essentially did not affect body-weight gain and survival rate (14).

Tab. 3.   Effect of DFMO in the drinking water (1% w/v) on the estab-
          lished skin papilloma.

Treatment period (weeks)	pa/mouse	pa $\geq$ 2mm	weight/mouse (gm)
Tap water 0	15.7	4.0	35.9
Tap water 7	15.7	4.6	38.8
DFMO 0	12.9	5.1	34.8
DFMO 7	7.0	2.0	33.2

DFMO in the drinking water inhibited 1-methyl-1-nitrosourea (NMU)-
induced mammary cancer in female Sprague-Dawley rats (39,40).   In this
experiment, DFMO (0.125, 0.25, or 0.5% w/v) treatment was started one
week post NMU administration and was continued during the entire period
(183 da) of the experiment.   DFMO at 0.125% and 0.50% dose levels significantly inhibited the cancer incidence.   These results indicate that
DFMO blocks the promotion stage of carcinogenesis.   Furthermore, DFMO
did not inhibit NMU-induced rat mammary cancer when given in conjunction with putrescine, implying that the inhibition of putrescine synthesis
is involved in the mechanism of inhibition of the induction of rat mammary
cancer by DFMO (39,40).   DFMO treatment has also been reported to prevent rat urinary bladder carcinogenesis (26).

Continuous DFMO treatment has been found to be more effective in
inhibiting azoxymethane-induced intestinal tumors in rats than DFMO medication for a short duration (25).   As observed with other carcinogenesis
models, combination of a low dose of DFMO with a low dose of the prostaglandin synthesis inhibitor piroxicam reduced intestinal tumor formation
more effectively than either a dose of DFMO or piroxicam alone (24).

SUMMARY AND CONCLUSION

Knowledge of the mechanisms of carcinogenesis is helpful for planning strategies and in the rational choice of agents for cancer prevention.
There is a great potential for intervention at the promotion step of human
carcinogenesis.   ODC induction is associated with the promotion stage of
carcinogenesis.   Consequently, DFMO may be a useful drug for cancer
prevention in humans.   Long-term medication with higher doses (9
$gm/m^2/da$) of DFMO has resulted in several toxic side effects, such as
thrombocytopenia and reversible ototoxicity.   However, doses of DFMO (<1
$gm/m^2/da$), selected by our in vitro human skin punch biopsy assay (16,
47), may be given for a longer period without appreciable toxicity and
should be evaluated in human cancer prevention trials.

ACKNOWLEDGEMENTS

This work was supported by U.S. Public Health Service grant CA
35368.

REFERENCES

1. Abeloff, M.D., S.T. Rosen, G.D. Luk, S.B. Baylin, M. Zeltzman, and A. Sjoerdsma (1986) Phase II trials of α-difluoromethylornithine, an inhibitor of polyamine synthesis, in advanced small cell lung cancer and colon cancer. Cancer Treat. Rep. 70:843-845.

2. Abeloff, M.D., M. Slavik, G.D. Luk, C.A. Griffin, J. Hermann, O. Blanc, A. Sjoerdsma, and S.B. Baylin (1984) Phase I trial and pharmacokinetic studies of α-difluoromethylornithine--An inhibitor of polyamine biosynthesis. J. Clin. Oncol. 2:124-130.

3. Boutwell, R.K. (1974) The function and mechanism of promoters of carcinogenesis. CRC Crit. Rev. Toxicol. 2:419-443.

4. Carpenter, J.R., and J.R. Fozard (1982) Enhancement of ovulation in the rat by α-difluoromethylornithine, an irreversible inhibitor of ornithine decarboxylase. Eur. J. Pharmacol. 80:263-266.

5. Danzin, C., M.J. Jung, J. Grove, and P. Bey (1979) Effect of α-difluoromethylornithine, an enzyme-activated, irreversible inhibitor of ornithine decarboxylase, on polyamine levels in rat tissues. Life Sci. 24:519-524.

6. Egner, P.A., and T.W. Kensler (1985) Effects of a biomimetic superoxide dismutase on complete and multistage carcinogenesis in mouse skin. Carcinogenesis 6:1167-1172.

7. Gilmour, S.K., N. Avdalovic, T. Madara, and T.G. O'Brien (1985) Induction of ornithine decarboxylase by 12-O-tetradecanoylphorbol-13-acetate in hamster fibroblasts. J. Biol. Chem. 260:16439.

8. Grove, J., J.R. Fozard, and P.S. Mamont (1981) Assay of α-difluoromethylornithine in body fluids and tissues by automatic amino acid analysis. J. Chromatogr. 223:409-416.

9. Gupta, C., B.R. Sonawane, S.J. Yaffe, and B.H. Shapiro (1980) L-Ornithine decarboxylase: An essential role in early mammalian embryogenesis. Science 208:505-508.

10. Heller, J.S., W.F. Fong, and E.S. Canellakis (1976) Induction of a protein inhibitor to ornithine decarboxylase by the end product of its reaction. Proc. Natl. Acad. Sci., USA 73:1858.

11. Hennings, H., R. Shores, M.L. Wenk, E.F. Spangler, R. Tarone, and S.H. Yuspa (1983) Malignant conversion of mouse skin tumors is increased by tumor initiators and unaffected by tumor promoters. Nature 304:67-71.

12. Hsieh, J.T., and A.K. Verma (1988) Involvement of protein kinase C in transcriptional regulation of ornithine decarboxylase gene expression by 12-O-tetradecanoylphorbol-13-acetate in T24 human bladder carcinoma cells. Arch. Biochem. Biophys. 262:326.

13. Kato, R., T. Nakadate, S. Yamamoto, and T. Sugimura (1983) Inhibition of 12-O-tetradecanoylphorbol-13-acetate-induced tumor promotion and ornithine decarboxylase activity by quercetin: Possible involvement of lipoxygenase inhibition. Carcinogenesis 4:1301-1305.

14. Kingsnorth, A.N., W.W.K. King, K.A. Diekema, P.P. McMann, J.S. Ross, and R.A. Malt (19) Inhibition of ornithine decarboxylase with 2-difluoromethylornithine: Reduced incidence of dimethylhydrazine-induced colon tumors in mice. Cancer Res. 43:2245-2249.

15. Kozumbo, W.J., J.L. Seed, and T.W. Kensler (1983) Inhibition by 2(3)-tert-butyl-4-hydroxyanisole and other antioxidants of epidermal ornithine decarboxylase activity induced by 12-O-tetradecanoylphorbol-13-acetate. Cancer Res. 43:2555.

16. Loprinzi, C.L., R.L. Love, T.M. Therneau, and A.K. Verma (1989) Inhibition of human skin ornithine decarboxylase activity by oral α-difluoromethylornithine. In Cancer Therapy and Control (in press).

17. Luk, G.D., and S.B. Baylin (1984) Ornithine decarboxylase as a biologic marker in familial colonic polyposis. N. Engl. J. Med. 311:80-83.

18. Luk, G.D., M.D. Abeloff, P.P. McCann, A. Sjoerdsma, and S.B. Baylin (1986) Long-term maintenance therapy of established human small cell variant lung carcinoma implants in athymic mice with a cyclic regimen of difluoromethylornithine. Cancer Res. 46:1849-1853.

19. Maegele, K.D., R.G. Alken, J. Grove, P.J. Schechter, and J. Koch-Weser (1981) Kinetics of α-difluoromethylornithine: An irreversible inhibitor of ornithine decarboxylase. Clin. Pharmacol. Ther. 30:210-217.

20. Metcalf, B.W., P. Bey, C. Danzin, M.J. Jung, P. Casara, and J.P. Vevert (1978) Catalytic irreversible inhibition of mammalian ornithine decarboxylase (EC 4.1.1.17) by substrate and product analogues. J. Am. Chem. Soc. 100:2551-2553.

21. Meyskens, F.L., E.M. Kingsley, T. Glagke, L. Loescher, and A. Booth (1986) A phase II study of α-difluoromethylornithine (DFMO) for the treatment of metastatic melanoma. Invest. New Drugs 4:257-262.

22. Nakadate, T., S. Yamamoto, M. Ishii, and R. Kato (1982) Inhibition of 12-O-tetradecanoylphorbol-13-acetate-induced epidermal ornithine decarboxylase activity by phospholipase A_2 inhibitors and lipoxygenase inhibition. Cancer Res. 42:2841.

23. Nakadate, T., S. Yamamoto, E. Aizu, and R. Kato (1986) Inhibition of 12-O-tetradecanoylphorbol-13-acetate-induced tumor promotion and epidermal ornithine decarboxylase activity in mouse skin by palmitoylcaritine. Cancer Res. 46:1589-1593.

24. Nigro, N.D., A.W. Bull, and M.E. Boyd (1986) Inhibition of intestinal carcinogenesis in rats: Effect of difluoromethylornithine with piroxicam or fish oil. J. Natl. Cancer Inst. 77:1309-1313.

25. Nigro, N.D., A.W. Bull, and M.E. Boyd (1987) Importance of the duration of inhibition on intestinal carcinogenesis by difluoromethylornithine in rats. Cancer Lett. 35:153-158.

26. Nowels, K., Y. Homma, J. Seidenfeld, and R. Oyasu (1986) Prevention of inhibitory effects of α-difluoromethylornithine on rat urinary bladder carcinogenesis by exogenous putrescine. Cancer Biochem. Biophys. 8:257-263.

27. Pegg, A.E. (1982) Polyamine metabolism and function. Am. J. Physiol. 243:C212-C221.

28. Pegg, A.E. (1986) Recent advances in the biochemistry of polyamines in eukaryotes. Biochem. J. 234:249-262.

29. Pegg, A.E. (1988) Polyamine metabolism and its importance in neoplastic growth and as a target for chemotherapy. Cancer Res. 48:759-774.

30. Pegg, A.E., H. Poso, K. Shuttleworth, and R.A. Bennet (1982) Effect of inhibition of polyamine synthesis on the content of decarboxylated s-adenosylmethionine. Biochem. J. 202:519-526.

31. Pegg, A.E., R.S. Wechter, R.S. Clark, L. Wiest, and B.G. Erwin (1986) Acetylation of decarboxylated S-adenosylmethionine by mammalian cells. Biochemistry 25:379-384.

32. Perchellet, J.P., and R.K. Boutwell (1981) Effect of 3-isobutyl-1-methylxanthine and cyclic nucleotides on the biochemical processes linked to skin tumor promotion by 12-O-tetradecanoylphorbol-13-acetate. Cancer Res. 41:3927-3935.

33. Perchellet, J.P., and R.K. Boutwell (1981) Effects of 3-isobutyl-1-methylxanthine and cyclic nucleotides on 12-O-tetradecanoylphorbol-

13-acetate-induced ornithine decarboxylase activity in mouse epidermis in vivo. Cancer Res. 41:3918-3926.

34. Russell, D.H. (1981) Posttranslational modification of ornithine decarboxylase by its product putrescine. Biochem. Biophys. Res. Commun. 99:1167.

35. Seely, J.E., J. Poso, and A.E. Pegg (1982) Effect of androgens on turnover of ornithine decarboxylase in mouse kidney. J. Biol. Chem. 257:7549.

36. Slaga, T.J., S.M. Fischer, K. Nelson, and G.L. Gleason (1980) Studies on the mechanism of skin tumor promotion: Evidence for several stages in promotion. Proc. Natl. Acad. Sci., USA 77:3659-3663.

37. Slotkin, T.A., A. Gignolo, W.L. Whitmore, P.A. Trepanier, G.A. Barnes, S.J. Weigel, F.J. Seidler, and J. Bartolome (1982) Impaired development of central and peripheral catecholamine neurotransmitter systems in preweanling rats treated with α-difluoromethylornithine, a specific inhibitor of ornithine decarboxylase. J. Pharmacol. Exp. Ther. 222:746-751.

38. Takigawa, M., A.K. Verma, R.C. Simsiman, and R.K. Boutwell (1983) Inhibition of mouse skin tumor promotion and of promoter-stimulated epidermal polyamine biosynthesis by α-difluoromethylornithine. Cancer Res. 43:3732-3738.

39. Thompson, H.J., L.D. Meeker, E.J. Herbst, A.M. Ronan, and R. Minocha (1985) Effect of concentration of D,L-2-α-difluoromethylornithine on murine mammary carcinogenesis. Cancer Res. 45:1170-1173.

40. Thompson, H.J., E.J. Herbst, L.D. Meeker, R. Minocha, A.M. Ronan, and R. Fite (1984) Effect of D,L-α-difluoromethylornithine on murine mammary carcinogenesis. Carcinogenesis 5:1649-1651.

41. Verma, A.K., and R.K. Boutwell (1980) Effects of dose and duration of treatment with the tumor-promoting agent, 12-O-tetradecanoylphorbol-13-acetate, on mouse skin carcinogenesis. Carcinogenesis (London) 1:271-276.

42. Verma, A.K., and R.K. Boutwell (1987) Inhibition of carcinogenesis by inhibitors of putrescine biosynthesis. In Inhibition of Polyamine Metabolism: Biological Significance and Basis for New Therapies, P.P. McCann, A.E. Peggy, and A. Sjoerdsma, eds. Academic Press, Inc., Orlando, Florida, pp. 249-258.

43. Verma, A.K., C.L. Ashendel, and R.K. Boutwell (1980) Inhibition by prostaglandin synthesis inhibitors of the induction of epidermal ornithine decarboxylase activity, the accumulation of prostaglandins, and tumor promotion caused by 12-O-tetradecanoylphorbol-13-acetate. Cancer Res. 40:308.

44. Verma, A.K., L. Duvick, and M. Ali (1986) Modulation of mouse skin tumor promotion by dietary 13-cis-retinoic acid and α-difluoromethylornithine. Carcinogenesis 7:1019-1023.

45. Verma, A.K., R.C. Pong, and D. Erickson (1986) Involvement of protein kinase C in ornithine decarboxylase gene expression in primary culture of newborn mouse epidermal cells and in skin tumor promotion by 12-O-tetradecanoylphorbol-13-acetate. Cancer Res. 46:6149.

46. Verma, A.K., B.G. Shapas, H.M. Rice, and R.K. Boutwell (1979) Correlation of the inhibition by retinoids of tumor promoter-induced mouse epidermal ornithine decarboxylase activity and of skin tumor promotion. Cancer Res. 39:419-425.

47. Verma, A.K., C.L. Loprinze, R.K. Boutwell, and P.P. Carbone (1985) In vitro induction of human skin ornithine decarboxylase by

the tumor promoter 12-O-tetradecanoylphorbol-13-acetate. J. Natl. Cancer Inst. 75:85-90.

48. Wattenberg, L.W. (1985) Chemoprevention of cancer. Cancer Res. 45:1-18.

49. Weekes, R.G., A.K. Verma, and R.K. Boutwell (1980) Inhibition by putrescine of the induction of epidermal ornithine decarboxylase activity and tumor promotion caused by 12-O-tetradecanoylphorbol-13-acetate. Cancer Res. 40:4013-4018.

50. Weeks, C.E., A.L. Herrman, F.R. Nelson, and T.J. Slaga (1982) α-Difluoromethylornithine, an irreversible inhibitor of ornithine decarboxylase, inhibits tumor promoter-induced polyamine accumulation and carcinogenesis in mouse skin. Proc. Natl. Acad. Sci., USA 79:6928-6932.

NEW ANTITUMOR PROMOTERS: (-)-EPIGALLOCATECHIN GALLATE AND SARCOPHYTOLS A AND B

H. Fujiki,[1] M. Suganuma,[1] H. Suguri,[1] K. Takagi,[1]
S. Yoshizawa,[1] A. Ootsuyama,[2] H. Tanooka,[2] T. Okuda,[3]
M. Kobayashi,[4] and T. Sugimura[5]

[1]Cancer Prevention Division
[2]Radiobiology Division
National Cancer Center Research Institute
Tsukiji 5-1-1, Chuo-ku, Tokyo 104, Japan

[3]Faculty of Pharmaceutical Sciences
Okayama University, Okayama 700, Japan

[4]Faculty of Pharmaceutical Sciences
Hokkaido University, Sapporo 060, Japan

[5]National Cancer Center, Chuo-ku, Tokyo 104, Japan

INTRODUCTION

Cancer development is a severe social and medical problem for human beings, because the occurrence of most cancers is associated with aging or the elongated lifespan of humans. Although recent advances in cancer diagnosis as well as cancer therapy are remarkable, we have to note that they have become more and more costly. We as basic cancer researchers are concerned about how we can contribute to solving this problem.

To prevent the development of human cancer, we think that it might be possible to use drugs or natural compounds which inhibit the stage of tumor promotion, as in the two-stage carcinogenesis experiment. In this study, inhibitors were found by experiments on mouse skin. We review here our recent study on new inhibitors of tumor promotion, like (-)-epigallocatechin gallate (EGCG) (13) and sarcophytols A and B (2).

MATERIALS AND METHODS

Chemicals

(-)-Epigallocatechin gallate (Fig. 1) was isolated from Japanese green tea leaves as reported previously (13). EGCG used for the experiment contained EGCG (85%), (-)-epicatechin (10%), and (-)-epicatechin gallate

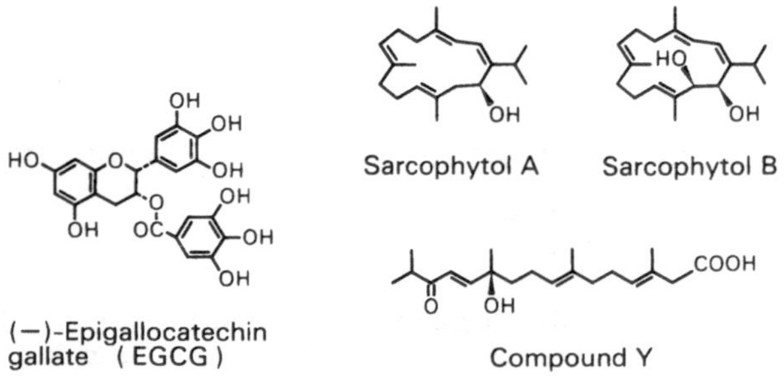

(−)-Epigallocatechin
gallate (EGCG)

Sarcophytol A   Sarcophytol B

Compound Y

Fig. 1.   Structures of new antitumor promoters.   Reprinted, in part, from Ref. 12, with permission.

(5%).   Sarcophytol A and sarcophytol B (Fig. 1) were isolated from a soft coral, Sarcophyton glaucum, collected off Ishigaki Island, Okinawa, Japan, as reported previously (4).   Compound Y, which is a secocembranoid carboxylic acid derivative (Fig. 1), was obtained by cleaving the cembrane-ring of sarcophytol A (2).   7,12-Dimethylbenz(a)anthracene (DMBA) was purchased (Sigma Chemical Co., St. Louis, Missouri).   Teleocidin was isolated from Streptomyces mediocidicus (1).

Animals

Seven-wk-old female CD-1 mice (Japanese Charles River Co., Ltd., Kanagawa, Japan) were used.   SHN mice were raised and mated at the National Cancer Center Research Institute, Tokyo, Japan.

Inhibition of Tumor Promotion in a Two-Stage Carcinogenesis Experiment

The experimental procedure was carried out as reported previously (2,13).   Briefly, initiation was achieved by a single application of either 50 μg or 100 μg of DMBA dissolved in 0.1 ml acetone to the skin of backs of 8-wk-old mice.   Beginning one week later, a control group received repeated applications of 2.5 μg of teleocidin dissolved in 0.1 ml acetone twice a week.   The experimental group received a limited amount of a test compound dissolved in 0.1 ml acetone, which was applied to the initiated area, before each treatment with teleocidin.   Each group consisted of 15 mice.   Inhibition of tumor promotion was estimated as reduction of the percentages of tumor-bearing mice, the average numbers of tumors per mouse, and delay in weeks required to achieve 50% tumor-bearing mice.

Inhibition of Spontaneous Mammary Tumor Development in SHN Mice

Virgin mice were used for the experiment (5).   The control group consisted of 11 mice, which were given tap water.   The experimental group consisted of 12 mice, which were given 3 ppm sarcophytol A in drinking water from 7 wk of age.   The development of spontaneous mammary tumors was tested by palpation.

RESULTS AND DISCUSSION

EGCG

A recent study revealed that 12-O-tetradecanoyl-phorbol-13-acetate (TPA)-type tumor promoters, such as TPA, teleocidin, and aplysiatoxin, activate calcium-activated, phospholipid-dependent protein kinase (protein kinase C), which serves as the phorbol ester receptor (1,8). The process of phosphorylation may lead to the activation of several cellular functions and subsequently to the tumor promotion in mouse skin (1,8). Based on this understanding, it is assumed that the most effective antitumor promoters should be inhibitors of protein kinase C or inhibitors of specific binding of [³H]TPA to the phorbol ester receptor. Palmitoyl-carnitine, a protein kinase C inhibitor, showed an inhibitory effect on tumor promotion of TPA in a two-stage carcinogenesis experiment (6,11). Strong binding inhibitors, except for phorbol dipropionate and phorbol dibutyrate, were not reported to be potent antitumor promoters (11).

Polyphenolic compounds like quercetin showed inhibitory effects on specific binding of [³H]TPA as well as activation of protein kinase C at high concentrations, due to interaction with membranes (3). Interestingly, EGCG dose-dependently inhibited the activation of protein kinase C induced by 2.2 µM of teleocidin (13). As we reported previously, its 50% inhibition was achieved by 1.4 µM EGCG, the amount of which was ten times more effective than that of quercetin. From the above-mentioned results, EGCG, like quercetin, was thought to have antitumor-promoting activity in mouse skin. EGCG and extracts of green tea were reported to inhibit the mutagenicity of Trp-P-1 and MNNG (9); thus, EGCG may inhibit initiation and promotion stages. Because EGCG is the main constituent of green tea infusion, it is worthwhile to study the antitumor-promoting activity of EGCG as one of the cancer chemopreventives for humans.

EGCG inhibited the tumor promotion by teleocidin in a two-stage carcinogenesis experiment (13). The control group, treated with a single application of 50 µg of DMBA and repeated applications of 2.5 µg of teleocidin twice a week, resulted in 53% of tumor-bearing mice at week 25. Five mg EGCG was applied topically to the same area of the skin before each treatment with teleocidin. The group treated with DMBA and teleocidin plus EGCG induced tumors only in 13% at week 25 (Fig. 2). EGCG also reduced the average numbers of tumors per mouse from 2.1 to 0.1 at week 25 (13). Penta-O-galloyl-β-D-glucose (5GG), isolated from tannic acid, is structurally similar to EGCG. We found that 5GG, like EGCG, showed inhibitory effects on tumor promotion by teleocidin (Horiuchi et al., ms. in prep.).

Because ingested as green tea by an oral route, EGCG was required for studying the inhibition of tumor promotion in the gastrointestinal tract. Although the results will be reported in more detail elsewhere, EGCG showed inhibition of carcinogenesis with N-ethyl-N'-nitro-N-nitroso-guanidine (ENNG) (Fujita et al., ms. in prep.). It is worthwhile to note that the antitumor promoter, like EGCG, is able to inhibit tumor development through inhibition of initiation and tumor promotion. We would like to stress that man receives EGCG in green tea every day in Japan, and its effects invite further epidemiological study.

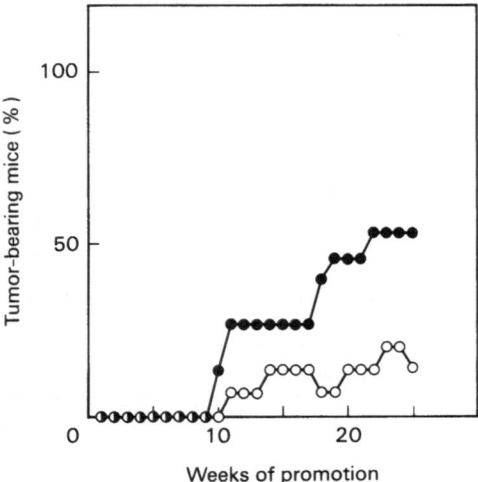

Fig. 2.   Inhibition with EGCG of tumor promotion by teleocidin.   DMBA
and teleocidin (●), and DMBA and teleocidin plus EGCG (o).
Reprinted from Ref. 12, with permission.

Sarcophytols A and B

Many compounds which inhibit tumor promotion in a two-stage carcinogenesis experiment have been reported.   However, there are few compounds which are not only nontoxic to animals, but also active in the
range of a few μgs per application.   As Fig. 1 shows, sarcophytols A and
B are chemically members of the cembrane-type diterpenes.   In 1985,
Saito et al. reported that two diastereoisomers of 2,7,11-cembratriene-
4,6-diol (α- and β-CBT) inhibit tumor promotion of TPA in mouse skin
(10).   Since α- and β-CBT were structurally similar to sarcophytols A
and B, we studied inhibition of tumor promotion by teleocidin in a two-
stage carcinogenesis experiment.

As Fig. 3 shows, the control group treated with DMBA and teleocidin
was the same as that in the above-mentioned experiment with EGCG.   Its
percentage of tumor-bearing mice was 53% at week 25.   Before each treatment with teleocidin, 1.6 μg of sarcophytol A, which corresponds to an
equimolar ratio of 2.5 μg of teleocidin, was applied to the skin of backs
of mice.   The treatment with 1.6 μg sarcophytol A induced tumors in only
7.1% at week 25 (2).   The sarcophytol A treatment reduced the average
numbers of tumors per mouse from 2.1 to 0.1 at week 25.   These results
clearly showed that 1.6 μg sarcophytol A was sufficient to effectively inhibit tumor promotion by teleocidin (Fig. 3).

To verify the inhibitory effect of sarcophytol A, the second experiment was repeated with a slightly modified experimental procedure in
order to confirm the results of the previous experiment.   In the second
experiment, the initiation was carried out by an application of 100 μg
DMBA, which was twice as much as that used in the previous experiment.
In addition to sarcophytol A, sarcophytol B with two hydroxyl groups and

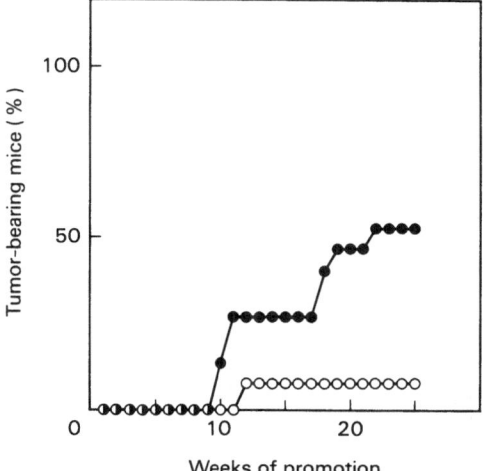

Fig. 3.   Inhibition with sarcophytol A of tumor promotion by teleocidin.
DMBA and teleocidin (●), and DMBA and teleocidin plus sarcophytol A (o). Reprinted from Ref. 2, with permission.

compound Y were included in the second experiment. The doses of all
the test compounds were equimolar to that of teleocidin per application.
As Fig. 4A and 4B show, sarcophytols A and B gave similar curves for
inhibition of tumor promotion by teleocidin. In particular, the onset of
tumor formation was delayed some weeks by treatments with sarcophytols
A and B. Also, the percentages of tumor-bearing mice in these groups
were lower than in the control group treated with DMBA and teleocidin
throughout the 20 wk. Sarcophytols A and B were effective in delaying
the weeks required for 50% of tumor-bearing mice and in reducing the

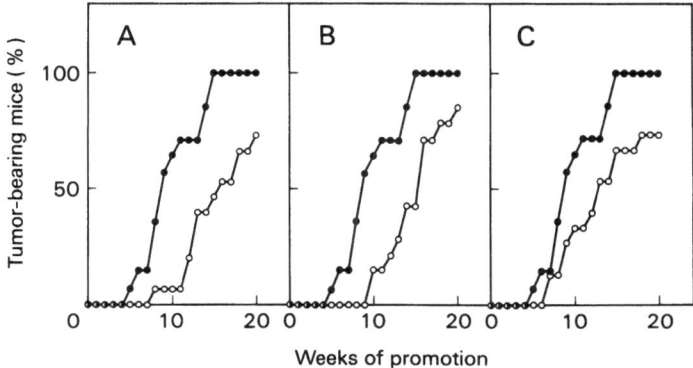

Fig. 4.   Inhibition with sarcophytol A, sarcophytol B, and compound Y
of tumor promotion. DMBA and teleocidin (●); and (A) DMBA
and teleocidin plus sarcophytol A (o), (B) DMBA and teleocidin
plus sarcophytol B (o), and (C) DMBA and teleocidin plus compound Y (o). Reprinted from Ref. 2, with permission.

average number of tumors per mouse. Compound Y had fewer apparent inhibitory effects than sarcophytols A and B (Fig. 4C). These results indicated that sarcophytols A and B were effective in inhibiting tumor promotion with an equimolar ratio to teleocidin.

Based on the evidence that [^{3}H]sarcophytol A appears to be distributed throughout the body of the mouse, sarcophytol A was given in drinking water to SHN mice. Nagasawa et al. reported that SHN mice bear spontaneous mammary tumors in almost 90% of virgin mice within 2 yr (5). As Fig. 5 shows, the experimental group bore the first tumor 1 mo later than the control group. Furthermore, the control group resulted in 82% of tumor-bearing mice in 16 mo. The experimental group induced only 58% of tumor-bearing mice in 16 mo. A solution of 3 ppm sarcophytol A was in a limited amount in drinking water due to the insolubility of sarcophytol A. These results suggested that sarcophytol A in the diet will provide more significant inhibition than sarcophytol A in drinking water. As an alternative, this experiment indicated the need of chemical modification of sarcophytols to increase solubility. It is of interest to note that sarcophytol A was effective in inhibiting the tumor development of spontaneous mammary tumors in SHN mice bearing Mouse Mammary Tumor Virus (MMTV).

Recently, sarcophytol A was found to inhibit the development of large bowel cancer of Fisher rats induced by N-methyl-N-nitrosourea (7). Also, the proliferation of a human hepatocellular carcinoma cell line, so-called Alexander cells, was inhibited dose-dependently by sarcophytols A and B. In this experiment, inhibition of release of α-fetoprotein was observed (Takagi et al., ms. in prep.). Sarcophytols A and B, which showed inhibition of tumor promotion by teleocidin in mouse skin, were effective in inhibiting tumor development in the above-mentioned experiments with viral or chemical carcinogenesis.

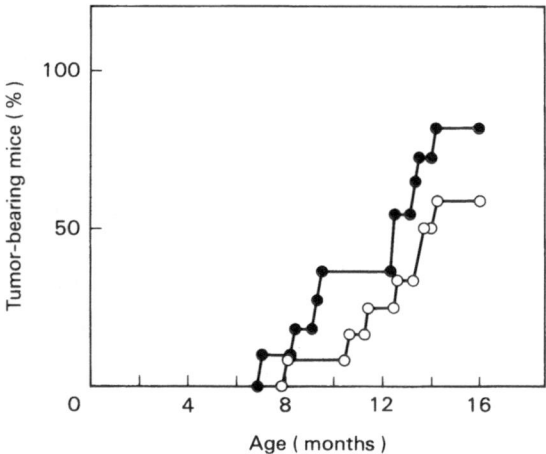

Fig. 5.   Inhibition with sarcophytol A of spontaneous mammary tumor development.   Group given sarcophytol A in drinking water (o), and control group (●).

The mechanism of action of sarcophytols A and B has not been well investigated. Because sarcophytol A did not share the phorbol ester receptor with TPA or teleocidin, or inhibit the activation of protein kinase C in vitro, we thought that sarcophytol A may not interact directly with the phorbol ester receptor (data not shown). A cytosolic fraction of mouse liver contains the specific binding protein for [^{3}H]sarcophytol A, although its binding affinity is still weak compared to that of the phorbol ester receptor. We are investigating the mechanism of action of sarcophytol A through this specific binding protein.

Sarcophytol A is not toxic to mice or rats in an acute toxicity test. Administration of 5 g/kg sarcophytol A p.o. did not cause any adverse effects. We are now conducting a chronic toxicity test for sarcophytol A. Sarcophytols A and B seem to be promising chemopreventives. In studying the process of two-stage carcinogenesis, we learned that initiation is a very rapid and irreversible process, but tumor promotion is a long and reversible process. We think that inhibition of tumor promotion by nontoxic chemopreventives might have a more practical consequence for humans than inhibition of initiation.

SUMMARY

EGCG, the main constituent of green tea, and sarcophytols A and B, isolated from a soft coral, inhibited tumor promotion by teleocidin in a two-stage carcinogenesis experiment on mouse skin. EGCG and sarcophytols showed inhibition of tumor development by chemical carcinogenesis. A possibility of developing these compounds as cancer chemopreventives for human beings is discussed.

ACKNOWLEDGEMENTS

This work was supported in part by Grants-in-Aid for Cancer Research from the Ministry of Education, Science and Culture and the Ministry of Health and Welfare for a Comprehensive Ten-Year Strategy for Cancer Control, Japan, and by grants from the Foundation for Promotion of Cancer Research, the Princess Takamatsu Cancer Research Fund, and the Smoking Research Foundation.

REFERENCES

1. Fujiki, H., and T. Sugimura (1987) New classes of tumor promoters: teleocidin, aplysiatoxin, and palytoxin. Adv. Cancer Res. 49:223-264.
2. Fujiki, H., M. Suganuma, H. Suguri, S. Yoshizawa, K. Takagi, and M. Kobayashi (1989) Sarcophytols A and B inhibit tumor promotion by teleocidin in two-stage carcinogenesis in mouse skin. J. Cancer Res. Clin. Oncol. 115:25-28.
3. Fujiki, H., T. Horiuchi, K. Yamashita, H. Hakii, M. Suganuma, H. Nishino, A. Iwashima, Y. Hirata, and T. Sugimura (1986) Inhibition of tumor promotion by flavonoids. In Plant Flavonoids in Biology and Medicine: Biochemical, Pharmacological, and Structure-Activity Relationships, V. Cody, E. Middleton, Jr., and J.B. Harborne, eds. Alan R. Liss, Inc., New York, pp. 429-440.

4.   Kobayashi, M., T. Nakayama, and H. Mitsuhashi (1979) Marine ter-
     penes and terpenoids. I. Structures of four cembrane-type diter-
     penes; Sarcophytol-A, sarcophytol-A acetate, sarcophytol-B, and
     sarcophytonin-A, from the soft coral, Sarcophyton glaucum. Chem.
     Pharm. Bull. 27:2382-2387.
5.   Nagasawa, H., R. Yanai, H. Taniguchi, R. Tokuzen, and W.
     Nakahara (1976) Two-way selection of a stock of swiss albino mice
     for mammary tumorigenesis: Establishment of two new strains (SHN
     and SLN). J. Natl. Cancer Inst. 57:425-430.
6.   Nakadate, T., S. Yamamoto, E. Aizu, and R. Kato (1986) Inhibition
     of 12-O-tetradecanoylphorbol-13-acetate-induced tumor promotion and
     epidermal ornithine decarboxylase activity in mouse skin by palmi-
     toylcarnithine. Cancer Res. 46:1589-1593.
7.   Narisawa, T., M. Takahashi, M. Niwa, Y. Fukaura, and H. Fujiki
     (1989) Inhibition of methylnitrosourea-induced large bowel cancer
     development in rats by sarcophytol A, a product from a marine soft
     coral Sarcophyton glaucum. Cancer Res. (in press).
8.   Nishizuka, Y. (1984) The role of protein kinase C in cell surface
     signal transduction and tumour promotion. Nature 308:693-698.
9.   Okuda, T., K. Mori, and H. Hayatsu (1984) Inhibitory effect of tan-
     nins on direct-acting mutagens. Chem. Pharm. Bull. 32:3755-3758.
10. Saito, Y., H. Takizawa, S. Konishi, D. Yoshida, and S. Mizusaki
     (1985) Identification of cembratriene-4,6-diol as antitumor-promoting
     agent from cigarette smoke condensate. Carcinogenesis 6:1189-1194.
11. Schmidt, R., and E. Hecker (1982) Simple phorbol esters as inhibi-
     tors of tumor promotion by TPA in mouse skin. In Carcinogenesis--
     A Comprehensive Survey, Vol. 7, E. Hecker, N.E. Fusenig, W.
     Kunz, F. Marks, and H.W. Thielmann, eds. Raven Press, New
     York, pp. 53-63.
12. Verma, A.K., R.-C. Pong, and D. Erickson (1986) Involvement of
     protein kinase C activation of ornithine decarboxylase gene expres-
     sion in primary culture of newborn mouse epidermal cells and in skin
     tumor promotion by 12-O-tetradecanoylphorbol-13-acetate. Cancer
     Res. 46:6149-6155.
13. Yoshizawa, S., T. Horiuchi, H. Fujiki, T. Yoshida, T. Okuda, and
     T. Sugimura (1987) Antitumor promoting activity of (-)-epigallocate-
     chin gallate, the main constituent of "tannin" in green tea. Phyto-
     therapy Res. 1:44-47.

CHEMOPREVENTION OF MAMMARY CANCER BY RETINOIDS

Richard C. Moon and Rajendra G. Mehta

Laboratory of Pathophysiology
IIT Research Institute
Chicago, Illinois 60616

INTRODUCTION

Chemoprevention of cancer centers on the early stages of the carcinogenic process. Chemopreventive agents as opposed to chemotherapeutic agents should either 1) inhibit or prevent initiation, 2) arrest or prevent the progression of premalignant cells, or 3) delay or prevent the development of invasive, malignant disease. Since, for the most part, initiation in the human population is unknown, chemoprevention studies in experimental animals have focused on the latter two processes. For an agent to achieve a chemopreventive role, it should enhance the physiological processes that protect the organism against the growth of abnormal cells with a potential of developing into an invasive cancer.

Since retinoids are involved in the general maintenance and/or enhancement of differentiation and cancer is a process in which loss of differentiation occurs, retinoids become principal agents targeted for their possible role as chemopreventive agents. As an example, an insufficiency of dietary vitamin A leads to a clinical syndrome manifested in growth retardation, degeneration of reproductive organs, and metaplasia and hyperkeratinization of epithelial tissues (9). Furthermore, animals deficient in vitamin A are also more susceptible to chemical carcinogens than are nondeficient animals (5,31). Such studies have been extended to show that exogenous retinoids can inhibit tumor induction in epithelia at several organ sites when administered during the preneoplastic period (28, 29), while several epidemiological investigations have noted an inverse relationship between vitamin A intake and risk for developing cancer (3, 23).

During the past ten years, the role of retinoids as chemopreventive agents has been examined in numerous in vivo and in vitro model systems. It is well documented that retinoids can reverse premalignant changes in the epithelium of mouse prostate glands in organ culture (4, 12) and suppress malignant transformation of cells in vitro, irrespective of whether the transformation is induced by ionizing radiation (8), chemical carcinogens (22), or transforming polypeptides (33). Moreover,

retinoids are also potent inhibitors of phorbol ester-induced tumor promotion (35). Inasmuch as retinoids inhibit several aspects of the carcinogenic process (transformation, metaplasia, promotion), and since the
majority of primary human cancers arise in epithelial tissues that depend
upon retinoids for normal cellular differentiation, a concerted effort has
been directed toward retinoid modulation of tumorigenesis in these tissues.

Although several reports have appeared relative to inhibition of carcinogenesis of the respiratory tract, skin, mammary gland, urinary bladder, intestine, stomach, esophagus, cervix, and pancreas by retinoids,
the majority of studies have dealt with chemoprevention of cancer of the
skin, mammary gland, and urinary bladder. Since carcinogenesis of the
mammary gland is the only process in which the chemopreventive activity
of retinoids either alone or in combination with other modifiers has been
studied in any detail, we shall briefly review experiments conducted in
our and other laboratories concerning the influence of retinoids on carcinogenesis of this organ.

ANIMAL MODELS FOR CHEMOPREVENTION

Several tumor models for various target organs are available to
study modulation of carcinogenesis by exogenous factors. The development of such tumor models should result in an animal tumor model exhib-
iting several of the following characteristics: 1) cancer development
in high numbers should be relatively rapid, preferably 6 to 12 mo;
2) initiation should be accomplished with a single carcinogen dose or, at
most, a few doses; 3) cancer should develop only in organ or tissue of
interest, i.e., exhibit target organ specificity; 4) tumors which develop
should be histologically comparable to the human counterpart; 5) tumors
should be invasive and/or metastatic; and 6) the carcinogen should induce little or no systemic toxicity. Several existing tumor models have
been used for experimental chemoprevention studies in our laboratory;
some offer specific advantages over others and are listed in Tab. 1.
Since the emphasis of this report centers on the chemoprevention of mammary cancer, a further description of the model systems used in such
studies appears warranted.

Mammary cancers can be selectively induced in rats by either 7,12-
dimethylbenz(a)-anthracene (DMBA) or N-methyl-N-nitrosourea (MNU),
and both DMBA- and MNU-induced mammary tumor models have been successfully utilized for chemoprevention studies. The MNU-induced tumor
model was originally described by Gullino et al. (7), and subsequently
modified in our laboratory (14,25). When DMBA is administered in a single intragastric dose at a concentration of 15 mg/ml sesame oil/rat, a
90-100% incidence of mammary tumors is obtained at 180 da post carcinogen, whereas a single intravenous injection of 50 mg MNU/kg body weight
(pH 5.0) also can induce 100% mammary tumor incidence in rats during a
6-month period. The majority of the cancers induced by these carcinogens are ovarian hormone dependent, with a small percentage of tumors
remaining as hormone dependent.

Earlier studies on chemoprevention by retinoids were conducted with
the DMBA-induced cancer model; however, the MNU-induced mammary carcinogenesis has remained a model of choice in our studies for several
reasons: 1) DMBA-induced tumors are encapsulated and do not invade

Tab. 1.   Animal tumor models used in the study of chemoprevention.

Species	Target	Carcinogen	Tumor type	Advantage
Rat	Mammary	MNU	Adenocarcinoma	Invasive, hormone responsive
Rat	Mammary	DMBA	Adenocarcinoma fibroadenoma	Hormone responsive, initiation/promotion
Mouse	Urinary bladder	OH-BBN	TCC	Invasive, highly aggressive
Rat	Urinary bladder	OH-BBN	Papilloma, TCC	Slow growing, histologic stages
Hamster	Tracheo-bronchial	MNU	SCC	Localization of tumor
Hamster	Lung	DEN	Papilloma, SCC, adenosquamous carcinoma	Histologic types
Mouse	Colon	Azoxymethane	Adenocarcinoma	Initiation/promotion
Rat	Colon	MAM	Adenocarcinoma	Single dose
Mouse	Skin	DMBA-TPA	Papilloma, SCC	Initiation/promotion

MNU = N-Methyl-N-nitrosourea
DMBA = 7,12-Dimethylbenz(a)anthracene
OH-BBN = N-Butyl-N-(4-hydroxybutyl)nitrosamine
DEN = Diethylnitrosamine
TPA = 12-O-Tetradecanoylphorbol-13-acetate
MAM = Methylazoxymethanol
TCC = Transitional cell carcinoma
SCC = Squamous cell carcinoma

or metastasize; 2) DMBA must be metabolized to an active form; and 3) DMBA induces a high incidence of adenomas and fibroadenomas. These complications do not arise in the MNU-induced cancer model.

Retinoid chemoprevention studies on "spontaneous" mammary cancer in mice have been conducted with mammary tumor virus (MTV) positive mice (13,39), except for the study in which C3H mice negative for the MTV were used to evaluate the effect of retinoids on development of pre-malignant hyperplastic alveolar nodules (28). Other studies in mice have involved mammary tumor induction in the BDF1 and GR strains with DMBA and the ovarian steroids, respectively.

RETINOIDS AND EXPERIMENTAL MAMMARY CARCINOGENESIS

The anticarcinogenic activity of retinoids has been investigated in all the tumor models described above. The majority of the in-depth work with a variety of retinoids has been conducted in the urinary bladder, mammary, and skin carcinogenesis models. Other model systems have been used infrequently.

The relative efficacy of nontoxic doses of several retinoids in the inhibition of mammary carcinogenesis induced in the rat by both DMBA and MNU has been investigated in our laboratory for the past ten years. Retinyl acetate and 4-hydroxyphenyl retinamide (4-HPR) are highly effective in reducing mammary cancer incidence and increasing the latency of induced mammary cancers (28). In addition, the number of mammary carcinomas is also significantly reduced by the administration of either of these retinoids. However, 13-cis-retinoic acid has little effect upon the appearance of MNU-induced mammary carcinomas; retinyl methyl ether is of intermediate efficacy, although the latter compound is extremely effective against 7,12-dimethylbenz(a)anthracene-induced mammary carcinogenesis (Tab. 2). Thus, it is readily apparent that minor alterations in the basic retinoid structure can significantly alter the activity of the molecule with respect to the inhibition of chemical carcinogenesis of the mammary gland.

The toxicity induced by a retinoid is of extreme importance in long term chemoprevention studies. As an example, retinyl acetate and 4-HPR are both effective inhibitors of chemical carcinogenesis of the rat mammary gland, but the patterns of metabolism and organ distribution of the two compounds are quite different. Chronic dietary administration of high doses of retinyl acetate results in an accumulation of retinyl esters in the liver, a process frequently accompanied by significant hepatic toxicity. On the other hand, dietary administration of 4-HPR results in a much higher level of retinoid in the mammary gland, but with relatively little liver accumulation (10,26). Moreover, 4-HPR is active both in mice and rats. Thus, on the basis of its organ distribution, it would appear that 4-HPR is preferable to retinyl acetate for use in the prevention of experimental breast cancer.

Because the exact time of carcinogen exposure in the human population cannot be determined, it is clinically important to determine how long after the carcinogenic insult that retinoid administration can be delayed and still maintain chemopreventive efficacy. Generally, retinoids are most effective in inhibiting mammary carcinogenesis when administered shortly after carcinogen treatment. However, McCormick and Moon (17) showed that retinoid treatment can be delayed in the rat mammary tumor model for as long as 16 wk after carcinogen administration and still retain its chemopreventive effectiveness, whereas in groups of animals in which retinoid treatment was initiated 20 wk after carcinogen administration, the ability of retinoids to inhibit carcinogenesis was lost. These findings indicate that retinoid administration can be delayed for as long as 4 mo but not for 5 mo. It was apparent, therefore, that a "critical" point exists in tumor development beyond which most retinoids are ineffective. Such a delay corresponds to approximately ten years in the human being.

In an effort to more closely simulate the clinical situation, we conducted two experiments in which retinoid treatment was not initiated until after the surgical removal of the first palpable tumor (18). Very little quantitative inhibition of mammary tumorigenesis was evident until approximately 50 da following tumor excision, after which a significantly reduced rate of tumor appearance was noted in the retinoid-treated group in comparison to control animals. These studies suggest that retinoids inhibit cancer by suppressing the progression of early lesions. If the retinoid offered any protection against the proliferation of fully established cancers, the effect would not have been delayed for 50 da after tumor excision. Thus, as noted earlier, there must be a crucial step(s)

Tab. 2.   Retinoids and in vivo mammary carcinogenesis.

Host/species	Carcinogen	Retinoid	Effect
Rat (S/D)	DMBA	Retinyl methyl ether	
Rat (S/D)		Retinyl acetate	Inhibition
Rat (Lewis)		Retinyl acetate	of
Rat (S/D)		N-(4-Hydroxyphenyl)- retinamide (HPR)	carcinogenesis
Rat (S/D)		Temaroten (RO 15-0778)	
Rat (S/D)	DMBA	Retinyl palmitate	No effect
Rat (S/D)	MNU	Retinyl acetate Retinyl methyl ether HPR Axerophthene All-trans-Retinoic acid 13-cis HPR Temaroten (RO 15-0778)	Inhibition of carcinogenesis

Rat (S/D)	MNU	13-cis-Retinoic acid
		Retinyl butyl ether
		N-Ethylretinamide
		Retinylidene dimedone
		Retinylidene Acetylacetone
		Etretinate (RO10-9359)
		Trimethylmethoxyphenyl analog
		   of retinyl methyl ether
		   (RO21-7925)
			No inhibition

Rat (Lewis)	Benzo(a)pyrene	Retinyl acetate	Inhibition of carcinogenesis
Rat (S/D)	MNU	HPR + ovariectomy Retinyl acetate + ovariectomy Retinyl acetate + CB154 HPR + tamoxifen Retinyl acetate + selenium	Enhanced inhibition
Rat (S/D)	MNU	HPR + MVE-2	No enhanced inhibition
Mouse (C_3H/A^{vy})	MTV	Retinyl acetate	No effect
Mouse (GR)	Hormones and MTV	Retinyl acetate	Enhancement of carcinogenesis

S/D = Sprague/Dawley
HPR = Hydroxyphenyl retinamide
DMBA = 7,12-Dimethylbenz(a)anthracene
MNU = N-Methyl-N-nitrosourea
MTV = Mammary tumor virus
MVE-2 = Maleic anhydride-divinyl ether co-polymer.

by which retinoid-responsive preneoplastic or early neoplastic lesions become retinoid resistant.

Although the evidence supporting the chemopreventive activity of
retinoids in mammary tumor models in rats is substantial, only a few
studies have been reported describing the use of retinoids in mammary
tumorigenesis in mice.   In the initial study, it was found that retinyl

acetate did not influence (neither inhibited nor enhanced) tumor incidence, latency, or tumor number in C_3H/A^{vy} female mice positive for the
mammary tumor virus (13).   However, using C_3H mice negative for the
MTV, it was found that the number of hyperplastic alveolar nodules
(putative precancerous lesion) developing in animals receiving dietary
4-HPR was significantly less than that of control mice receiving the
placebo diet, while retinyl acetate did not affect noduligenesis in these
experiments (28).   Later studies indicated that 4-HPR inhibited tumor
development in C_3H-MTV positive mice (39).   On the other hand, Welsch
et al. (37) have reported an enhancement of tumor development in nulliparous and multiparous mice of the GR/A strain fed a diet supplemented
with retinyl acetate.

COMBINATION CHEMOPREVENTION OF MAMMARY CANCER

Even though retinoids can significantly inhibit experimental carcinogenesis of various organs, a totally effective retinoid that can reduce
cancer incidence to zero is yet to be developed.   It is, however, possible
to enhance the inhibition achieved by retinoids if a combination of retinoids with other growth modifiers is employed.

Recent studies by several investigators have demonstrated such interaction between retinoids and other modifiers of mammary carcinogenesis.   In most cases, combined treatment affords greater protection against
mammary carcinogenesis than either treatment alone.   Carcinogen-induced
rat mammary cancer models are subject to inhibition by both retinoids and
modification of host hormonal status.   It is well known that ovarian hormone-dependent tumors regress following ovariectomy of the tumor-bearing animal. Similarly, if animals are ovariectomized shortly after carcinogen administration, only the ovarian hormone-independent tumors appear,
and cancer incidence is low.   The combination of ovariectomy (2 wk postcarcinogen administration) and retinyl acetate (Fig. 1) results in a synergistic inhibition of tumor incidence and multiplicity (27). Similar results
were obtained with 4-HPR.

Synergistic inhibition has also been demonstrated in the MNU-induced mammary carcinogenesis model by concomitant administration of
retinyl acetate and 2-bromo-a-ergocryptine, an inhibitor of pituitary prolactin secretion (36).   Because the blood prolactin levels of the retinyl
acetate-treated rats were similar to those of control animals, the enhanced
combination effect was probably not due to a further suppression of prolactin secretion but to an effect at the level of the mammary parenchymal
cell.   In a more recent study (30), it was demonstrated that tamoxifen
and 4-HPR, when used in combination following excision of the first
tumor, were much more effective in inhibiting the appearance of subsequent mammary cancers than was either agent alone (Fig. 2).

Although hormonal modification of experimental mammary tumorigenesis is well established, the evidence cited above indicates that the retinoids also effectively alter mammary tumorigenesis.   These data suggest
the existence of populations of preneoplastic and/or neoplastic cells dis-
playing differential sensitivity to the retinoids and hormones.   Whether
retinoids preferentially suppress the growth of hormone-dependent cell
populations, reverse the neoplastic potential of these cells, or induce
terminal differentiation of preneoplastic cells, as has been demonstrated in
C3H 10T1/2 cells (22), is presently unknown.

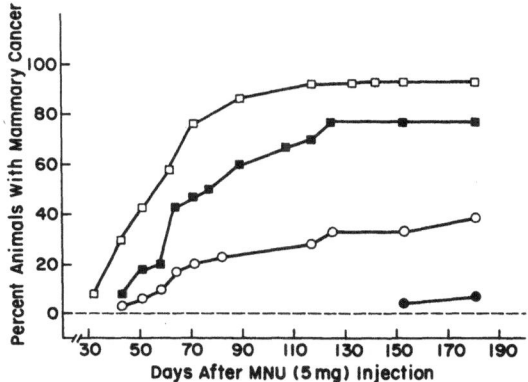

Fig. 1.   Influence of ovariectomy and retinyl acetate on the multiplicity of mammary adenocarcinoma in rats treated with N-methyl-N-nitrosourea. Sprague-Dawley female rats were injected i.v. with N-methyl-N-nitrosourea (50 mg/kg) at 50 da of age. Animals received either placebo or retinyl acetate (1 mmol/kg) supplemented diet at 7 da postcarcinogen, and two groups of the animals were bilaterally ovariectomized at 14 da postcarcinogen. Reprinted from Ref. 27, with permission from Plenum Press.

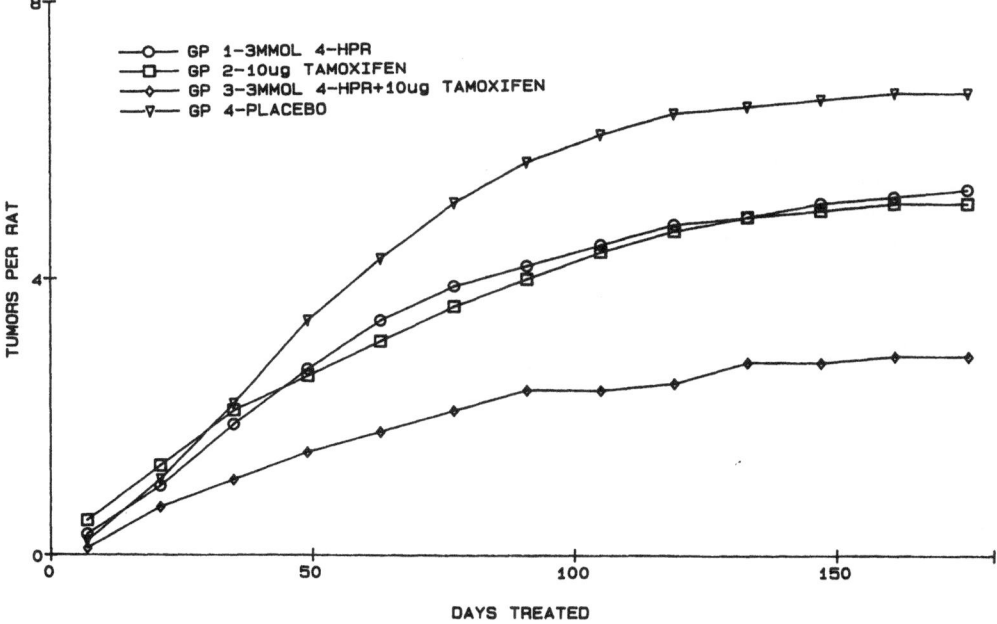

Fig. 2.   Effect of combined treatment with N-(4-hydroxyphenyl)retinamide (3 mmol/kg diet) plus tamoxifen (10 µg/rat thrice weekly, s.c.) on number of N-methyl-N-nitrosourea-induced mammary cancers following excision of the primary cancer in female Sprague-Dawley rats.

Combination chemoprevention has also been demonstrated with reti- noids and other agents that inhibit development of mammary cancer. Thompson et al. (32) were the first to show an enhanced inhibition of MNU-induced rat mammary carcinogenesis with retinyl acetate and seleni- um. The effect was confirmed by Ip and Ip (11) using the DMBA-in- duced mammary tumor model. Although both groups of workers found that the combined effect of retinyl acetate and selenium was substantially greater than the effect of either treatment alone, both studies were com- plicated by the significant reduction in food intake and body weight gain in animals receiving these chemopreventive agents. Attempts to combine modalities for prevention of mammary cancer are not always successful. For example, HPR and MVE-2 (a maleic anhydride-divinyl ether copoly- mer), an immunostimulatory agent, are both effective inhibitors of MNU- induced mammary carcinogenesis in rats. However, the combination of HPR and MVE-2 was no more effective in inhibiting cancer than either agent alone (16).

In recent years we have extended the combination chemoprevention approach to other organ sites. The results have met with mixed success. For example, in the OH-BBN-induced urinary bladder tumor model, the combination of 4-HPR with selenium and α-tocopherol enhanced the inhibi- tion caused by 4-HPR alone; however, in the DEN-induced lung cancer model, there was no enhanced suppression of carcinogenesis (unpubl. re- sults). Several of the combination studies in organs other than mammary gland are in progress in our laboratory; however, it is too early to spec- ulate on the outcome of these experiments.

TARGET ORGAN AND SPECIES SPECIFICITY

As indicated above and in Tab. 2, not all retinoids are effective in- hibitors of mammary carcinogenesis. Modification of the basic structure of the retinoid molecule has yielded a range of biologically active analogs. The majority of these modifications have occurred at the polar terminal group, but changes in the aromatic ring and side chain have also led to highly active compounds, some of which impart less toxicity than the basic molecule. Not only has toxicity been altered by changes in the basic retinoid structure, but such alterations can also have dramatic ef- fects upon organ distribution and metabolism and, hence, the cancer pre- ventive activity of the retinoid.

For instance, the synthetic retinoid 13-cis-retinoic acid is highly effective in the inhibition of two-stage skin tumorigenesis (35) and uri- nary bladder cancer induced in rats (1,6) and mice (2) by the carcino- gens N-butyl-N-(4-hydroxybutyl) nitrosamine (OH-BBN) or N-methyl-N- nitrosourea (MNU). However, by contrast, the compound has little cancer inhibitory activity in the rat mammary gland. The trimethylmethox- yphenyl (TMMP) analog of ethyl retinoate is highly effective against mouse skin carcinogenesis (35), although this compound is ineffective against either bladder carcinogenesis in mice or mammary cancer in rats (28). On the other hand, retinyl acetate is extremely active in the rat mammary cancer model (15,24), but exhibits little chemopreventive protec- tion against two-stage skin tumorigenesis (35) or mammary carcinogenesis in mice (13).

MECHANISM OF RETINOID CHEMOPREVENTION

Although the mechanism(s) by which retinoids inhibit mammary carcinogenesis is unknown, some insight into the process has been gained from the effect of retinoids on the mammary gland per se. Both 4-HPR and retinyl acetate exert an antiproliferative effect on the mammary epithelium. This is exemplified by the significant inhibition of ductal branching and the end bud proliferation of the glands of rats fed the retinoids chronically in the diet (26). Retinyl acetate has been also shown to inhibit chemical carcinogen-induced increases in mammary gland DNA synthesis in rats (19), and the induction by carcinogens of terminal ductal hyperplasias, a putative precancerous lesion (15).

It is interesting to note that the synthetic retinoid, trimethylmethoxyphenyl analog of retinyl methyl ether, which is ineffective in inhibiting carcinogen-induced mammary tumorigenesis, is also ineffective in suppressing mammary DNA synthesis (19). Furthermore, the addition of 4-HPR or retinoic acid to organ culture of mouse mammary glands (21) inhibits prolactin-induced increases in DNA synthesis, which is reflected in a decreased structural differentiation in such glands. In addition, recent in vivo studies in C_3H mice also suggest an antiproliferative effect for 4-HPR in that hyperplastic alveolar noduligenesis is reduced in animals maintained on a diet supplemented with the retinoid (28). These effects on the mammary gland are probably not mediated via an influence on host hormonal levels, since retinoid administration has little effect upon either circulating prolactin levels (36) or normal ovarian function (24). Moreover, the additive or synergistic effect of the retinoid plus hormonal manipulation in the combination studies cited above would also appear to substantiate this view.

At present, it is speculative to suggest that the interaction of the retinoid (or a metabolite) with the nucleus results in altered genomic expression. However, there are numerous reports which indirectly support such a view. For example, retinoids inhibit tumor promoter-induced ornithine decarboxylase activity (34), carcinogen-induced DNA synthesis (19), and growth factor-induced transformation (33). Recent studies (20) of RNA polymerase activity of mammary tumor nuclei are also suggestive of such an effect: nuclei isolated from mammary cancers preincubated with retinoic acid exhibited reduced RNA polymerase activity compared to tissues incubated under similar conditions without the retinoid. Furthermore, the nuclei which were preincubated with mammary cytosol containing retinoic acid-receptor complex also showed reduced RNA polymerase activity, as compared with that of nuclei incubated with either buffer or with free retinoic acid. Activity of both RNA polymerase I and II was reduced as a result of retinoid treatment. These results indicate that retinoid may be active at the chromatin level, and that retinoic acid-retinoic acid receptor complexing may be an important step in the mediation of retinoid action in the mammary tumor cell.

REFERENCES

1. Becci, P.J., H.J. Thompson, C.J. Grubbs, C.C. Brown, and R.C. Moon (1979) Effect of delay in administration of 13-cis-retinoic acid on the inhibition of urinary bladder carcinogenesis in the rat. Cancer Res. 39:3141-3144.

2. Becci, P.J., H.J. Thompson, J.M. Strum, et al. (1981) N-butyl-N(4-hydroxybutyl)nitrosamine-induced urinary bladder in C57BL/6 x DBA/2F$_1$ mice as a useful model for study of chemoprevention of cancer with retinoids. Cancer Res. 41:927-932.
3. Bjelke, E. (1975) Dietary vitamin A and human lung cancer. Int. J. Cancer 15:561-565.
4. Chopra, D.P., and C.J. Wilkoff (1976) Inhibition and reversal by β-retinoic acid of hyperplasia induced in cultured mouse prostate tissue by 3-methylcholanthrene or N-methyl-N-nitrosoguanidine. J. Natl. Cancer Inst. 56:583- 589.
5. Cohen, S.M., J.F. Wittenberg, and G.T. Bryan (1976) Effect of avitaminosis A and hypervitaminosis A on urinary bladder carcinogenesis of N-[4-(5-nitrofuryl)-2-thozolyl)]formamide. Cancer Res. 36:2334-2339.
6. Grubbs, C.J., R.C. Moon, R.A. Squire, G.M. Farrow, S.F. Stinson, D.G. Goodman, C.B. Brown, and M.B. Sporn (1977) 13-cis-retinoic acid: inhibition of bladder carcinogenesis induced in rats by N-butyl-N-(4-hydroxbutyl)nitrosamine. Science 198:743-744.
7. Gullino, P.M., H.M. Pettigrew, and F.H. Grantham (1975) N-Nitrosomethylurea was mammary gland carcinogen in rats. J. Natl. Cancer Inst. 54:401-414.
8. Harisiadis, L., R.C. Miller, E.J. Hall, and C. Borek (1978) A vitamin A analogue inhibits radiation-induced oncogene transformation. Nature 274:486-487.
9. Hicks, R.M. (1983) The scientific basis for regarding vitamin A and its analogs as anticarcinogenic agents. Proc. Nutr. Soc. 42:83-93.
10. Hultin, T.A., C.M. May, and R.C. Moon (1986) N-(4-hydroxyphenyl)-all-trans-retinamide pharmacokinetics in female rats and mice. Drug Metab. Dispos. 14:714-717.
11. Ip, C., and M.M. Ip (1981) Chemoprevention of mammary tumorigenesis by a combined regimen of selenium and vitamin A. Carcinogenesis 2:915-918.
12. Lasnitzki, I. (1976) Reversal of methylcholanthrene-induced changes in mouse prostates in vitro by retinoic acid and its analogs. Br. J. Cancer 34:239-248.
13. Maiorana, A., and P. Gullino (1980) Effect of retinyl acetate on the incidence of mammary carcinomas and hepatomas in mice. J. Natl. Cancer Inst. 64:655-663.
14. McCormick, D.L., C.B. Adamowski, A. Fiks, and R.C. Moon (1981) Lifetime dose response relationship for mammary tumor induction by a single administration of N-methyl-N- nitrosourea. Cancer Res. 41:1690-1694.
15. McCormick, D.L., F.J. Burns, and R.E. Albert (1981) Inhibition of benzo(a)pyrene-induced mammary carcinogenesis by retinyl acetate. J. Natl. Cancer Inst. 66:559-564.
16. McCormick, D.L., P.J. Becci, and R.C. Moon (1982) Inhibition of mammary and urinary bladder carcinogenesis by a retinoid and a maleic anhydride-divinyl ether copolymer (MVE-2). Carcinogenesis 3:1473-1477.
17. McCormick, D.L., and R.C. Moon (1982) Influence of delayed administration of retinyl acetate on mammary carcinogenesis. Cancer Res. 42:2639-2643.
18. McCormick, D.L., Z.L. Sowell, C.A. Thompson, and R.C. Moon (1983) Inhibition by retinoid and ovariectomy of additional primary malignancies in rats following surgical removal of the first mammary cancer. Cancer 51:594-599.

19. Mehta, R.G., and R.C. Moon (1981) Inhibition of DNA synthesis by retinyl acetate durnig chemically-induced mammary carcinogenesis. Cancer Res. 40:1109-1111.

20. Mehta, R.G., M.E. Hawthorne, and R.C. Moon (1988) Effect of all-trans-retinoic acid on nuclear RNA polymerase activity in chemically-induced rat mammary tumors. Cancer Lett. 42:1-5.

21. Mehta, R.G., W.L. Cerny, and R.C. Moon (1983) Retinoid inhibition of prolactin-induced development of the mammary gland in vitro. Carcinogenesis 4:23-26.

22. Merriman, R.L., and J.S. Bertram (1979) Reversible inhibition by retinoids of 3-methylcholanthrene-induced neoplastic transformation in C3H/10T1/2 CL8 cells. Cancer Res. 39:1661-1666.

23. Mettlin, C., S. Graham, and M. Swanson (1979) Vitamin A and lung cancer. J. Natl. Cancer Inst. 62:1435-1438.

24. Moon, R.C., C.J. Grubbs, and M.B. Sporn (1976) Inhibition of 7,12-dimethylbenz(a)anthracene-induced mammary carcinogenesis by retinyl acetate. Cancer Res. 36:2626-2630.

25. Moon, R.C., C.J. Grubbs, M.B. Sporn, and D.G. Goodman (1977) Retinyl acetate inhibits mammary carcinoma induced by N-methyl-N-nitrosourea. Nature 267:620-621.

26. Moon, R.C., H.J. Thompson, P.J. Becci, C.J. Grubbs, R.J. Gander, D.L. Newton, J.M. Smith, S.R. Phillips, W.R. Henderson, L.T. Mullen, C.C. Brown, and M.B. Sporn (1979) N-(4-hydxoxyphenyl)retinamide, a new retinoid for prevention of breast cancer in the rat. Cancer Res. 39:1339-1346.

27. Moon, R.C., and R.G. Mehta (1982) Retinoid binding in normal and neoplastic mammary tissue. In Hormones and Cancer, W.W. Leavit, ed. Plenum Press, New York, pp. 231-249.

28. Moon, R.C., D.L. McCormick, and R.G. Mehta (1983) Inhibition of carcinogenesis by retinoids. Cancer Res. 43:2469s-2475s.

29. Moon, R.C., and L. Itri (1984) Retinoids and cancer. In The Retinoids, M.B. Sporn, A.B. Roberts, and D.S. Goodman, eds. Academic Press, Orlando, Florida, pp. 327-371.

30. Ratko, T.A., C.J. Detrisac, N.M. Dinger, C.F. Thomas, G.J. Kelloff, and R.C. Moon (1989) Chemopreventive synergy of combined retinoid and tamoxifen treatment following surgical excision of a primary mammary cancer in female rats. Cancer Res. (submitted for publication).

31. Rogers, A.E., B.J. Herndon, and P.M. Newberne (1973) Induction by dimethylhydrazine of intestinal carcinoma in normal rats fed high and low levels of vitamin A. Cancer Res. 33:1003-1009.

32. Thompson, H.J., L.D. Meeker, and P.J. Becci (1981) Effect of combined selenium and retinyl acetate treatment on mammary carcinogenesis. Cancer Res. 41:1413-1416.

33. Todaro, G.J., J.E. DeLarco, and M.B. Sporn (1978) Retinoids block phenotypic cell transformation produced by sarcoma growth factor. Nature 276:272.

34. Verma, A.K., and R.K. Boutwell (1977) Vitamin A acid (retinoic acid), a potent inhibitor of 12-0-tetradecanoyl-phorbol-13-acetate-induced ornithine decarboxylase activity in mouse epidermis. Cancer Res. 37:2196-2201.

35. Verma, A.K., B.G. Shapas, H.M. Rice, and R.K. Boutwell (1979) Correlation of the inhibition by retinoids of tumor promoter-induced mouse epidermal ornithine decarboxylase activity and of skin tumor promotion. Cancer Res. 39:419-425.

36. Welsch, C.W., C.K. Brown, M. Goodrich-Smith, J. Chiusano, and R.C. Moon (1980) Synergistic effect of chronic prolactin suppression

and retinoid treatment in the prophylaxis of N-methyl-N-nitrosourea-induced mammary tumorigenesis in female Sprague-Dawley rats. Cancer Res. 40:3095-3098.

37. Welsch, C.W., M. Goodrich-Smith, C.C. Brown, and N. Crowe (1981) Enhancement by retinyl acetate of hormone induced mammary tumorigenesis in female GR/A mice. J. Natl. Cancer Inst. 67:935-938.

38. Welsch, C.W., M. Goodrich-Smith, C.K. Brown, D. Mackie, and D. Johnson (1982) 2-Bromo-α-ergocryptine (CB 154) and tamoxifen induced suppression of the genesis of mammary carcinoma in female rats treated with 7,12-dimethylbenz(a)anthracene (DMBA): A comparison. Oncology 39:88-92.

39. Welsch, C.W., J.V. DeHoog, and R.C. Moon (1983) Inhibition of mammary tumorigenesis in nulliparous C3H mice by chronic feeding of the synthetic retinoid, N-(4-hydroxyphenyl) retinamide. Carcinogenesis 4:1185-1187.

SUPPRESSION OF TUMOR PROMOTION BY INHIBITORS

OF POLY(ADP)RIBOSE FORMATION

Walter Troll, Seymour Garte, and Krystyna Frenkel

Institute of Environmental Medicine
New York University Medical Center
New York, New York 10016

ABSTRACT

Tumor promoters, such as phorbol esters or hormones, cause many biological effects which may contribute to the expression of cancer. The mechanism of cancer expression may have a common theme. One method of learning about this common mechanism is the identification of chemicals that interfere with tumor development. That there is actually a common theme between very different substances, such as inflammatory skin tumor promoters and estradiol causing breast cancer, was shown by the fact that both skin and breast cancers are suppressed by the same agents, e.g., protease inhibitors and retinoids. In addition to skin and breast, protease inhibitors suppress colon, bladder, and liver cancers. The substances that crossed over in suppressing many varieties of cancer were found to inhibit oxygen radical formation by tumor promoter-activated neutrophils and ras oncogene expression in NIH 3T3 cells. Poly(ADP)ribose polymerase (PADPR polymerase) may serve as the connecting link between oxygen radicals that cause its activation and oncogene expression. PADPR polymerase is inhibited by retinoids, antioxidants, and some protease inhibitors. Benzamide, an inhibitor of PADPR polymerase, is also a chymotrypsin inhibitor which suppresses oxygen radical formation by tumor promoter-activated neutrophils. The inhibition of PADPR polymerase causes the expulsion of some oncogenes from NIH 3T3 cells at definite times after oncogene transfection. Further work is required to find what are the contributions of PADPR polymerase to tumor promotion and of its inhibitors to suppression of oncogene expression.

INTRODUCTION

The contribution of genetic effects to carcinogenesis has been extensively investigated and significant progress has led to methods for identifying carcinogens in our environment. These carcinogenic agents were so numerous that it appeared difficult to rationalize the relatively low cancer occurrence in certain populations and the evolutionary stability of our species without contemplating the presence of major anticarcinogenic components in our environment. Epidemiological data on the occurrence of cancer throughout the world identified vegetarians as having lower incidence of many cancers in comparison to meat (fat)-consuming populations. The highest rates of breast, colon, and prostatic cancers were noted in countries where the people consumed a characteristically "western" diet high in meat content (e.g., the United States). The lowest levels were observed in populations that excluded meat and dairy products from their diet for economic reasons (e.g., Japan and Thailand) or religious dictates (e.g., Seventh-Day Adventists). Since such diets consist mainly of plants and their products, these findings suggest that plants contain anticarcinogenic agents (1,7,32).

Cellular DNA can be modified by a variety of agents, a process that may result in the creation of a new cell species. Such cells, by successfully competing with normal cells, can form tumors that further develop into malignant cells and are capable of metastasis to other sites in the host. The first demonstration of the multistage nature of carcinogenesis was shown by Berenblum in skin cancer (3). Inflammatory croton oil, when painted on the backs of mice, caused tumors only when the mice were pretreated with a minute dose of a carcinogen that by itself would be insufficient to cause cancer. This carcinogen putatively gives rise to a cell type that is indistinguishable from normal cells until exposure to a promoting agent converts it to a tumor cell. Many promoters (e.g., phorbol esters, which are the active compounds in croton oil) were identified as derivatives of 12-0-tetradecanoyl-phorbol-13-acetate (TPA) (21,41). Fujiki et al. identified other types of promoting agents in our environment, including teleocidin, aplysiatoxin, and okadaic acid (16).

In contrast to initiation, which requires only one dose, many applications of a promoting agent are needed to produce tumors. Thus, promotion requires multiple or prolonged exposure to the agent before tumor growth becomes inevitable. Tumor development has been divided into three phases: stage I--conversion; stage II--promotion; and stage III--progression to malignant neoplasm (3-5,26,34). The concept of multistage carcinogenesis in mouse skin led to identification of promoting agents that contribute to the development of other cancers as well. The determination of the site where the tumor occurs facilitated identification of ovarian hormones as tumor promoters. The hormones estradiol and testosterone are considered possible causative agents in breast and prostatic cancers, respectively. Since the initiators of these common human cancers are not known, the identification of agents as tumor promoters depends on their ability to produce cancer at specific sites where receptors binding them are present (28,33). Such cancer in humans appears to be preventable by nutritional components that inhibit tumor promotion. Similar considerations have led to the tentative identification of bile acids as contributors to colon cancer (36). That phenobarbital promotes liver cancer has been demonstrated in animal experiments (36).

Studies on the mechanism of tumor promotion have revealed a number of common characteristics that are shared by the seemingly unrelated agents that contribute to cancer. These include protein kinase C activation, proteases, as well as induction of oxygen radicals and PADPR polymerase (30,31,37,39). The most interesting common aspect of tumor promotion at certain sites has come from the observations that the same anticarcinogenic compounds that are capable of counteracting experimental tumor promotion by TPA in mouse model, counteract breast and colon cancers in animals and also, as shown by epidemiological studies, in humans (36). After it was noted that croton oil or its purified principle, TPA, induced proteases in mouse skin, it was hypothesized that proteases may be involved in tumor promotion. To test this possibility, synthetic protease inhibitors were applied to mouse skin and were found to block the promoting action of TPA. Low doses (1-10μg) of tosyl-L-lysine chloromethyl ketone, tosyl-phenylalanine-chloromethyl ketone, and the competitive substrate tosyl-L-arginine methyl ester specifically counteracted promotion. The number and incidence of tumors decreased and latent periods increased. The inhibition of tumor promotion by protease inhibitors was confirmed by applying leupeptin to mouse skin (24). Feeding a raw soybean diet rich in protease inhibitors not only suppressed tumors in mouse skin initiated with 4-nitroquinoline-N-oxide and promoted by TPA, but also suppressed breast tumors induced by X-ray irradiation in Sprague-Dawley rats and spontaneous liver cancer in C_3H mice (39). Feeding animals leupeptin suppressed rat mammary tumors induced by 7,12-dimethylbenz(a)anthracene, whereas ε-aminocaproic acid or Bowman-Birk soybean inhibitor inhibited colon cancer induced by 1,2-dimethylhydrazine (36). The epidemiological studies of Correa showed decreased breast, prostatic, and colon cancers in populations consuming seeds rich in protease inhibitors (e.g., rice, soybeans, chick-peas) (9).

A similar crossover of anticarcinogenic agents that were active in the mouse skin system was evidenced with retinoids which also suppressed breast cancer (42). Bryostatin, a modifier of protein kinase C that had been shown to inhibit tumor promotion by phorbol esters in SENCAR mice, was found to inhibit leukemia in humans (23,29). These apparent common anticarcinogenic actions also have recently been observed with sarcophytols A and B, cembrane-type diterpenes isolated from the soft coral Sarcophyton glaucum. These compounds were shown to inhibit teleocidin-mediated tumor promotion in mouse skin. When fed to mice, sarcophytol A inhibited tumor promotion by TPA, aplysiatoxin, and okadaic acid in large bowel carcinogenesis, as well as spontaneous mammary and liver tumors (17). These compounds are reminiscent of retinoids in structure but appear more effective in suppressing a variety of tumors and exhibit only negligible toxicity.

Recent developments have stimulated a growing interest in the role of oxygen radicals and particularly of hydrogen peroxide in tumor promotion. Hydrogen peroxide and organic peroxides have been shown to be tumor promoters in SENCAR mice (35). Tumor promoters, including phorbol esters and indole alkaloids, induce a respiratory burst in polymorphonuclear leukocytes (PMNs), resulting in the formation of superoxide anion radicals ($\cdot O_2^-$), hydrogen peroxide (H_2O_2), hydroxyl radicals ($\cdot OH$), and singlet oxygen (2,11,12,16,25,27). Phorbol derivatives that are inactive as tumor promoters (i.e., phorbol, phorbol diacetate, or 4-0-methyl-TPA) also fail to elicit production of $\cdot O_2^-$ and H_2O_2 by PMNs (13,20). Interestingly, chemopreventive protease inhibitors and retinoids suppress formation of $\cdot O_2^-$ as well as H_2O_2 by TPA-activated PMNs (14, 38). Thus, the ability of the above agents to inhibit production of active oxygen species may be responsible for their chemopreventive properties in respect to suppression of tumor promotion and cancer development.

The induction of PADPR polymerase is of specific interest among the many possible contributions of oxygen radicals to tumor promotion. For example, it has been observed that inhibitors of PADPR polymerase cause deletion of oncogenes from transfected NIH 3T3 cells (30). Benzamide, an inhibitor of PADPR polymerase, has been shown to cause a loss of exogenously-supplied H-ras genes from NIH 3T3 cells, which results in morphologically-normal flat cells. Benzamide, 3-aminobenzamide, and nicotinamide were all found to be protease inhibitors with a preference for chymotrypsin. They also inhibited TPA-induced formation of oxygen radicals by human neutrophils (Ref. 40; Tab. 1 and 2).

Naturally-occurring chymotrypsin-inhibiting protease inhibitors (e.g., potato inhibitor I) have been shown to effectively block H_2O_2 formation by TPA-activated human neutrophils, and perhaps they are also inhibitors of poly(ADP)ribosylation (14). Antipain, another protease inhibitor, has been shown to be a PADPR polymerase inhibitor as well (8). The contribution of PADPR polymerase to promotion becomes of particular interest when the actions of protease inhibitors, retinoids, and antioxidants are examined in their role of interfering with PADPR formation.

Garte et al. (18) have demonstrated that H-ras oncogene-induced transformation can be inhibited by leupeptin, antipain, ε-aminocaproic acid, and α_1-antitrypsin. Inhibition of cell transformation occurred only when antipain was added to NIH 3T3 cells three to nine days after transfection with ras, which suggests cell proliferation as the sensitive zone for suppression of ras expression. Similarly, addition of 3-aminobenzamide, the poly(ADP-ribose)transferase inhibitor, less than three days after transfection was ineffective in inhibiting transformation of NIH 3T3 cells by the ras, v-raf, and v-mos oncogenes (10). Such results suggest a central role for poly(ADP)ribosylation in oncogene expression--a possible step in tumor promotion. The demonstration that PADPR inhibitors cause deletion of exogenously-supplied ras with concomitant reversion of cells from transformed to morphologically-normal flat cells also points to poly(ADP)ribosylation as being involved in the transformation process.

Tab. 1. Effect of various inhibitors on hydrolysis of [^{3}H]casein by chymotrypsin and trypsin.

Inhibitor	Concentration of inhibitor (mM)	Inhibition (%)	
		Chymotrypsin[a]	Trypsin[a]
Nicotinamide	10	7	6
	20	22	12
	40	53	14
	60	52	40
Benzamide	10	57	24
	20	64	37
	40	76	44
	60	88	64
3-Aminobenzamide	10	34	6
	20	50	27
	40	69	31
	60	72	40

[a] Twenty nanograms chymotrypsin or trypsin per assay.

Tab. 2.   Inhibition of superoxide anion formation in human neutrophils.

Concentration of inhibitor	Inhibition (%)		
	Nicotinamide	Benzamide	3-Aminobenzamide
2 mM	18	23	15
5 mM	23	34	48
10 mM	35	84	73

Luminol, a potent inhibitor of PADPR polymerase, also induced loss of exogenous ras sequence and formation of the flat cells (30). It is of interest to speculate if other chemopreventive anticarcinogens are active through repressing PADPR polymerase. For example, it was shown that all-trans-retinoic acid is effective in suppressing ras-induced transformation (19). It may be that retinoids inhibit formation of PADPR by interfering with activation of PMNs to form oxygen radicals, which by causing DNA damage are responsible for inducing poly(ADP)ribosylation. This could be the connection of antioxidants to PADPR polymerase inhibition. The inhibition of PADPR polymerase may not be the only mechanism of action by anticarcinogenic agents, since ε-aminocaproic acid and α_1-trypsin inhibitor have been shown to suppress the induction of DNA polymerase α, a necessary enzyme for carcinogen-mediated DNA amplification (6,22). The suppression of ras-oncogene expression demonstrated by these particular protease inhibitors is more likely due to this latter effect, since they are unlikely prospects for being inhibitors of PADPR polymerase (8,14,18).

More work is required to permit measurement of the chemopreventive agents in our environment and elucidation of the mechanisms of their action (43). A possible relationship of protease inhibitors and antioxidants as contributors to the prevention of PADPR formation that results in deletion of oncogenes awaits further investigation.

ACKNOWLEDGEMENTS

Special thanks are expressed to Drs. Takashi Sugimura and Minako Nagao for discussing their recent findings relating PADPR to oncogene expression, and to Dr. Hirota Fujiki for making available his recent data on chemopreventive agents. Supported in part by the Foundation for Promotion of Cancer Research, Tokyo, Japan, by the National Institute of Environmental Health Sciences grants ES 00260 and P42 ES 048995, and by Public Health Service grant CA 37858, awarded by the National Cancer Institute, Department of Health and Human Services.

REFERENCES

1. Armstrong, B., and R. Doll (1975) Environmental factors and cancer incidence and mortality in different countries, with special reference to dietary factors. Int. J. Cancer 15:617-631.
2. Badwey, J.A., and M.L. Karnovsky (1980) Active oxygen species and the functions of phagocytic leukocytes. Ann. Rev. Biochem. 49:695-726.

3.   Berenblum, I. (1941) The cocarcinogenic action of croton resin. Cancer Res. 1:44-48.
4.   Boutwell, R.K. (1964) Some biological aspects of skin carcinogenesis. Prog. Exp. Tumor Res. 4:207-250.
5.   Boutwell, R.K. (1983) Diet and anticarcinogenesis in the mouse skin two-stage model. Cancer Res. 43(Suppl.):2465s-2468s.
6.   Bürkle, A., T. Meyer, H. Hilz, and H. zur Hausen (1987) Enhancement of N-methyl-N'-nitro-N-nitrosoguanidine-induced DNA amplification in a Simian virus 40-transformed Chinses hamster cell line by 3-aminobenzamide. Cancer Res. 47:3632-3636.
7.   Carroll, K.K. (1975) Experimental evidence of dietary factors and hormone-dependent cancers. Cancer Res. 35:3374-3383.
8.   Cleaver, J.E., M.J. Banda, W. Troll, and C. Borek (1986) Some protease inhibitors are also inhibitors of poly(ADP-ribose) polymerase. Carcinogenesis 7:323-325.
9.   Correa, P. (1981) Epidemiological correlations between diet and cancer frequency. Cancer Res. 41:3685-3690.
10. Diamond, A.M., C.J. Der, and J.L. Schwartz (1989) Alterations in transformation efficiency by. the ADPRT-inhibitor 3-aminobenzamide are oncogene specific. Carcinogenesis 10:383-385.
11. Fantone, J.C., and P.A. Ward (1982) Role of oxygen-derived free radicals and metabolites in leukocyte-dependent inflammatory reactions. Am. J. Pathol. 107:397-418.
12. Formisano, J., W. Troll, and T. Sugimura (1983) Superoxide response induced by indole alkaloid tumor promoters. Ann. NY Acad. Sci. 407:429-431.
13. Frenkel, K., and K. Chrzan (1987) Hydrogen peroxide formation and DNA base modification by tumor promoter-activated polymorphonuclear leukocytes. Carcinogenesis 8:455-460.
14. Frenkel, K., K. Chrzan, C. Ryan, R. Wiesner, and W. Troll (1987) Chymotrypsin-specific protease inhibitors decrease H_2O_2 formation by activated human polymorphonuclear leukocytes. Carcinogenesis 8:1207-1212.
15. Narisawa, T., M. Takahashi, M. Niwa, Y. Fukaura, and H. Fujiki (1989) Inhibition of methylnitrosourea-induced large bowel cancer development in rats by Sarcophytols A, a product from a marine soft-coral Sarcophyton glaucum. Cancer Res. 49:3287-3289.
16. Fujiki, H., M. Mori, M. Nakayasu, T. Terda, T. Sugimura, and R.E. Moore (1981) Indole alkaloids: Dihydroteleocidin B, teleocidin, and lyngbyatoxin A as members of a new class of tumor promoters. Proc. Natl. Acad. Sci., USA 78:3872-3876.
17. Fujiki, H., M. Suganuma, H. Suguri, S. Yoshizawa, K. Takagi, and M. Kobayashi (1989) Sarcophytols A and B inhibit tumor promotion by teleocidin in two-stage carcinogenesis in mouse skin. J. Cancer Res. Clin. Oncol. 115:25-28.
18. Garte, S.J., D.D. Currie, and W. Troll (1987) Inhibition of H-ras oncogene transformation of NIH 3T3 cells by protease inhibitors. Cancer Res. 47:3159-3162.
19. Garte, S.J., D. Currie, J. Motz, and W. Troll (1988) Retinoic acid inhibits transformation of NIH 3T3 cells by the human H-ras oncogene. Proc. Am. Assoc. Cancer Res. 29:140.
20. Goldstein, B.D., G. Witz, M. Amoruso, D.S. Stone, and W. Troll (1981) Stimulation of human polymorphonuclear leukocyte superoxide anion radical production by tumor promoters. Cancer Lett. 11:257-262.
21. Hecker, E. (1968) Cocarcinogenic principles from the seed oil of

Croton tiglium and from other euphorbiaceae. Cancer Res. 28:2338-2349.

22. Heilbronn, R., J.R. Schlehofer, A.O. Yalkinoglu, and H. zur Hausen (1985) Selective DNA amplification induced by carcinogens (initiators): Evidence for a role of proteases and DNA polymerase α. Int. J. Cancer 36:85-91.

23. Hennings, H., P.M. Blumberg, G.R. Pettit, C.L. Herrald, R. Shores, and C.H. Yuspa (1981) Bryostatin 1 an activator of protein kinase C inhibits tumor promotion by phorbol esters in SENCAR mouse skin. Carcinogenesis (Lond.) 8:1343-1346.

24. Hozumi, M., M. Ogawa, T. Sugimura, T. Takeuchi, and H. Umezawa (1972) Inhibition of tumorigenesis in mouse skin by leupeptin, a protease inhibitor from Actinomycetes. Cancer Res. 32:1725-1729.

25. Janoff, A., A. Klassen, and W. Troll (1970) Local vascular changes induced by the cocarcinogen, phorbol myristate acetate. Cancer Res. 30:2568-2571.

26. Kinzel, V., G. Fürstenberger, H. Loehrke, and F. Marks (1986) Three-stage tumorigenesis in mouse skin: DNA synthesis as a prerequisite for the conversion stage induced by TPA prior to initiation. Carcinogenesis 7:779-782.

27. Klebanoff, S.J. (1980) Oxygen metabolism and the toxic properties of phagocytes. Ann. Intern. Med. 93:480-489.

28. Knight, W.A., R.B. Livingstone, E.J. Gregory, and W.L. McGuire (1977) Estrogen receptors as an independent prognostic factor for early recurrence in breast cancer. Cancer Res. 37:4669-4671.

29. Kraft, A.S., F. William, G.R. Pettit, and M.B. Lilly (1989) Varied differentiation responses of human leukemias to bryostatin 1. Cancer Res. 49:1287-1293.

30. Nakayasu, M., H. Shima, S. Aonuma, H. Nakagama, M. Nagao, and T. Sugimura (1988) Deletion of transfected oncogenes from NIH 3T3 transformants by inhibitors of poly(ADP-ribose)polymerase. Proc. Natl. Acad. Sci., USA 85:9066-9070.

31. Nishizuka, Y. (1984) The role of protein kinase C in cell surface signal transduction and tumor promotion. Nature 308:693.

32. Phillips, R.L. (1975) Role of life style and their habits in risk of cancer among Seventh-Day Adventists. Cancer Res. 35:3513-3522.

33. Pollard, M., and P.H. Luckert (1986) Promotional effects of testosterone and dietary fat on the development of autochronous prostate cancer in rats. Cancer Lett. 32:223-227.

34. Slaga, T.J., S.M. Fischer, K. Nelson, and G.L. Gleason (1980) Studies on mechanism of action of anti-tumor-promoting agents: Evidence for several stages in promotion. Proc. Natl. Acad. Sci., USA 77:3659-3663.

35. Slaga, T.J., A.J.P. Klein-Szanto, L.L. Triplett, L.P. Yotti, and J.E. Trosko (1981) Skin tumor promoting activity of benzoyl peroxide, a widely used free radical-generating compound. Science 213:1023-1025.

36. Troll, W. (1989) Cancer prevention by inhibitors of tumor promotion. Proc. Am. Assoc. Cancer Res. 30:684-685.

37. Troll, W., and R. Wiesner (1985) The role of oxygen radicals as a possible mechanism of tumor promotion. Ann. Rev. Pharmacol. Toxicol. 25:509-528.

38. Troll, W., K. Frenkel, and G. Teebor (1984) Free oxygen radicals: Necessary contributors to tumor promotion and carcinogenesis. In Cellular Interactions by Environmental Tumor Promoters, H. Fujiki, E. Hecker, R.E. Moore, T. Sugimura, and I.B. Weinstein, eds.

Japan Scientific Societies Press, Tokyo/VNU Science Press BV, Utrecht, pp. 207–218.

39. Troll, W., K. Frenkel, and R. Wiesner (1984) Protease inhibitors as anticarcinogens. J. Natl. Cancer Inst. 73:1245–1250.

40. Troll, W., R. Wiesner, and K. Frenkel (1987) Anticarcinogenic action of protease inhibitors. In Advances in Cancer Research, Vol. 49, G. Klein and S. Weinhouse, eds. Academic Press, Inc., New York, pp. 265–283.

41. Van Duuren, B.L., and L. Orris (1965) The tumor-enhancing principles of Croton tiglium. Cancer Res. 25:1871–1875.

42. Verma, A.K., B.G. Shapas, H.M. Rice, and R.K. Boutwell (1979) Correlation of the inhibition by retinoids of tumor promoter-induced mouse epidermal ornithine decarboxylase activity and of skin tumor promotion. Cancer Res. 39:419–425.

43. Wattenberg, L.W. (1985) Chemoprevention of cancer. Cancer Res. 45:1–8.

ANTIMUTAGENIC ACTIVITY OF VITAMINS

IN CULTURED MAMMALIAN CELLS

Yukiaki Kuroda

Department of Ontogenetics
National Institute of Genetics
Mishima, Shizuoka 411, Japan

Cultured mammalian cell systems are useful for detecting quantitatively the mutagenicity and antimutagenicity of chemicals under strictly controlled conditions. These systems have some specific characteristics which are not seen in microbial systems.

CHARACTERISTICS OF CULTURED MAMMALIAN CELL SYSTEMS

By using cultured human diploid cells, the author detected the mutagenicities of some fungal toxins, food additives, amino acid pyrolysates, and other chemical mutagens, and evaluated their genetic toxicities for human cells (Tab. 1).

In these systems, the mutagenicity of some carcinogens whose mutagenicity is not detected in microbial systems can be detected as gene mutations, chromosome aberrations, sister chromatid exchanges (SCEs), unscheduled DNA syntheses, DNA strand breaks, and so on (2). We have detected the mutagenicities of hexamethylphosphoramide, o-toluidine, benzene, safrole, and phenobarbital by using 6-thioguanine ($\overline{6}$TG) resistance as a genetic marker in Chinese hamster V79 cells (27). In these systems, 8-azaguanine (8AG)-resistant and 6TG-resistant mutations are used to detect base changes and frameshifts of DNA nucleotides as well as various types of deletions of DNA. Similarly, ouabain (OVA)-resistant mutations are used to detect only base-change and frameshift mutations, which correspond well to reverse mutations in Salmonella typhimurium in the Ames test systems (Tab. 2) (28).

Aniline is a typical example of the usefulness of this system for detecting the mutagenicity of chemicals which are not detected in microbial systems. The mutagenicity of this chemical was not detected in S. typhimurium TA98, TA100, TA1535, or TA1537 (31), chromosome aberrations in Chinese hamster Don (1) and CHL cells (17), SCEs (1) and transformations in Syrian hamster BHK cells (39) (Tab. 3). However, aniline induced various sarcomas, fibromas, and cytomas in rats administered 0.3% or 0.6% aniline in their food (Tab. 4) (10). It was found that aniline

Tab. 1.   The activity of various chemicals to induce 8AG-resistant mutations in cultured human diploid cells.

Chemical	LD_{50} (μmole-hr/ml)	Induced mutation frequency		Reference
		At LD_{50} (x10 [5])	(x10 [5]/μmole-hr/ml)	
STC	0.004	10.2	2,550	Kuroda (24)
AF2	0.08	58.7	734	Kuroda (21)
Phloxine	0.07	24.3	347	Kuroda (22)
Trp-P-1	0.06	10.0	167	Kuroda (25)
Trp-P-2	0.02	2.8	140	Kuroda (26)
Glu-P-1	2	2.7	1.35	Kuroda (26)
EMS	20	24.5	1.23	Kuroda (20)
Sodium bisulfite	20	11.7	0.59	Kuroda (23)
Quercetin	1.32	0.66	0.5	Kuroda (unpubl.)

STC: Sterigmatocystin; AF2: trans-2-(2-furyl)-3-(5-nitro-2-furyl) acrylamide or furyl-
furamide;   phloxine: disodium   9-(3',4',5',6'-tetrachloro-o-carboxyphenyl)-6-hydroxy-
2,4,5,7-tetrabromo-3-isoxanthine; EMS: ethyl methanesulfonate.

Tab. 2.   Difference in inducibility of mutations among different genetic
markers by various chemicals in Chinese hamster V79 cells.

Mutagen	Ames test	Genetic markers		
		$8AG^r$	$6TG^r$	OUA^r
Ethyl methanesulfonate	+	+	+	+
Methyl methanesulfonate	+	+	+	+
MNNG	+	+	+	+
Trp-P-1	+	+		+
Nitrobenzene	–	+		+
Aniline	–	+		–
o-Chloroaniline		+		–
m-Chloroaniline		+		–
Hexachlorobenzene		+		–
Chloroform		+		–

Tab. 3. Mutagenicity and carcinogenicity of aniline in various test systems.

Test	Strain	Result	Reference
Mutation in Ames test	Salmonella TA 98 TA 100 TA 1535 TA 1537	– – – –	MacCann et al. (31)
Chromosome aberration	Chinese hamster Don CHL	– –	Abe and Sasaki (1) Ishidate and Shima (17)
SCE	Chinese hamster Don	–	Abe and Sasaki (1)
Cell transformation	Syrian hamster BHK	–	Styles (39)
Carcinogenesis in rats	Fischer 344	+	NIH Publication No. 78-1385 (10)
Carcinogenesis in mice	C57B16 X C3H F1	–	

induced 8AG-resistant mutations without S-9 mix, but not OUA-resistant mutations with or without S-9 mix, in Chinese hamster V79 cells. This suggests that aniline may induce some deletion-type mutations in mammalian cells (Fig. 1) (28).

In these systems, the treatment time of chemicals can be controlled only by the exchange of medium containing mutagens and carcinogens. The effects of more than two chemicals can be determined when they are present simultaneously, or at different times from each other. The combined effects of more than two mutagens or carcinogens on cell survival and mutation induction can be determined. Some chemicals act

Tab. 4. Carcinogenicity of aniline in rats (10).

	Control	Aniline HCL	
		0.3%	0.6%
No of rats tested	25	50	50
No of rats with			
Spleen			
Sarcoma	0	4(18 %)	2 (4 %)
Fibroma	0	7(14 %)	6 (13 %)
Fibrosarcoma	0	3(6 %)	7 (15 %)
Hemangiosarcoma	0	19(38 %)	20 (43 %)
Pituitary adenoma	2(9 %)	7(16 %)	4 (12 %)
Adrenal pheochromocytoma	1(4 %)	5(10 %)	11 (25 %)

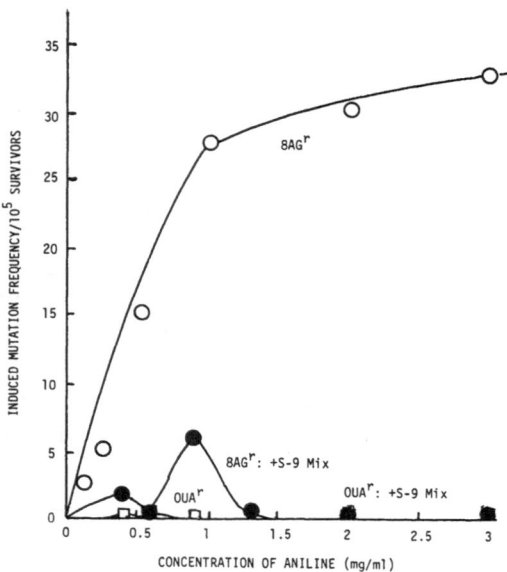

Fig. 1.   Effect of aniline on mutation inductions in Chinese hamster V79 cells (28).

synergistically with other chemicals, other chemicals act additively, and still other chemicals are competitive with each other on mutation induction of cells.   When ethyl methanesulfonate (EMS) at a concentration producing an LD_{50} was combined with methyl methanesulfonate (MMS) at various concentrations, some synergistic effects of both chemicals on the induction of 6TG-resistant mutations were found, compared with the theoretical values calculated from the mutation frequencies obtained by single chemicals (Fig. 2) (18).

However, when the cytotoxic effects of both chemicals were considered, the combined effects were not synergistic, but additive.   When cells were treated with the two chemicals successively at different times with an interval incubation in normal medium for 3 hr, the induced mutation frequency decreased, suggesting the operation of a repair mechanism for mutational damage after treatment with the first chemical (Fig. 3) (19).   The procedure for simultaneous or differential treatment of two chemicals can be utilized to examine the effects of antimutagens/anticarcinogens on mutagens/carcinogens in mammalian cells.   In the present study, the antimutagenic effects of vitamins on 6TG-resistant mutations induced by EMS were examined in Chinese hamster V79 cells.

MATERIALS AND METHODS

Cells and Culture

The cell line used was the V79 strain of Chinese hamster lung cells, isolated by Ford and Yerganian (11) from a normal young adult male Chinese hamster.   The cells were grown in large mass, distributed in many small ampules, and frozen at -80°C.   Before use, cells were thawed

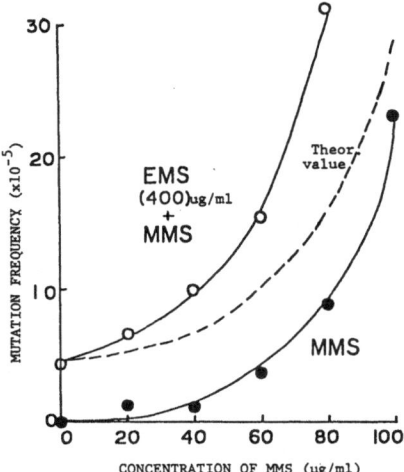

Fig. 2. Combined effects of EMS and MMS on mutation induction of Chinese hamster V79 cells (18).

at 38°C, suspended in fresh medium, and cultured in GHAT medium (3 x 10^{-6} M glycine, 10^{-4} M hypoxanthine, 4 x 10^{-7} M amethopterine, and 1.6 x 10^{-5} M thymidine) at 38°C for 24 hr, to remove the pre-existing 6TG-resistant cells in the cell population. During further subcultivation in normal medium, the appearance of spontaneous 6TG-resistant mutations was occasionally noted. For this reason, stock cultures were discarded at the tenth subculture and replaced from frozen stocks. The culture

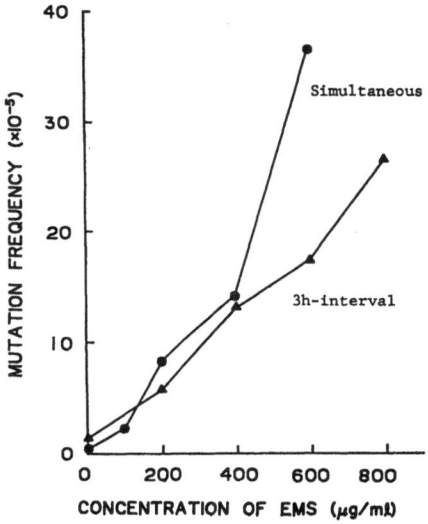

Fig. 3. Effects of interval incubation between treatments with EMS and MMS (400 μg/ml) on mutation induction of Chinese hamster V79 cells (19).

medium was Eagle's minimal essential medium (Nissui Seiyaku Co., Tokyo) supplemented with 10% fetal bovine serum (Gibco Laboratory, Grand Island, New York). The cells were cultured in 60-mm plastic contour petri dishes (Lux Science Corp., Newbury Park, California; catalog No. 5216) in 5 ml of medium under 5% CO_2 and 95% air at 38°C.

Mutagen and Vitamins

EMS used as a mutagen throughout the present experiments was a product of Aldrich Chemical Co., Inc., Milwaukee, Wisconsin. Vitamin C (L-ascorbic acid) was purchased from Wako Pure Chemical Industries, Ltd., Osaka. Derivatives of vitamin C used were dehydro-vitamin C (dehydroascorbic acid; Aldrich Chemical Co., Inc.) and iso-vitamin C (D-isoascorbic acid; Wako Pure Chemical Industries, Ltd.). The chemical structures of vitamin C and its derivatives are shown in Fig. 4.

Vitamin C and its derivatives were directly dissolved in Hanks' solution, adjusted to pH 7.2, and sterilized by filtration through a millipore filter. Like other vitamins, vitamin A (vitamin A acetate in oil) and vitamin E (DL-α-tocopherol) were purchased from Wako Pure Chemical Industries, Ltd. Vitamins A and E were dissolved in ethyl alcohol, filtered through a millipore filter, and added to Hanks' solution at a final concentration of 2% ethyl alcohol.

Cytotoxicity Assay

Cells were dissociated from monolayer cultures by treatment with 0.25% trypsin (Difco, 1:250) solution for 10 min at 38°C. The cytotoxicity assays of vitamins were carried out by determining colony formation of cells as shown in Fig. 5. Triplicate inocula of 10^2 cells in 5 ml of normal medium in 60-mm petri dishes were incubated for 20 hr at 38°C, during which time most cells became attached to the surface of the dishes. Then cells were treated with EMS and/or vitamins at various concentrations in Hanks' solution at 38°C for 3 hr.

After treatment of cells with EMS and/or vitamins, the cells were rinsed twice with Hanks' solution and incubated in normal medium for 6 da at 38°C. The cell colonies formed in petri dishes were fixed with methanol, and stained with May Grünwald-Giemsa solution. Colonies containing more than 50 cells were scored under a binocular microscope, and the colony-forming activity was calculated as the average number of colonies found as a percentage of the number of cells initially inoculated.

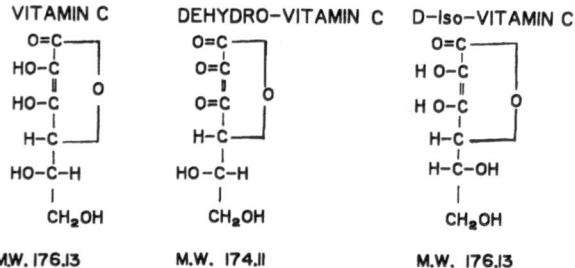

Fig. 4.   The chemical structures of vitamin C and its derivatives.

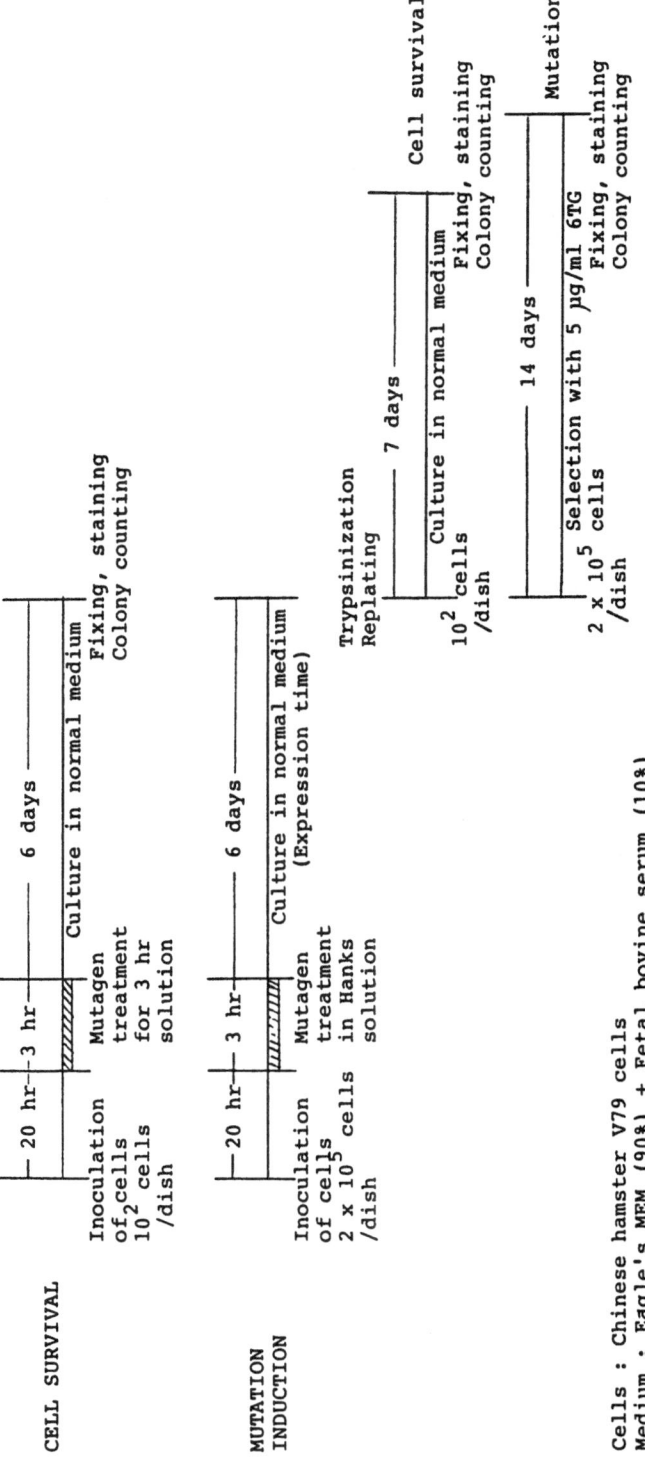

Cells : Chinese hamster V79 cells
Medium : Eagle's MEM (90%) + Fetal bovine serum (10%)

Fig. 5.   Procedure for detecting the cytotoxicity and mutagenicity of chemicals in Chinese hamster V79 cells.

The effects of EMS and vitamins are expressed as fractions (surviving fractions) of the colony-forming activity of untreated control cultures.

Mutagenicity and Antimutagenicity Assay

The mutagenicity and antimutagenicity of mutagen and antimutagen were determined by the "replating method" (27) (Fig. 5). The inocula of 2×10^5 cells in 10 ml of normal medium in 100-mm plastic petri dishes (Lux Science Corp.; catalog No. 5211) were incubated for 20 hr at 38°C and then treated with EMS and/or vitamins at almost the same concentrations as those used for the cytotoxicity assay for 3 hr. Then the cells were rinsed twice with Hanks' solution and incubated in normal medium for a mutation expression time of 6 da. The cells were then treated with trypsin, and 2×10^5 cells were replated each in five 100-mm petri dishes in medium containing 6TG (5 µg/ml) (Sigma Chemical Company). After incubation for 14 da, the cell colonies formed were fixed and stained and the number of 6TG-resistant colonies was scored. In parallel experiments, duplicate inocula of 10^2 cells in 60-mm petri dishes were incubated in normal medium for 7 da. The colony-forming activity of replated cells was determined as described above. The observed number of 6TG-resistant mutant colonies was corrected for the decrease in the colony-forming activity of cells in normal medium. The number of induced mutants was calculated by subtracting the number of colonies in untreated control cultures from those in treated cultures. The induced mutation frequency was expressed as the number of induced mutants per 10^5 colony-forming cells.

EFFECTS OF VITAMIN C AND ITS DERIVATIVES

Vitamin C is contained in Japanese green tea, fruits such as strawberry, pineapple, and orange, and vegetables such as cabbage, spinach, and pimentos in great quantity. It is effective in preventing scurvy, and its antioxidant activity is useful for detoxification of some drugs and for protection against rancidity in foodstuffs. In the present experiment, the effects of vitamin C and its derivatives on cytotoxicity and mutagenicity of EMS were compared in Chinese hamster V79 cells.

Effects of Vitamin C and Its Derivatives on EMS-Induced Cytotoxicity

When cells were treated with EMS alone for 3 hr, the cell survivals calculated from the number of colonies formed are shown in Tab. 5.

The cell survival decreased as the concentrations of EMS increased. At a concentration of 1,000 µg/ml of EMS, the cell survival was 9% (Fig. 6). The LD_{50} value of EMS calculated from the survival curve of cells was 541 µg/ml.

Vitamin C and its derivatives showed only weak or no cytotoxicity at less than 100 µg/ml for vitamin C and at less than 50 µg/ml for dehydro-vitamin C and iso-vitamin C (Tab. 6).

At higher concentrations, cell survival dropped rapidly. The LD_{50} values of vitamin C, dehydro-vitamin C, and iso-vitamin C were 132 µg/ml, 269 µg/ml, and 59 µg/ml, respectively, indicating that iso-vitamin C had the most cytotoxicity, and dehydro-vitamin C had the least cytotoxicity.

Tab. 5.   Effect of vitamin C and its derivatives on EMS-induced cytotoxicity of Chinese hamster V79 cells.

Concentration	Colony-forming activity (%)			
of EMS (μg/ml)	None	With vitamin C (100 μg/ml)	With dehydro vitamin C (100 μg/ml)	With iso- vitamin C (50 μg/ml)
0	100	100	100	100
100	72	129	69	106
200	70	99	56	99
400	68	139	29	94
600	44	126	13	66
800	29	129	4	16
1,000	9	58	0	13

The colony-forming activity of cells that were treated with EMS and vitamin C or its derivatives simultaneously for 3 hr is shown in Tab. 5.

In the presence of vitamin C, the cytotoxicity of EMS was markedly reduced. When vitamin C was added to EMS solution at a concentration of 100 μg/ml, the colony-forming activity of cells was maintained or was a little bit higher at concentrations less than 800 μg/ml of EMS and the

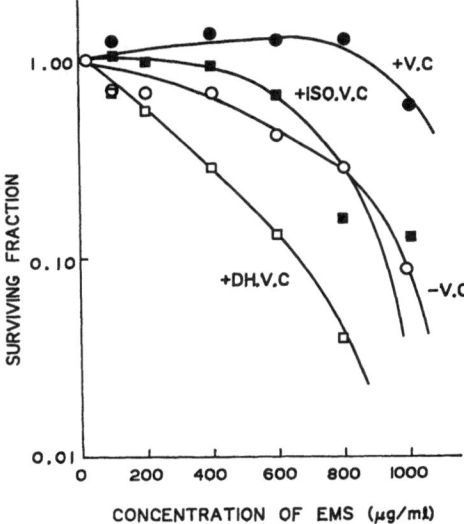

Fig. 6.   Effects of vitamin C and its derivatives on EMS-induced cytotoxicity in Chinese hamster V79 cells.

Tab. 6.   Effects of vitamin C and its derivatives on cell survival of
          Chinese hamster V79 cells.

Concentrations	Colony-forming activity (%)		
of vitamins (µg/ml)	Vitamin C	Dehydro-vitamin C	Iso-vitamin C
0	100	100	100
0.1	100	80	117
1	100	80	99
5	92	71	91
10	86	75	105
50	85	109	99
100	79	13	2
300	0	44	0

LD_{50} value of EMS was more than 1,000 µg/ml. On the other hand, in
the presence of dehydro-vitamin C at a concentration of 100 µg/ml, the
cytotoxicity of EMS was enhanced and the LD_{50} of EMS became 234 µg/ml.
When iso-vitamin C was present together with EMS at the time of treatment, the cytotoxicity of EMS slightly decreased. In the presence of
iso-vitamin C at a concentration of 50 µg/ml, the LD_{50} of EMS was 639
µg/ml. The survival curves of EMS in the presence of vitamin C or its
derivatives are shown in Fig. 6.

Effects of Vitamin C and Its Derivatives on EMS-Induced Mutations

When cells were treated with EMS alone for 3 hr, the frequencies of
6TG-resistant mutations increased as the concentrations of EMS increased,
as shown in Tab. 7. At a concentration of 1,000 µg/ml, EMS induced
6TG-resistant mutations at a frequency of 87.5×10^{-5}. On the other
hand, vitamin C, dehydro-vitamin C, or iso-vitamin C alone had no detectable activities in inducing 6TG-resistant mutations (Tab. 8).

The frequencies of 6TG-resistant mutations induced by EMS when
cells were treated with EMS and vitamin C or its derivatives simultaneously for 3 hr are shown in Tab. 7.

In the presence of vitamin C at a concentration of 100 µg/ml, EMS-
induced mutations are reduced significantly to about one-third or one-
fourth. Dehydro-vitamin C at a concentration of 100 µg/ml also decreased
the EMS-induced mutations to about one-half or one-third. Iso-vitamin C
at a concentration of 50 µg/ml decreased the EMS-induced mutations to a
similar frequency to that of dehydro-vitamin C. The frequencies of 6TG-
resistant mutations induced by EMS in the presence of vitamin C or its
derivatives are compared in Fig. 7.

It was reported that vitamin C at concentrations of higher than 10^{-3}
M (= 176.12 µg/ml) was mutagenic and induced chromosome aberrations
and unscheduled DNA synthesis (14,38). It is also mutagenic in the

Tab. 7.   Effects of vitamin C and its derivatives on 6TG-resistant mutations in Chinese hamster V79 cells.

Concentration	Induced mutation frequency (per 10^5 survivors)		
of vitamins	Vitamin C	Dehydro-vitamin C	Iso-vitamin C
(µg/ml)			
0	0	0	0
0.1	0.2	0	0.1
1	0	0	0.2
5	0	0	0
10	0.5	0	0.3
50	0.7	0	1.5
100	0	0	0

Ames Salmonella test (14).   At concentrations lower than 10^{-3} M, vitamin
C had no such effects.   In the present experiments, vitamin C at concentrations lower than 100 µg/ml (= 5.7 X 10^{-4} M) was not so cytotoxic and
inhibited EMS-induced mutations markedly in Chinese hamster V79 cells.
Dehydro-vitamin C and iso-vitamin C had more cytotoxic effect than
vitamin C and were less active in reducing EMS-induced mutations than
vitamin C.

Tab. 8.   Effects of vitamin C and its derivatives on EMS-induced mutations in Chinese hamster V79 cells.

Concentration	Induced mutation frequency (per 10^5 survivors)			
of EMS (µg/ml)	Without	With	With	With
	vitamin C	vitamin C	dehydro-vitamin	iso-vitamin
		(100µg/ml)	C (100µg/ml)	C (50µg/ml)
0	0	0	0	0
100	3.0	2.5	1.5	3.6
200	11.0	1.2	3.5	5.6
400	19.5	7.7	13.9	15.4
600	36.0	12.3	13.4	18.1
800	43.0	12.2	26.7	20.7
1,000	87.5	24.7	32.7	25.0

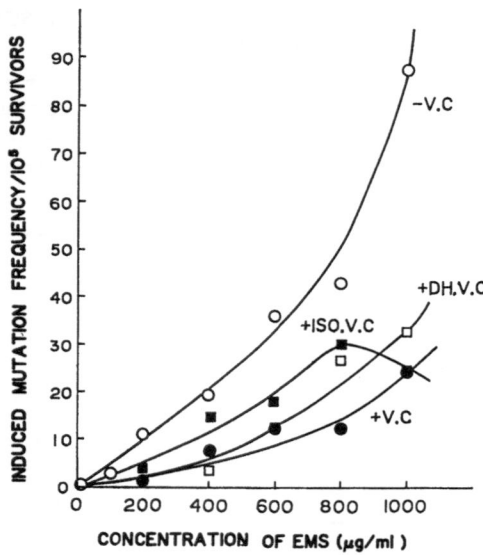

Fig. 7.   Effects of vitamin C and its derivatives on EMS-induced mutations in Chinese hamster V79 cells.

Mechanisms of Inhibition of Mutations by Vitamin C

        The mechanisms by which vitamin C inhibits EMS-induced mutations
were examined by experiments in which vitamin C was administered at different times from that of EMS.

        Figure 8 presents a protocol of such experiments.   In all experi-
ments, the concentration of EMS was 1,000 μg/ml and that of vitamin C
was 50 μg/ml.   In the previous experiments described above, cells were
treated with vitamin C together with EMS simultaneously.   In the present
experiments, cells were previously treated with vitamin C for 3 hr, rinsed

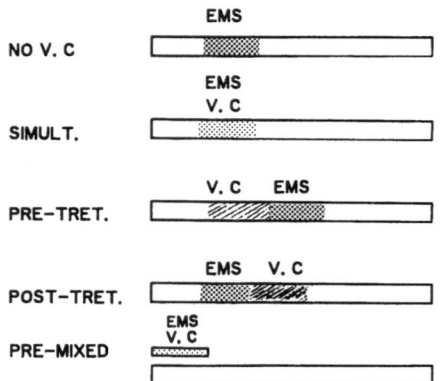

Fig. 8.   Protocol for testing effects of various treatments with vitamin C
on EMS-induced mutations.

twice in Hanks' solution, and then treated with EMS for another 3 hr. In the second experiment, cells were previously treated with EMS for 3 hr, rinsed in Hanks' solution, and then treated with vitamin C for 3 hr. In the last experiment in this series, a mixture of EMS and vitamin C was previously incubated without cells at 38°C for 3 hr, and then it was added to the medium in which cells were cultured and incubated for 3 hr. Results are shown in Tab. 9.

The addition of vitamin C (100 µg/ml) simultaneously with EMS had a marked effect in reducing the mutation frequencies, as described above. Pretreatment with vitamin C (50 µg/ml) had no detectable effects on EMS-induced mutation frequencies. Post-treatment with vitamin C (50 µg/ml) was also not effective in reducing the EMS-induced mutation frequencies, but enhanced slightly the mutation frequencies induced by EMS at concentrations of 400 to 800 ug/ml. However, preincubating a mixture of EMS and vitamin C had a marked effect in reducing the EMS-induced mutation frequencies to the frequencies obtained in the simultaneous treatment with vitamin C and EMS. Figure 9 shows the comparison of these various treatments with vitamin C on EMS-induced mutations.

These results indicate that vitamin C is effective in reducing EMS-induced mutations only when it is present together with EMS in the same solution. This suggests that vitamin C has a desmutagenic activity for EMS. Vitamin C may react directly with EMS and inactivate its mutation-inducing activity in Chinese hamster V79 cells.

EFFECTS OF VITAMIN E

It is known that vitamin E functions as an intracellular antioxidant and stabilizes cell membrane systems. In the present experiment, the

Tab. 9. Effects of various treatments with vitamin C on EMS-induced mutations in Chinese hamster V79 cells.

Concentration	Induced mutation frequency (per 10^5 survivors)				
of EMS (µg/ml)	Without V. C	With V. C (100µg/ml)	With pre-treatment with V. C (50µg/ml)	With Post-treatment with V. C (50µg/ml)	Treatment with a Mixture of EMS-V. C
0	0	0	0.2	0.4	0
100	3.0	2.5	12.0	10.6	2.2
200	11.0	1.2	7.0	18.0	4.9
400	19.5	7.7	19.0	31.9	15.1
600	36.0	12.3	45.0	61.0	20.3
800	43.0	12.2	69.0	77.7	30.6
1,000	87.5	24.7	67.0	92.3	26.0

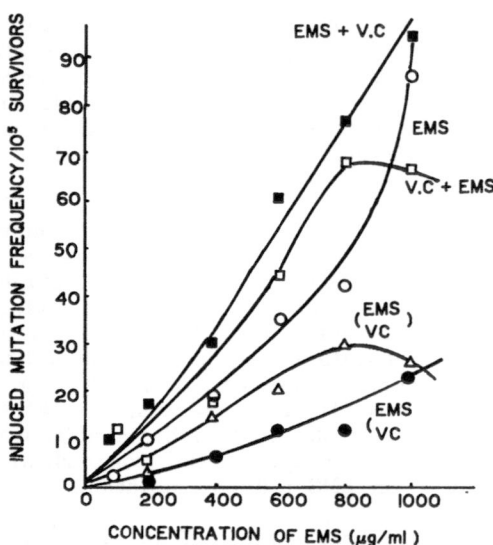

Fig. 9.   Comparison of various treatments with vitamin C on EMS-
         induced mutations in Chinese hamster V79 cells.

effects of vitamin E on the cytotoxicity and mutations induced by EMS
were examined.

Effect of Vitamin E on EMS-Induced Cytotoxicity

The cytotoxicity of vitamin E is shown in Tab. 10.   Hanks' solution
used for treatment of vitamin E contained ethyl alcohol at a final concentration of 2%, which reduced the cell survival to 71% of control cultures
treated with Hanks' solution.   Figure 9 indicates that vitamin E had no

Tab. 10.   Effect of vitamin E on cell survival of Chinese hamster V79
          cells.

Concentration of vitamin E (µg/ml)	Number of cells inoculated	Number of colonies formed	Surviving fraction
0	100	109	1.00
1	100	100	0.92
10	100	89	0.82
50	100	95	0.87
100	100	109	1.00
500	100	94	0.86
1,000	100	59	0.54

significant effect on cell survival of Chinese hamster V79 cells at concentrations lower than 500 µg/ml. The LD_{50} value of vitamin E was more than 1,000 µg/ml.

When cells were treated with 800 µg/ml EMS for 3 hr in the presence of vitamin E at various concentrations, the cell survival calculated from the colony-forming activities of cells is shown in Tab. 11.

EMS at a concentration of 800 µg/ml reduced cell survival to 19% of that of the control culture. In the presence of vitamin E at various concentrations, EMS-induced cytotoxicity was markedly enhanced, indicating that vitamin E and EMS had an additive cytotoxic effect.

Effects of Vitamin E on EMS-Induced Mutations

Vitamin E alone had no detectable mutagenicity in inducing 6TG-resistant mutations in Chinese hamster V79 cells. Table 12 shows that vitamin E did not induce mutations even at as high a concentration as 1,000 µg/ml (1 mg/ml). When cells were treated with EMS in the presence of vitamin E at various concentrations, EMS-induced mutations were markedly enhanced, as shown in Tab. 13.

These results indicate that vitamin E enhanced the frequencies of 6TG-resistant mutations induced by EMS. EMS at a concentration of 800 µg/ml induced 26 mutants per 10^5 survivors. When the concentrations of vitamin E added to Hanks' solution were increased, the EMS-induced mutation frequencies rose to 73 per 10^5 survivors at a concentration of 100 µg/ml of vitamin E.

To examine the mechanism of the enhancing effect of vitamin E on mutations induced by EMS, cells were treated with vitamin E at various concentrations, rinsed in Hanks' solution, and then treated with EMS. The frequency of 6TG-resistant mutations obtained in this experiment is shown in Tab. 14.

Tab. 11. Effect of vitamin E on EMS-induced cytotoxicity of Chinese hamster V79 cells.

Concentration of EMS (µg/ml)	Concentration of vitamin E (µg/ml)	Number of colonies formed	Surviving fraction
0	0	109	1.00
800	0	21	0.19
800	5	11	0.10
800	10	9	0.08
800	50	4	0.04
800	100	5	0.05
800	500	4	0.04
800	1,000	0	0

Tab. 12. Effect of vitamin E on 6TG-resistant mutations in Chinese hamster V79 cells.

Concentration of vitamin E (µg/ml)	cell survival	Number of 6TG-resistant mutants	Induced mutation frequency per 10^5 survivors
0	1.00	0.2	0
1	1.01	0.5	0.2
10	1.09	0.6	0.2
50	1.10	0	0
100	0.96	0	0
500	0.99	0.5	0.2
1,000	0.88	0	0

These data indicate that the pretreatment with vitamin E had no detectable effect in modifying the frequencies of EMS-induced mutations even when cells were pretreated with vitamin E at a concentration of 1,000 µg/ml. These results suggest that vitamin E had a marked enhancing effect on EMS-induced mutations and that this enhancing effect was not produced by pretreatment with vitamin E.

Tab. 13. Effect of vitamin E on EMS-induced mutations in Chinese hamster V79 cells.

Concentration of EMS (µg/ml)	Concentration of vitamin E (µg/ml)	Cell survival	Number of 6TG-resistant mutants	Induced mutation frequency per 10^5 survivors
0	0	1.36	0	0
800	0	0.91	51	26
800	1	0.85	88	44
800	5	0.60	120	60
800	10	0.70	100	50
800	50	0.65	143	72
800	100	0.76	146	73

Tab. 14. Effect of pretreatment with vitamin E on EMS-induced mutations in Chinese hamster V79 cells.

Concentration of vitamin E (μg/ml)	Concentration of EMS (μg/ml)	Cell survival	Number of 6TG-restant mutants	Induced mutation frequency per 10^5 survivors
0	0	1.18	0.2	0
0	600	0.77	127.0	63.4
1	600	0.70	143.6	71.7
10	600	0.89	139.9	69.9
50	600	1.06	83.7	41.8
100	600	0.97	107.8	53.8
500	600	0.88	122.1	61.0
1,000	600	1.05	80.7	40.3

EFFECT OF VITAMIN A

Vitamin A (retinol) is also a vitamin which is known for its various physiological activities in higher animals. The effect of vitamin A on EMS-induced cytotoxicity was examined. Vitamin A alone had no cytotoxic effects at concentrations less than 100 μg/ml. At more than 500 μg/ml, vitamin A was cytotoxic (Tab. 15). The LD_{50} value of vitamin A was 302 μg/ml. Cells were treated with EMS at various concentrations in the presence of vitamin A at a concentration of 100 μg/ml.

Tab. 15. Effect of vitamin A on cell survival of Chinese hamster V79 cells.

Concentration of vitamin A (μg/ml)	Number of cells inoculated	Number of colonies formed	Surviving fraction
0	100	107	1.00
1	100	97	0.91
10	100	94	0.88
50	100	120	1.12
100	100	130	1.21
500	100	38	0.36
1,000	100	0	0

The effect of vitamin A on survival (colony-forming ability) of cells treated with EMS is shown in Tab. 16. This data indicates that vitamin A had no marked effect on EMS-induced cytotoxicity in Chinese hamster V79 cells.

Vitamin A alone did not induce 6TG-resistant mutations in Chinese hamster V79 cells (Tab. 17).

In the presence of vitamin A at a concentration of 100 µg/ml, EMS-induced mutation frequencies rose markedly as the concentrations of EMS increased. The results are shown in Tab. 18. When the frequencies of EMS-induced mutations are compared in the absence (second column in Tab. 7) and the presence of vitamin A (Tab. 18), the frequencies of induced mutations in both experiments are similar to each other, although vitamin A slightly reduced EMS-induced mutations at concentrations less than 600 µg/ml of EMS.

DISCUSSION

Vitamin C at high concentrations is cytotoxic and mutagenic as described in earlier reports, inducing DNA-strand breakages (33) and SCEs (12, 37, 42). Vitamin C ($\geq 10^{-4}$ M) enhanced the numbers of SCEs induced by an enzymatic oxygen radical-generating system (xanthine oxidase plus hypoxanthine) in cultured Chinese hamster ovary cells (42).

On the other hand, vitamin C at concentrations as low as 5×10^{-6} M decreased the frequency of SCEs induced by trenimon in human peripheral blood lymphocytes with and without addition of rat-liver S-9 mix (13).

Vitamin C was found to inhibit mutagenicity of dimethyl-nitrosamine (14; 29), sterigmatocystin (29), β-propiolactone (36), N-methyl-N'-nitro-N-nitrosoguanidine (14,34), and N-hydroxy-2-acetylaminofluorene (40).

Tab. 16. Comparison of effects of EMS on cell survival in the absence and presence of vitamin A in Chinese hamster V79 cells.

Concentration of EMS (µg/ml)	Number of cells inoculated	without vitamin A		with vitamin A (100 µg/ml)	
		Number of colonies	Surviving fraction	Number of colonies	surviving fraction
0	100	100	1.00	108	1.00
100	100	72	0.72	90	0.83
200	100	70	0.70	94	0.87
400	100	68	0.68	63	0.58
600	100	44	0.44	25	0.23
800	100	29	0.29	12	0.11
1,000	100	9	0.09	3	0.03

Tab. 17. Effect of vitamin A on 6TG-resistant mutations in Chinese hamster V79 cells.

Concentration of vitamin A (μg/ml)	Cell survival	Number of 6TG-resistant mutants	Induced mutation frequency per 10^5 survivors
0	0.78	0	0
1	0.93	0.2	0.1
10	0.79	0.3	0.2
50	0.83	0.5	0.3
100	0.90	0	0

The effect of vitamin C in inhibiting mutagenicity of a variety of carcinogens was attributed to its antioxidant action.

Many studies also have been carried out on the anticarcinogenic activity of vitamin C. Sodium ascorbate inhibited tumor induction in rats treated with morpholine and sodium nitrite or with nitrosomorpholine (32). Vitamin C is an effective protector against the carcinogenic activity of ionizing radiations and chemical agents. It is effective in affecting 3-methyl-cholanthrene-induced cell transformation in C3H/10T 1/2 mouse-embryo fibroblast cells in culture (35). The amount of vitamin C which we ingest in our food may be 20-500 μg/kg of body weight. This amount corresponds to about 1 to 25 x 10^{-4} M. In the present experiment, the

Tab. 18. Effect of vitamin A on EMS-induced mutations in Chinese hamster V79 cells.

Concentration of EMS (μg/ml)	Concentration of vitamin A (μg/ml)	Cell survival	Number of 6TG-resistant mutants	Induced mutation frequency per 10^5 survivors
0	0	1.17	1.0	0
0	100	0.84	1.9	1.0
100	100	0.82	7.7	3.9
200	100	0.78	7.7	3.9
400	100	0.78	24.1	12.1
600	100	0.66	38.5	14.3
800	100	0.64	96.6	48.3
1,000	100	0.59	100.2	75.1

concentration of vitamin C that inhibited EMS-induced mutations in Chinese hamster V79 cells was less than 100 µg/ml (= 5.7 x 10^{-4} M).

It was shown that vitamin C is effective in reducing the EMS mutagenesis only when it is present in Hanks' solution together with mutagens during treatment with mutagens. Most effects of vitamin C are assumed to be due to its desmutagenic activity.

As for vitamin E, a few studies on its role in mutagenesis and carcinogenesis have been reported. It was found that vitamin E (α-tocopherol) reduced the mutagenic effect of malonaldehyde and β-propiolactone, both direct-acting mutagens (36).

The trenimon-induced SCEs in human peripheral blood lymphocytes decreased with simultaneous addition of vitamin E in the mutagen-containing medium (13). The effective concentration of vitamin E was 5 x 10^{-5} M. In the present experiment, vitamin E had no inhibitory effect on the mutations induced by EMS, but at concentrations up to 100 µg/ml (= 2.3 x 10^{-4} M), vitamin E effected a marked enhancement of the mutation frequency in Chinese hamster V79 cells. It is interesting to note that vitamin E was remarkably different from the other antioxidant, vitamin C, in inhibiting EMS-induced mutations.

On the other hand, the antimutagenic actions of vitamin A have been more extensively investigated. Vitamin A (retinol) inhibited the mutagenicity induced by aflatoxin B_1 (4,6), 2-aminofluorene, 2-acetylaminofluorene (7), orthoaminoazotoluene (8,41), protein pyrolysate (9), dimethylnitrosamine (DMN), and diethylnitrosamine (DEN) (16) in the Salmonella/liver microsome assays. For some procarcinogen-induced mutagenesis, it was suggested that vitamin A may exert its antimutagenic activities by inhibiting certain forms of the cytochrome P450 isoenzymes required for activation of procarcinogens (16). In cultured mammalian cells, vitamin A inhibited growth of tumor cell lines (30), suppressed carcinogen-induced cell transformation (3), and antagonized the effect of tumor promoters (5). It was suggested that anticarcinogenic effects of vitamin A may result from inhibition of the promotion and/or expression step of carcinogenesis (16). For mutagenicity by chemicals in mammalian cells, vitamin A inhibited SCEs induced by cyclophosphamide, aflatoxin B_1 (15), DMN, and DEN (16) in Chinese hamster V79 cells. In the present experiment, vitamin A slightly reduced EMS-induced 6TG-resistant mutations in Chinese hamster V79 cells. The concentrations used for inhibition of mutagen-induced SCEs were 4 to 32 µg/ml (16). For inhibition of EMS-induced mutations, higher concentrations of vitamin A were effective in the present experiment.

In the present studies, EMS was used as a typical mutagen to induce 6TG-resistant mutations in Chinese hamster V79 cells. Effective and non-effective vitamins examined here for inhibiting EMS-induced cytotoxicity and 6TG-resistant mutations have inhibitory activities to various other mutagens with different action mechanisms which remain to be examined in future experiments.

SUMMARY

Cultured mammalian cell systems are useful for examining the quantitative effects of mutagens and antimutagens on cell survival and gene mutations and the mechanisms of the interaction of two chemicals in the process of mutation induction. In the present article, the antimutagenic effects of vitamins C, E, and A, and derivatives of vitamin C on EMS-induced 6TG-resistant mutations in Chinese hamster V79 cells were examined.

Vitamin C was most effective in inhibiting EMS-induced cytotoxicity and 6TG-resistant mutations. In the presence of vitamin C at a concentration of 100 µg/ml, EMS-induced mutations were reduced to about one-third or one-fourth of those in control cultures treated with EMS alone. Dehydro-vitamin C and iso-vitamin C also inhibited EMS-induced mutations to about one-half or one-third of the control level. The fact that vitamin C was effective in reducing EMS-induced mutations when EMS was previously incubated together with vitamin C for 3 hr suggests that vitamin C may react directly with EMS as a desmutagen and thus inactivate its mutation-inducing activity in Chinese hamster V79 cells.

Vitamin E had an additive cytotoxic effect on EMS-induced cytotoxicity. This vitamin enhanced the frequencies of 6TG-resistant mutations induced by EMS. Pretreatment with vitamin E before treatment with EMS resulted in no detectable effect in modifying the EMS-induced mutations. On the contrary, vitamin A markedly enhanced EMS-induced mutation frequencies.

ACKNOWLEDGEMENTS

This work was supported in part by a Grant-in-Aid for Environmental Science from the Ministry of Education, Science and Culture, Japan. The author is indebted to Dr. Hideyuki Furukawa, Meijyo University, Nagoya, for providing samples of vitamin C derivatives and for helpful discussions. The author wishes to thank Miss Y. Takada for her technical assistance during this work. This is contribution No. 1787 from the National Institute of Genetics, Mishima, Japan.

REFERENCES

1. Abe, S., and M. Sasaki (1977) Chromosome aberrations and sister chromatid exchanges in Chinese hamster cells exposed to various chemicals. J. Natl. Cancer Inst. 58:1635-1641.
2. Ashby, J., F.J. de Serres, M. Ishidate, Jr., B.H. Margolin, B.E. Matter, and M.D. Shelby, eds. (1985) Evaluation of Short-term Tests for Carcinogenesis: Report of the International Programme on Chemical Safety Collaborative Study on In Vitro Assays, Elsevier Science Publisher, Amsterdam, Oxford, New York, 752 pp.
3. Bertram, J.S., and J.E. Marner (1985) Inhibition by retinoids of neoplastic transformation in vitro: Cellular and biochemical mechanisms. In Ciba Foundation Symposium 113, Retinoids, Differentiation and Disease, Pitman, London, pp. 29-41.
4. Bhattacharya, R.K., A.R. Francis, and T.K. Shetty (1987) Modifying role of dietary factors on the mutagenicity of aflatoxin B_1: In vitro effect of vitamins. Mutat. Res. 188:121-128.
5. Boutwell, R.K., A.K. Verma, M. Takigawa, C.L. Loprinzi, and P.P. Cabone (1985) Retinoids as inhibitors of tumor promotion. In Retinoids: New Trends in Research and Therapy, J.H. Saunat, ed. Karger, Basel, pp. 83-96.

6. Busk, L., and U.G. Ahlborg (1980) Retinol (vitamin A) as an inhib-
    itor of the mutagenicity of aflatoxin B_1. Toxicol. Lett. 6:243-249.
7. Busk, L., and U.G. Ahlborg (1982) Retinol (vitamin A) as a modi-
    fier of 2-aminofluorene and 2-acetyl-aminofluorene mutagenesis in the
    Salmonella/microsome assay. Arch. Toxicol. 49:169-174.
8. Busk, L., and U.G. Ahlborg (1982) Retinoids as inhibitors of ortho-
    amino-azotoluene-induced mutagenesis in the Salmonella/liver micro-
    some test. Mutat. Res. 104:225-231.
9. Busk, L., U.G. Ahlborg, and L. Albanus (1982) Inhibition of pro-
    tein pyrolysate mutagenicity by retinol (vitamin A). Food Chem.
    Toxicol. 20:535-539.
10. Carcinogenesis Testing Program (1978) Report on the Bioassay of
    Aniline Hydrochloride for Possible Carcinogenicity, U.S. Department
    of Health, Education and Welfare, National Institute of Health, DHEW
    Publication No. 78-1385, pp. 1-53.
11. Ford, D.K., and G. Yerganian (1958) Observations on the chromo-
    somes of Chinese hamster cells in tissue culture. J. Natl. Cancer
    Inst. 21:393-425.
12. Galloway, S.M., and R.B. Painter (1978) Vitamin C is positive in the
    DNA synthesis inhibition and sister-chromatid exchange tests.
    Mutat. Res. 60:321-327.
13. Gebhart, E., H. Wagner, K. Griziwok, and H. Behnsen (1985) The
    action of anticlastogens in human lymphocyte cultures and their
    modification by rat-liver S9 mix. II. Studies with vitamins C and E.
    Mutat. Res. 149:83-94.
14. Guttenplan, J.B. (1977) Inhibition by L-ascorbate of bacterial muta-
    genesis induced by two N-nitroso compounds. Nature 268:368-370.
15. Huang, C.C., J.L. Hsueh, H.H. Chen, and T.R. Butt (1982) Reti-
    nol (vitamin A) inhibits sister chromatid exchanges and cell cycle
    delay induced by cyclophosphamide and aflatoxin B in Chinese ham-
    ster V79 cells. Carcinogenesis 3:1-5.
16. Huang, C.C. (1987) Retinol (vitamin A) inhibition of dimethylnitros-
    amine (DMN) and diethylnitrosamine (DEN) induced sister-chromatid
    exchanges in V79 cells and mutations in Salmonella/microsome assays.
    Mutat. Res. 187:133-140.
17. Ishidate, Jr., M., and S. Odashima (1977) Chromosome tests with
    134 compounds in Chinese hamster cells in vivo--A screening for
    chemical carcinogens. Mutat. Res. 48:337-354.
18. Kojima, H., H. Konishi, and Y. Kuroda (1986) Combined effects of
    chemicals on mammalian cells in culture. I. Effects of methyl meth-
    anesulfonate (MMS) and ethyl methanesulfonate (EMS) on the muta-
    tion induction (abstract). Mutat. Res. 164:272.
19. Kojima, H., H. Konishi, and Y. Kuroda (1987) Combined effects of
    chemicals on mammalian cells in culture. Effect of order and inter-
    vals of administration of methyl methanesulfonate and ethyl methane-
    sulfonate (abstract). Mutat. Res. 182:364.
20. Kuroda, Y. (1974) Mutagenesis in cultured human diploid cells. II.
    Chemical induction of 8-azaguanine-resistant mutations. Japan. J.
    Genet. 49:389-398.
21. Kuroda, Y. (1975) Mutagenesis in cultured human diploid cells. III.
    Induction of 8-azaguanine-resistant mutations by furylfuramide.
    Mutat. Res. 30:229-238.
22. Kuroda, Y. (1975) Mutagenesis in cultured human diploid cells. IV.
    Induction of 8-azaguanine resistant mutations by phloxine, a muta-
    genic red dye. Mutat. Res. 30:239-248.
23. Kuroda, Y. (1977) Induction of 8-azaguanine-resistant mutations by

sulfite in cultured embryonic human diploid cells. Ann. Rep. Natl. Inst. Genet. Japan 27:39-40.

24. Kuroda, Y. (1979) Induction of 8-azaguanine resistant mutations by sterigmatocystin in cultured embryonic human diploid cells. Ann. Rep. Natl. Inst. Genet. Japan 29:38-39.

25. Kuroda, Y. (1979) Mutagenic activity of tryptophan pyrolysis products on embryonic human diploid cells in culture. Ann. Rep. Natl. Inst. Genet. Japan 29:39.

26. Kuroda, Y. (1981) Mutagenic activity of Trp-P-2 and Glu-P-1 on embryonic human diploid cells in culture. Ann. Rep. Natl. Inst. Genet. Japan 31:45-46.

27. Kuroda, Y., A. Yokoiyama, and T. Kada (1985) Assays for the induction of mutations to 6-thioguanine resistance in Chinese hamster V79 cells in culture. In Evaluation of Short-Term Tests for Carcinogenesis: Report of the International Programme on Chemical Safety Collaborative Study on In Vitro Assays, J. Ashby, F.J. de Serres, M. Ishidate, Jr., B.H. Margolin, B.E. Matter, and M.D. Shelby, eds. Elsevier Science Publisher, Amsterdam, Oxford, New York, pp. 537-542.

28. Kuroda, Y. (1986) Genetic and chemical factors affecting chemical mutagenesis in cultured mammalian cells. In Antimutagenesis and Anticarcinogenesis Mechanisms, D.M. Shankel, P.E. Hartman, T. Kada, and A. Hollaender, eds. Plenum Press, New York and London, pp. 359-375.

29. Lo, L.W., and H.F. Stich (1978) The use of short-term tests to measure the preventive action of reducing agents on formation and activation of carcinogenic nitroso compounds. Mutat. Res. 57:57-67.

30. Lotan, R. (1985) Mechanism of inhibition of tumor cell proliferation by retinoids. Studies with cultured melanoma cells. In Retinoids: New Trends in Research and Therapy, J.H. Saunat, ed. Karger, Basel, pp. 9-105.

31. MacCann, J., E. Choi, E. Yamasaki, and B.N. Ames (1975) Detection of carcinogens as mutagens in the Salmonella/microsome test: Assay of 300 chemicals. Proc. Natl. Acad. Sci., USA 72:5135-5139.

32. Mirvish, S.S., A.F. Pelfrene, H. Garcia, and P. Shubik (1976) Effect of sodium ascorbate on tumor induction in rats treated with morpholine and sodium nitrite, and with nitrosomorphiline. Cancer Lett. 2:101-108.

33. Morgan, A.R., R.L. Cone, and T.M. Elgert (1976) The mechanism of DNA strand breakage by vitamin C and superoxide and the protective roles of catalase and superoxide dismutase. Nucl. Acids Res. 3:1139-1149.

34. Rosin, M.P., and H.F. Stich (1979) Assessment of the use of the Salmonella mutagenesis assay to determine the influence of antioxidants on carcinogen induced mutagenesis. Int. J. Cancer 23:722-727.

35. Rosin, M.P., A.R. Peterson, and H.F. Stich (1980) The effect of ascorbate on 3-methylcholanthrene-induced cell transformation in C3H/10Tl/2 mouse-embryo fibroblast cell cultures. Mutat. Res. 72:533-537.

36. Shamberger, R.J., C.L. Corlett, K.D. Beaman, and B.L. Kasten (1979) Antioxidants reduce the mutagenic effect of malonaldehyde and β-propiolactone, Part IX. Antioxidants and cancer. Mutat. Res. 66:349-355.

37. Speit, G., M. Wolf, and W. Vogel (1980) The SCE-inducing capacity
     of vitamin C: Investigation in vitro and in vivo. Mutat. Res.
     78:273-278.
38. Stich, H.F., J. Karim, J. Koropatnick, and L. Lo (1976) Mutagenic
     action of ascorbic acid. Nature 26:722-724.
39. Styles, J.A. (1978) Mammalian cell transformation in vitro. Br. J.
     Cancer 37:931-936.
40. Thorgeirsson, S.S., S. Sasaki, and P.J. Wirth (1980) Effect of as-
     corbic acid on the in vitro mutagenicity and in vivo covalent binding
     of N-hydroxy-2-acetylaminofluorene in the rat. Mutat. Res. 70:395-
     398.
41. Victorin, K., L. Busk, and U.G. Ahlborg (1987) Retinol (vitamin A)
     inhibits the mutagenicity of o-aminoazotoluene activated by liver
     microsomes from several species in the Ames test. Mutat. Res.
     179:41-48.
42. Weitberg, A.B., and S.A. Weitzman (1985) The effect of vitamin C
     on oxygen radical-induced sister-chromatid exchanges. Mutat. Res.
     144:23-26.
43. Weitberg, A.B. (1987) Antioxidants inhibit the effect of vitamin C on
     oxygen radical-induced sister-chromatid exchanges. Mutat. Res.
     191:53-56.

MICROCELL-MEDIATED CHROMOSOME TRANSFER: A STRATEGY

FOR STUDYING THE GENETICS AND MOLECULAR PATHOLOGY

OF HUMAN HEREDITARY DISEASES WITH ABNORMAL RESPONSES

TO DNA DAMAGE

Errol C. Friedberg,[1] Karla Henning,[1] Clare Lambert,[2]
Paul J. Saxon,[2] Roger A. Schultz,[2] Gurbax S. Sekhon,[3]
and Eric J. Stanbridge[4]

[1]Departments of Pathology and Genetics
Stanford University School of Medicine
Stanford, California 94305

[2]Division of Human Genetics
University of Maryland
Baltimore, Maryland 21201

[3]Cytogenetics Laboratory
University of Wisconsin
Madison, Wisconsin 53705

[4]Department of Microbiology and Molecular Genetics
University of California
Irvine, California 92717

INTRODUCTION

Recent progress in molecular biology has begun to facilitate an understanding of many hereditary human diseases at the molecular level. Not surprisingly, attention has been largely focused on diseases that are particularly prevalent and which extract a major toll in terms of human morbidity and mortality. A general experimental strategy that has been pursued in many laboratories is to first map the position of the disease locus of interest in the human genome (frequently by demonstrating genetic linkage to known restriction fragment length polymorphisms), and to identify closely linked markers which can be used as probes to physically delineate a defined region of the genome (19). Such probes can then be used to screen genomic libraries in order to isolate recombinant plasmids containing inserts from the region of interest.

The localization of a disease locus to a relatively defined region of the three billion base pairs in the human genome represents a significant technical tour de force. Nonetheless, the precise delineation of a specific single gene of interest remains a time consuming and challenging endeavor, because in most cases thus far studied the mutant cells do not have selectable phenotypes. Hence, the presumption of successful gene cloning must necessarily rely on criteria such as appropriate (i.e., tissue-specific) expression, or on sequence comparisons between wild-type and mutant alleles.

Several rare human hereditary diseases do offer the advantage of a selectable phenotype in cell culture and hence are especially attractive for gene cloning by phenotypic complementation. Among these are xeroderma pigmentosum (XP), ataxia telangiectasia (AT), Fanconi's anemia (FA), and Cockayne's syndrome (CS), all of which are diseases characterized at the cellular level by abnormal sensitivity to killing by agents known to damage DNA (9). The incidence of these diseases is rare compared to diseases such as cystic fibrosis or muscular dystrophy. Nonetheless, they are associated with crippling infirmities and/or significant predisposition to neoplasia, and constitute important members of the family of human hereditary disease states.

HUMAN HEREDITARY DISEASES WITH ABNORMAL SENSITIVITY TO DNA DAMAGE

Xeroderma Pigmentosum

XP is the only member of this group of diseases in which cellular hypersensitivity to killing by DNA damaging agents is well correlated with defective DNA repair (9). Experiments in numerous laboratories have demonstrated defective excision of helix-distorting bulky base adducts from DNA, such as those produced by exposure of cells to UV radiation at $\sim$254 nm, or to a variety of bulky chemicals (9). This latter group includes well-characterized carcinogens such as polycyclic aromatic hydrocarbons and aromatic amines. Individuals homozygous for the genetic defect are extremely sensitive to UV rays from the sun and manifest severe photosensitivity of the skin, eyes, and anterior region of the tongue. Consistent with the role of DNA repair in protecting the genome from mutations in somatic cells, XP sufferers have an extremely high incidence of cancers of photosensitive areas of the body (9).

The genetic aspects of this disease are extremely complex. The XP defect is recessive; hence, fusion of fibroblasts from different XP individuals is associated with restoration of normal DNA repair in the hetero-dikaryons, and has facilitated the identification of 10 genetic complementation groups. Individuals from any single complementation group are clinically indistinguishable (9).

This disease is also inherited as an autosomal recessive, but it has been estimated that in heterozygous individuals the AT gene may contribute significantly to the incidence of cancer in individuals under the age of 40 (30,31). Like XP cells, AT cells are extremely sensitive to killing, in this case by ionizing radiation and by so-called X-ray-mimetic chemical agents (21,28). Additionally, AT cells manifest abnormal kinetics of DNA synthesis (20) and also accumulate in the G2/M phase of the cell cycle after exposure to X-rays (26). The biochemical basis for these phenotypes is unknown. Equally mysterious is the complex clinical phenotype of the disease. Patients typically have severe cerebellar ataxia, telangiectasia (abnormal dilatation of small vessels), immune deficiency and an increased incidence of neoplasms of the lymphoreticular system (21). Treatment of the neoplastic state with ionizing radiation leads to severe radiation toxicity (21).

Like XP, the genetic aspects of this disease are complex. The detection of genetic complementation groups is technically more difficult. Nonetheless, based on complementation of defective kinetics of DNA synthesis after exposure to ionizing radiation, at least four complementation groups have been identified in AT (18). One of these (complementation group A) has recently been mapped to the q arm of chromosome 11 (12).

Fanconi's Anemia

FA is characterized clinically by pancytopenia and diverse congenital abnormalities (7). A high incidence of leukemia and of other neoplasms has been reported in association with this disease (24). At the cellular level, FA is characterized by marked sensitivity to killing by agents known to produce intrastrand cross-links in DNA. However, no specific defect in the repair of DNA of these lesions has been firmly established (8,11,15).

Cockayne's Syndrome

This disease manifests with dwarfism plus a variety of other congenital anomalies (3,4). Like XP cells, those from CS individuals are abnormally sensitive to killing by UV radiation (1). Unlike XP, however, there is no evidence of a DNA repair defect, and the pathogenesis of this cellular sensitivity remains totally unexplained. Three genetic complementation groups have been identified (16,32).

PHENOTYPIC COMPLEMENTATION BY DNA TRANSFECTION

The phenotype of increased sensitivity to killing by DNA-damaging agents suggests that molecular cloning of wild-type genes defective in these diseases would be facilitated by screening genomic DNA for complementing sequences. Such experiments are predicated on a number of assumptions.

(i)   The complementing genes must be small enough to integrate as functionally intact units following standard DNA transfection procedures.

(ii) The efficiency of DNA transfection and the frequency of stable integration of complementing sequences must be sufficiently high to facilitate the logistics of screening large numbers of transfected cells.

(iii) The frequency of spontaneous reversion to resistance to killing must be significantly lower than that of stable complementation.

(iv)   Stable (presumably nonhomologous) integration of a single copy
       of the complementing gene must be associated with levels of
       expression sufficient to provide unequivocal phenotypic comple-
       mentation.

Phenotypic complementation following DNA transfection has proven to
be extremely effective using immortalized UV- or X-ray-sensitive rodent
cell lines as recipients, and has facilitated the isolation of several human
genes presumably involved in cellular responses to damage produced by
radiation (10,14,35,36). However, for reasons that are not understood,
one or more of the requirements listed above are not satisfactorily met
following transfection of human XP, AT, or CS lines. To date only a
single example of such complementation in XP-A cells has been reported
(33). The frequency of primary transfectants identified in this study
directly supports the contention that human cells are highly inefficient
recipients for stable transfection of the genes of interest.

PHENOTYPIC COMPLEMENTATION BY MICROCELL-MEDIATED CHROMOSOME TRANSFER (MMCT)

As evidenced from the results of somatic cell fusions in XP and in
AT, the introduction of wild-type genes carried on chromosomes is associated with high levels of phenotypic complementation (18,29). Furthermore, techniques have been developed whereby single human chromosomes
can be introduced and maintained in recipient cells of interest (22).
Such a general experimental strategy is an attractive one for investigating
human diseases with selectable phenotypes, for several
reasons.

(i)   Phenotypic complementation of cells from different genetic com-
       plementation groups by different human chromosomes provides
       direct and independent evidence for the genetic complexity of
       diseases such as XP and AT and, in the case of the latter, can
       facilitate more specific definition of this complexity.

(ii)   The isolation of complementing single human chromosomes with
       deletions and/or rearrangements of the complementing region
       can facilitate refined subchromosomal mapping of the gene of
       interest.

(iii) Mapping the gene of interest can facilitate the identification of
       linked DNA probes, particularly in regions of the genome ex-
       tensively mapped with such markers. Such probes can be used
       to screen recombinant plasmids representing a limited region of
       the human genome for inserts containing complementing se-
       quences.

Microcell-Mediated chromosome Transfer

Mammalian cells in culture can be stimulated to form micronuclei by
treatment with colcemid. Subsequent treatment of cells with cytochalasin
followed by low-speed centrifugation results in enucleation of the micronuclei, and the resulting microcells can be harvested and size-selected by
ultrafiltration. The microcells can then be fused to recipient human cells
of interest, resulting in transfer of one or more human chromosomes (22)
(Fig. 1).

Selection for the transfer of human chromosomes and their stable retention in recipient cells is facilitated by prior tagging of the human donor chromosomes with a dominant selectable marker such as the E. coli neo or gpt gene. Such chromosomal tagging is readily achieved by stable integration of a plasmid carrying the selectable marker into human cell lines used as the source of donor chromosomes (13,22) (Fig. 1). Micronucleation of human cells is less efficient than with mouse cells (6), and for this reason mouse cells have been most extensively used for refining the technology of MMCT.

In order to generate microcells carrying tagged human chromosomes, human donor cells are first fused to mouse cells to form mouse-human hybrids (Fig. 1). Such hybrids segregate human chromosomes and ideally, when grown under selection, will retain a single tagged human chromosome, generating a monochromosomal human-mouse hybrid for use with MMCT (Fig. 1). Typically, however, hybrids derived by fusion of mouse and human cells retain more than one tagged human chromosome, and monochromosomal derivatives must be derived by MMCT into mouse cells and screening these for the presence of a single human chromosome (Fig. 1). Monochromosomal hybrids can also be derived by rescue of the complementing chromosome from a recipient human cell line. Such rescue is most efficiently effected by fusion of the complemented cell to mouse cells, followed by MMCT into mouse cells from the resulting human-mouse hybrids (Fig. 1).

Monochromosomal hybrids can be maintained by selection for the marker carried on the single human chromosome. The goal of this and of other laboratories is to generate panels of such hybrids carrying different dominant selectable markers representative of the entire human karyotype. Such panels will be of obvious utility in exploring the genetics and molecular biology of any cell type with a selectable phenotype.

PHENOTYPIC COMPLEMENTATION OF XP CELL
WITH SINGLE HUMAN CHROMOSOMES

Mixed pools of polychromosomal human-mouse hybrids, as well as clonal populations of several monochromosomal hybrids carrying neo- or gpt-tagged human chromosomes have been used for MMCT into immortalized XP cells from genetic complementation groups A (25), D (R.A. Schultz and E.C. Friedberg, unpubl. data), and F (23). In all cases, phenotypic complementation of sensitivity to UV radiation was observed and was attributed to the passage of a single chromosome. In all cases, the complementing chromosome was maintained in a monochromosomal hybrid, thereby facilitating its cytogenetic and molecular characterization.

Complementation of XP-A cells is close to wild-type levels of UV resistance and is accompanied by extensive complementation of defective nucleotide excision repair, as determined by repair synthesis using a standard autoradiographic assay (2,25). The complementing chromosome is rearranged, and hence identification of the chromosome carrying the XP-A locus was not immediately possible. Detailed cytogenetic characterization of the chromosome has demonstrated a translocation of the region 9q22.2-9q34.3 to the centromere of chromosome 11 [designated t(9,11) (9q22.2→9q34.3::11cen→11qter) (K. Henning, R.A. Schultz, G.S. Sekhon, and E.C. Friedberg, unpubl. data). Transfer of chromosome 11 into XP-A cells has no phenotypic effect, and a derivative of the

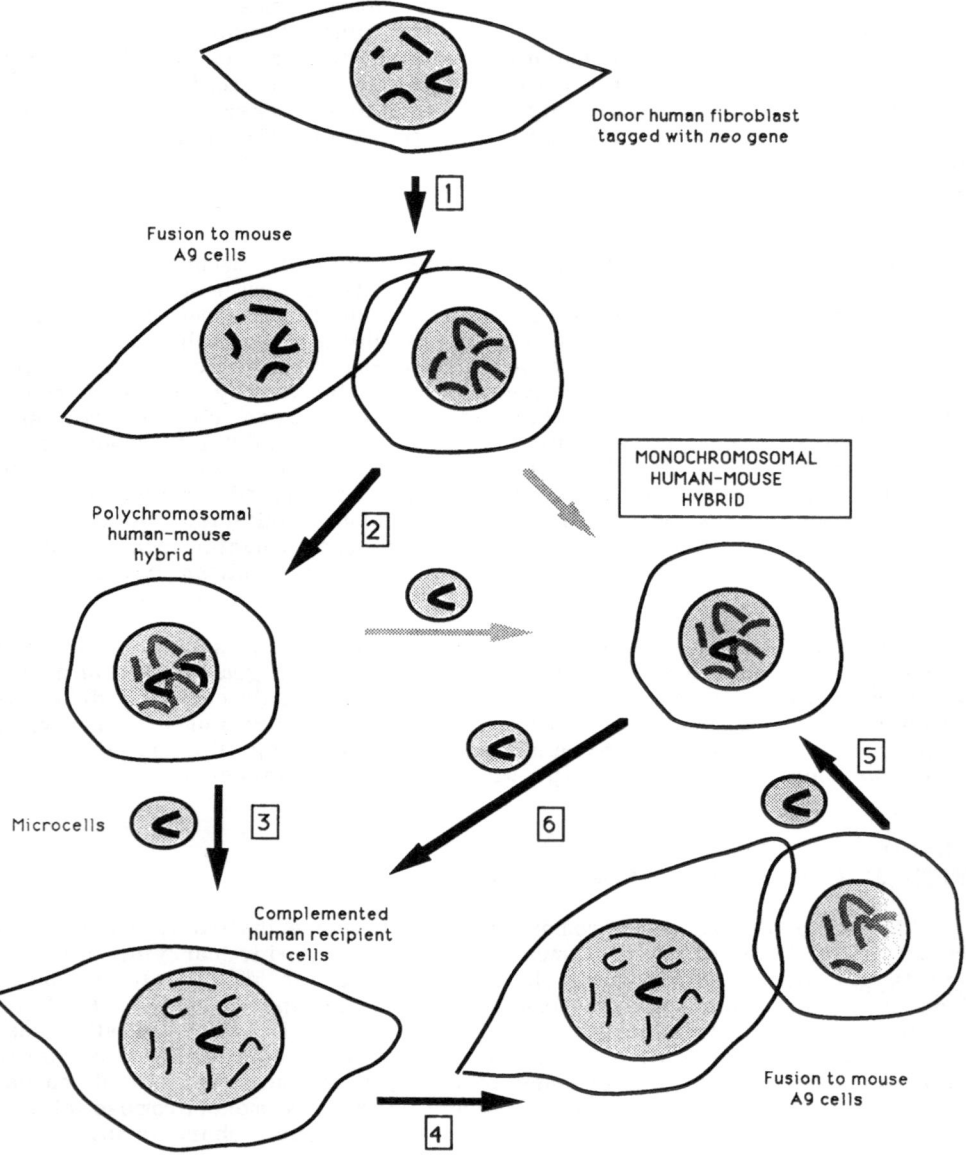

Donor human fibroblast
tagged with *neo* gene

Fusion to mouse
A9 cells

1

Polychromosomal
human-mouse
hybrid

2

MONOCHROMOSOMAL
HUMAN-MOUSE
HYBRID

5

Microcells

3

6

Complemented
human recipient
cells

Fusion to mouse
A9 cells

4

translocated chromosome carrying only 11q also does not alter the UV sensitivity of these cells. Thus, we tentatively conclude that the XP-A locus is carried on 9q. More refined subchromosomal localization of this gene is in progress.

Complementation of UV sensitivity and of sensitivity to killing by the UV-mimetic chemical 4-nitroquinoline-1-oxide is observed following MMCT of chromosome 15 into XP-F cells (23). Additionally, XP-F cells carrying this chromosome show increased levels of repair synthesis (23). Enhanced DNA repair capacity has also been demonstrated by measuring

←

Fig. 1.   Schematic representation of the transfer of single human chromosomes by microcell mediated chromosome transfer (MMCT).
Human cells (dark, bold chromosomes) in which chromosomes
have been tagged with a genetic marker (such as the E. coli
neo gene), are fused to mouse cells (light, bold chromosomes)
[1]. The hybrids segregate many of the human chromosomes
but eventually stabilize in culture carrying varying numbers of
human chromosomes (polychromosomal human-mouse hybrids)
[2]. Occasionally, segregation of human chromosomes after cell
fusion results in the retention of only a single tagged human
chromosome (monochromosomal human-mouse hybrid). In practice, such monochromosomal hybrids are more typically generated by MMCT of single human chromosomes from the polychromosomal hybrids to mouse A9 cells. Polychromosomal or monochromosomal hybrids can be used to generate microcells for transfer
of human chromosomes to recipient cells of interest [3,6].
When a complementing chromosome is detected, it can be res-
cued by fusion of the complemented cells to mouse cells [4],
followed by MMCT from the resulting polychromosomal hybrids
to generate monochromosomal hybrids carrying the complementing chromosome [5]. Rescue can be confirmed by MMCT back
to the recipient cell of interest [6].

reactivation of the plasmid-borne E. coli chloramphenicol acetyltransferase
(cat) gene following transfection of cells with UV-irradiated plasmid (23).
All these phenotypes are complemented partially following passage of a
single copy of chromosome 15. Extensive screening for other human
chromosomes that complement XP-F cells has been negative. Thus, at the
present time we are pursuing the hypothesis that partial complementation
reflects a gene dosage effect and that further copies of the complementing
gene are required to observe more extensive complementation (23).

Complete complementation of the phenotypes of XP-D cells is associated with the passage of a second rearranged chromosome. At present
this chromosome is the subject of intensive cytogenetic and molecular
characterization. The chromosomes that complement XP-A and XP-F cells
are genetically specific; i.e., phenotypic effects are only observed in
cells from the genetic complementation groups indicated and not in other
XP cell types.

PHENOTYPIC COMPLEMENTATION OF AT CELLS
WITH A SINGLE HUMAN CHROMOSOME

In addition to ionizing radiation, AT calls are also abnormally sensitive to killing by the X-ray-mimetic chemical streptonigrin (34,27). A
single human chromosome complements this sensitivity fully in a cell line
designated AT5BI, belonging to genetic complementation group D (C.
Lambert, R.A Schultz, and E. C. Friedberg, unpubl. data). Additionally, passage of this chromosome is associated with complementation of (i)
the abnormal kinetics of semiconservative DNA synthesis observed in AT
cells exposed to ionizing radiation; (ii) accumulation of cells at the G2/M

phase of the cell cycle following exposure to this form of radiation; (iii) the extended S phase observed in unirradiated AT cells (5,18); and (iv) the enhanced levels of chromosomal breakage observed after treatment of cells with bleomycin. These results suggest that all these phenotypes are associated with a single genetic defect in cells from complementation group D. In contrast, others have reported segregation of the X-ray-sensitive and abnormal DNA synthesis phenotypes in revertants of AT cells (17).

Cytogenetic and molecular characterization of the complementing chromosome has identified it as human chromosome 22 (C. Lambert, R.A. Schultz, and E.C. Friedberg, unpubl. data). As indicated earlier, recent studies by others have localized the AT-A gene to chromosome 11 (12). Clearly, continued chromosome transfers into immortalized AT cell lines and continued linkage studies in AT families have the potential for a complete definition of the genetic complexity of this disease.

CONCLUSIONS

The technique of MMCT has proven to be an informative experimental tool for approaching the genetics and, more remotely, the molecular biology of mutant cells with defective responses to DNA damage. Screening recipient cells for successful transfer of complementing human chromosomes is relatively facile once libraries of human-mouse hybrids carrying suitably tagged human chromosomes are established. Obviously, with each successful effort and with the elaboration of a new monochromosomal hybrid, the potential of realizing the goal of a complete panel of hybrids is increased, thus greatly facilitating screening the full human karyotype in subsequent experiments.

The road from a single human chromosome to a single human gene is a long one. However, as emphasized in the Introduction, the delineation of a specific region of a chromosome is a rational strategy for reducing the complexity of the human genome prior to molecular cloning of a gene of interest. The rapid development of a comprehensive physical map of the human genome, the emergence of refined tools for isolating defined regions of the genome (in the order of 2-3 megabases), the development of vectors that can accommodate inserts of this magnitude, and the powerful selection provided by the use of cells from the diseases discussed in this article suggest that the cloning and characterization of human genes defective in these diseases is a reasonable expectation in the relatively near future.

ACKNOWLEDGEMENTS

This manuscript is dedicated to the memory of Drs. Alexander Hollaender and Tsuneo Kada for their contributions to the field of mutagenesis. These studies were supported by research grant CA 44247 from the USPHS (ECF), and CA 19401 from the USPHS and 1475-AR from the Council for Tobacco Research (EJS).

REFERENCES

1. Andrews, A.D., F.W. Yoder, S.F. Barrett, and J.H. Robbins (1976) Cockayne's syndrome fibroblasts have decreased rates of colony

forming ability but normal rates of unscheduled DNA synthesis after ultraviolet irradiation. Clin. Res. 24: 624A.

2. Cleaver, J.E. (1983) Xeroderma pigmentosum. In The Metabolic Basis of Inherited Disease, J.B. Stanbury, J.B. Wyngaarden, D.S. Fredricksen, J.S. Goldstein, and J.S. Brown, eds. McGraw-Hill, New York, pp. 1227-1248.

3. Cockayne, E.A. (1936) Dwarfism with retinal atrophy and deafness. Arch. Dis. Childhood 11:1-8.

4. Cockayne, E.A. (1946) Dwarfism with retinal atrophy and deafness. Arch. Dis. Childhood 21:52-54.

5. Cohen, M.M., and S.J. Simpson (1980) Growth kinetics of ataxia telangiectasia lymphoblastoid cells. Evidence for a prolonged S period. Cytogent. Cell Genet. 28:24-33.

6. Crenshaw, A.H., J.W. Shay, and L.R. Murrell (1981) Colcemid-induced micronucleaction in cultured human cells. J. Ultrastr. Res. 75:179-186.

7. Fanconi, G. (1967) Familial constitutional panmyelopathy. Fanconi's anaemia (FA). I. Clinical Aspects. Semin. Hematol. 4:233-240.

8. Fornace, Jr., A., A.B. Little, and R.R. Weichselbaum (1977) DNA repair in a Fanconi's anaemia. Biochim, Biophys. Acta 561:99-109.

9. Friedberg, E.C. (1985) DNA Repair, W.H. Freeman, New York, pp. 505-540.

10. Friedberg, E.C., C. Backendorf, J. Burke, A. Collins, L. Grossman, J.H. Hoeijmakers, A.R. Lehmann, E. Seeberg, G.P. van der Schans, and A.A. van Zeeland (1987) Molecular aspects of DNA repair. Mutat. Res. 184:67-86.

11. Fujiwara, Y., M. Tatsumi, and M.S. Sasaki (1977) Cross-link repair in human cells and its possible defect in Fanconi's anaemia cells. J. Molec. Biol. 113:635-649.

12. Gatti, R.A., I. Berkel, E. Boder, G. Braedt, P. Charmley, P. Concannon, F. Ersoy, T. Foroud, N.G.J. Jaspers, K. Lange, G. M. Lathrop, M. Leppert, Y. Nakamura, P. O'Connell, M. Paterson, W. Salser, O. Sanal, J. Silver, R.S. Sparkes, E. Susi, D.E. Weeks, S. Wei, R. White, and F. Yoder (1988) Localization of an ataxia-telangiectasia gene to chromosome 11q22-23. Nature 336:577-580.

13. Graham, F.L., and A.J. van der Eb (1973) A new technique for the assay of infectivity of human adenovirus 5 DNA. Virology 52:456-467.

14. Hoeijmakers, H.J.H. (1987) Characterization of genes and proteins involved in excision repair of human cells. J. Cell Sci. Suppl. 6:111-125.

15. Kaye, J., C. Smith, and P.C. Hanawalt (1980) DNA repair in human cells containing photoadducts of 8-methoxypsoralen or angelicin. Cancer Res. 40:696-702.

16 Lehmann, A.P. (1982) Three complementation groups in Cockayne's Syndrome. Mutat. Res. 106:347-356.

17. Lehmann, A.R., C.F. Arlett, J.F. Burke, M.H.L. Green, M.R. James, and J.E. Lowe (1986) A derivative of an ataxia telangiectasia (AT) cell line with normal radiosensitivity but AT- like inhibition of DNA synthesis. Int. J. Radiat. Biol. 49:639-643.

18. Murnane, J.P., and R.B. Painter (1982) Complementation of defects in DNA synthesis in unradiated and unirradiatsd ataxia telangiectasia cells. Proc. Natl. Acad. Sci., USA 79:1960-63.

19. Orkin, S.H.(1986) Reverse genetics and human disease. Cell 47: 845-850.

20. Painter, R.B. (1985) Altered DNA synthesis in irradiated and unirradiated ataxia telangiectasia cells. In Ataxia Telangiectasia: Genet-

ics, Neuropathology, and Immunology of a Degenerative Disease of Childhood, R.A. Gatti and M. Swift, eds. Alan R. Liss, Inc., New York. pp. 89-100.

21. Paterson, M.C., and P.J. Smith (1979) Ataxia telangiectasia: An inherited human disorder involving hypersensitivity to ionizing radiation and related DNA-damaging chemicals. Ann. Rev. Genet. 13: 291-318.

22. Saxon, P.J., E.S. Srivatsan, G.V. Leipzig, J.H. Sameshima, and E.J. Stanbridge (1985) Selective transfer of individual human chromosomes to recipient cells. Molec. Cell. Biol. 5:140-146.

23. Saxon, P.J., R.A. Schultz, E.J. Stanbridge, and E.C. Friedberg (1989) Human chromosome 15 confers partial complementation of phenotypes to xeroderma pigmemtosum group F cells. Am. J. Human Genet. (in press).

24. Schroeder, T.M., D. Tilger, J. Krueger, and F. Volger (1976) Formal genetics of Fanconi's anaemia. Human, Genet. 32:257-288.

25. Schultz, R.A., P.J. Saxon, T.W. Glover, and E.C. Friedberg (1987) Microcell-mediated transfer of a single human chromosome complements xeroderma pigmentosum group A fibroblasts. Proc. Natl. Acad. Sci., USA 84:4176-4179.

26. Scott, D., and F. Zampetti-Bosseler (1982) Cell-cycle dependence of mitotic delay in X-irradiated normal and ataxia telangiectasia fibroblasts. Int. J. Radiat. Biol. 42:679-683.

27. Shiloh, Y., E. Tabor, and Y. Becker (1983) Abnormal response of ataxia telangiectasia cells to agents that break the deoxyribose moiety of DNA via a targeted free radical mechanism. Carcinogenesis 4:1317-1322.

28. Shiloh, Y., E. Tabor, and Y. Becker (1985) In vitro phenotype of AT fibroblast strains: Clues to the nature of the "AT DNA lesion" and the molecular defect in AT. In Ataxia Telangiectasia: Genetics, Neuropothology, and Immunology of a Degenerative Disease of Childhood, R.A. Gatti and M. Swift, eds. Alan R. Liss, Inc., New York, pp. 111-121.

29. Stefanini, M., W. Keijzer, A. Westerfield, and D. Bootsma (19S5) Interspecies complementation analysis of xeroderma pigmentosum and UV-sensitive Chinese hampster cells. Exp. Cell. Res. 161:373-380.

30. Swift, M. (1982) Disease predisposition of AT heterozygotes. In Ataxia Telangiectasia: A Cellular and Molecular Link Between Cancer, Neuropathology, and Immune Deficiency, B.A. Bridges and D.G. Harnden, eds. J. Wiley and Sons, New York. pp. 355-361.

31. Swift, M., L. Sholman, M. Perry, and C. Chase (1976) Malignant neoplasms in families of patients with ataxia-telangiectasia. Cancer Res. 36:209-215.

32. Tanaka, K., K. Kawai, Y. Kumahara, M. Ikenaga and Y. Okada (1981) Genetic complementation groups in Cockayne's Syndrome. Somat. Cell. Genet. 7:445-447.

33. Tanaka, K., I. Satokata, and Y. Okada (1989) Molecular cloning of the mouse DNA repair gene which can complement the defect in xeroderma pigmentosum complementation group A by DNA transfection methods. EMBO J. (in press).

34. Taylor, A.M.R., H.B. Laher, and G.R. Morgan (1985) Unscheduled DNA synthesis induced by streptonigrin in ataxia telangiectasia fibroblasts. Carcinogenesis 6:945-947.

35. van Duin, M., J. de Wit, H. Odijk, A. Westerveld, A. Yasui, M. Koken, H.J.H. Hoeijmakers, and D. Bootsma (1986) Molecular characterization of the human excision repair gene ERCC-1: cDNA cloning and amino-acid homology with the yeast DNA repair gene RAD10. Cell 44:913-923.

36. Weber, C.A., E.P. Salazar, S.A. Stewart, and L.H. Thompson (1988) Molecular cloning and biochemical characterization of a human gene, ERCC-2, that corrects the nucleotide excision repair defect in CHO UV5 cells. Molec. Cell. Biol. 8:1137-1146.

GENETIC ANALYSES OF CELLULAR FUNCTIONS REQUIRED

FOR UV MUTAGENESIS IN ESCHERICHIA COLI

John R. Battista, Takehiko Nohmi,
Caroline E. Donnelly, and Graham C. Walker

Biology Department
Massachusetts Institute of Technology
Cambridge, Massachusetts 02139

ABSTRACT

In Escherichia coli, most UV and chemical mutagenesis is not a passive process and requires the participation of the umuD and umuC gene products. However, the molecular mechanism of UV mutagenesis is not yet understood and the roles of the UmuD and UmuC proteins have not been elucidated. The umuDC operon is induced by UV irradiation and regulated as part of the SOS response. Genetic evidence now indicates that RecA-mediated cleavage activates UmuD for its role in mutagenesis. The COOH-terminal fragment of UmuD is both necessary and sufficient for this role. The RecA protein appears to have a third role in UV mutagenesis besides mediating the cleavage of LexA and UmuD at the time of SOS induction. In addition, we have obtained evidence which indicates that the GroEL and GroES proteins also play a role in UV mutagenesis. Similarities of the amino acid sequence of UmuD to the sequence of gene 45 protein of bacteriophage T4 and of the sequence of UmuC to those of the gene 44 and gene 62 proteins suggest possible roles for UmuD and UmuC in mutagenesis that are supported by preliminary evidence.

INTRODUCTION

Studies of Escherichia coli have indicated that much of the mutability of this organism by UV and by many chemical agents is due to the existence of a system that processes damaged DNA in such a way that mutations result. Experiments of Weigle (21) provided the first clue that, at least in E. coli, the process of UV mutagenesis requires the participation of host functions and that the expression of one or more of these functions was inducible. Yet, despite many years of study, the molecular mechanism of UV and chemical mutagenesis in E. coli has not been elucidated.

Genetic analyses have led to the identification of two host genes, umuD and umuC, whose function is required for UV mutagenesis (9,20).

Mutations in either of these genes render E. coli largely nonmutable with UV and a variety of other chemical mutagens. In other words, the mutations themselves are antimutagenic. It is interesting, in this context, that numbers of bacteria are naturally nonmutable with UV and various agents (20). The umuD and umuC genes are organized in an operon whose expression is regulated by the SOS circuitry (20). In response to an SOS-inducing treatment, the RecA protein becomes activated and then mediates a proteolytic cleavage of LexA at its Ala^{84}-Gly^{85} bond (20), apparently by facilitating a specific autodigestion of LexA (10). Slilaty and Little (18) have recently suggested that hydrolysis of the LexA Ala-Gly bond proceeds by a mechanism similar to that of serine proteases, with Ser^{119} acting as a nucleophile and Lys^{156} as an activator. LexA shares homology with the repressors of bacteriophages lambda, 434, P22, and ϕ80, and cleavage of these proteins appears to occur by an analogous mechanism (6,17). The cleavage site of all these proteins is an Ala-Gly bond except for the ϕ80 repressor, which has a Cys-Gly cleavage site (6).

The umuD and umuC genes encode proteins of approximately 15 and 47 kDa. An evolutionarily diverged but functionally analogous operon is present on the plasmid pKM101 which is present in the Ames Salmonella strains used for detecting mutagens (15). The deduced amino acid sequences of the UmuD and MucA proteins are 41% homologous and those of the UmuC and MucB are 55% homologous. Our observation that each gene requires its cognate for biological function has led us to propose previously that the UmuD and UmuC proteins physically interact (14).

Despite the progress in understanding aspects of the regulation and processing of UmuD and UmuC, their biochemical role in UV and chemical mutagenesis has remained elusive, as has a biochemical demonstration of the mechanism of UV and chemical mutagenesis. Our observation that overexpression of the UmuDC operon results in a cold-sensitive block to DNA replication has led us to hypothesize that UmuD and UmuC interact with components of the replication apparatus (12). The concept that an altered polymerase/replication apparatus might be involved in UV mutagenesis was proposed by Witkin (22). This hypothesis was later expanded by Radman's suggestion (16) that SOS-regulated proteins were involved in this process. Extensive analyses of DNA sequence changes resulting from UV or chemical mutagenesis have indicated that $umuD^{+}C^{+}$-dependent mutagenesis is targeted, and have therefore supported the concept that a key event in such mutagenesis is misincorporation of bases opposite noncoding or potentially miscoding lesions (20).

On the basis of a set of physiological experiments involving photoreactivation of umuC cells, Bridges and Woodgate (3) have proposed a two-step model for mutagenesis: a $recA^{+}$-dependent, $umuC^{+}$-independent step in which an incorrect base (or bases) is inserted opposite a premutagenic lesion, and a subsequent $umuC^{+}$-dependent step in which chain elongation is continued past the misincorporated base. Based on in vitro studies of the behavior of DNA polymerase III holoenzyme on damaged templates, Livneh has proposed that UmuD and UmuC and possibly other SOS-induced proteins may act by helping the polymerase to reinitiate after terminating and dissociating at the site of a lesion (11) or by making the polymerase more processive and thus facilitating bypass (8).

UmuD SHARES HOMOLOGY WITH LexA AND PHAGE REPRESSORS

We recently reported that UmuD and MucA share homology with the carboxyl-terminal regions of LexA and the repressors of lambda, 434, P22, and ϕ80 (14). This led us to hypothesize that UmuD and MucA might interact with activated RecA and that this interaction could result in a proteolytic cleavage of these proteins that would activate or unmask their function required for mutagenesis. The putative cleavage site of UmuD is the Cys^{24}-Gly^{25} bond. A RecA-mediated cleavage of UmuD has now been shown to occur both in vivo (19) and in vitro (4), and work from our lab discussed below has shown that the purpose of this cleavage is to activate UmuD for its role in mutagenesis (13).

HOMOLOGY OF UmuD TO LexA AND PHAGE REPRESSORS HAS FUNCTIONAL SIGNIFICANCE

We used site-directed mutagenesis of an $umuD^+C^+$ plasmid to create certain umuD mutations that were analogous to lexA or lambda repressor mutations that block both RecA-mediated cleavage and autodigestion, and found that all these umuD mutations caused major reductions in the ability of UmuD to function in UV mutagenesis. Changing the Gly^{25} residue of the putative Cys^{24}-Gly^{25} UmuD cleavage site to Glu or Lys largely abolished the ability of UmuD to function in UV mutagenesis. A Gly→Glu change at the Ala-Gly cleavage site of lambda repressor completely blocks cleavage, while a corresponding Gly→Asp change in LexA largely blocks cleavage. In addition, we found that changing either Ser^{60} or Lys^{97} to Ala also greatly reduced UmuD's ability to function in UV mutagenesis. Slilaty and Little (18) have shown that changes of the corresponding Ser^{119} and Lys^{156} residues of LexA to Ala completely block cleavage.

CLEAVED UmuD IS FUNCTIONAL IN MUTAGENESIS

To test more directly the hypothesis that cleavage of UmuD is important for mutagenesis, we constructed a umuD mutant of a $umuD^+C^-$ plasmid in which overlapping termination (TGA) and initiation (ATG) codons were introduced at the site in the umuD sequence that corresponds to the putative cleavage site (13). The plasmid carrying this engineered form of UmuD encodes two polypeptides rather than one. These two polypeptides are virtually the same as those that would result from cleavage at the Cys^{24}-Gly^{25} bond of UmuD. When a plasmid carrying this engineered umuD encoding two polypeptides was introduced into a nonmutable umuD44 strain, it restored the UV mutability of the cell to that of a $umuD^+$ gene product in mutagenesis.

This result strongly indicates that at least one of the products resulting from cleavage of the UmuD at its Cys^{24}-Gly^{25} bond is capable of carrying out the role of the umuD gene product in mutagenesis. Furthermore, it rules out the possibility that the purpose of UmuD cleavage is to inactivate UmuD. A plasmid that encoded only the polypeptide corresponding to the small NH_2-terminal fragment of UmuD failed to complement the UV nonmutability of a umuD44 strain, whereas a plasmid that encoded only the large COOH-terminal polypeptide made the strain more UV-mutable than a plasmid carrying $umuD^+$. These results strongly suggest that the COOH-terminal cleavage product of UmuD is both necessary and sufficient for the role of UmuD in UV mutagenesis.

THE COOH-TERMINAL POLYPEPTIDE OF UmuD RESTORES MUTABILITY TO A lexA(Def) recA430 STRAIN

To test the physiological significance of UmuD cleavage, we introduced plasmids carrying either the engineered umuD encoding two polypeptides or the COOH-terminal polypeptide of UmuD into a lexA71::Tn5(Def) recA430 strain. The recA430 mutation has differential effects on RecA's ability to mediate proteolytic cleavage (20). The RecA430 protein fails to mediate the cleavage of lambda repressor, mediates the cleavage of LexA with reduced efficiency, and mediates the cleavage of φ80 repressor normally. recA430 strains are UV nonmutable. The introduction of the plasmid encoding the two engineered UmuD polypeptides partially restored the UV mutability of this strain, while the plasmid encoding only the COOH-terminal UmuD polypeptide restored the UV mutability of the strain to that of a lexA71::Tn5(Def) recA$^+$ strain carrying a umuD$^+$ plasmid.

The restoration of UV mutability to the recA430 strains observed when we circumvented the need for UmuD cleavage strongly indicates that the primary cause for the UV nonmutability of recA430 derivatives is a result of an inability to mediate the cleavage of UmuD. It furthermore implies that the purpose of RecA-mediated cleavage of UmuD is to activate UmuD for its role in mutagenesis. Thus, it appears that RecA carries out two mechanistically related roles in UV and chemical mutagenesis: (i) transcriptional derepression of the umuDC operon by mediating the cleavage of LexA; and (ii) post-translational activation of UmuD by mediating its cleavage.

A THIRD ROLE FOR RecA IN MUTAGENESIS

In contrast to the situation discussed above, the introduction of a plasmid encoding the COOH-terminal UmuD polypeptide did not suppress the nonmutability of a lexA(Def) recA1730 mutant (5). The recA1730 mutation impairs the ability of RecA to mediate the cleavage of LexA but not of lambda repressor. recA1730 is dominant to recA$^+$ with respect to the nonmutability phenotype but recessive to recA$^+$ with respect to all other functions tested. Taken together, these observations suggest that the RecA protein plays a third role in UV mutagenesis besides mediating the cleavage of LexA and UmuD.

EVIDENCE THAT THE GroEL AND GroES PROTEINS PLAY A ROLE In UV MUTAGENESIS

We have recently observed that the E. coli heat-shock proteins, GroEL and GroES, appear to play roles in UV mutagenesis. This line of experimentation grew out of our previous observation that overexpression of the umuDC operon in a lexA(Def) strain results in cold-sensitive growth and that DNA replication is blocked upon a shift from 42° to 30°C (12). This cold sensitivity can be suppressed by various mutations affecting genes involved in the heat-shock response; recent results indicate that mutations in groEL and groES are particularly effective suppressors. These results suggest that the products of umuDC may interact with the replication fork and thus raise the possibility that GroEL and GroES might play a role in facilitating this interaction. This led us to the discovery that groEL and groES mutants of E. coli are greatly reduced in their UV mutability. It is possible that they may exert their function by playing some type of molecular chaperone role with respect to UmuD and/or UmuC, since the nonmutability of these groEL and groES mutants can be suppressed by increased production of UmuD and UmuC.

SEQUENCE SIMILARITIES OF UmuD AND UmuC TO DNA ACCESSORY PROTEINS OF T4 BACTERIOPHAGE

We have observed that the amino acid sequences of UmuD and UmuC share limited similarity with the DNA accessory proteins 45, 44, and 62 of T4 bacteriophage. UmuD aligns with accessory protein 45 (gp45), and UmuC aligns with accessory proteins 44 and 62 (gp 44 and gp 62) (2). The three DNA accessory proteins function as a complex and together increase the processivity of T4 DNA polymerase (1). The similarities between UmuD and UmuC and these DNA accessory proteins suggest the possibility that the target of the UmuD and UmuC proteins may be the replication apparatus of E. coli when it has been stalled by a damaged template. Such a model is consistent with suggestions by Bridges and Woodgate (3) that umuDC function involves translesion bypass of DNA damage, and by Hevroni and Livneh (8) that UmuD and UmuC make the polymerase more processive.

Our approach has concentrated on developing a functional link between the DNA accessory proteins and UmuD and UmuC. This is possible because the accessory proteins are well-characterized biochemically. The gp44 and gp62 proteins form a complex that exhibits a ssDNA-dependent ATPase activity which is stimulated by gp45. We have demonstrated that the UmuC protein, like the gp44/62 complex, binds to ssDNA. Further, binding to ssDNA appears to require ATP. UmuC binds equally well in the presence or absence of UmuD or UmuD*, and there is preliminary evidence that the UmuD* protein will bind, albeit slightly, to ssDNA that has UmuC bound to it. Neither UmuD nor UmuD* binds ssDNA in the absence of UmuC.

ACKNOWLEDGEMENTS

This work was supported by Public Health Service grants CA21615 and GM28988, awarded by the National Cancer Institute and National Institute of General Medical Sciences, respectively. J.R.B. and C.E.D. were supported by postdoctoral fellowships from the National Institutes of Health and American Cancer Society, respectively.

REFERENCES

1. Alberts, B.M. (1984) The DNA enzymology of protein machines. Cold Spring Harbor Symp. Quant. Biol. 49:1-12.
2. Battista, J.R., T. Nohmi, C.E. Donnelly, and G.C. Walker (1988) Role of UmuD and UmuC in UV and chemical mutagenesis. In Mechanisms and Consequences of DNA Damage Processing, E.C. Friedberg and P. Hanawalt, eds. Alan R. Liss, Inc., New York, pp. 455-459.
3. Bridges, B.A., and R. Woodgate (1984) Mutagenic repair in Escherichia coli X. The umuC gene product may be required for replication past pyrimidine dimers but not for the coding error in UV mutagenesis. Molec. Gen. Genet. 196:364-366.

4. Burckhardt, S.E., R. Woodgate, R.H. Scheuermann, and H. Echols (1988) UmuD mutagenesis protein of Escherichia coli: Overproduction, purification, and cleavage by RecA. Proc. Natl. Acad. Sci., USA 85:1811-1815.

5. Dutreix, M., P.E. Moreau, A. Bailone, F. Galibert, J. Battista, G.C. Walker, and R. Devoret (1989) New recA mutations that dissociate the various RecA protein activities in Escherichia coli: Evidence for an additional role for RecA protein in UV mutagenesis. J. Bacteriol. 171:2415-2423.

6. Eguchi, Y., T. Ogawa, and H. Ogawa (1988) Cleavage of phage φ80 cI repressor by RecA protein. J. Molec. Biol. 202:565-573.

7. Elledge, S.J., and G.C. Walker (1983) Proteins required for ultraviolet light and chemical mutagenesis: Identification of the products of the umuC locus of E. coli. J. Molec. Biol. 164:175-192.

8. Hevroni, D., and Z. Livneh (1988) Bypass and termination at apurinic sites during replication of single-stranded DNA in vitro: A model for apurinic site mutagenesis. J. Biol. Chem. 85:5046-5050.

9. Kato, T., and Y. Shinoura (1977) Isolation and characterization of mutants of Escherichia coli that are deficient in induction of mutations by ultraviolet light. Molec. Gen. Genet. 156:121-131.

10. Little, J.W. (1984) Autodigestion of LexA and phage λ repressors. Proc. Natl. Acad. Sci., USA 81:1375-1379.

11. Livneh, Z. (1986) Mechanism of replication of ultraviolet-irradiated single stranded DNA by DNA polymerase III holoenzyme of Escherichia coli: Implications for SOS mutagenesis. J. Biol. Chem. 261:9526-9533.

12. Marsh, L., and G.C. Walker (1985) Cold sensitivity induced by overproduction of UmuDC in Escherichia coli. J. Bacteriol. 162:155-161.

13. Nohmi, T., J.R. Battista, L.A. Dodson, and G.C. Walker (1988) RecA- mediated cleavage activates UmuD for mutagenesis: Mechanistic relationship between transcriptional derepression and posttranslational activation. Proc. Natl. Acad. Sci., USA 85:722-737.

14. Perry K.L., S.J. Elledge, B.B. Mitchell, L. Marsh, and G.C. Walker (1985) umuDC and mucAB operons whose products are required for UV light- and chemical-induced mutagenesis: UmuD, MucA, and LexA proteins share homology. Proc. Natl. Acad. Sci., USA 82:4331-4335.

15. Perry, K.L., and G.C. Walker (1982) Identification of plasmid-(pKM101)-coded proteins involved in mutagenesis and UV resistance. Nature (London) 300:278-281.

16. Radman, M. (1974) Phenomenology of an inducible mutagenic DNA repair pathway in Escherichia coli: SOS repair hypothesis. In Molecular and Environmental Aspects of Mutagenesis, L. Prakash, F. Sherman, M. Miller, C. Lawrence, and H.W. Tabor, eds. Charles C. Thomas, Publisher, Springfield, Illinois, pp. 128-142.

17. Sauer, R.T., R.R. Yocum, R.F. Doolittle, M. Lewis, and C.O. Pabo (1982) Homology among DNA-binding proteins suggests use of a conserved super-secondary structure. Nature 298:447-451.

18. Slilaty, S.N., and J.W. Little (1987) Lysine-156 and serine-119 are required for LexA repressor cleavage: A possible mechanism. Proc. Natl. Acad. Sci., USA 84:3987-3991.

19. Shinagawa, H., H. Iwasaki, T. Kato, and A. Nakata (1988) RecA protein-dependent cleavage of UmuD protein and SOS mutagenesis. Proc. Natl. Acad. Sci., USA 85:1806-1810.

20. Walker, G.C. (1984) Mutagenesis and inducible responses to deoxy-

ribonucleic acid damage in _Escherichia coli_. _Microbiol. Rev._ 48:60–93.

21. Weigle, J.J. (1953) Induction of mutation in a bacterial virus. _Proc. Natl. Acad. Sci., USA_ 39:628–636.

22. Witkin, E. (1969) Ultraviolet-induced mutation and DNA repair. _Microbiol. Rev._ 23:487–514.

POSITION OF A SINGLE ACETYLAMINOFLUORENE ADDUCT

WITHIN A MUTATIONAL HOT SPOT IS CRITICAL

FOR THE RELATED MUTAGENIC EVENT

Dominique Burnouf, Patrice Koehl, and Robert P.P. Fuchs

Groupe de Cancérogénèse et de Mutagénèse
  Moleculaire et Structurale
IBMC du CNRS
67084 Strasbourg Cedex, France

ABSTRACT

2-Acetylaminofluorene, a potent rat liver carcinogen, which binds primarily to C_8 of guanines, has been shown to induce mainly frameshift mutations in the bacteria Escherichia coli. Mutations occur at specific sequences, known as mutation hot spots, of which two types may be considered. First, repetitive sequences, where deletions of a single unit occur (GGGGG→GGGG). Second, the so-called NarI site, $^5{'}$GGCGCC$^{3'}$, where only -2-bp deletions are observed ($G_1G_2CG_3CC$→GGCC). Mutagenesis within repetitive sequences is dependent on the $UmuCD^+$ gene functions, whereas mutagenesis in the NarI site is not. These differences in the genetic requirements of mutagenesis at these hot spots suggest that two different pathways operate. In order to precisely determine the actual involvement of each of the three premutagenic lesions that may form in the NarI site in the course of the mutational process, we designed a single adduct mutagenesis experiment, and found that AAF binding to the G_3 induced only a -2 frameshift mutation event. This result will be discussed in terms of local DNA conformation.

INTRODUCTION

There is now an increasingly large collection of data emphasizing the interrelationship between alterations of a genome and the development of degenerative diseases like cancer or metabolic disorders (2,14-16,21,24, 26). Presumably, different kinds of endogenous or exogenous factors may provoke alterations of the genetic material, and among them, environmental factors, mainly chemicals and radiations, are the most studied because they are responsible for the largest proportion of cancers.

Most chemical mutagens bind covalently to DNA and form a variety of premutagenic lesions that are processed by cellular error-free repair mechanisms, such as excision repair or recombinational repair (25). Some of the adducts escape these repair processes, however, and are converted into mutations, either by direct miscoding or by cellular processes that are induced when the replication fork is blocked. This block in the replication processes triggers in <u>Escherichia coli</u> the induction of the SOS regulon. The SOS response contains several co-regulated genes, all under the control of the $lexA^+$- and $recA^+$-encoded proteins, whose gene products are involved in DNA repair or in the mutagenic processing of adducts in DNA (25).

A general strategy for studying mutagenic mechanisms is to determine the forward mutation spectrum observed within a gene that has been randomly modified with a mutagen. In our laboratory, we have developed such a forward mutation assay based on the inactivation of the tetracycline-resistant gene of the plasmid pBR322 (8). This method was shown to be very sensitive and led us to determine the mutation spectra induced by several chemicals--the antitumoral drug cis-diaminodichloroplatinum (4) and some aromatic amines (1,8,12; G. Hoffman and R.P.P. Fuchs, ms. in prep.).

One of these molecules that we studied more extensively is the ultimate carcinogen N-acetoxy-N-2-acetylaminofluorene (N-Aco-AAF), which is used as a model for the strong rat liver carcinogen N-2-acetylaminofluorene (AAF). N-Aco-AAF binds covalently to DNA, mainly at position C_8 of guanine (13). The mutation spectrum was determined by use of the tetracycline-resistance gene inactivation assay (8), and it was found that AAF adducts mainly induce frameshift mutations (more than 90%) (8,12). These mutations have been detected in a small number of sites, the so-called mutation hot spots. They occur within two types of sequences: (i) runs of guanines; and (ii) the sequence GGCGCC, which is the recognition sequence of the restriction enzyme NarI. The fact that the average reactivity of the guanines within these sites is equal to the average reactivity of any guanine in DNA (7) rules out a purely chemical explanation for the presence of hot spots of mutagenesis at these sites.

The mutations in the run of guanines are primarily -1 frameshift mutations and can be related to a slippage-type mechanism, as described by Streisinger and co-workers (22,23), in which one strand of the DNA can misalign with respect to the other strand during replication or repair synthesis. Mutations in the NarI sequence are -2 frameshift mutations. Mutagenesis processes observed at both sites do not seem to follow the same mechanism, in that the two pathways have different genetic requirements (3,12). Briefly, mutagenesis in the repetitive sequences is dependent on $umuDC^+$ gene function, whereas mutagenesis in the NarI sequence is not, although this pathway also requires a LexA-controlled function (10).

Due to the nature of the sequence of the NarI site, $G_1G_2CG_3CC$, three different molecular events, each involving the deletion of two contiguous base pairs (i.e., G_2C, CG_3, G_3C), can give rise to the observed end point (GGCC). We have used a single-adduct mutagenesis strategy to investigate further the mechanism of AAF-induced mutations within the NarI sequence.

In this paper, we describe the construction of a series of plasmids, all bearing a single AAF adduct located in a NarI site. These monomodified plasmids were used in mutagenesis experiments in <u>E. coli</u>.

MATERIALS AND METHODS

Strain, Plasmids, and Medium

Plasmid DNA was grown in E. coli strain JM103 and purified on cesium chloride gradients, as described previously (9). Plasmid pEMBL8(-) (6) is a gift from Dr. G. Cesarini (EMBL Laboratory, Heidelberg, FRG). Luria broth (LB) contained ampicillin (100 µg/ml), 5-bromo-4-chloro-3-indolyl-β-D-galactopyranoside (X-gal) (50 µg/ml), and isopropyl-β-D-thiogalactopyranoside (IPTG) (60 µg/ml).

DNA Restriction/Modification Enzymes

All enzymes were purchased from New England Biolabs (Beverly, Massachusetts), except topoisomerase I (BRL Laboratories) and HincII (IBI). Enzymes were used as specified by the manufacturers. Preparation of single-stranded plasmids and sequencing of the mutants were done using a Sequenase kit (UCB).

Transformation Protocol

Induction of SOS functions prior to the transformation step was achieved by UV irradiation of the host E. coli JM103 in $MgSO_4$ at fluences ranging from 50 to 70 J/m^2. Cells were made competent by $CaCl_2$ treatment, as previously described (12). Cells were plated on ampicillin plates immediately after the heat shock without any expression period to ensure the independence of the mutant clones.

Preparation of AAF Monomodified Oligonucleotides

The synthesis, purification, and analysis of monomodified oligonucleotides has been extensively described elsewhere (9).

Recovery and Analysis of Mutants

Transformed bacteria were plated on LB plates containing ampicillin, IPTG, and X-gal. Potential -2 frameshift mutants were detected as blue colonies among a majority of white colonies. The blue color is ascribable to the restoration of the lacZ gene function when two base pairs that had been built into the pSM14 plasmids as part of a 14-mer are deleted. Plasmids were prepared from the blue colonies and retransformed into JM103 cells. Each plasmid was digested by ClaI, NarI and by ClaI, HincII restriction enzymes. A -2 frameshift mutation within the NarI site was considered confirmed when a plasmid from a blue colony had neither a NarI nor a HincII site. The ClaI, HincII digestion was introduced to identify any pEMBL8(-) plasmids that were carried over during the preparation of the plasmids containing single adducts. Plasmids with the restriction phenotype ClaS HindR NarR were scored as mutants.

GENERAL STRATEGY FOR UNIQUE SITE MUTAGENESIS*

The general strategy for the construction of genomes containing a single AAF adduct in the NarI site is described in Fig. 1. Two related

---

*This method has been extensively described elsewhere (5,9).

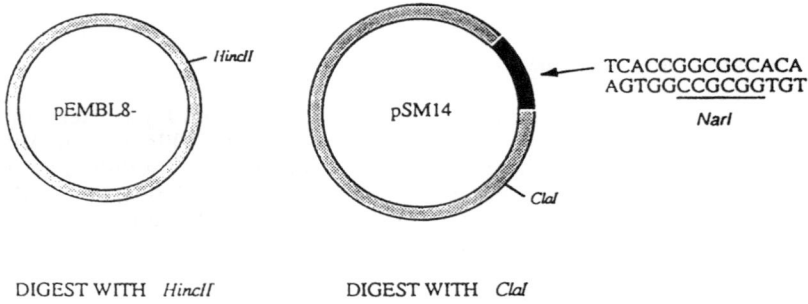

DIGEST WITH *HincII*   DIGEST WITH *ClaI*

MIX THE LINEAR FORMS OF pEMBL and pSM14

DENATURE AND RENATURE
TO FORM GAPPED-DUPLEX

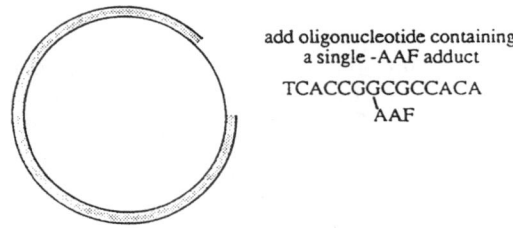

HYBRIDIZE AND LIGATE

ISOLATE COVALENTLY CLOSED CIRCLES
ON CESIUM CHLORIDE GRADIENTS

Fig. 1.   Strategy for the construction of pSM14 plasmids containing single AAF adducts on one of the guanine residues of the NarI sequence. Reprinted from Ref. 9, with permission.

plasmids, previously linearized, are used to reconstitute a circular duplex
molecule containing a single-stranded gap, in which an oligonucleotide,
modified or not, is hybridized and ligated. The covalently-closed circular
molecules resulting from a double ligation event are then purified on cesium chloride gradients and used in mutagenesis experiments in E. coli.

    The construction of the plasmid is such that it enables us to detect
-2 frameshift mutations. We used a phenotypic ·selection based on the
LacZ α-complementation assay. The parent plasmid, pEMBL8(-) (6), encodes part of the β-galactosidase gene and produces blue colonies when
propagated in an adequate host on X-gal plates. The cloning of the
14-bp oligonucleotide d(TCACCGGCGCCACA) in the HincII site of the
polylinker region of pEMBL8(-) produces a +2 frameshift in the original
reading frame of the β-galactosidase gene. Bacteria transformed with this
new plasmid, pSM14, give white clones when plated on X-gal plates. The
oligonucleotide contains a NarI site, which will be the target for the
mutagenesis assay. Any -2 frameshift event that will occur within the
NarI site will restore the reading frame of the LacZ gene, thus giving
blue colonies (Fig. 2).

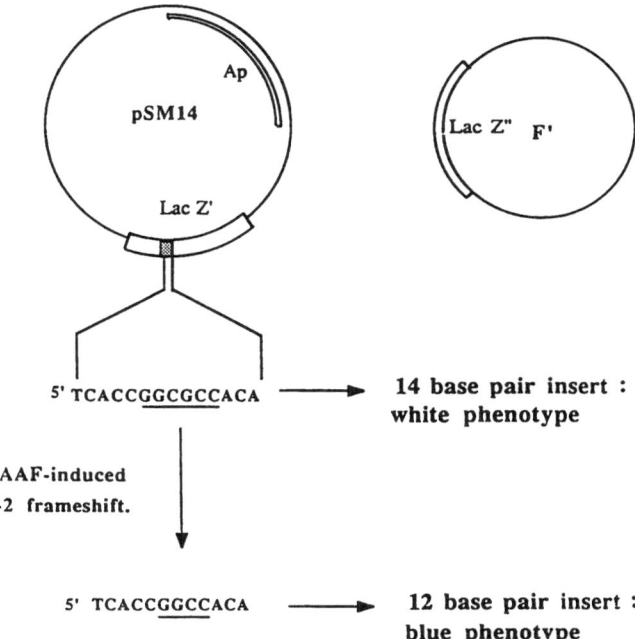

Fig. 2.   Design of pSM14 as a probe for detecting -2 frameshift mutations within the NarI site. Reprinted from Ref. 9, with permission.

RESULTS AND DISCUSSION

Purification and Characterization of
the Monomodified Oligonucleotides

        The 14 mer oligonucleotide 5'd(TCACCGGCGCCACA) was modified by
N-Aco-AAF, as described elsewhere (9), and purified by two successive
reverse-phase HPLC experiments. Several reaction products resulted
from the in vitro modification of the 14 mer by the carcinogen: one
tri-adducted, three di-adducted, and three mono-adducted oligonucleotides. Figure 3A shows the HPLC purification of the crude reaction mixture. A first run of purification allowed the molecules to be separated on
the basis of the modification level, as revealed by UV spectroscopy (9).
At this step, the three mono-adducted 14 mer eluted in overlapping pics
(Fig. 3A). A second set of purification was necessary to separate the
three positional isomers (Fig. 3B, C, and D). After this ultimate purification, the exact position of the AAF adduct on the oligonucleotide and
the purity of each monomodified oligonucleotide were checked using a T4
DNA polymerase assay. This enzyme, when incubated without triphospho-
nucleotides, exhibited a 3'→5' exonucleolytic activity, which was blocked
when the enzyme encountered an AAF adduct (7). Figure 4 shows the
result of the digestion experiment analyzed on a 20% acrylamide gel. Each
band represents the length of a DNA molecule, corresponding to the position where the enzyme stops. Band b shows the digest profile of a pool
of the three mono-adducted molecules; the c, d, and e slots reveal the
purity of each monomodified 14 mer, each of which is considered to be
greater than 90%.

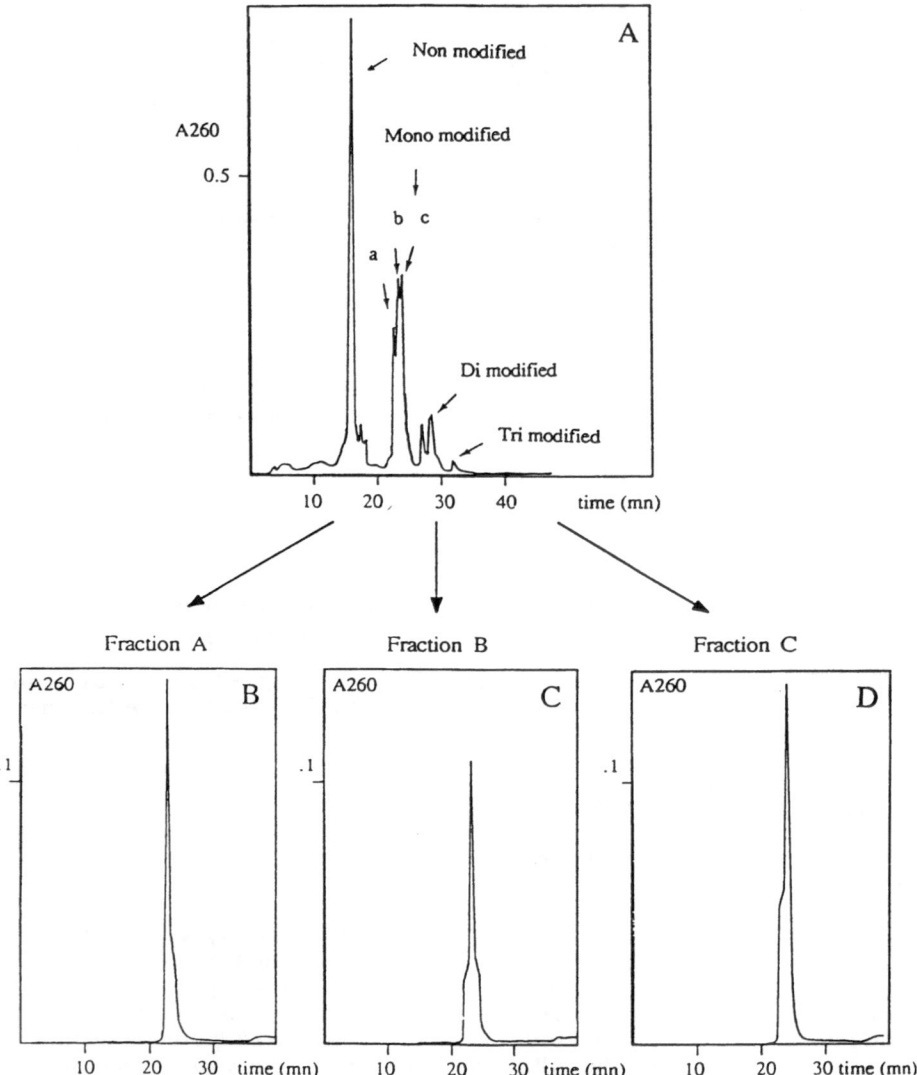

Fig. 3.   An HPLC profile (panel A) of the crude mixture obtained after
          reaction of N-Aco-AAF with d(TCACCGGCGCCACA). The pic
          of mono-adducted oligonucleotides was collected in three dif-
          ferent fractions, noted (a), (b), and (c), which were then re-
          purified by HPLC, as presented in panels B, C, and D, re-
          spectively. Reprinted from Ref. 9, with permission.

Construction of Monomodified Plasmids

As mentioned above, pSM14 plasmid results from the cloning of a 14
mer duplex, 5'd(TCACCGGCGCCACA)3', which contains a NarI site, in
the unique HincII restriction site of plasmid pEMBL8(-). The orientation
of the oligonucleotide was checked by sequencing. Plasmids pEMBL8(-)

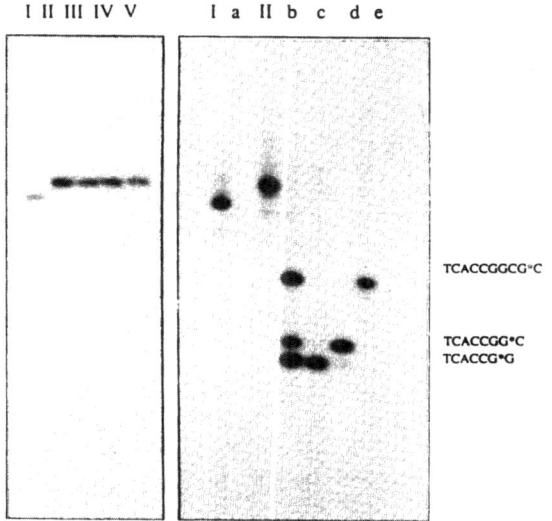

Fig. 4.   T4 DNA polymerase analysis of the purified mono-adducted iso-
           mers of the 14 mer d(TCACCGGCGCCACA). All nucleotides
           were ^{32}P 5' end-labeled. The analysis was performed on a 20%
           polyacrylamide sequencing gel. Left panel: Lane I shows the
           nonmodified oligonucleotide, and lane II shows a pool of mono-
           adducted oligonucleotides. The AAF adduct is responsible for a
           slightly slower migration of mono-adducted 14 mer (lanes II,
           III, IV, and V). The mono-adducted isomers are not resolved
           on the sequencing gel (same migration in lanes II, III, IV, and
           V). Right panel: Undigested DNA is shown in lanes I and II,
           respectively, designated as in the left panel. Lane a shows
           that the unmodified oligomer is completely digested by T4 DNA
           polymerase. Lanes b, c, d, and e show the DNA fragments
           produced when the 3'→5' exonuclease activity of the T4 DNA
           polymerase is blocked one nucleotide 3' of the AAF adduct,
           hence revealing the position of the modified guanine in the 14
           mer (7). These lanes show the digestion products obtained
           with the pool (lane b) and the three mono-adducted oligonu-
           cleotides (lanes c, d, and e). Reprinted from Ref. 9, with
           permission.

and pSM14 were linearized by the restriction enzymes HincII and ClaI, respectively. The two double-stranded plasmids were then mixed, heat-
denatured, and renatured. The renatured DNA mixture contains both the
linear forms of the two plasmids and two species of gapped duplex molecules. Only one of the two species will have a gap with the sequence
complementary to the oligonucleotide. In a typical experiment, we estimate that in the renatured mixture, at least 10% of the DNA molecules
are gapped duplexes with the adequate structure.

     The unmodified 14 mer and the three mono-adducted position isomers
were 5'-phosphorylated, and the ^{32}P-labeled molecules were purified from
the unphosphorylated molecules on a 20% acrylamide gel. These molecules

were added to the crude gapped-duplex mixture, and hybridization and ligation steps were performed, as described above (see "Materials and Methods").

The formation of a double ligation product was certified by the appearance, on an ethidium bromide-containing agarose gel, of a new band which migrates as covalently-closed plasmids. Autoradiograms of the gel showed that these plasmids integrated a radioactively-labeled material. In order to use them in mutagenesis experiments, we purified the covalently-closed molecules by centrifugation in CsCl gradients containing ethidium bromide.

Only the AAF-G_3 Lesion Induces a Mutational Event

The reconstructed covalently-closed plasmids, designated as pSM14-1, pSM14-2, and pSM14-3, each containing a single AAF adduct on the guanine 1, 2, and 3, respectively, of the NarI site, were used to transform SOS-induced or -uninduced JM103 bacteria. Mutant clones were selected on the basis of a phenotypic assay (Fig. 2) and confirmed by restriction analysis, namely, by the loss of the unique NarI restriction site or by sequencing.

In previous experiments (11), heteroduplex molecules were constructed to investigate the segregation of each strand as a function of its modification level. It was concluded that the presence of blocking lesions in one strand leads to the uncoupling of the two polymerase complexes, and that the modified strand is specifically lost.

From this observation, we expected a bias to be introduced in the unique site mutagenesis experiments because of the specific loss of the monomodified strand. To avoid such a problem in our experiment, we irradiated the pSM14 parental plasmid before the denaturation-renaturation step, at a UV dose giving about 10 to 12 pyrimidine dimers per plasmid. UV irradiation of the complementary strand did not interfere with our mutation assay (Tab. 1). This shows that the background mutation frequency introduced by the chemistry of the oligonucleotide and the engineering of the gapped duplex is quite low.

As expected from our previous results with randomly modified plasmids (8,12), the NarI mutant pathway is strongly SOS-dependent (Tab. 1).

As a main result, we found that only -AAF modification at the G_3 position induces the -2 frameshift event that characterizes the NarI mutation pathway, with a frequency of about 10^{-2}, when the SOS functions are induced. No mutant clones were recovered with the constructions that carry the -AAF modification at positions G_1 or G_2. Modification at G_2 could potentially trigger the -2 event. Indeed, three different molecular events (i.e., deletion of G_2C, CG_3, or G_3C) could have given rise to the observed end point (GGCGCC→GGCC). We suggest that AAF-induced mutagenesis at NarI site hot spots is related to local conformational changes that are caused by the binding of an AAF moiety at the G_3 residue of these peculiar sequences. Such a local structure is not induced upon binding of -AAF to G_1 or G_2.

Tab. 1. Mutation frequencies of the AAF-induced -2 frameshift event GGCGCC→GGCC in the NarI site.

Plasmids	pSM14	pSM14-1	pSM14-2	pSM14-3
No SOS induction	ND	<19 (0/510)	<9 (0/1,119)	4 (1/2,315)
With SOS induction (50 to 70 J/m^2)	<1 (0/10,080)	<6 (0/1,650)	<10 (0/978)	102 (17/1670)

All plasmids bear UV lesions (5 to 6) in the strand complementary to the AAF-modified one. No AAF lesion was introduced in the control plasmid pSM14. The plasmids pSM14-1, pSM14-2, and pSM14-3 have a unique AAF adduct on G_1, G_2, and G_3,respectively. Mutation frequencies are expressed in 10^{-4}. Terms in parentheses give the number of mutant clones over the number of total transformants. Mutagenesis experiments were done in SOS-induced JM103 cells. ND, not done.

Mutagenesis at the NarI Site Is Related to an Unusual DNA Conformation

Several studies demonstrate that binding of AAF to alternating pyrimidine-purine polynucleotides induces a B→Z transition (17-20,27), the driving force being the rotation of the modified guanine from its anti to syn conformation as a result of AAF binding to the C_8 position. Moreover, it has been suggested that AAF-modified d(CpG) dinucleotide, in a poly-d(GC).poly-d(GC), may adopt a Z conformation (17). We propose that modification of the G_3 at a NarI sequence induced such a local transition from B to Z conformation. Such a transition would not happen when AAF binds to G_1 or G_2, on the basis that their corresponding flanking regions would not adapt to the formation of a B/Z junction. Little is known about the structural requirements for the formation of B/Z junctions. These requirements may be critical in the formation of the local Z form. We suggest that this conformational change to Z DNA is recognized by the proteins that process the lesion, and that AAF mutagenesis occurs during a repair type of processing of the lesion on double-stranded DNA rather than during the course of replication.

NarI Site Mutagenesis Is a Model for a More General Mutation Pathway

The NarI mutation pathway is not restricted to the NarI site; related sequences of alternating CG show the same characteristics (Ref. 3; R. Schaaper, N. Koffel-Schwartz, and R.P.P. Fuchs, pers. comm.). That is, these sequences are mutational hot spots for AAF-induced mutagenesis; mutations are -2-bp deletions and occurred independently of the UmuDC functions. Moreover, this pathway presents the same characteristics as the reversion of the his502 allele in Salmonella typhimurium. In the Ames tester strains TA1538 and TA98, reversion of this allele occurs by the loss of two base pairs in a (GC)$_4$ track, independently of the

MucAB genes, which are functionally homologous to UmuDC. Thus, it seems that the NarI mutation pathway is part of a broader mutational pathway, which could be involved, in different organisms, in the processing of unusual structures induced in specific sequences.

REFERENCES

1. Bichara, M., and R.P.P. Fuchs (1985) DNA binding and mutation spectra of the carcinogen N-2-aminofluorene in Escherichia coli. A correlation between the conformation of the premutagenic lesion and the mutation specificity. J. Molec. Biol. 183:341-351.
2. Bos, J.L., M. Verlaan-de Vries, C.J. Marshall, G.H. Veeneman, J.H. van Boom, and A.J. van der Eb (1986) A human gastric carcinoma contains a single mutated and an amplified normal allele of the Ki-ras oncogene. Nucl. Acids Res. 14:1209-1217.
3. Burnouf, D., and R.P.P. Fuchs (1985) Construction of frameshift mutation hot spots within the tetracycline resistance gene of pBR322. Biochimie 67:385-389.
4. Burnouf, D., M. Daune, and R.P.P. Fuchs (1987) Spectrum of cis-platin-induced mutations in Escherichia coli. Proc. Natl. Acad. Sci., USA 84:3758-3762.
5. Burnouf, D., P. Koehl, and R.P.P. Fuchs (1989) Single adduct mutagenesis: Strong effect of the position of a single acetylamino-fluorine adduct within a mutation hot spot. Proc. Natl. Acad. Sci., USA (in press).
6. Dente, L., G. Cesarini, and R. Cortese (1984) pEMBL: A new family of single stranded plasmids. Nucl. Acids Res. 11:1645-1655.
7. Fuchs, R.P.P. (1984) DNA binding spectrum of the carcinogen N-acetoxy-N-2-acetylaminofluorene significantly differs from the mutation spectrum. J. Molec. Biol. 177:173-180.
8. Fuchs, R.P.P., N. Schwartz, and M.P. Daune (1981) Hot spots of frameshift mutations induced by the ultimate carcinogen N-acetoxy-N-2-acetylaminofluorene. Nature 294:657-659.
9. Koehl, P., D. Burnouf, and R.P.P. Fuchs (1989) Construction of plasmids containing a unique acetylaminofluorene adduct located within a mutation hot spot; a new probe for frameshift mutagenesis. J. Molec. Biol. (in press).
10. Koffel-Schwartz, N., and R.P.P. Fuchs (1988) Genetic control of AAF induced mutagenesis at alternating sequences: An additional role of RecA. Molec. Gen. Genet. 215:306-311.
11. Koffel-Schwartz, N., G. Maenhaut-Michel, and R.P.P. Fuchs (1987) Specific strand loss in N-2-acetylaminofluorene modified DNA. J. Molec. Biol. 193:651-659.
12. Koffel-Schwartz, N., J.M. Verdier, M. Bichara, A.M. Freund, M. Daune, and R.P.P. Fuchs (1984) Carcinogen-induced mutation spectrum in wild-type, uvrA and umuC strains of Escherichia coli. J. Molec. Biol. 177:33-51.
13. Kriek, E., J.A. Miller, U. Juhl, and E.C. Miller (1967) 8-(N-2-fluorenylacetamido) guanosine, an arylamidation reaction product of guanosine and the carcinogen N-acetoxy-N-2-fluorenylacetamide in neutral solution. Biochemistry 6:177-182.
14. Malhotra, S.B., K.A. Hart, H.J. Klamut, N.S.T. Thomas, S.E. Bodrug, A.H.M. Burghes, M. Bobrow, P.S. Harper, M.W. Thompson, P.N. Ray, and R.G. Worton (1988) Frame-shift deletions in patients with Duchenne and Becker muscular dystrophy. Science 242:755-759.

15. Orkin, S.H., T.-C. Cheng, S.E. Antonarakis, and H.H. Kazazian, Jr. (1985) Thalassemia due to a mutation in the cleavage-polyadenylation signal of the human β-globin gene. EMBO J. 4:453-456.

16. Premkumar Reddy, E., R.K. Reynolds, E. Santos, and M. Barbacid (1982) A point mutation is responsible for the acquisition of transforming properties by the T24 human bladder carcinoma oncogene. Nature 300:149-152.

17. Sage, E., and M. Leng (1980) Conformation of poly(dG-dC) modified by the carcinogens N-acetoxy-N-acetyl-2-aminofluorene and N-hydroxy-N-2-aminofluorene. Proc. Natl. Acad. Sci., USA 77:4597-4601.

18. Sage, E., and M. Leng (1981) Conformation of poly(dG-dC) modified by the carcinogen N-acetoxy-N-acetyl-2-aminofluorene. Nucl. Acids Res. 9:1241-1249.

19. Santella, R.M., D. Grunberger, S. Broyde, and B.E. Hingerty (1981) Z-DNA conformation of N-2-acetylaminofluorene modified poly(dG-dC).poly(dG-dC) determined by reactivity with anti cytidine antibodies and minimized potential energy calculations. Nucl. Acids Res. 9:5459-5467.

20. Santella, R.M., D. Grunberger, I.B. Weinstein, and A. Rich (1981) Induction of the Z conformation in poly(dG-dC).poly(dG-dC) by binding of N-2-acetylaminofluorene to guanine residues. Proc. Natl. Acad. Sci., USA 78:1451-1455.

21. Schmale, H., and D. Richter (1984) Single base deletion in the vasopressin gene is the cause of diabetes insipidus in Brattleboro rats. Nature 308:705-709.

22. Streisinger, G., and J. Owen (Emrich) (1985) Mechanisms of spontaneous and induced frameshift mutation in bacteriophage T4. Genetics 109:633-659.

23. Streisinger, G., Y. Okada, J. Emrich, J. Newton, A. Tsugita, E. Terzaghi, and M. Inouye (1966) Frameshift mutations and the genetic code. Cold Spring Harbor Symp. Quant. Biol. 31:77-89.

24. Tabin, C.J., S.M. Bradley, C.I. Bargmann, R.A. Weinberg, A.G. Papageorge, E.M. Scolnick, R. Dhar, D.R. Lowy, and E.H. Chang (1982) Mechanism of activation of a human oncogene. Nature 300:143-149.

25. Walker, G.C. (1984) Mutagenesis and inducible responses to deoxyribonucleic acid damage in Escherichia coli. Microbiol. Rev. 48:60-93.

26. Weinstein, I.B. (1988) The origins of human cancer: Molecular mechanisms of carcinogenesis and their implications for cancer prevention and treatment. Cancer Res. 48:4135-4143.

27. Wells, R.D., J.J. Miglietta, J. Klysik, J.E. Larson, S.M. Stirdivant, and W. Zaccharias (1982) Spectroscopic studies on acetylaminofluorene-modified $(dT-dG)_n.(dC-dA)_n$ suggest a left-handed conformation. J. Molec. Biol. 257:10166-10171.

FIDELITY OF ANIMAL CELL DNA POLYMERASES α AND δ

AND OF A HUMAN DNA REPLICATION COMPLEX

David C. Thomas, John D. Roberts,
Mary P. Fitzgerald, and Thomas A. Kunkel

Laboratory of Molecular Genetics
National Institute of Environmental Health Sciences
Research Triangle Park, North Carolina 27709

ABSTRACT

We are investigating the mechanisms by which mutations are produced or avoided during DNA synthesis. Using in vitro fidelity assays, we have defined the error frequency and mutational specificity of the replicative animal cell DNA polymerases (α and δ). With DNA polymerase α or the four-subunit DNA polymerase α–DNA primase complex, neither of which contains detectable associated exonuclease activity, the fidelity of the polymerization step is low relative to spontaneous mutation rates in vivo. DNA polymerase δ is much more accurate, partly due to proofreading by the 3'→5' exonuclease activity associated with this polymerase. These fidelity studies have been extended to the replication apparatus present in extracts of human HeLa cells. The replication complex is highly accurate, suggesting that additional fidelity components are operating in the extract during bidirectional, semiconservative replication of double-stranded DNA. Nevertheless, in highly sensitive reversion assays, base substitution errors can be readily detected at frequencies greater than the estimated rate of spontaneous mutation in vivo. This suggests that fidelity components may be missing and/or that human cells depend heavily on postreplicative repair processes to correct replication errors.

INTRODUCTION

The extremely accurate manner in which animal cells replicate and maintain their genome is essential for the survival of the organism and the species. Spontaneous mutation rates in eukaryotes have been estimated to be 10^{-10} to 10^{-12} errors per base pair per generation (19), which is at least as low as that estimated for prokaryotes (3). It is reasonable to expect that this high accuracy reflects the fidelity of DNA synthesis by the enzymes that synthesize new DNA during replication, repair, and recombination. Indeed, there is considerable genetic and biochemical evidence that DNA polymerases in both prokaryotes and eukaryotes play a

central role in controlling mutation rates (14,15). For this reason we and
others have been investigating the fidelity of DNA synthesis by purified
DNA polymerases.

There are four classes of DNA polymerases in higher eukaryotes,
designated α, β, δ, and γ and classified according to size, cellular
location, response to inhibitors, antigenicity, template preference, and
putative function (reviewed in Ref. 5). Two forms of DNA polymerase δ
have been reported, a PCNA (proliferating cell nuclear antigen)-depend-
ent and a PCNA-independent form, the latter designated DNA polymerase
δII (reviewed in Ref. 2). The functions of these polymerases and their
relationship (if any) are uncertain. However, a variety of data (extensively reviewed in Ref. 2, 4, 17, and 18) suggest that both DNA polymerase α and DNA polymerase δ function in replication of the nuclear
genome.

One current model that emerges from the existing data to describe
the participation of these two polymerases at a replication fork is depicted
in Fig. 1 (from Ref. 18). DNA polymerase α is proposed to be responsible for replication of the lagging-strand at the replication fork, while
DNA polymerase δ has been proposed as the leading-strand polymerase.
Although the model is still hypothetical, it provides a useful framework in
which to describe our studies of the fidelity of DNA synthesis catalyzed
in vitro by purified DNA polymerases α and δ and by an SV40-origin-
dependent, semiconservative, double-stranded, bidirectional DNA replication apparatus present in extracts of human HeLa cells.

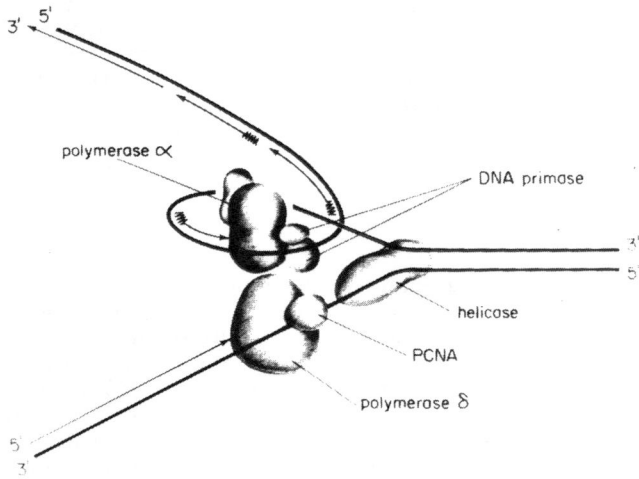

Fig. 1.   A hypothetical view of the eukaryotic replication fork. This
         model suggests how DNA polymerase α-primase and DNA polym-
         erase δ may synthesize the lagging and leading strands, re-
         spectively, in a coordinated manner. Reprinted from Ref. 18
         with permission.

METHODS

Gap-filling Assays

The DNA substrate for fidelity assays with purified polymerases is an M13mp2 molecule with a gap of single-stranded DNA that contains as a mutational target the lacZ α-complementation sequence (6). Following gap-filling synthesis by the DNA polymerase, which is monitored by running an aliquot of the reaction products on an agarose gel, the DNA is transfected into an appropriate E. coli host strain. The forward mutation assay uses the wild-type lacZ α sequence, which yields dark-blue plaques on X-gal plates due to α-complementation of β-galactosidase activity (6). Errors committed by the polymerase during gap-filling synthesis are detected as lighter blue or colorless plaques due to loss of the nonessential α-complementation function. In this assay, a variety of mutations at many different sites can be scored, including all 12 base substitution errors, frameshifts, deletions, and complex errors. To determine the type of error made, DNA is purified from cultures obtained from independent mutant plaques and the sequence of each mutant is determined.

In addition to the forward mutation assay, a base substitution reversion assay has been developed using an altered form of the gapped M13mp2 substrate. In this assay (11), the gap contains a single-base change (G→A in the viral, template strand) at position 89 in the lacZ α coding sequence. This yields a TGA opal codon which results in a colorless plaque phenotype. Following gap-filling synthesis, eight out of nine possible single-base substitutions at the TGA codon can be scored as blue revertants. The proportion of blue to total plaques (the reversion frequency) represents the base substitution error frequency of a single round of DNA synthesis. Because of the much smaller target in this assay (only three bases), the spontaneous revertant background is much lower than is the spontaneous mutant frequency in the forward assay. This allows fidelity estimates to be made for highly accurate synthesis reactions. Furthermore, in this assay the error rate can be easily modulated by creating dNTP pool imbalances in the reaction mixture, which is necessary for forcing errors under the most accurate synthesis conditions.

Replication Assays

To measure the fidelity of DNA replication in extracts of human cells, we chose an SV40-based system. This is a useful model system for studying replication in mammalian cells because synthesis is initiated at a unique site, the SV40 origin, proceeds in a bidirectional, semiconservative mode, converting a double-stranded circular template into double-stranded daughter molecules, and is carried out almost entirely by host cellular replication proteins, requiring only one viral protein, the SV40 large tumor antigen (1,12,13). A minimal list of host factors required for this system is DNA polymerase α-primase complex, topoisomerases, DNA ligase, single-strand binding protein, PCNA, and possibly DNA polymerase δ.

In order to measure fidelity with the M13mp2-based system, the SV40 origin of replication was cloned into M13mp2 just to the left of the lacZ α-complementation target at the unique Ava II site (Fig. 2A, from Ref. 10). This substrate can be used to measure fidelity for bidirectional, semiconservative DNA replication carried out by the extracts. A second substrate was also constructed (Fig. 2B) using Ava II-digested

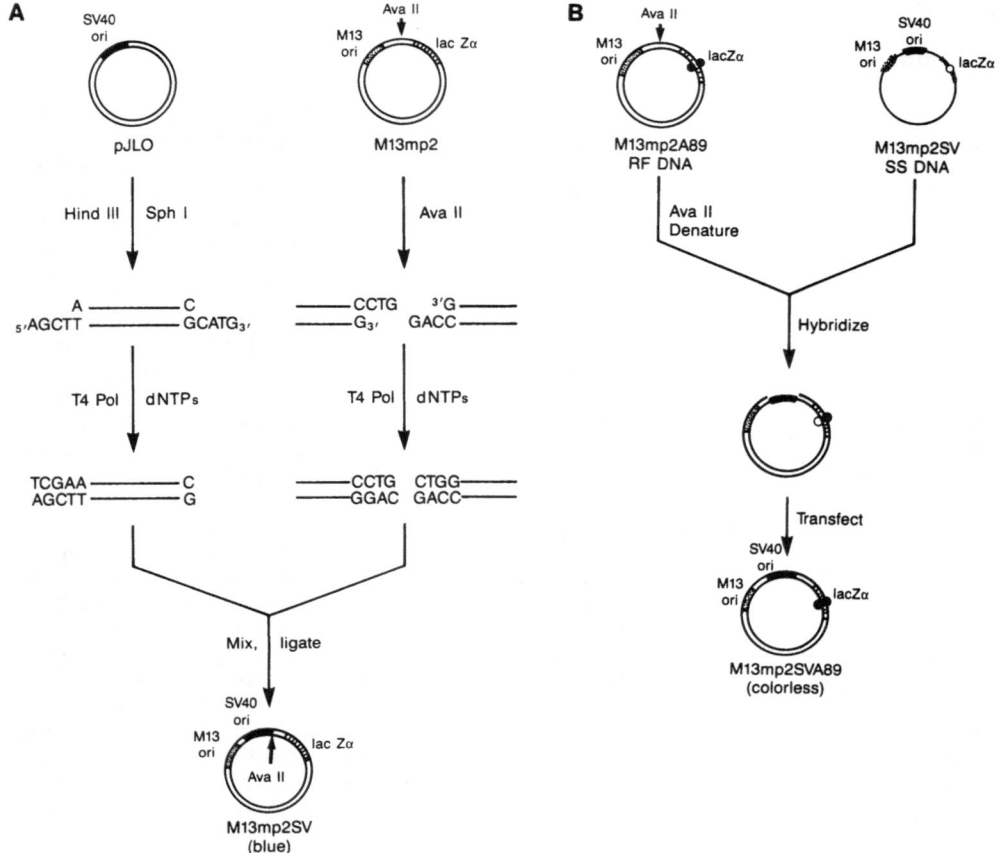

M13mp2 duplex DNA containing the G→A base change at position 89 to hybridize to the single-stranded viral form of the SV40 origin-containing construct containing a wild-type lacZ α sequence. Following transfection, colorless plaques were isolated and the resulting DNA was used for a reversion assay in the SV40 replication system, similar to the A89 reversion assay used for purified polymerases in the gap-filling reaction.

Replication reactions are incubated for 6 hr as described (16), after which the extent of the reaction is monitored by precipitation of an aliquot of the ^{32}P-labeled product on a glass fiber filter. The remaining sample is digested with Dpn I to determine the extent of semiconservative replication, as unreplicated products, which are fully methylated, are digested by this enzyme. The digested sample, now containing only fulllength hemimethylated or unmethylated replicated molecules, is transfected into a mismatch-repair deficient E. coli strain (NR9162, mutS) in order to eliminate methyl-directed mismatch repair of mutant replication products. The resulting mutants are scored and sequenced.

Fig. 2.   Construction of SV40-origin-containing M13mp2 substrates for replication fidelity assays. (A) Construction of M13mp2SV for the forward mutation assay. The Hind III-Sph I fragment of SV40 (nucleotides 5171-128 on the SV40 map) obtained from plasmid pJLO (12) was inserted at the unique Ava II site in M13mp2 by blunt end ligation of the appropriate fragments. Following transfection of the ligation products into competent E. coli S90C cells (6), M13 phage that produced blue plaques was purified. The presence of the SV40 origin was examined by digestion of RF DNA with Sty I, an enzyme that cuts once within the SV40 origin but not within M13mp2. The lacZ α region and the SV40 origin and flanking sequences were sequenced to confirm that no base changes had occurred during construction. Reprinted from Ref. 10 with permission. (B) Scheme for introducing A89 mutation in lacZ α into M13mp2SV for the opal codon reversion assay. Single-stranded circular, viral M13mp2SV DNA was hybridized with denatured M13mp2A89 RF DNA previously digested with Ava II. The gapped heteroduplex hybridization products were used to transfect competent cells. Colorless (white) plaques were picked, and the resulting DNA sequenced to confirm that no changes had occurred in the lacZ α and SV40 origin regions. This molecule, M13mp2SVA89, contains an opal codon (TGA) due to the G→A base change at position 89 in the α-complementation coding sequence. Other mutations can be introduced into M13mp2SV using this procedure.

RESULTS AND DISCUSSION

Fidelity of Purified Eukaryotic DNA Polymerases α

Results for DNA polymerase α purified from several sources are shown in Tab. 1, using the forward mutation assay. When purified by conventional chromatographic techniques, the catalytic subunit of Pol α is partially proteolyzed (the typical molecular weight is ∿120,000 to 140,000), and the enzyme preparation has no associated DNA primase activity. When purified by immunoaffinity chromatography, a four-subunit complex is obtained containing both DNA polymerase and DNA primase activity. The polymerase catalytic subunit is less proteolyzed (e.g., the typical molecular weight is primarily 180,000). The enzyme preparations shown in Tab. 1 are devoid of detectable 3'→5' exonucleolytic proofreading activity.

Each of these enzyme preparations generates mutants during gap-filling DNA synthesis at a frequency well above the spontaneous background mutant frequency of the assay. DNA sequence analysis of mutants demonstrates that three classes of errors are produced: single-base substitutions, one-base frameshifts, and deletions of two or more bases. The error rates per detectable nucleotide polymerized are shown in Tab. 1 for two of these three classes of errors. Given the number and different sources of alpha polymerases examined, the results are remarkably similar. Error rates are generally in the range of 1 per 10,000 bases polymerized (although wide variations in error rates are seen for individual sites and subclasses of base-substitution and frameshift errors, reviewed

Tab. 1.   Fidelity of DNA polymerase α and DNA polymerase α-DNA prim-
          ase complexes.

Enzyme Source	Mutant Frequency	Error Rate	
		Base Substitutions	Frameshifts
	(x 10^{-4})		
DNA polymerase α			
Calf thymus	130	1/ 4,300	1/ 9,300
HeLa cell	80	---------	---------
Mouse myeloma	120	---------	---------
Chick embryo	80	1/ 4,700	1/33,000
DNA polymerase α-DNA primase			
KB cell	160	1/ 3,100	1/ 8,500
HeLa cell	130	---------	---------

Data are taken from references 7 and 16.

in Ref. 9).   Neither DNA polymerase α nor the DNA polymerase α-DNA
primase complex is highly accurate relative to the estimated rate of
spontaneous mutation in vivo.   Clearly, additional factors must be involved to assure the accurate replication and maintenance of genetic
information.

Fidelity of DNA Polymerase δII

Studies with prokaryotic DNA polymerases indicate that one likely
candidate for enhancing the fidelity of DNA synthesis in eukaryotes is
proofreading by polymerase-associated 3'→5' exonuclease activity.   In
collaboration with Sabatino and Bambara (11), we therefore measured the
fidelity of DNA polymerase δII, the PCNA-independent form of Pol δ,
which contains a nondissociable 3'→5' exonuclease.   Synthesis by this
enzyme is highly accurate, having a base-substitution error rate in the
opal codon reversion assay of less than one error per million bases incorporated (Tab. 2).   When AMP is included in the reaction to inhibit the
proofreading exonuclease, the error rate of DNA polymerase δII is increased more than ten-fold, suggesting that proofreading enchances fidelity by at least this amount.   This is a minimum estimate, since even 5 mM
AMP does not completely abolish proofreading activity.   As with prokaryotic enzymes, the extent to which proofreading enhances fidelity probably
varies over a wide range, depending on the neighboring base sequence
and composition of the mispair.   In the future, it may be possible to define the specificity of proofreading by DNA polymerases using the forward
mutation assay.

Fidelity of Replication

It is obvious from the data in Tab. 1 and 2 that DNA polymerases α
and δ have very different fidelity.   This is interesting relative to the
model for their specific roles in replication (Fig. 1).   If DNA polymerase
δ is participating in replication, the fidelity of replication may be an

Tab. 2. Base substitution fidelity of DNA polymerases α and δII in the opal codon reversion assay.

Enzyme	Reaction conditions		Error rate
	[dNTP]	AMP	
Calf thymus Pol α-primase	500 μM	none	1/8,000
Calf thymus Pol δII[a]	100 μM	none	1/1,300,000
	200 μM	5 mM	1/120,000

[a]Synthesis reactions were performed with a 10-fold excess of dGTP, dCTP and dTTP over dATP and the error rates were calculated to reflect this imbalance.

Data are taken from references 8 and 11.

average of the α and δ polymerases operating at the replication fork. Alternatively, the fidelity of replication may be higher if the low fidelity of polymerase α is increased by interactions with accessory proteins that make the DNA polymerase α-dependent steps more accurate.

As a first step to address these questions, we have developed a fidelity assay system to measure the accuracy of replication complexes-- that is, extracts or reconstituted enzyme preparations that are capable of semiconservative, bidirectional replication of double-stranded DNA molecules into two daughter molecules. We have cloned the SV40 origin of replication into M13mp2 and, with the addition of SV40 large T antigen, have demonstrated that efficient replication of this molecule can be achieved using extracts prepared from HeLa cells. As determined in the forward mutation assay, replication by this extract has an estimated error rate of <1 error per 150,000 nucleotides incorporated. This fidelity value is considerably higher than the immunoaffinity-purified DNA polymerase α-primase complex from HeLa cells, which has an error rate in the forward mutation assay of 1/5,000.

A second comparison can be obtained with the opal codon reversion assay, which is much more sensitive than the forward assay (i.e., the background mutant frequency is 100-fold lower). With this assay, the base substitution error rate during replication ranges from 1/110,000 to <1/1,500,000, depending on the mispair formed at the TGA codon. By comparison, the base substitution error rate of DNA polymerase α-primase is 1/8,000, while that of pol δII is 1/1,300,000 (or 1/120,000 with proofreading partially inhibited). Thus, the extract is more accurate than DNA polymerase α-primase, about as accurate as DNA polymerase δII, and not nearly as accurate as the estimated spontaneous mutation rate in vivo.

There are several possible explanations for these observations. Since replication fidelity is greater than that of purified DNA polymerase α, if, as suggested by a wealth of data and the model in Fig. 1, polymerase α is indeed responsible for at least some of the replicative DNA synthesis, then it is possible that additional fidelity components are operating during replication. Likely candidates for improving accuracy include: (i) accessory proteins that alter the polymerase binding affinity for dNTPs, primer, or template; (ii) single-stranded DNA binding

Tab. 3.   Fidelity of semiconservative double-stranded DNA replication by
           HeLa cell extract.

Reaction Conditions	Mutant Frequency	Error Rate
Forward Mutation Assay		
Equal dNTPs, 100 µM	3.1×10^{-4}	$\leq 1/150,000$
Opal Codon Reversion Assay		
Equal dNTPs, 100 µM	1×10^{-5}	$\sim 1/180,000$
20 x dGTP and dCTP	14×10^{-5}	$1/86,000$
From DNA sequence analysis of revertants:		
First position mispairs:	A·dCTP; T·dGTP	$1/450,000$
	A·dGTP; T·dCTP	$\leq 1/1,500,000$
Third position mispairs:	T·dGTP; A·dCTP	$1/110,000$
	T·dCTP; A·dGTP	$1/710,000$

Data are taken from reference 16. The background mutant frequency of
uncopied DNA has been subtracted from the values shown.

proteins; and (iii) a proofreading exonuclease. DNA polymerase δ is
known to utilize proofreading to increase fidelity, though a role for this
polymerase in replication is uncertain.

Regardless of the involvement of DNA polymerase δ or a high fidelity
form of DNA polymerase α, the base substitution fidelity of the extract,
while high, is still not consistent with the very low estimated rate of
spontaneous mutagenesis in vivo. This finding suggests that fidelity
components may be missing or inoperative and/or that repair processes
contribute a large portion of the accuracy of replication in vivo. Initial
data demonstrate that some form of mismatch repair is indeed functioning
in the extract to correct single-base mismatches in defined heteroduplex-
es. However, the extent of this repair is insufficient to explain the high
fidelity of the extract.

The speculation that DNA polymerase α-primase and DNA polymerase
δ operate on opposite strands at the replication fork poses the obvious
question of whether there is a difference in the fidelity of leading-
versus lagging-strand replication. We are also interested in the question
of whether proofreading is occurring during replication in these extracts,
and we want to examine further the extent and specificity of mismatch
repair in this system. These issues, and a search for proteins other
than the polymerases that may enhance accuracy, are currently being
pursued using M13mp2 DNA substrates.

REFERENCES

1. Ariga, H., and S. Sugano (1983) Initiation of simian virus 40 DNA
    replication in vitro. J. Virol. 48:481-491.

2. Bambara, R.A., T.W. Myers, and R.D. Sabatino (1989) DNA polymerase δ. In The Eucaryotic Nucleus: Molecular Biochemistry and Macromolecular Assemblies, P. Strauss and S. Wilson, eds. Telford Press, Caldwell, New Jersey (in press).

3. Drake, J.W. (1969) Comparative rates of spontaneous mutation. Nature 221:1132.

4. Focher, F., E. Ferrari, S. Spadari, and U. Hubscher (1988) Do DNA polymerases δ and α act coordinately as leading and lagging strand replicases? FEBS Lett. 229:6-10.

5. Fry, M., and L.A. Loeb (1986) Animal Cell DNA Polymerases, CRC Press, Boca Raton, Florida.

6. Kunkel, T.A. (1985) The mutational specificity of DNA polymerase-β during in vitro DNA synthesis. Production of frameshift, base substitution, and deletion mutants. J. Biol. Chem. 260:5787-5796.

7. Kunkel, T.A. (1985) The mutational specificity of DNA polymerases-α and γ during in vitro DNA synthesis. J. Biol. Chem. 260:12866-12874.

8. Kunkel, T.A., R.A. Bambara, K. Bebenek, J.D. Roberts, R.D. Sabatino, M.P. Smith, and A. Soni (1988) Analysis of mutational mechanisms with eukaryotic DNA polymerases. In Mechanisms and Consequences of DNA Damage Processing, E.C. Friedberg and P.C. Hanawalt, eds. Alan R. Liss, Inc., New York, pp. 521-528.

9. Kunkel, T.A., and K. Bebenek (1988) Recent studies of the fidelity of DNA synthesis. Biochim. Biophys. Acta 951:1-15.

10. Kunkel, T.A., K. Bebenek, J.D. Roberts, M.P. Smith, and D.C. Thomas (1989) Analysis of fidelity mechanisms with eukaryotic DNA replication and repair proteins. Genome (in press).

11. Kunkel, T.A., R.D. Sabatino, and R.A. Bambara (1987) Exonucleolytic proofreading by calf thymus DNA polymerase δ. Proc. Natl. Acad. Sci., USA 84:4865-4869.

12. Li, J.J., and T.J. Kelly (1984) Simian virus 40 DNA replication in vitro. Proc. Natl. Acad. Sci., USA 81:6973-6977.

13. Li, J.J., and T.J. Kelly (1985) Simian virus 40 DNA replication in vitro: Specificity of initiation and evidence for bidirectional replication. Molec. Cell. Biol. 5:1238-1246.

14. Loeb, L.A., and T.A. Kunkel (1982) Fidelity of DNA synthesis. Ann. Rev. Biochem. 52:429-457.

15. Loeb, L.A., and M.E. Reyland (1987) Fidelity of DNA synthesis. In Nucleic Acids and Molecular Biology, Vol. 1, F. Eckstein and D.M.J. Lilley, eds. Springer-Verlag, Berlin, pp. 157-173.

16. Roberts, J.D., and T.A. Kunkel (1988) Fidelity of a human cell DNA replication complex. Proc. Natl. Acad. Sci., USA 85:7064-7068.

17. So, A.G., and K.M. Downey (1988) Mammalian DNA polymerases α and δ: Current status in DNA replication. Biochemistry 27:4591-4595.

18. Stillman, B. (1988) Initiation of eukaryotic DNA replication in vitro. BioEssays 9:56-60.

19. Wabl, M., P.D. Burrows, A. von Gabain, and C. Steinberg (1985) Hypermutation at the immunoglobulin heavy chain locus in a pre-B-cell line. Proc. Natl. Acad. Sci., USA 82:479-482.

MOLECULAR MECHANISMS OF REPLICATIONAL FIDELITY

IN ESCHERICHIA COLI

Hisaji Maki, Masahiro Akiyama, Takashi Horiuchi,
and Mutsuo Sekiguchi

Departments of Biochemistry and Molecular Biology
Faculty of Medicine
Kyushu University, Fukuoka 812, Japan

ABSTRACT

DNA polymerase III holoenzyme is responsible for chromosomal DNA synthesis in Escherichia coli and seems to be a major determinant of the fidelity of replication of this organism. Among ten different subunits of the holoenzyme, the α subunit, encoded by the dnaE gene, has a polymerase activity, while the ε subunit, encoded by the dnaQ gene, is a proofreader with a 3'-5' exonuclease activity. Using poly(dA)/oligo(dT)$_{20}$ as a template-primer, misincorporation of dGMP, dCMP, and dAMP by the α subunit and exonucleolytic editing of those mispairs by the ε subunit were investigated. When the polymerization reaction was performed with the α subunit, dCMP and dGMP but not dAMP were misincorporated. This would suggest that the polymerase might have a base-selecting function to avoid dA:dA mispairing. A subassembly of the DNA polymerase III consisting of α, ε, and θ subunits misincorporated only dGMP. This would imply that the proofreading function of the ε subunit may correct the dC:dA but not the dG:dA mispair. Addition of a protein encoded by the mutT gene, defects of which cause AT to CG transversions in vivo, diminished the misincorporation of dGMP onto poly(dA) template by the α subunit. A dGTPase activity was associated with the MutT protein. The significance of the dGTPase activity in the prevention of dG:dA mispairing is discussed.

INTRODUCTION

Most spontaneous mutations are base substitutions resulting from mispairings occurring during DNA replication (21). Many factors are involved in the formation of mispairs and conversion into the base substitution (10). Mispairs are generated as errors of the action of DNA polymerase. The mispairs are examined and corrected by the proofreading function present in the DNA polymerase, and most of the mispairs are corrected in this step. Thus, DNA polymerase seems to play crucial roles in controlling the fidelity of DNA replication.

In Escherichia coli, the DNA polymerase III holoenzyme is a major replicative enzyme and responsible for maintaining the replicational fidelity (12). This enzyme has an asymmetric dimeric structure with ten different subunits (17). Each half catalyzes the leading- or lagging-strand DNA synthesis at the replication fork, and both contain a catalytic core of the polymerase which consists of α, ε, and θ subunits (18). This subassembly, called pol III core, has both polymerase activity and proofreading capacity. With the isolated α and ε subunits, these enzymatic functions of pol III core could be shown to be attributable to these two subunits; the α subunit possesses the polymerase activity but neither 3'-5' nor 5'-3' exonuclease activity (15), and the 3'-5' exonuclease activity resides in the ε subunit (22). In contrast to other prokaryotic polymerases, two different polypeptides of pol III core account for polymerization on the one hand and the 3'-5' exonuclease function on the other; together they provide the editing function (16).

Mutator mutants which show elevated levels of spontaneous mutation rate have aided in the identification of factors involved in this mutagenesis, and the molecular mechanisms of these processes are now better understood. Among the 9 mutator genes identified in E. coli, dnaE, dnaQ, and mutT may be involved in the fidelity of replication (9,11,25). Other mutator genes were assigned to postreplicative mismatch repair systems. The dnaE and dnaQ genes encode the α and ε subunits of DNA polymerase III, respectively, and mutations that occurred in these genes cause conditional lethal mutator activity, thereby reflecting the fact that the editing exonuclease activity resides in the ε subunit. Recently, a dnaE^{ts} mutant was isolated with a mutator activity comparable to that of the dnaQ mutants (H. Maki, T. Horiuchi, and M. Sekiguchi, unpubl. results). The mutT mutants exhibit an increased frequency of a unidirectional AT to CG transversion (6), and this type of transversion is apparently not enhanced in other mutator mutants. DNA replication seems to be required for expression of the mutator effects of the mutT mutant as well as of the dnaQ and dnaE strains. Based on the finding that the temperature-sensitive phenotype of dnaE mutations is partially suppressed by the mutT mutation, it was suggested that the mutT gene product might be interacting with the DNA polmerases III (6). However, the mutT gene does not seem to be essential for cell growth (5).

We used a gel electrophoresis assay to analyze formation of mispairs by the α subunit, exonucleolytic editing of such mispairs by the ε subunit, and function of the MutT protein. Our results suggested that the α subunit might have a base selection function. We found that dG:dA mispairs were little eliminated by the 3'-5' exonuclease associated with the pol III and that the MutT protein prevented the formation of dG:dA mispairs.

SPECIFICITY OF BASE SELECTION AND EXONUCLEOLYTIC EDITING
BY DNA POLYMERASE III

Using purified preparations of the α subunit and pol III core, we explored replicational errors and the specificity of proofreading in DNA synthesis in vitro. Figure 1 shows a schematic diagram of the assay for formation of mispairs in vitro. Oligo(dT)$_{20}$ was annealed to the poly(dA) with an average gap size of 20 nucleotides, and misincorporation of nucleotides onto the primer-template was examined. Either one of the nucleoside triphosphates labeled with α-^{32}P and the polymerase were

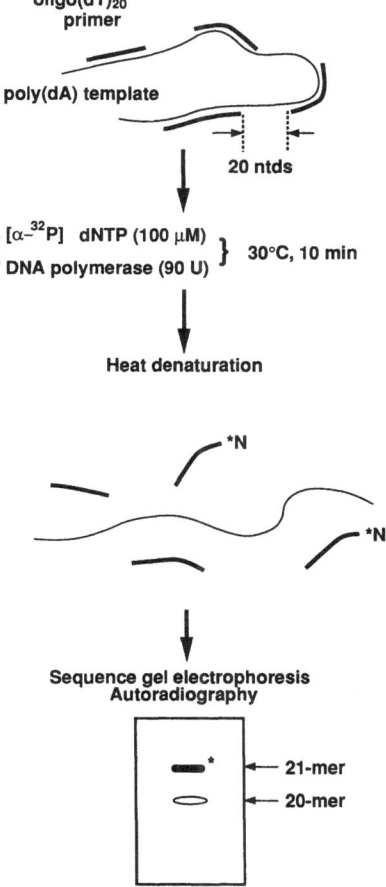

Fig. 1.   A schematic diagram of the assay for formation of mispairs in vitro.

added to the reaction mixture containing the template, and incubated at 30°C for 10 min.   After the reaction was terminated, DNA was extracted, denatured, and applied to a DNA sequencing gel electrophoresis.   Misincorporated nucleotides were detected as bands corresponding to 21-mer on the autoradiogram.

As shown in Fig. 2, the α subunit misincorporated dGMP and dCMP but not dAMP, while the core misincorporated only dGMP.   These observations suggest that dG:dA and dC:dA mispairs can be readily formed during the polymerase action and that the proofreading capacity can eliminate the dC:dA but not the dG:dA mispairs.

From these observations, we can depict several aspects of polymerase and proofreading functions.   First, the polymerase seems to have a base selection function, which excludes dA:dA mispairing.   This is consistent with the in vivo observation that the dC:dA mispairs are generated more frequently than the dA:dA mispairs (8).   Second, differing from the in

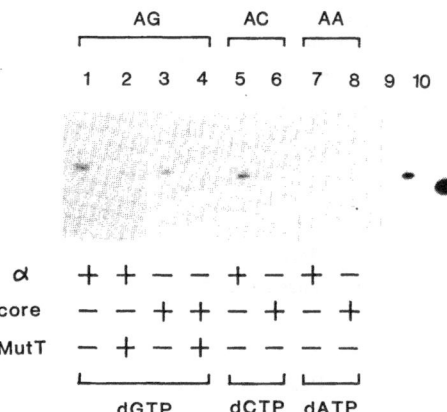

Fig. 2.   Misincorporation of dGMP and other nucleotides with poly(dA) as a template. The reaction was carried out, as indicated in Fig. 1, with 90 units of the α subunit (for lanes 1, 2, 5, and 7) or the pol III core (for lanes 3, 4, 6, and 8). Samples for lanes 2 and 4 were with 25 ng of the purified MutT protein. Lanes 1-4: dGMP misincorporation. Lanes 5-6: dCMP misincorporation. Lanes 7-8: dAMP misincorporation. The samples were run on 12% polyacrylamide gels containing 8 M urea, and nucleotides misincorporated were detected on the autoradiogram. In lanes 9 and 10, ^{32}P-labeled oligo(dT)$_{21}$ and oligo-(dT)$_{20}$ were run, respectively.

vivo data, the dG:dA mispairs are formed as frequently as dC:dA mispairs. Using synthetic DNAs specially designed for a misincorporation assay, Sloane et al. (23) estimated the accuracy of DNA synthesis in vitro catalyzed by the α subunit; misinsertion frequencies of dG, dC, and dA the opposite to dA were 5.2×10^{-4}, 7.1×10^{-5}, and 1.3×10^{-5}, respectively. Furthermore, the proofreading capacity of the polymerase is incompetent to eliminate the dG:dA mispairs. This means that the editing exonuclease does have substrate specificity. Fersht and Knill-Jones (7) arrived at a similar conclusion by demonstrating that the relative frequencies of mispairs formed in vitro by the pol III holoenzyme were dG:dA > dA:dA = dC:dA = dT:dG > dT:dC.

Since the dG:dA mispairs are not corrected by the mismatch repair (19), a frequent generation and a poor correction of dG:dA mispairs would result in the occurrence of a high frequency of AT to CG transversion. However, the frequency of this transversion in E. coli is low, and it seems probable that the cell comes equipped with a certain mechanism(s) to prevent formation of the dG:dA mispairs or to eliminate such pairs. The mutT gene product might be involved in this process.

PURIFICATION AND CHARACTERIZATION OF MutT PROTEIN

We cloned the mutT gene and determined the DNA sequence. Based on these data, we identified the product as a polypeptide of about 15,000 daltons (1). To examine whether the MutT protein has the capacity to suppress dG:dA mispairing, we purified this protein (2). From cells overproducing the MutT protein, a nearly homogeneous preparation of the protein was obtained. Since we knew little of the enzymatic activity associated with this protein, we followed the protein in the SDS gel throughout the purification.

The MutT protein apparently possesses no DNA polymerase activity nor affects the rate of polymerase reaction when added to the E. coli DNA polymerase III core enzyme. However, on incubation of the MutT protein with DNA and dNTPs in the polymerase assay, formation of a small amount of dGMP did occur. GTP and dGDP were also hydrolyzed, with lower rates, as compared with the hydrolysis of dGTP. No other nucleoside monophosphate was formed under these conditions. Except for Mg^{++}, other components for the DNA polymerase reaction were not required for the dGTPase activity.

Bhatnagar and Bessman (4) reported the presence of a deoxynucleoside triphosphatase activity in their MutT protein preparation. They found that the protein hydrolyzed all four deoxynucleoside triphosphates, although the rates of hydrolysis for dATP, dCTP, and dTTP were 2-30% of that for dGTP. We found that the MutT protein degrades dGTP but not other deoxynucleoside triphosphates. This discrepancy may relate to assay conditions; they used a buffer at pH 9.0 for the assay, whereas we carried out the reaction at pH 7.5.

PREVENTION OF MISINCORPORATION OF dGMP BY THE MutT PROTEIN

Effects of the MutT protein on the formation of dG:dA mispairs in vitro were also examined. The purified MutT protein clearly inhibited the dG:dA mispairing. When the MutT protein was filtered through a Suparose-12 column, the activity required to inhibit the mispairing coincided well with the peak of the protein. This inhibition was specifically observed with the dG:dA mispairing, and preformed dG:dA mispairs were not eliminated by the MutT protein. This suggests that the MutT protein acts before or during the DNA synthesis to prevent the formation of dG:dA mispairs.

To elucidate the correlation of activities of the MutT protein, we quantitatively measured both reactions catalyzed by the MutT protein (Fig. 3). On incubation with increasing amounts of MutT protein, the amount of dGTP in the reaction mixture decreased while that of dGMP increased. In parallel, the rate of misincorporation of dGMP decreased, but this suppression of misincorporation occurred when a large amount of dGTP was present in the reaction mixture. Therefore, it is unlikely that the prevention of misincorporation is caused by loss of the dGTP substrate. Presumably, some mutagenic form of dGTP that can pair with dA is specifically degraded by the MutT protein.

According to a model proposed by Topal and Fresco (24), the syn form of dGMP can pair with a tautomer of dAMP within the Watson-Crick geometry. We propose the model for the MutT protein action shown in Fig. 4. If the MutT protein preferentially hydrolyzes the syn-formed dGTP, the prevention of formation of the dG:dA mispair could be explained. Supporting evidence for this model is that the apparent Km for the dGTPase activity of MutT protein exceeded 1 mM, thereby suggesting that the proper substrate of the MutT protein might not be the major component, the anti-formed dGTP. Since conversion between the anti and

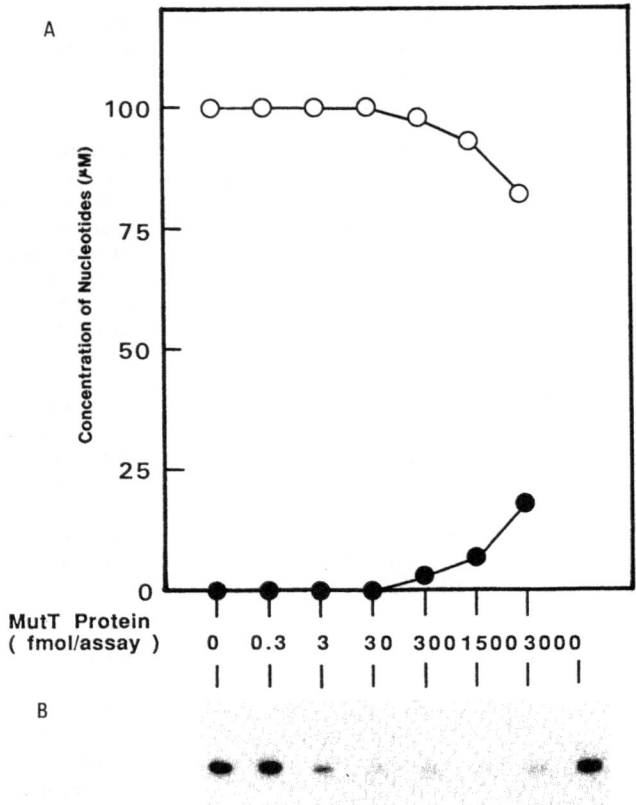

Fig. 3.   Prevention of misincorporation of dGMP by MutT protein.   As-
          says for the misincorporation of dGMP were carried out with 90
          units of the α subunit and various amounts of the MutT pro-
          tein.   Aliquots were subjected to PEI-cellulose chromatography
          to determine contents of dGTP and free dGMP; the remaining
          was used for the misincorporation assay.   Samples analyzed
          contained 0, 0.33, 3.3, 33, 330, 1,500, 3,300 fmol of MutT pro-
          tein.   (A) Conversion of dGTP to dGMP by the MutT protein.
          o, amount of dGTP; *, amount of dGMP.   (B) An autoradiogram
          showing the misincorporation of dGMP onto a poly(dA) template.

the syn forms seems to be nearly free (24), we assume that hydrolysis
occurs at the site of DNA replication.   Since the occurrence of G:C to
C:G transversion, which is also predictable from the Topal-Fresco model,
is not stimulated in mutT mutants, the dG:dG mispairs might be excluded
by other mechanisms, such as base selection by the α subunit.

MULTISTEPS IN ERROR PREVENTION

     As the accuracy required for cellular DNA replication is too high to
be achieved by a single step of reaction, several consecutive reactions

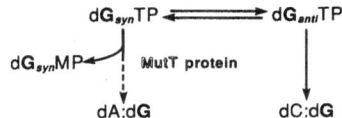

Fig. 4. Model for MutT protein action.

are needed to reduce the rate of formation of error during DNA replication. Here, discrimination of unfavorable precursors in the nucleotide pool, selection of appropriate nucleotides during polymerization, elimination of mismatched bases by proofreading, and correction by postreplicational mismatch repair systems are involved. E. coli cell has at least three pathways for the postreplicational mismatch repair: the mutH, mutL, mutS, and uvrD-dependent pathway (13), the very short patch (VSP) repair pathway (14), and the mutY-dependent pathway (3).

The specificity and intensity of the error-prevention systems in vivo can be estimated from spectrum analyses of apontaneous mutations in wild-type and several mutator strains. All types of transitions and transversions, except for GC to CG, were generated in the wild-type strain with almost equal frequencies of about 10^{-10}/base/replication (21). The spectra of spontaneous mutations in strains lacking a capacity for the post-replicative mismatch repair would reflect the spectra of mispairs which originated during the polymerization and which evaded the proofreading if the replicational errors were corrected by only this type of repair system. In mutH, mutL, mutS, and uvrD-dependent mismatch repair-deficient strains, frequencies of transitions are enhanced about 200-fold, whereas the increase of those of transversions is a few-fold, if any (20). Therefore, this mismatch repair system corrects only dA:dC and dG:dT mispairs, and such mispairs seem to be predominant in the newly replicated DNA.

Similarly, the spectra obtained with a mutD5 mutant defective in the proofreading capacity would provide information regarding the extent of base selection of pol III holoenzyme in vivo (8). Of 42 base substitutions found in the proofreading-deficient strain, 31 were transitions. Eight AT to TA and three GC to TA transversions were found, but AT to CG and GC to CG transversions were not evident. From these results, the frequencies of transitions were estimated to be 10^{-6} to 10^{-5}, while those for transversions were less than 10^{-7} to 10^{-6}. These values are close to those obtained in vitro for frequencies of formation of dC:dA and dA:dA mispairs (23), hence, the base selection function of the α subunit is probably a major determinant for the frequencies of formation of these types of mispairs in E. coli. As expected from the in vitro data, the proofreading capacity of pol III apparently removes mispairs causing the transitions, dA:dC and dG:dT, and also those due to misincorporation of dAMP onto dA or dG residue of the template strand.

The rare occurrence of AT to CG and GC to CG transversions in a proofreading-deficient strain suggests that the dG:dA and dG:dG mispairs might be less frequently formed than other types of mispairs during polymerization. However, from the in vitro studies, it appeared that the polymerase does form the dG:dA mispairs with a frequency comparable to

Tab. 1.   Prevention of replicational errors.

Mispairing		Replication error		Proofreading		Mismatch repair
Transition	AC, GT	10^{-6}	→	10^{-8}	→	10^{-10}
Transversion	GA, AA	10^{-7}	→	10^{-10}		
	AG, GG	10^{-10}				

The lanes underlined represent those on the template strand.

that for the dC:dA mispairs, and that the editing function does not operate to remove the mispairs. We found that in vitro, the MutT protein
prevented the formation of dG:dA mispairs during or before the DNA
polymerase III began to function.

All these observations taken together lead to the proposal that there
are three classes of mispairs that are corrected, each in a different manner (Tab. 1). The first class is dA:dC and dG:dT mispairs which would
result in transition mutations. These occur most frequently during the
polymerase action and are corrected by both proofreading and the mutH,
mutL, mutS, and uvrD-dependent mismatch repair system. The second
class is mispairs of coming dA on template dG or dA residue. These
arise with a moderate frequency and are efficiently corrected by the
proofreading function. The base selection function of the α subunit might
strictly differentiate the dA:dA mispairing. The mutY-dependent mismatch repair pathway is also known to convert the dA:dG to dC:dG pairs
(3). The third class is mispairs of the dG on template dA and dG residues. Occurrence of this class during the DNA synthesis is as frequent
as that of the first class. Neither proofreading nor mismatch repair is
effective to correct these types of mispairs. The MutT protein is required to avoid dG:dA mispairing.

ACKNOWLEDGEMENTS

We thank M. Ohara for helpful comments. This work was supported
by grants from the Ministry of Education, Science and Culture, Japan.

REFERENCES

1.   Akiyama, M., T. Horiuchi, and M. Sekiguchi (1987) Molecular clon-
     ing and nucleotide sequence of the mutT mutator of Escherichia coli
     that causes A:T to C:G transversion. Molec. Gen. Genet. 206:9-16.
2.   Akiyama, M., H. Maki, M. Sekiguchi, and T. Horiuchi (1989) A spe-
     cific role of MutT protein to prevent dG:dA mispairing in DNA rep-
     lication. Proc. Natl. Acad. Sci., USA (in press).
3.   Au, K.G., M. Cabrera, J.H. Miller, and P. Modrich (1988) Escheri-
     chia coli mutY gene product is required for specific A:T-C:G mis-
     match correction. Proc. Natl. Acad. Sci., USA 85:1963-1966.

4. Bhatnagar, S.K., and M.J. Bessman (1988) Studies on the mutator gene, mutT of Escherichia coli: Molecular cloning of the gene, purification of the gene product and identification of a novel nucleoside triphosphatase. J. Biol. Chem. 263:8953-8957.

5. Conrad, S.E., K.T. Dussik, and E.C. Siegel (1976) A phage mu-1 induced mutation to mutT in Escherichia coli. J. Bacteriol. 125:1018-1025.

6. Cox, E.C. (1973) Mutator gene studies in Escherichia coli: The mutT gene. Genetics 73(Suppl.):67-80.

7. Fersht, A.R., and J.W. Knill-Jones (1981) DNA polymerase accuracy and spontaneous mutation rates: Frequencies of purine-purine, purine-pyrimidine, and pyrimidine-pyrimidine mismatches during DNA replication. Proc. Natl. Acad. Sci., USA 78:4251-4255.

8. Fowler, R.G., R.M. Schaaper, and B.W. Glickman (1986) Characterization of mutational specificity within the lacI gene for a mutD5 mutator strain of Escherichia coli defective in 3'-5' exonuclease (proofreading) activity. J. Bacteriol. 167:130-137.

9. Horiuchi, T., H. Maki, and M. Sekiguchi (1978) A new conditional lethal mutator (dnaQ49) in Escherichia coli. Molec. Gen. Genet. 163:277-283.

10. Horiuchi, T., H. Maki, and M. Sekiguchi (1989) Mutators and fidelity of DNA replication. Bull. Inst. Pasteur (in press).

11. Konrad, E.B. (1978) Isolation of an Escherichia coli K-12 dnaE mutation as a mutator. J. Bacteriol. 133:1197-1202.

12. Kornberg, A. (1980) DNA Replication, W.H. Freeman and Co., San Francisco.

13. Lahue, R.S., S.-S. Su, K. Welsh, and P. Modrich (1986) Analysis if methyl-directed mismatch repair in vitro. UCLA Symp. Molec. Cell. Biol. 47:125-134.

14. Lieb, M. (1987) Bacterial genes mutL, mutS, and dcm participate in repair of mismatches at 5-methylcytosine sites. J. Bacteriol. 169:5241-5246.

15. Maki, H., and A. Kornberg (1985) The polymerase subunit of DNA polymerase III of Escherichia coli. II. Purification of the α subunit, devoid of nuclease activities. J. Biol. Chem. 260:12987-12992.

16. Maki, H., and A. Kornberg (1987) Proofreading by DNA polymerase III of Escherichia coli depends on cooperative interaction of the polymerase and exonuclease subunits. Proc. Natl. Acad. Sci., USA 84:4389-4392.

17. Maki, H., S. Maki, and A. Kornberg (1988) DNA polymerase III holoenzyme of Escherichia coli. IV. The holoenzyme is an asymmetric dimer with twin active sites. J. Biol. Chem. 263:6570-6578.

18. McHenry, C.S., and W. Crow (1979) DNA polymerase III of Escherichia coli: Purification and identification of subunits. J. Biol. Chem. 254:1748-1753.

19. Radman, M., and R. Wagner (1986) Mismatch repair in Escherichia coli. Ann. Rev. Genet. 20:523-538.

20. Schaaper, R.M., and R.L. Dunn (1987) Spectra of spontaneous mutations in Escherichia coli strains defective in mismatch correction: The nature of in vivo DNA replication errors. Proc. Natl. Acad. Sci., USA 84:6220-6224.

21. Schaaper, R.M., B.N. Danforth, and B.W. Glickman (1986) Mechanisms of spontaneous mutagenesis: An analysis of the spectrum of spontaneous mutation in the Escherichia coli. J. Molec. Biol. 189:273-284.

22. Scheuermann, R.H., and H. Echols (1984) A separate editing exo-
     nuclease for DNA replication: The ε subunit of Escherichia coli DNA
     polymerase III holoenzyme. Proc. Natl. Acad. Sci., USA 81:7747-
     7751.
23. Sloane, D.L., M.F. Goodman, and H. Echols (1988) The fidelity of
     base selection by the polymerase subunit of DNA polymerase III holo-
     enzyme. Nucleic Acids Res. 16:6465-6475.
24. Topal, M.D., and J.R. Fresco (1976) Complementary base pairing
     and origin of substitution. Nature (London) 263:285-289.
25. Treffers, H.P., V. Spinelli, and N.O. Belser (1954) A factor (or
     mutator gene) influencing mutation rates in Escherichia coli. Proc.
     Natl. Acad. Sci., USA 40:1064-1071.

PERMANENT CONVERSION OF NIH3T3 CELLS TRANSFORMED BY
ACTIVATED c-Ha-ras, c-Ki-ras, N-ras, OR c-raf, AND OF HUMAN
PANCREATIC ADENOCARCINOMA CONTAINING ACTIVATED c-Ki-ras TO
APPARENTLY NORMAL CELLS BY TREATMENT WITH THE ANTIBIOTIC
AZATYROSINE

Nobuko Shindo-Okada,[1] Osamu Makabe,[2]
Hikaru Nagahara,[1] and Susumu Nishimura[1]

[1]Biology Division
National Cancer Center Research Institute
Tsukiji 5-1-1, Chuo-ku, Tokyo 104, Japan
[2]Central Research Laboratories
Meiji Seika Kaisha, Ltd.
Morooka, Kohoku-ku, Yokohama 222, Japan

An antibiotic isolated from Streptomyces chibanensis, L-β-(5-hydroxy-2-pyridyl)-alanine, named azatyrosine (1) (Fig.1), inhibited growth of NIH3T3 cells transformed by the activated human c-Ha-ras gene at 500 µg/ml concentration, but did not significantly inhibit the growth of normal NIH3T3 cells. Surprisingly, on treatment with azatyrosine, most of the transformed cells were converted to cells that apparently behave similarly to normal cells (Fig. 2). These apparently normal cells, named revertant cells, could grow in the presence of azatyrosine, and stopped growing when they reached confluency (Fig. 3), and their normal phenotype persisted during prolonged culture in the absence of azatyrosine. The revertant cells could not grow in soft agar (Fig. 4), and scarcely proliferated in nude mice.

The human c-Ha-ras gene present in transformed NIH3T3 cells was found to be still present in the revertant cells, and was expressed to the same extent as in the original transformed cells, producing the same amount of activated p21 (Fig. 5).

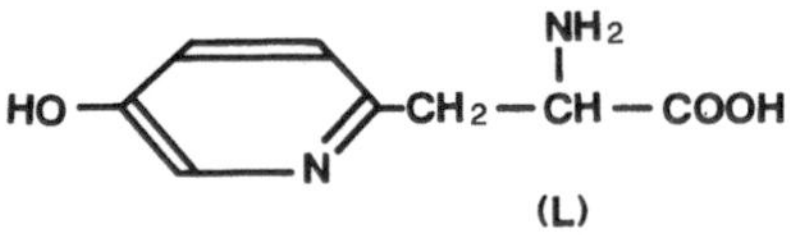

$$HO-\underset{\underset{N}{}}{\bigcirc}-CH_2-\underset{\underset{(L)}{}}{\overset{\overset{NH_2}{|}}{CH}}-COOH$$

Fig. 1.   Structure of antibiotic azatyrosine, L-β-(5-hydroxy-2-pyridyl)-alanine.

a) **after 6 days**   b) **after 13 days**

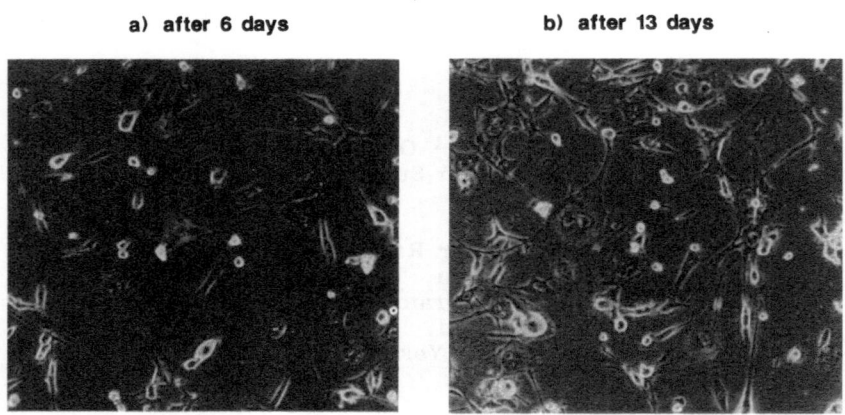

Fig. 2.   Morphological change of NIH3T3 cells transformed by activated
c-Ha-<u>ras</u> by treatment with azatyrosine.

transformed NIH3T3   revertant cells   revertant cells   normal NIH3T3
      cells   (clone1)   (clone2)   cells

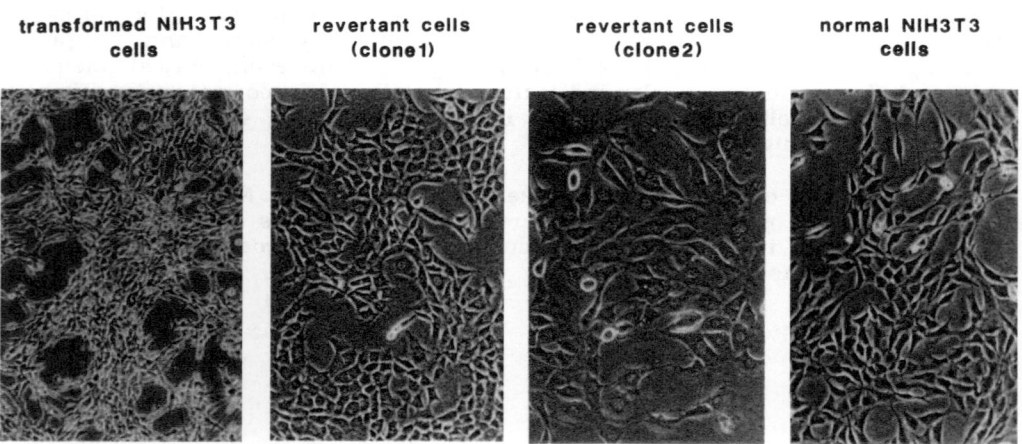

Fig. 3.   Comparison of appearance of cloned revertant cells as with
those of the original transformed NIH3T3 cells and normal
NIH3T3 cells.

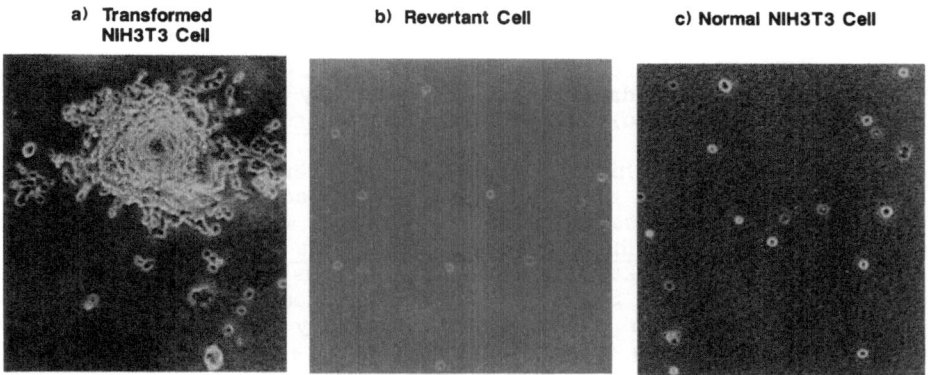

a) Transformed NIH3T3 Cell   b) Revertant Cell   c) Normal NIH3T3 Cell

Fig. 4.   Inability of revertant cells to form colonies in soft agar; (a) NIH3T3 cells transformed by activated c-Ha-ras; (b) revertant cells; (c) normal NIH3T3 cells.

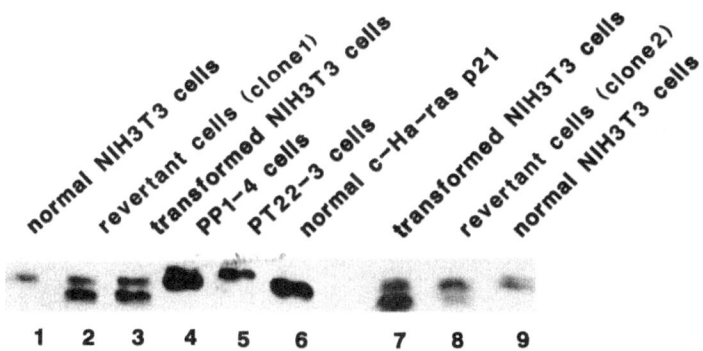

Fig. 5.   Expression of activated c-Ha-ras p21 in revertant cells.

Treatment with azatyrosine caused similar conversion of NIH3T3 cells transformed by activated c-Ki-ras, N-ras, or c-raf to apparently normal cells, but NIH3T3 cells transformed by hst or ret were not exclusively converted by azatyrosine.   Human pancreatic adenocarcinoma cells, which are known to contain an amplified activated c-Ki-ras gene and amplified c-myc gene (2,3), were also converted to flat and giant revertant cells by treatment with azatyrosine.

REFERENCES

1. Inouye, S., T. Shomura, T. Tsuruoka, Y. Ogawa, H. Watanabe, J.
    Yoshida, and T. Niida (1975) L-β-(5-hydroxy-2-pyridyl)-alanine and
    L-β-(3-hydroxyureido)-alanine from Streptomyces. Chem. Pharm.
    Bull. 23:2669-2677.
2. Yamada, H., T. Yoshida, H. Sakamoto, M. Terada, and T. Sugimura
    (1986) Establishment of a human pancreatic adenocarcinoma cell line
    (PSN-1) with amplification of both c-myc and activated c-Ki-ras by a
    point mutation. Biochem. Biophys. Res. Comm. 140:167-173.
3. Yamada, H., H. Sakamoto, M. Taira, S. Nishimura, Y. Shimosato,
    M. Terada, and T. Sugimura (1986) Amplification of both c-Ki-ras
    with a point mutation and c-myc in a primary pancreatic cancer and
    its metastatic tumors in lymph nodes. Jpn. J. Cancer Res. (Gann)
    77:370-375.

ONCOGENES OF STOMACH CANCERS

M. Terada, Y. Hattori, T. Yoshida, H. Sakamoto,
O. Katoh, A. Wada, J. Yokota, and T. Sugimura

Genetics Division
National Cancer Center Research Institute
1-1 Tsukiji 5-chome, Chuo-ku
Tokyo, 104 Japan

INTRODUCTION

Although stomach cancer is a common malignant disease in Japan as well as elsewhere in the world, little is known about the genetic changes that may be associated with the development of tumors. Therefore, we examined the transforming activities of 58 samples of DNA from stomach cancers and noncancerous portions of stomach mucosa. We have detected and identified a novel transforming gene, HST1, in three of these samples (22). The HST1 gene was also subsequently identified as a transforming gene in three stomach cancers (11), three hepatomas (17,43), one colon cancer (11), and in Kaposi's sarcoma (6). Thus, the HST1 gene is at present the most frequently encountered transforming gene other than those of the RAS gene family.

A molecular genetic method with probes revealing restriction fragment-length polymorphism (RFLP) has become available to detect allelic deletion or loss of heterozygosity at various chromosomal loci. This method allowed us to study gene loss at specific chromosomal loci in various cancers by directly analyzing genomic DNAs obtained from surgical specimens without prior culture in vitro. In this paper, we summarize our studies on HST1 and on LOH at specific chromosomal loci in stomach cancers.

TRANSFECTION ASSAYS OF DNA SAMPLES FROM STOMACH CANCERS

DNAs were isolated from 21 stomach cancers, 16 lymph node metastases of stomach cancers, and 21 samples from a noncancerous region of stomach mucosa obtained from 26 patients at the time of surgery. These DNA samples were tested for their transforming activity to NIH 3T3 cells by DNA transfection assay.

In each transfection passage, 60 μg of DNA was used for two recipient cultures containing 5-10 x 10^5 cells per culture. As a positive control, DNA from a1-1 cells, secondary transformants of NIH 3T3 cells induced by transfection of DNA from a T24 bladder carcinoma line, was used in each transfection assay. Only those experimental results in which a1-1 cellular DNA gave more than 0.5 foci per μg of DNA were evaluated. The transfection assay was repeated at least four times for each DNA sample.

Three DNA samples out of a total of 58 showed transforming activity by transfection. Two samples, nos. 361 and 363, were derived from a primary stomach tumor and a noncancerous stomach mucosa, respectively, of the same patient. The third DNA sample, no. 51, was extracted from a lymph-node metastasis of mucinous adenocarcinoma in stomach cancer of another patient (22). It should be noted that activated RAS genes have not been detected in the present study, indicating that a point-mutational activation of the RAS oncogene family is not frequently associated with stomach cancer. The DNA samples from the primary transformants all induced secondary transformants upon transfection. All of the transformants grew well in 0.3% soft agar and were highly tumorigenic when injected subcutaneously into nude mice.

MOLECULAR CLONING OF HST1 cDNA AND ITS CHARACTERIZATION

One of the secondary transformants induced by transfection of no. 361 DNA was designated T361-2nd-1 and was chosen for cloning of portions of the transforming gene. DNA fragments containing an exon of the transforming gene were identified. In northern blot hybridization, these probes hybridized to 3.3 kb poly(A)$^+$ RNA from T361-2nd-1 cells. All of the primary and secondary transformants induced by DNA samples nos. 361, 363, and 51, containing the DNA fragments hybridized to these probes. These data confirmed that the same transforming gene was responsible for acquisition of the transforming activity of three different DNA samples (22).

A cDNA library in λgt10 was constructed from poly(A)$^+$ RNA from T361-2nd-1 cells. Four cDNA clones were isolated and were shown to contain overlapped inserts by restriction mapping. One of these clones, which contained the 3.2-kb insert, encompassed nearly the full length of the major transcript of the transforming gene by northern blot analysis of poly(A)$^+$ RNA from T361-2nd-1 cells (28). It was also shown that all of the primary and the secondary transformants induced by the nos. 361, 363, and 51 DNA samples contained DNA fragments of human origin hybridized to this cDNA clone, confirming the results that the transformants induced by the three different samples of DNA all contained the same transforming gene.

This cDNA clone rendered a transforming activity to NIH 3T3 cells when placed downstream of the SV40 early promoter as an expression plasmid pKOc1. Sequence analysis revealed two open reading frames, termed ORF1 and ORF2, with a coding capacity of more than 150 amino acids. Of these two, only ORF1 was found to possess the transforming activity by transfection of mutants expressing either ORF1 or ORF2. The deduced amino acid sequence of ORF1 was not identical to any of the known gene products in the data bases, and so was designated HST1 as a novel transforming protein. The gene was officially registered in human gene nomenclature as HSTF1 (heparin-binding secretory transforming factor 1) (38).

SEQUENCE ANALYSIS AND TRANSFORMING ACTIVITY OF THE HST1 GENE

The human and mouse HST1 genes were cloned by screening of genomic. libraries in cosmid with the cDNA fragment of the human HST1
gene. The BamHI-SalI 6,181-base genomic fragment from human leukemic
leukocyte DNA was subcloned and sequenced in full (40). By alignment
of the genomic and cDNA sequences, the coding regions of these clones
were shown to be completely identical. The human and mouse genomic
fragments of HST1 had as potent a transforming activity to NIH 3T3 cells
at an efficiency of 100-500 foci/pmol as the human HST1 genomic fragment
in the T361-2nd-1 transformant (23,39).

HST1 PRODUCT

The sequence analysis of the human HST1 cDNA and genomic DNA
predicted that the product was composed of 206 amino acid residues with
a characteristic N-terminal signal peptide. The protein was 40-50% homologous to human basic and acidic fibroblast growth factors (FGFs), mouse
Int2 protein, and human FGF5 in selected regions (40,45). Two cysteine
residues were conserved among this family, but a classical signal peptide
was present only in the HST1- and FGF5-encoded proteins (45). The
heparin-binding domains of the basic and acidic FGFs appeared to be conserved in the HST1 protein.

The human HST1 protein was synthesized and secreted by silkworm
cells (16). The protein was purified to a single band of 18 kDa by
heparin-affinity chromatography and reversed-phase HPLC. The N-terminus of this protein was sequenced, confirming that it was the processed
HST1 protein. When added exogenously to the culture medium of NIH 3T3
cells, the recombinant HST1 protein stimulated thymidine uptake and induced morphological transformation at a concentration as low as 0.1-1
ng/ml.

Thymidine incorporation by primary culture of human umbilical endothelial cells was also stimulated by the protein. The protein supported
colony formation of NIH 3T3 cells and NRK 49F cells in soft agar in a
dose-dependent manner.

CO-AMPLIFICATION OF HST1 AND INT2

HST1 was mapped to the human chromosome band 11q13.3, the same
site as INT2 (38). Using pulsed-field gel electrophoresis (PFGE) and
cosmid cloning, the physical distance between HST1 and INT2 was determined to be as short as 35 kbp (35) (Fig. 1). These genes were on the
same DNA strand in the same transcriptional orientation, with INT2
located 5' upstream of HST1. In the mouse genome, too, both genes were
found to be less than 20 kbp apart.

The HST1 gene was amplified sporadically in some human cancers, up
to a 20-fold degree. The amplification of HST1 was, however, observed
in more than 40% of esophageal cancers. In our study, INT2 was always
amplified together with HST1 to the same degree (30,32,38).

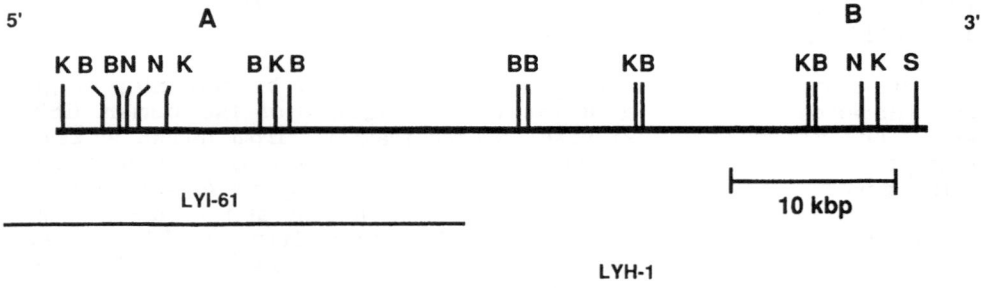

Fig. 1.   A restriction map of the INT2-HST1 region on chromosome
11q13.   HST1 is located 3' downstream of INT2 on the same
DNA strand. LYI-61 and LYH1 are overlapping cosmid clones.
INT2 (A) and HST1 (B) regions are also indicated. K, KpnI;
B, BamHI; N, NotI; S, SalI.

EXPRESSION OF HST1 IN EMBRYOS AND IN GERM CELL TUMORS

Northern blot analysis revealed that HST1 is rarely expressed, with
the eXception of germ cell tumors; the gene was transcribed in five out
of nine surgical specimens of human testicular germ cell tumors, a cell
line derived from human immature teratoma, NCC-IT, and a mouse terato-
carcinoma cell line, F9 (41,42).   When F9 cells were induced to differentiate to parietal endodermal cells in vitro, the transcript of Hst1
decreased dramatically.   As described by others, Int2 was detectably
transcribed only in the differentiated F9 cells in which expression of Hst1
was turned off (41).

RFLP ANALYSIS ON STOMACH CANCER

To clarify whether loss of heterozygosity on certain specific chromosomal loci is actually involved in the development of stomach cancer, we
performed a molecular genetic analysis with RFLP probes, and obtained
information on chromosomal abnormalities at 25 loci on 18 autosomal chromosomes, except chomosomes 2, 4, 7, and 21 in 30 patients with stomach
cancer (35).   In the present study, loss of heterozygosity was observed
only in five of 30 cases with stomach cancer.   It was infrequently detected at ten loci on seven different chromosomes, and not at all at the
other 14 loci on 11 chromosomes.   Thus, in stomach cancer, loss of
chromosomal heterozygosity occurred infrequently even at loci often
deleted in several other types of cancers including lung cancer
(27,32,36).

Heterozygosity at several loci tested in this study is frequently lost
in specific types of tumors.   These include chromosome 1p in endocrine
neoplasia (15); 3p in renal cell carcinoma (44) and small cell carcinoma of
the lung (18,19); 5q in colorectal carcinoma (26); 11p in Wilms' tumor
(5,7,12,13,20,21), breast cancer (1,29), and bladder cancer (9); 13q in
retinoblastoma (2-4,10) and breast cancer (14); 17p in colorectal tumor
(8); and 22q in acoustic neuroma and meningioma (24,25).   We have also
shown that loss of heterozygosity on chromosomes 3p, 13q, and 17p
occurs in nearly 100% of the small cell carcinomas of the lung and, in the

case of chromosome 3p, in adenocarcinomas of the lung (36). Besides these loci, incidence of loss of heterozygosity at the other loci was relatively high in lung cancer, ranging from 9% on chromosomes 5 and 14q to 33% on chromosome 12p. In addition, we previously showed that loss of the Y chromosome occurred in three of 21 stomach cancers and in two of 12 lung cancers (33). In the present study, we analyzed the same loci as previously tested in lung cancer for possible loss of chromosomal heterozygosity in stomach cancer. Loss of heterozygosity was observed in only five of 30 cases, and it was detected on seven of 17 chromosomes with comparatively low frequency. These results indicate that, in contrast to lung cancer, loss of heterozygosity at these chromosomal loci is infrequent in stomach cancer and that there may be no specific locus commonly deleted in this type of cancer, though the possibility remains that loss of heterozygosity might occur frequently at some chromosomal loci other than those tested in this study.

ACKNOWLEDGEMENTS

This work was supported in part by a Grant-in-Aid for a Comprehensive 10-Year Strategy for Cancer Control from the Ministry of Health and Welfare of Japan and a Grant-in-Aid from the Ministry of Health and Welfare of Japan.

REFERENCES

1. Ali, I.U., R. Lidereau, C. Theillet, and R. Callahan (1987) Reduction to homozygosity of genes on chromosome 11 in human breast neoplasia. Science 238:185-188.
2. Benedict, W.F., E.S. Srivatsan, C. Mark, A. Banerjee, R.S. Sparkes, and A.L. Murphree (1987) Complete or partial homozygosity of chromosome 13 in primary retinoblastoma. Cancer Res. 47:4189-4191.
3. Cavenee, W.K., M.F. Hansen, M. Nordenskjold, E. Kock, I. Maumenee, J.A. Squire, R.A. Phillips, and B.L. Gallie (1985) Genetic origin of mutations predisposing to retinoblastoma. Science 228:501-503.
4. Cavenee, W.K., T.P. Dryja, R.A. Phillips, W.F. Benedict, R. Godbout, B.L. Gallie, A.L. Murphree, L.C. Strong, and R.L. White (1983) Expression of recessive alleles by chromosomal mechanisms in retinoblastoma. Nature (London) 305:779-784.
5. Dao, D.D., W.T. Schroeder, L.Y. Chao, H. Kikuchi, L.C. Strong, V.M. Riccardi, S. Pathak, W.W. Nichols, W.H. Lewis, and G.F. Saunders (1987) Genetic mechanisms of tumor-specific loss of 11p DNA sequences in Wilms tumor. Am. J. Hum. Genet. 41:202-217.
6. Delli Bovi, P., A.M. Curatola, F.G. Kern, A. Greco, M. Ittmann, and C. Basilico (1987) An oncogene isolated by transfection of Kaposi's sarcoma DNA encodes a growth factor that is a member of the FGF family. Cell 50:729-737.
7. Fearon, E.R., B. Vogelstein, and A.P. Feinberg (1984) Somatic deletion and duplication of genes on chromosome 11 in Wilms' tumours. Nature (London) 309:176-178.
8. Fearon, E.R., S.R. Hamilton, and B. Vogelstein (1987) Clonal analysis of human colorectal tumors. Science 238:193-197.
9. Fearon, E.R., A.P. Feinberg, S.H. Hamilton, and B. Vogelstein (1985) Loss of genes on the short arm of chromosome 11 in bladder cancer. Nature (London) 377-380.

10. Hansen, M.F., A. Koufos, B.L. Gallie, R.A. Phillips, O. Fodstad, A. Brogger, T. Gedde-Dahl, and W.K. Cavenee (1985) Osteosarcoma and retinoblastoma: A shared chromosomal mechanism revealing recessive predisposition. Proc. Natl. Acad. Sci., USA 82:6216-6220.

11. Koda, T., A. Sasaki, S. Matsushima, and M. Kakimuna (1987) A transforming gene, hst, found in NIH3T3 cells transformed with DNA from three stomach cancers and a colon cancer. Jpn. J. Cancer Res. (Gann) 78:325-328.

12. Koufos, A., M.F. Hansen, N.G. Copeland, N.A. Jenkins, B.C. Lampkin, and W.K. Cavenee (1985) Loss of heterozygosity in three embryonal tumours suggests a common pathogenetic mechanism. Nature (London) 316:330-334.

13. Koufos, A., M.F. Hansen, B.C. Lampkin, M.L. Workman, N.G. Copeland, N.A. Jenkins, and W.K. Cavenee (1984) Loss of alleles at loci on human chromosome 11 during genesis of Wilms' tumour. Nature (London) 309:170-172.

14. Lundberg, C., L. Skoog, W.K. Cavenee, and M. Nordenskjold (1987) Loss of heterozygosity in human ductal breast tumors indicates a recessive mutation on chromosome 13. Proc. Natl. Acad. Sci., USA 84:2372-2376.

15. Mathew, C.G.P., B.A. Smith, K. Thorpe, Z. Wong, N.J. Royle, A.J. Jeffreys, and B.A.J. Ponder (1987) Deletion of genes on chromosome 1 in endocrine neoplasia. Nature (London) 328:524-526.

16. Miyagawa, K., H. Sakamoto, T. Yoshida, Y. Yamashita, Y. Mitsui, M. Furusawa, S. Maeda, F. Takaku, T. Sugimura, and M. Terada (1988) hst-1 transforming protein: Expression in silkworm cells and characterization as a novel heparin-binding growth factor. Oncogene 3:383-389.

17. Nakagama, H., S. Ohnishi, M. Imawari, H. Hirai, F. Takaku, H. Sakamoto, M. Terada, M. Nagao, and T. Sugimura (1987) Identification of transforming genes as hst in DNA samples from two human hepatocellular carcinomas. Jpn. J. Cancer Res. (Gann) 78:651-654.

18. Naylor, S.L., B.E. Johnson, J.D. Minna, and A.Y. Sakaguchi (1987) Loss of heterozygosity of chromosome 3p markers in small-cell lung cancer. Nature (London) 329:451-454.

19. Naylor, S.L., J. Minna, B. Johnson, and A.Y. Sakaguchi (1984) DNA polymorphisms confirm the deletion in the short arm of chromosome 3 in small cell lung cancer. Am. J. Human Genet. 36(Suppl.): 35.

20. Orkin, S.H., D.S. Goldman, and S.E. Sallan (1984) Development of homozygosity for chromosome 11p markers in Wilms' tumor. Nature (London) 309:172-174.

21. Reeve, A.E., P.J. Housiaux, R.J.M. Grander, W.E. Chewings, R.M. Grindley, and L.J. Millow (1984) Loss of a Harvey ras allele in sporadic Wilms' tumour. Nature (London) 309:174-176.

22. Sakamoto, H., M. Mori, M. Taira, T. Yoshida, S. Matsukawa, K. Shimizu, M. Sekiguchi, M. Terada, and T. Sugimura (1986) Transforming gene from human stomach cancers and a noncancerous portion of stomach mucosa. Proc. Natl. Acad. Sci., USA 83:3997-4001.

23. Sakamoto, H., T. Yoshida, M. Nakakuki, H. Odagiri, K. Miyagawa, T. Sugimura, and M. Terada (1988) Cloned hst gene from normal human leukocyte DNA transforms NIH3T3 cells. Biochem. Biophys. Res. Commun. 151:965-972.

24. Seizinger, B.R., R.L. Martuza, and J.F. Gusella (1986) Loss of genes on chromosome 22 in tumorigenesis of human acoustic neuroma. Nature (London) 322:644-647.

25. Seizinger, B.R., S. de la Monte, L. Atkins, J.F. Gusella, and R.L.

Martuza (1987) Molecular genetic approach to human meningioma: Loss of genes on chromosome 22. Proc. Natl. Acad. Sci., USA 84:5419-5423.

26. Solomon, E., R. Voss, V. Hall, W.F. Bodmer, J.R. Jass, A.J. Jeffreys, F.C. Lucibello, I. Patel, and S.H. Rider (1987) Chromosome 5 allele loss in human colorectal carcinomas. Nature (London) 328:616-619.

27. Suzuki, T., J. Yokota, H. Mugishima, I. Okabe, M. Ookuni, T. Sugimura, and M. Terada (1989) Frequent loss of heterozygosity on chromosome 14q in neuroblastoma. Cancer Res. 49:1095-1098.

28. Taira, M., T. Yoshida, K. Miyagawa, H. Sakamoto, M. Terada, and T. Sugimura (1987) cDNA sequence of human transforming gene hst and identification of the coding sequence required for transforming activity. Proc. Natl. Acad. Sci., USA 84:2980-2984.

29. Theillet, C., R. Lidereau, C. Escot, P. Hutzell, M. Brunet, J. Gest, J. Schlom, and R. Callahan (1986) Loss of c-H-ras-1 allele and aggressive human primary breast carcinomas. Cancer Res. 46:4776-4781.

30. Tsuda, T., H. Nakatani, T. Matsumura, K. Yoshida, E. Tahara, T. Nishihira, H. Sakamoto, T. Yoshida, M. Terada, and T. Sugimura (1988) Amplification of the hst-1 gene in human esophageal carcinomas. Jpn. J. Cancer Res. (Gann) 79:584-588.

31. Tsutsumi, M., J. Yokota, T. Kakizoe, K. Koiso, T. Sugimura, and M. Terada (1989) Loss of heterozygosity on chromosomes 1p and 11p in sporadic pheochromocytoma. J. Natl. Cancer Inst. 81:367-370.

32. Tsutsumi, M., H. Sakamoto, T. Yoshida, T. Kakizoe, K. Koiso, T. Sugimura, and M. Terada (1988) Coamplification of the hst-1 and int-2 genes in human cancers. Jpn. J. Cancer Res. (Gann) 79:428-432.

33. Wada, M., J. Yokota, H. Mizoguchi, M. Terada, and T. Sugimura (1987) Y chromosome abnormality in human stomach and lung cancer. Jpn. J. Cancer Res. (Gann) 78:780-783.

34. Wada, M., J. Yokota, H. Mizoguchi, T. Sugimura, and M. Terada (1988) Infrequent loss of chromosomal heterozygosity in human stomach cancer. Cancer Res. 48:2988-2992.

35. Wada, A., H. Sakamoto, O. Katoh, T. Yoshida, P.F.R. Little, T. Sugimura, and M. Terada (1988) Two homologous oncogenes, HST1 and INT2, are closely located in human genome. Biochem. Biophys. Res. Commun. 157:828-835.

36. Yokota, J., M. Wada, Y. Shimosato, M. Terada, and T. Sugimura (1987) Loss of heterozygosity on chromosomes 3, 13 and 17 in small-cell carcinoma and on chromosome 3 in adenocarcinoma of the lung. Proc. Natl. Acad. Sci., USA 84:9252-9256.

37. Yokota, J., Y. Tsukada, T. Nakajima, M. Gotoh, Y. Shimosato, N. Mori, Y. Tsunokawa, T. Sugimura, and M. Terada (1989) Loss of heterozygosity on the short arm of chromosome 3 in carcinoma of the uterine cervix. Cancer Res. (in press).

38. Yoshida, M.C., M. Wada, H. Satoh, T. Yoshida, H. Sakamoto, K. Miyagawa, J. Yokota, T. Koda, M. Kakimura, T. Sugimura, and M. Terada (1988) Human HST1 (HSTF1) gene maps to chromosome band 11q13 and coamplifies with the INT2 gene in human cancer. Proc. Natl. Acad. Sci., USA 85:4861-4864.

39. Yoshida, T., H. Sakamoto, K. Miyagawa, P.F.R. Little, M. Terada, and T. Sugimura (1987) Genomic clone of hst with transforming activity from a patient with acute leukemia. Biochem. Biophys. Res. Commun. 142:1019-1024.

40. Yoshida, T., K. Miyagawa, H. Odagiri, H. Sakamoto, P.F.R. Little,

M. Terada, and T. Sugimura (1987) Genomic sequence of hst, a transforming gene encoding a protein homologous to fibroblast growth factors and the int-2-encoded protein. Proc. Natl. Acad. Sci., USA 84:7305-7309.

41. Yoshida, T., H. Muramatsu, T. Muramatsu, H. Sakamoto, O. Katoh, T. Sugimura, and M. Terada (1988) Differential expression of two homologous and clustered oncogenes, Hst1 and Int-2, during differentiation of F9 cells. Biochem. Biophys. Res. Commun. 157:618-625.

42. Yoshida, T., M. Tsutsumi, H. Sakamoto, K. Miyakawa, S. Teshima, T. Sugimura, and M. Terada (1988) Expression of the HST1 oncogene in human germ cell tumors. Biochem. Biophys. Res. Commun. 155:1324-1329.

43. Yuasa, Y., and K. Sudo (1987) Transforming genes in human hepatomas detected by a tumorigenicity assay. Jpn. J. Cancer Res. (Gann) 78:1036-1040.

44. Zbar, B., H. Brauch, C. Talmadge, and M. Linehan (1987) Loss of alleles of loci on the short arm of chromosome 3 in renal cell carcinoma. Nature (London) 327:721-724.

45. Zhan, X., B. Bates, X. Hu, and M. Goldfarb (1988) The human FGF-5 oncogene encodes a novel protein related to fibroblast growth factors. Molec. Cell. Biol. 8:3487-3495.

ONCOGENIC POTENTIAL AND NORMAL FUNCTION OF THE PROTO-ONCOGENES ENCODING PROTEIN-TYROSINE KINASES

Tadashi Yamamoto, Tetsu Akiyama, Kentaro Semba,
Yuji Yamanashi, Kazushi Inoue, Yukinori Yamada,
Jun Sukegawa, and Kumao Toyoshima

Department of Oncology
Institute of Medical Science
University of Tokyo
Tokyo 108, Japan

INTRODUCTION

A number of protein-tyrosine kinases, including the cellular counterpart of the src gene product of Rous sarcoma virus, have been identified in mammalian cells and are suggested to be important in growth and/or differentiation of cells. Approximately half of the protein-tyrosine kinases are integral membrane proteins and are, in many cases, receptors for polypeptide growth factors (13). They are the receptors for epidermal growth factor (EGF), insulin, insulin-like growth factor-1, platelet-derived growth factors, and colony-stimulating factor-1. Cellular oncogenes such as met, erbB-2/neu, trk, and ret are also included in this group.

The remaining protein-tyrosine kinases, typified by the 60-kDa c-src protein, are membrane-associated proteins but lack transmembrane and extracellular domains. Genes encoding proteins of this group are called members of the src family and include src, yes, fgr, fyn, lck, lyn, hck, and possibly tkl (32). Comparison of the amino acid sequences of the protein products of the src gene family revealed that the entire sequences are highly conserved within this family, except for the amino-terminal regions, of about 75 residues, which are variable. The significance of the divergent segment has not presently been established for certain. It may contain recognition sequences for interaction with other proteins, namely substrates, regulators, or binding proteins, and thereby give each family member unique properties (8). The abl and fps genes may also fall in this group, although primary structures of their gene products are less homologous to the other src-like proteins. Several lines of evidence show that these src-like proteins are often expressed in cells committed to differentiate into specific lineages (8).

Many of the genes encoding the protein-tyrosine kinases, both receptor-type and nonreceptor-type, are either induced by retroviruses as oncogenes or activated by point mutation or rearrangement. This suggests that subversion of cell growth and/or differentiation via altered expression of protein-tyrosine kinases results in cellular transformation. For the past several years, we have been studying cellular genes encoding protein-tyrosine kinases to establish mechanisms of cellular transformation and their normal functions. In this chapter we present our studies on genes encoding nonreceptor-type and receptor-type protein-tyrosine kinases.

HUMAN yes AND ITS HOMOLOGS

The human cellular yes gene, c-yes-1, is a cellular counterpart of the yes oncogene of avian sarcoma virus. The amino acid sequence of the product of the viral yes gene (p90^{yes}) is partly very similar to that of the src gene product pp60^{src} of Rous sarcoma virus, and p90^{yes} exerts protein-tyrosine kinase activity (16,18). Thus, the yes gene is considered to be the most intimate cognate of the src gene. We first analyzed chromosomal yes gene by Southern blotting of DNAs extracted from human-mouse hybridoma cell clones using viral yes DNA as a probe. Under relaxed conditions, many bands were observed, and some of them could be clearly assigned to distinct chromosomes (28). A gene library constructed from human placental DNA was screened for further analysis of these bands. After restriction mapping and partial sequencing of the v-yes DNA hybridizing clones, the c-yes gene was identified and was assigned to chromosome 18q21.3 by use of a specific intron probe (46). In addition to the bands of c-yes, a 1.9-kbp EcoRI fragment also hybridized well with probe prepared from either the 3' end, middle, or 5' end of the v-yes gene, suggesting that the 1.9-kbp EcoRI fragment represented a processed gene. Actually, sequence data proved that this gene did not have an intron, but had one Alu repetitive sequence within the coding region and a stop codon in its frame. We believe that this gene is a pseudogene, which we named c-yes-2 and assigned to chromosome 22 (30). A third gene was found to be most similar to the v-fgr gene of feline sarcoma virus. This gene was proved to be the same as the src-2 gene, which was then renamed c-fgr gene and assigned to chromosome 1q34-36 (24). The human c-fgr gene does not contain an actin-like sequence upstream of the kinase domain (14,23), unlike the v-fgr gene, which is a trichimeric gene formed possibly by recombination of the viral genome and the γ-actin and c-fgr genes. A fourth gene was identified as the c-src gene located on chromosome 20q12-q13 (19). Further analysis of v-yes DNA-hybridizing clones suggested that there might be other related genes besides those described above (see below). We then analyzed cDNA clones of these yes-related genes.

The c-yes cDNA clones were obtained from poly(A)$^{+}$ RNA of human embryo fibroblasts. Sequence analysis of the clones showed that they contained inserts corresponding to nearly full-length human c-yes mRNA, which could encode a polypeptide of 543 amino acids with a relative molecular weight (M_r) of 60,801 (33). The predicted amino acid sequence of the protein, which closely resembles pp60^{c-src}, has no apparent membrane-spanning region or suspected ligand-binding domain. Comparison of the sequences of c-yes (33) and v-yes (18) revealed that the v-yes gene contains most of the c-yes coding sequence, except for the region encoding its extreme carboxyl terminus. The region missing from the

v-yes protein is the part that is highly conserved in cellular gene products of the src gene family and appears to be important for regulation of
its own kinase activity (see Ref. 8 for review).

During analysis of the c-yes gene and its mRNA, we found two novel
yes-related genes, termed fyn (fgr- and yes-related novel gene) (29) and
lyn (lck- and yes-related novel gene) (42). The fyn gene was first detected by screening a genomic library. Three of 26 independent isolates
gave a different restriction map from clones of c-yes-1, c-yes-2, c-src,
and c-fgr, suggesting that these three represented a novel v-yes-related
gene. The nucleotide sequence of one putative exon of these clones was
highly homologous to the corresponding portion of the v-yes gene (79%
homology), and the splicing junction was also identical to that of exon 8
of c-src (4,29). This gene was assigned to human chromosome 6 (45).
It was found to express a 2.8-kb mRNA in a variety of human embryonic
tissues and cell lines, and its expression was especially high in embryonic
brain. Analysis of cDNA clones revealed that this gene encoded a 537-
amino acid polypeptide with a relative molecular mass of 60,761.

Another yes-related novel gene, lyn, was isolated by screening a
cDNA library made from human placental RNA with v-yes DNA as a
probe. The open reading frame predicted from the nucleotide sequence
could encode a polypeptide of 512 amino acid residues with a calculated
molecular weight of 58,574. Although this gene was isolated with a v-yes
probe, the predicted structure of its product showed highest homology
with mouse T-lymphocyte-specific tyrosine kinase p56^{lck} (20,37). The
lyn gene was mapped on human chromosome 8q13-qter by Southern blot
analysis with sorted human chromosomes obtained from a karyotypically
normal cell line and from a cell line containing the chromosome translocation t(2;8) (q37; q13) (42). Expression of 3.2-kb lyn mRNA was relatively high in placenta and liver, low in brain, lung, and kidney, and
scarcely detectable in cultured fibroblasts (42).

The nucleotide sequence of c-fgr cDNA clones obtained from
poly(A)$^+$ RNA of a B lymphocyte cell line, IM-9, was also determined, revealing that human c-fgr mRNA could encode a polypeptide of 529 amino
acids with a calculated molecular weight of 59,478 (14). Northern blot
hybridization analysis using a c-fgr-specific sequence showed that the
c-fgr mRNA was expressed at a higher level in the liver than in the
brain, lung, or kidney of a human fetus. Expression of c-fgr mRNA in
the liver of an adult mouse was low (data not shown). These data suggest that c-fgr expression is specific in hematopoietic cells.

The predicted amino acid sequence of the c-yes-1, fyn, lyn, and
c-fgr gene products showed that they were very similar to that of the
c-src protein (Fig. 1), which suggested that they were protein-tyrosine
kinases. This was confirmed by an immune-complex kinase assay using
antibodies specific to each gene product and γ-^{32}P-ATP. An example that
demonstrates tyrosine kinase activity of p56^{lyn} is shown in Fig. 2. In
addition, these proteins start with the dipeptide Met-Gly. The src gene
product p60^{c-src} has a Met-1 which was cleaved and replaced with myris-
tic acid in the amide linkage to Gly-2 (see Ref. 8 for review). Since the
presence of Gly-2 seems to be critical for myristylation, the invariance of
Gly-2 at the amino termini of the gene products of c-yes-1, fyn, lyn, and
c-fgr means it is likely that these proteins are myristylated. The invar-
iant Gly-2 is followed by a totally diverged sequence of 70 to 80 residues

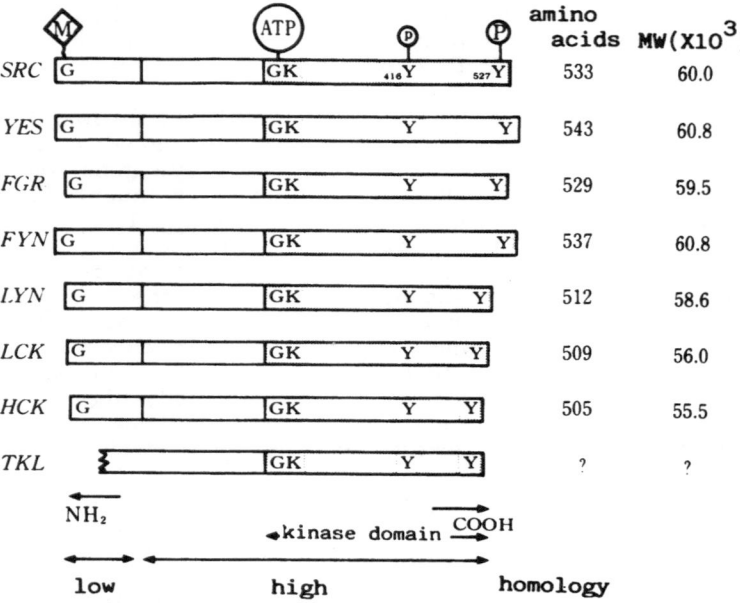

Fig. 1.   Schematic illustration of the protein products of the src-like
           genes.

before sequence similarities are evident.   The significance of the divergent sequences is presently unknown.   However, they are characteristic
of each gene product and may contain recognition sequences for interaction with substrates or regulatory binding proteins.   Following the variable region, the remaining 400 amino-acid sequences are quite constant.
This region includes the kinase domain as well as the SH2 and SH3
domains which are homologous to the noncatalytic portion of p150^{abl},
p90^{c-fps}, and phospholipase A$_2$ (see Ref. 8 for review).

      It should be noted that Tyr-527 in pp60^{c-src} is important for regu-
lating its tyrosine kinase activity, and tyrosine residues located proximal
to the carboxyl termini of c-yes-1, fyn, lyn, and c-fgr correspond well
to Tyr-527 of pp60^{c-src}.   Since v-src, v-yes, and v-fgr all have altered
carboxy-terminal sequences (i.e., Tyr-527 and its homolog, together with
the flanking sequence, were replaced by foreign sequences), it is postulated that similar alteration of the corresponding sequences of the fyn and
lyn proteins may activate their tyrosine kinase activities, which may result in cellular transformation.

Genomic Structures

      The proto-src genes of chickens and humans have been studied ex-
tensively.   They both have 11 introns and 12 exons, and their exon-
intron junctions are all precisely conserved (4,34).   In the analysis of
the human c-fgr gene, every intron-exon junction sequenced so far (exons 4 to 12) was found to be identical to that of the c-src gene.   This
identity suggests that these two proto-oncogenes evolved by duplication
of a single ancestral gene after completion of the organization of the
exon-intron structure.   This duplication probably occurred, at the latest,

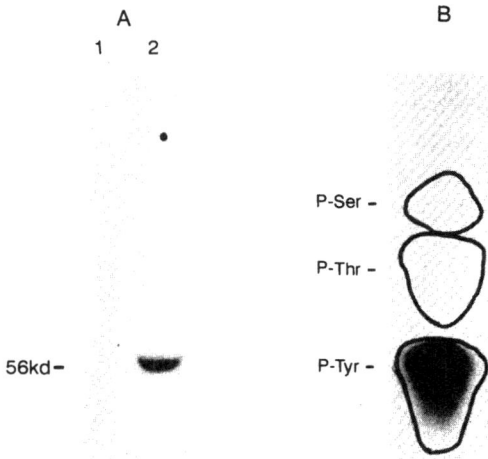

Fig. 2.   Autophosphorylation on tyrosine residue of $p56^{lyn}$. (A) L10 cells, human lyn cDNA transfectants, were lysed and the extracts were immunoprecipitated with antibody to the lyn gene products (lane 2) and normal rabbit serum (lane 1). The immunoprecipitates were incubated with γ-^{32}P-ATP, and the reaction mixtures were analyzed on SDS-polyacrylamide gel. (B) Phosphoamino acid analysis of the $p56^{lyn}$ labeled in vitro.

before the genealogical divergence of birds and mammals. Only a few junctions of c-yes-1, fyn, and lyn have so far been sequenced, and they all have been found to be completely identical with those of c-src, indicating that these genes are also descendants of the same ancestral gene as c-src.

Specialized Function of the yes Homologs

To date, the presence of three additional members of the src-like protein have been reported. Thus, the mammalian genome possesses at least eight cellular genes (c-src, c-yes, c-fgr, fyn, lyn, hck, lck, tkl) that encode proteins highly homologous to $p60^{c-src}$ (see Ref. 8 for review; Fig. 1). To rationalize the presence of a set of genes whose products were structurally similar to each other, having protein-tyrosine kinase activity, we assumed that each member of this family was expressed and that each functions in a cell-type-specific manner. This possibility was examined by northern blot hybridization of RNAs from various tissues (Fig. 3A) and by immunohistochemistry. As shown in Fig. 3, the level of lyn mRNA was higher in the fetus liver than in other tissues. The c-yes mRNA was expressed ubiquitously. The fyn mRNA was detected at a high level in the fetus brain, and the c-fgr mRNA (data not shown) was detected at a high level in the fetus liver. Also, when RNAs from cancerous cell lines were analyzed (Fig. 3B), patterns of expression were different among the c-yes, fyn, and c-fgr genes.

Since fetus liver is rich in hematopoietic cells, we assumed that expression of the c-fgr gene and the lyn gene may be associated with

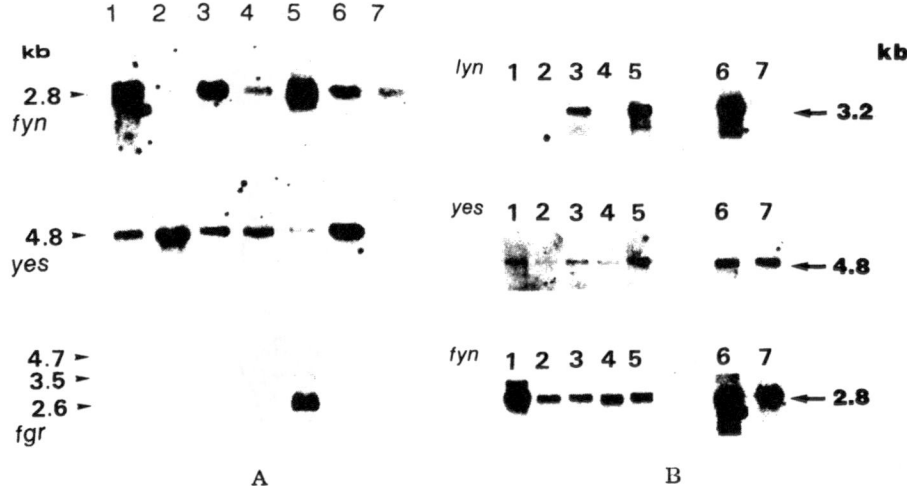

Fig. 3.   Levels of RNA transcripts of the yes homologs. Northern blot
hybridization of RNAs from normal human cells (A) and cancerous cell lines (B). (A) RNAs are from the brain (lane 1),
lung (lane 2), liver (lane 3), and kidney (lane 4) of a human
fetus, from placenta at term (lanes 5 and 6), and from human
embryo fibroblasts TIG-1 cells (lane 7).   (B) RNAs are from
placenta at term (lane 1), pancreas carcinoma UCVA cells
(lane 2), TIG-1 cells (lane 3), chronic myelogenous leukemia
k562 cells (lane 4), Epstein-Barr virus-immortalized B lymphocytes IM9 (lane 5), squamous carcinoma cells A431 (lane 6), and
follicular lymphoma cells FL-18 (lane 7).

hematopoiesis. In fact, expression of the lyn mRNA was almost confined
to the spleen of adult mice. Less expression of this mRNA was seen in
the thymus. Further analysis of hematopoietic cell lines revealed that lyn
was expressed in B lymphocytes but not in T lymphocytes, except for
HTLV-1 producer T cells (39). Consistent with these observations, western blotting analysis with anti-lyn gene product antibodies showed that
the lyn gene product $p56^{lyn}$ was expressed in tonsils where B lymphocytes were located. In addition, the western blot revealed that $p56^{lyn}$
was also expressed in macrophage/monocytes and platelets (39). The role
of $p56^{lyn}$ in these different cell types is not known.

Hybridization histochemistry with c-fgr DNA probe showed that the
c-fgr gene was expressed in large granular lymphocytes and monocytes/-
macrophages, both of which are important for natural immunity (K. Inoue
et al., ms. in prep.), in addition to B lymphocytes immortalized with
Epstein-Barr virus (EBV). These data suggest a specific role of the
c-fgr protein in effector cells in the development of of natural immunity.

Regulation of c-yes and fyn expression is currently under investigation. Our preliminary data suggest that, besides in the brain, fyn was
expressed in T lymphocytes, and that expression of the fyn gene was
elevated in T cells of murine lymphoproliferative diseases (T. Katagiri et
al., ms. in prep.).

MOLECULAR CHARACTERIZATION OF THE c-erbB-2 GENE

Identification of the c-erbB-2 Gene

The retroviral oncogene erbB is derived from the chicken EGF receptor gene (10,38). In addition to the EGF receptor gene, we found another v-erbB-related gene, c-erbB-2, in the human genome. Southern blot hybridization of DNAs from A431 epidermal carcinoma cells and placenta using a v-erbB probe showed that at least two EcoRI fragments, of 6.4 kb and 13 kb, respectively, were not amplified in A431 cells in which the gene encoding the EGF receptor is amplified. This suggested that besides the EGF receptor gene, there might be another v-erbB-related gene, which was tentatively termed c-erbB-2.

To examine this possibility, we searched a human genomic library for a v-erbB-related sequence and isolated one clone, λ107, whose restriction map differs from those of DNA clones representing the EGF receptor gene (27). Thus, we concluded that this clone represents the c-erbB-2 gene. The transcript of the c-erbB-2 gene (4.6 kb long) was found in various cells, which suggested that the c-erbB-2 gene was functional (11). By hybridizing both sorted chromosomes and metaphase spreads with the c-erbB-2 probe, we mapped the c-erbB-2 locus on human chromosome 17 at q21 (11).

Poly(A)$^+$ RNA was prepared from MKN-7 cells, a gastric cancer cell line in which the c-erbB-2 gene is amplified and overexpressed. A cDNA library was constructed from MKN-7 mRNA using Okayama-Berg cloning vector, and a 4,480-bp-long nucleotide sequence was determined from c-erbB-2 cDNA clones obtained from this library. This sequence contained a coding sequence of 1,255 amino acid residues. The primary translation product of the c-erbB-2 gene was calculated to have a relative molecular mass of 137,895 (40).

Inspection of the predicted amino acid sequence reveals that the c-erbB-2 protein is very similar to the EGF receptor (Fig. 4), suggesting that the c-erbB-2 protein is a receptor for an unknown growth factor. The putative extracellular domain shows 44% homology in amino acid sequence with the ligand-binding domain of the EGF receptor. A striking similarity is the presence of two cysteine-rich regions, in which the spatial distribution of cysteine residues is virtually identical with that in the EGF receptor. However, ^{125}I-labeled EGF did not bind to the c-erbB-2 protein at the surface of MKN-7 cells (1). A sequence of 260 amino acids of the intracellular domain of the c-erbB-2 protein is highly homologous with the kinase domain of the EGF receptor (80% homology). Thus, the c-erbB-2 protein could have tyrosine kinase activity.

Rabbit antibodies were raised against a synthetic peptide corresponding to the predicted 14 amino acid residues at the carboxy-terminus of the predicted c-erbB-2 protein. These antibodies immunoprecipitated a 185-kDa glycoprotein from a MKN-7 cell. An immune-complex kinase assay showed that the c-erbB-2 protein was associated with protein-tyrosine kinase activity (Ref. 1; Fig. 5). Three tyrosine residues near the carboxyl-terminus of the c-erbB-2 protein, which are at positions equivalent to those of the autophosphorylated tyrosines of the EGF receptor, may also be the sites of autophosphorylation.

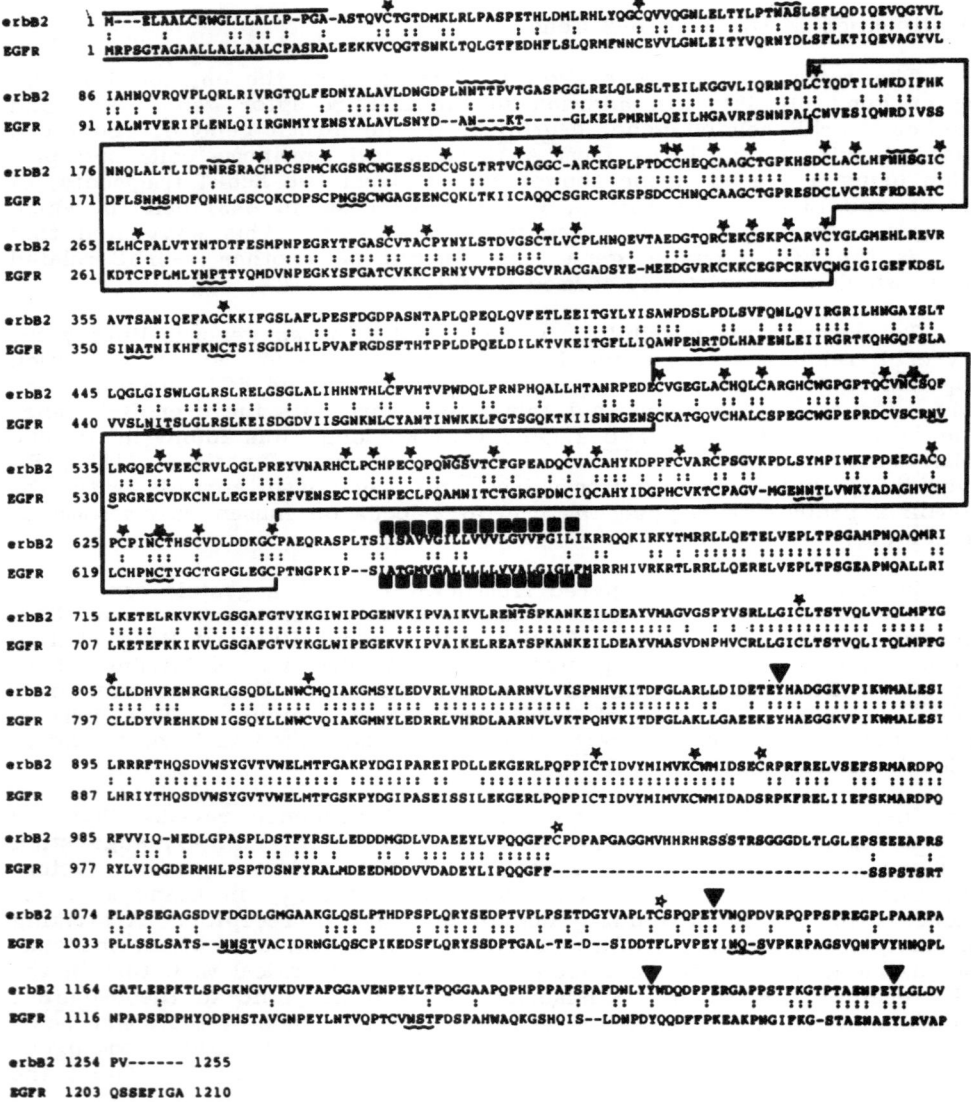

Fig. 4.   Alignment of the amino acid sequences of the c-erbB-2 protein
and EGF receptor. Identities in the sequences are marked by
two dots between the two lines. Predicted transmembrane regions are represented by dotted black bars; the possible N-
linked glycosylation sites by wavy lines. Horizontal lines
indicate signal peptides, and stars, cysteine residues in the
sequence; solid stars, cysteine common to the two proteins.
Major sites of tyrosine phosphorylation of the EGF receptor
are conserved in the c-erbB-2 sequence and are shown by tri-
angles.

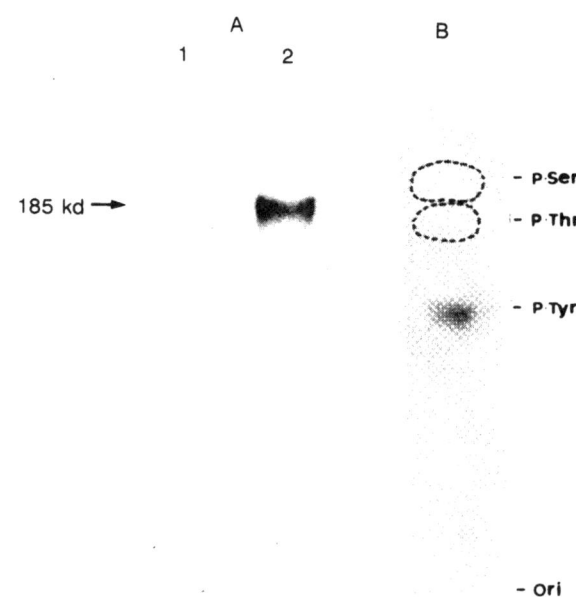

Fig. 5.   Autophosphorylation on tyrosine residue of p185 c-erbB-2.
(A) The lysates of a gastric cancer cell line (MKN-7 cells) were
subjected to immune-complex kinase assay as described in legend to Fig. 2. (B) Phosphoamino acid analysis of p185$^{c-erbB-2}$
labeled in vitro.

Phosphorylation of the c-erbB-2 Protein Induced by EGF

Treatment of MKN-7 cells with EGF or TPA stimulated phosphorylation of the c-erbB-2 protein (Ref. 2; Fig. 6a). EGF induced a rapid
increase in phosphotyrosine (Fig. 6b), followed by relatively gradual
increases in phosphoserine and phosphothreonine. Inhibition of EGF
binding to the EGF receptor by preincubation with anti-EGF receptor
antibody almost completely abolished the EGF-dependent tyrosine phosphorylation of the c-erbB-2 gene product as well as that of the EGF
receptor itself. In addition, the concentration of EGF required for
half-maximum phosphorylation of the EGF receptor was virtually identical
with that of the c-erbB-2 gene product (data not shown). From these
data, we conclude that activation of EGF receptor tyrosine kinase activity
upon EGF binding to its receptor leads to the phosphorylation of the
related c-erbB-2 gene product on tyrosine residues (Ref. 15; Fig. 6c).
Thus, EGF is expected to modulate presumptive functions of the c-erbB-2
gene product. Similar results have been reported by other investigators
(17,31).

Distribution of the c-erbB-2 Protein in Fetal Epithelial Cells

Using the c-erbB-2 antibodies, various fetal tissues from three
human abortuses at 9, 14, and 24 wk of gestation, were studied immuno-
histologically by the ABC method and immunochemically by western blot
analysis to determine the distribution of the c-erbB-2 gene product (22).

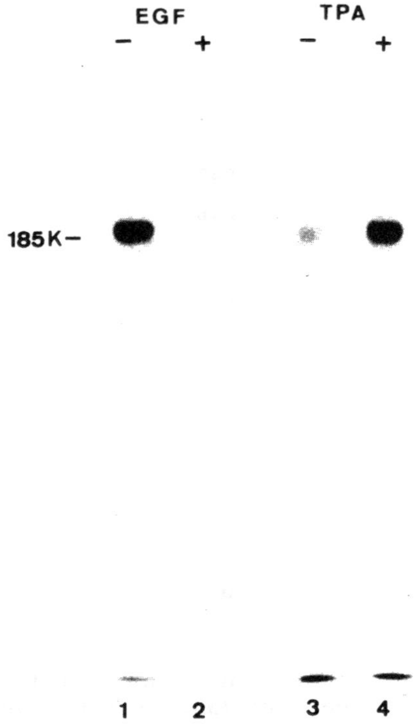
EGF
TPA
− + − +
185K−
1 2 3 4

a

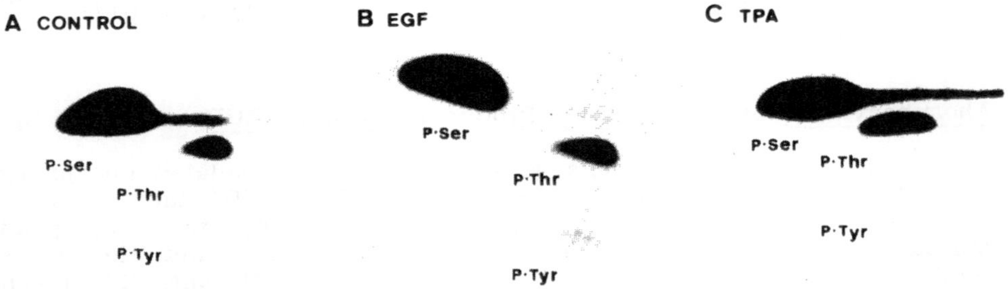
A CONTROL
P·Ser
P·Thr
P·Tyr
B EGF
P·Ser
P·Thr
P·Tyr
C TPA
P·Ser
P·Thr
P·Tyr

b

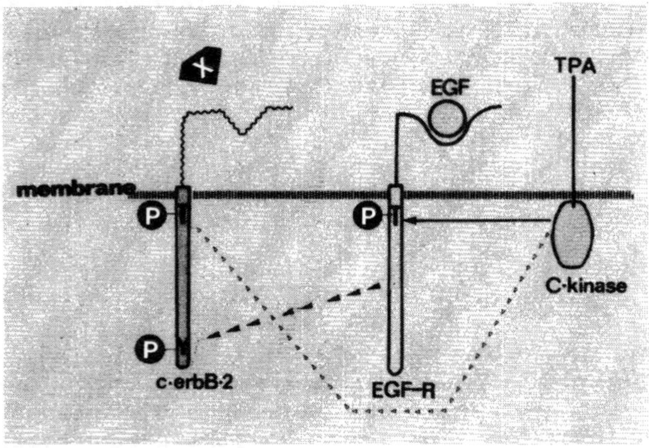
X
EGF
TPA
membrane
P
P
P
C·kinase
c·erbB·2
EGF–R

c

Fig. 6.   Phosphorylation of the c-erbB-2 protein stimulated by EGF and
            TPA. (a) MKN-7 cells were labeled with ^{32}P-orthophosphate for
            7 hr (lanes 1 and 3) and subsequently treated with EGF (50
            ng/ml) or TPA (50 ng/ml) for 15 min (lanes 2 and 4). Immuno-
            precipitate with anti-c-erbB-2 was analyzed on SDS polyacryl-
            amide gels. (b) Phosphoamino acid analysis of the c-erbB-2
            protein. The MKN-7 cells were labeled as described in (A)
            without EGF or TPA (A) and with EGF (B) or TPA (C) for the
            final 3 min of labeling. Phosphorylated 185-kDa protein was
            prepared and subjected to acid hydrolysis. Phosphoamino acids
            were resolved by two-dimensional electrophoresis at pH 1.9 and
            pH 3.5 on cellulose thin-layer plate. (c) Schematic illustration
            of "cross talk" between the surface proteins.

A heavy immune reaction was observed on the membrane of most of
the epithelial cells examined (Fig. 7), including the transitional gastrointestinal tract, the renal tubuli, the bronchi and pancreas, and the
stratified epithelium of the oral cavity, trachea, and esophagus in this
gestational period. A much more intense reaction was observed on the
basolateral side than on the apical side of these cells. The immune reaction was not observed in the liver, adrenal gland, striated and smooth
muscles, brain, endothelium, or fibroblasts. Western blot analysis confirmed increased expression of the c-erbB-2 gene product in fetal kidney
and intestine but not in the brain. As the immune reaction was not observed in normal adult epithelial cells, except for those in the renal
tubuli, this product may be a membrane-associated receptor protein which
regulates growth or differentiation of fetal epithelium.

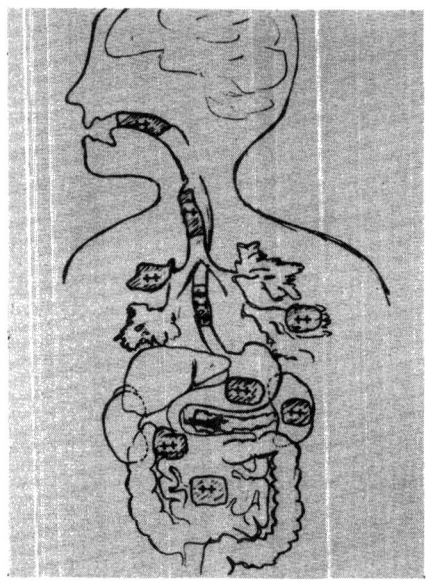

Fig. 7.   Tissue distribution of the c-erbB-2 protein. Embryonal epithel-
            ial tissues expressing the c-erbB-2 protein are denoted ++.

Overexpression of the c-erbB-2 Gene in Cancer Cells

There is accumulating evidence that overexpression of proto-oncogenes is associated with human tumors and presumably plays an important part in the neoplastic process (3). Since transformation of chicken cells by avian erythroblastosis virus is apparently due to increased expression of a truncated form of the EGF receptor (10,38), qualitatively and/or quantitatively abnormal expression of the EGF receptor gene in human cells may be involved in some stage of tumorigenesis. Therefore, we examined human tumors for aberrant expression of the EGF receptor gene as well as the c-erbB-2 gene.

In the initial experiments, we tested a total of 118 fresh samples of human malignant tumors of 25 different types for amplification of the EGF receptor and c-erbB-2 genes (43). Hybridization analysis of high-molecular-weight DNAs from these tumors after their digestion with restriction endonucleases showed that neither the EGF receptor gene nor the c-erbB-2 gene was amplified or rearranged grossly in 14 sarcomas, 20 leukemias, or 4 malignant lymphomas examined. However, amplification of the EGF receptor gene was seen in 2 of 14 squamous cell carcinomas and in 1 of 66 adenocarcinomas. This is consistent with a previous observation of ours and other investigators that EGF receptors are expressed at high levels on the cell surface of some squamous cell carcinomas (9,41). In contrast, the c-erbB-2 gene was amplified in 6 of 66 adenocarcinomas, but not in 14 squamous cell carcinomas. The 6 samples that showed amplification of the c-erbB-2 gene were from adenocarcinomas of the salivary gland (1 case), the stomach (2 cases), the breast (2 cases), and the kidney (1 case). We then examined over 400 fresh samples of human malignant tumors of various types and found that about 20% of the cases of breast cancer and stomach cancer, respectively, carry amplification of the c-erbB-2 gene (44,47; Y. Yamada et al., unpubl. data); amplification was frequently seen in schirrhous carcinomas of the breast and tubular adenocarcinomas of the stomach (Fig. 8). On the other hand, immunohistological and immunochemical analyses of the cancerous tissues

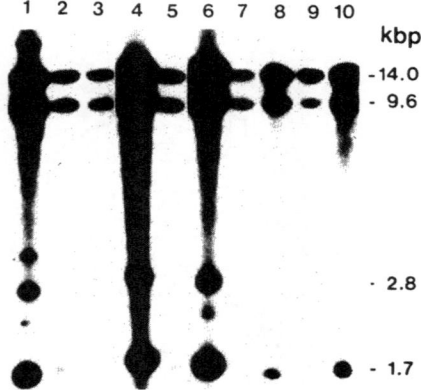

Fig. 8.   Amplification of the c-erbB-2 gene in tubular adenocarcinomas in the stomach. Hind III digests of DNAs from seven fresh tumors (lanes 1-7) and three cell lines (lanes 8-10) were probed with ^{32}P-labeled c-erbB-2 cDNA.

revealed that in about 40% of the cases of breast cancer, the c-erbB-2 gene product is overexpressed (Y. Yamada et al., unpubl. data). These observations suggest that aberrant expression of the c-erbB-2 gene is associated with induction and/or progression of malignancy in glandular epithelium of the breast and stomach.

Therefore, we examined the correlation between c-erbB-2 expression and the clinical activity of the breast cancer. As shown in Tab. 1, amplification of the c-erbB-2 gene was correlated with the stage of the tumor and with lymph node metastasis. When the protein level was examined, overexpression of the c-erbB-2 protein was correlated with the degree of lymph node metastasis (P=0.001) and with the stage of the tumor (P=0.01) (Y. Yamada et al., ms. in prep.). Moreover, using DNAs extracted from formalin-fixed, paraffin-embedded tissue blocks in 176 consecutive patients, association of c-erbB-2 amplification with ten-year survival was examined. As shown in Fig. 9, amplification of c-erbB-2 correlated well with poorer patient prognosis (35). These results emphasize that curative operation at the early stage of disease without lymph node metastasis is very important for patients with breast cancer carrying overexpressed c-erbB-2.

Association of Transforming Ability with Elevated Tyrosine-Kinase Activity

Complementary DNAs of normal or variously mutated c-erbB-2 genes were expressed under the control of the SV40 promoter and their transforming potential was investigated. The normal c-erbB-2 cDNA failed to transform NIH 3T3 cells, whereas a mutant encoding glutamic acid instead of valine at position 659 within the transmembrane domain, as is the case of the rat neu oncogene (6), exhibited transforming activity (Fig. 10A). The protein-tyrosine kinase activity of the mutated c-erbB-2 protein was enhanced relative to the normal c-erbB-2 protein both in vivo and in vitro (5,21,26; T. Akiyama et al., unpubl. data; see also Fig. 10B).

Tab. 1. Correlation between amplification of c-erbB-2 and characteristics of primary breast cancer (modified from Ref. 47).

Tumor characteristic	No. Studied	No. with c-erbB-2 amplification	P
Stage I and II	34	2	0.01
Stage III and IV	24	7	
Axillary nodes			
Negative	21	1	0.08
Positive	37	8	

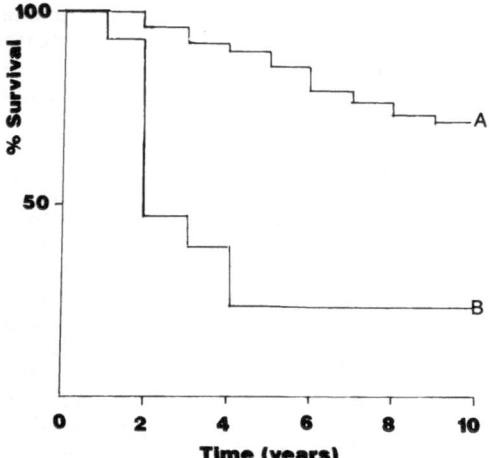

Fig. 9.   Actuarial survival curves of breast carcinoma patients. Comparison of patients with c-erbB-2 amplification (B) with patients without c-erbB-2 amplification (A).

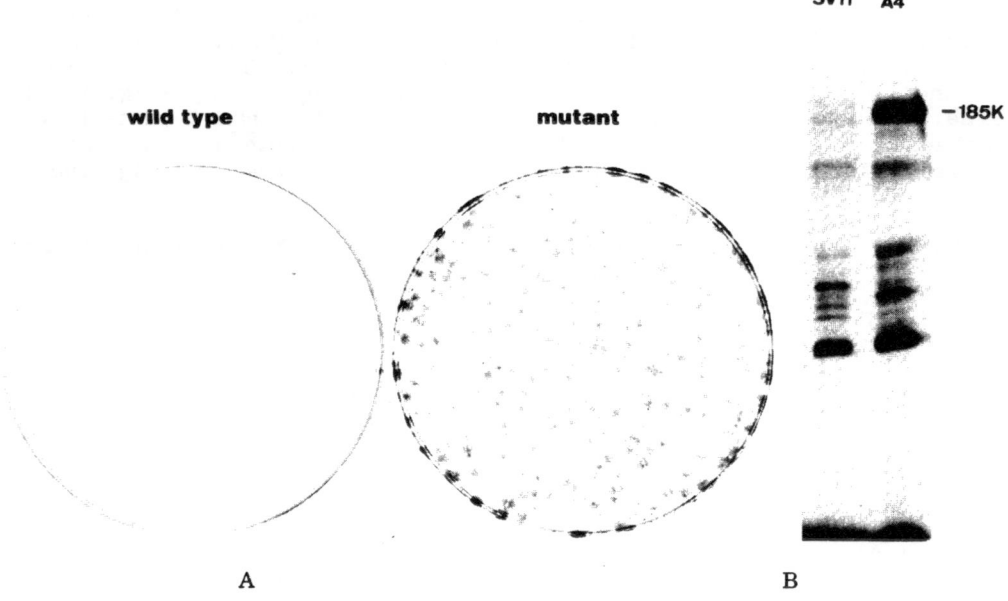

Fig. 10.   Elevated protein-tyrosine kinase activity of the onocogenic c-erbB-2 protein.   (A) Normal and oncogenically-mutated c-erbB-2 cDNAs were transfected to NIH 3T3 cells. Foci were induced only by the mutated c-erbB-2.   (B) Extracts NIH3T3 cells expressing normal c-erbB-2 (SV11 cells) and oncogenically-mutated c-erbB-2 (A4 cells) were immunoprecipitated with anti-c-erbB-2 antibodies.   Immunoprecipitates were subjected to an in vitro kinase assay, as described in the legend to Fig. 2.

The tyrosine residue at position 1248, located proximal to the carboxyl terminus of the activated c-erbB-2 protein, was phosphorylated at the elevated level. When a further mutation was introduced at the ATP binding site of the tyrosine kinase domain of the activated c-erbB-2 protein (i.e., Lys at position 753 was replaced with Met), the transforming ability of the gene product was abolished (T. Akiyama et al., ms. in prep.). These results indicate that the transforming potential of the c-erbB-2 gene is closely related to the elevated tyrosine kinase activity of the gene product. Our preliminary data suggest that a sequence of 230 amino acid residues at the carboxyl-terminus, which include tyrosine at position 1248, is important in the regulation of the c-erbB-2 transforming ability.

CONCLUSION

Among 24 known retroviral oncogenes, ten encode protein-tyrosine kinases, suggesting that activities of both receptor-type and nonreceptor-type protein-tyrosine kinases (Fig. 11) play a part in cellular transformation. Consistent with this notion, several cellular oncogenes (neu/c-erbB-2, met, trk, ret, sam) that encode protein-tyrosine kinases have been reported to be activated in tumors by either point mutation or rearrangement. Amplification of the EGF receptor gene/c-erbB-1 and c-erbB-2 is also suggested to be associated with induction and/or progression of human cancer. However, all the oncogenes of the protein-tyrosine kinase shown to be associated with human malignancy are of the receptor-type. Nonreceptor-type proto-oncogenes, such as c-src, c-yes, c-fgr, fyn, and lyn, have not been shown to be active in naturally-occurring cancers. We also could not detect any amplification of c-yes, fyn, and lyn in analyses of more than 200 samples of various human tumors.

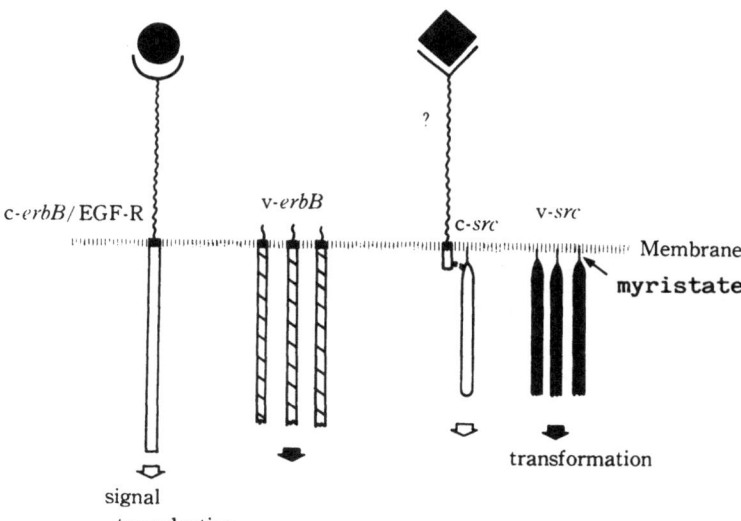

Fig. 11.   Schematic illustration of protein-tyrosine kinases of receptor-type and nonreceptor-type.

Accumulating evidence shows that nonreceptor-type kinases are expressed in differentiated hematopoietic cells (8; T. Yamamoto et al., unpubl. data). For example, c-src is expressed in macrophages and platelets, c-fgr in monocytes/macrophages and large granular lymphocytes, fyn in T cells, lyn in platelets, macrophages/monocytes, and B-lymphocytes, hck in granulocytes, and lck in T lymphocytes. Therefore, it would be reasonable to postulate that expression of nonreceptor-type kinase may be important in signal transduction required for the differentiated phenotype of the cells and may not be involved in growth control of normal cells. Recent evidence that the lck gene product interacts with the CD4 molecule suggests that lck is involved in signal transduction of T lymphocytes.

Nevertheless, unusual expression of nonreceptor-type tyrosine kinase would subvert the differentiated phenotype, which may result in abnormal growth of the cells. Recently, in collaboration with Katagiri and his colleagues, we observed elevated expression of a protein-tyrosine kinase $p59^{fyn}$ in $CD4^- CD8^-$ T lymphocyte of mutant mice which carried the autosomal recessive mutant gene lpr (lymphoproliferative) and suffered from lymphadenopathy and autoimmune disease (T. Katagiri et al., ms. in prep.). It should also be pointed out that the c-kit gene encoding receptor-type tyrosine kinase is suggested to play a role in the embryonic development of mice; this raises a possibility that the c-kit gene product is important in cell-cell communication during differentiation. Thus, detailed analysis of signaling mechanisms via protein-tyrosine kinases would help us to understand regulation of growth, differentiation, and malignant transformation of the cell.

ACKNOWLEDGEMENTS

A part of our study described in this chapter was performed in collaboration with J. Yokota, M.J. Cline, S. Hirohashi, and their colleagues.

REFERENCES

1. Akiyama, T., C. Sudo, H. Ogawara, K. Toyoshima, and T. Yamamoto (1986) The product of the human c-erbB-2 gene: A 185-kilodalton glycoprotein with tyrosine kinase activity. Science 232:1644-1646.
2. Akiyama, T., T. Saito, H. Ogawara, K. Toyoshima, and T. Yamamoto (1988) Tumor promoter and epidermal growth factor stimulate phosphorylation of the c-erbB-2 gene product in human adenocarcinoma cells. Molec. Cell. Biol. 8:1019-1026.
3. Alitaro, K., and M. Schwab (1986) Oncogene amplification in tumor cells. Adv. Cancer Res. 47:235-281.
4. Anderson, S.K., C.P. Griffs, A. Tanaka, H.-J. Kung, and D.J. Fujita (1985) Human cellular src gene: Nucleotide sequence and derived amino acid sequence of the region coding for the carboxyl-terminal two-thirds of $pp60^{c-src}$. Molec. Cell. Biol. 5:1122-1129.
5. Bargmann, C.I., and R.A. Weinberg (1988) Increased tyrosine kinase activity associated with the protein encoded by the activated neu oncogene. Proc. Natl. Acad. Sci., USA 85:5394-5398.
6. Bargmann, C.I., M.-C. Hung, and R.A. Weinberg (1986) Multiple independent activations of the neu oncogene by a point mutation altering the transmembrane domain of p185. Cell 45:649-657.

7. Chabot, B., D.A. Stephenson, V.M. Chapman, P. Besmer, and A. Bernstein (1988) The proto-oncogene c-kit encoding a transmembrane tyrosine kinase receptor maps to the mouse W locus. Nature 335:88-89.

8. Cooper, J.A. (1989) The src-family of protein tyrosine kinase. In Peptides and Protein Phosphorylation, B. Kemp and P.F. Alewood, eds. CRC Press, Inc., London (in press).

9. Cowley, O., J.A. Smith, B. Gusherson, F. Hender, and B. Ozanne (1984) The amount of EGF receptor is elevated on squamous cell carcinomas. Cancer Cells 1:5-10.

10. Downward, J., Y. Yarden, E. Mayes, G. Scrace, N. Totty, P. Stockwell, A. Ullrich, J. Schlessinger, and M.D. Waterfield (1984) Close similarity of epidermal growth factor receptor and v-erbB oncogene protein sequences. Nature 307:521-527.

11. Fukushige, S., K. Matsubara, M. Yoshida, M. Sasaki, T. Suzuki, K. Semba, K. Toyoshima, and T. Yamamoto (1986) Localization of a novel v-erbB-related gene, c-erbB-2, on human chromosome 17 and its amplification in a gastric cancer cell line. Molec. Cell. Biol. 6:955-958.

12. Geissler, E.N., M.A. Ryan, and D.E. Housman (1988) The dominant-white spotting (W) locus of the mouse encodes the c-kit protoonco-genes. Cell 55:185-192.

13. Hanks, S.K., A.M. Quinn, and T. Hunter (1988) The protein kinase family: Conserved features and deduced phylogeny of the catalytic domains. Science 241:42-52.

14. Inoue, K., S. Ikawa, K. Semba, J. Sukegawa, T. Yamamoto, and K. Toyoshima (1987) Isolation and sequencing of cDNA clones homologous to the v-fgr oncogene from a human B lymphocyte cell line, IM9. Oncogene 1:301-304.

15. Kadowaki, T., M. Kasuga, K. Tobe, F. Takaku, E. Nishida, H. Sakai, S. Koyasu, I. Yahara, K. Toyoshima, T. Yamamoto, and T. Akiyama (1987) A Mr=190,000 glycoprotein, a major substrate of the epidermal growth factor receptor tyrosine kinase in KB cells is the product of the human c-erbB-2 gene. Biochem. Biophys. Res. Commun. 144:699-704.

16. Kawai, S., M. Yoshida, K. Segawa, H. Sugiyama, R. Ishizaki, and K. Toyoshima (1980) Characterization of Y73, a newly isolated avian sarcoma virus: A unique transforming gene and its product, a phosphopolyprotein with protein kinase activity. Proc. Natl. Acad. Sci., USA 77:6199-6209.

17. King, C.R., I. Borrello, F. Bellot, P. Comoglio, and J. Schlessinger (1988) EGF binding to its receptor triggers a rapid tyrosine phosphorylation of the erbB-2 protein in the mammary tumor cell line SK-BR-3. EMBO J. 7:1647-1651.

18. Kitamura, N., A. Kitamura, K. Toyoshima, Y. Hirayama, and M. Yoshida (1982) Avian sarcoma virus Y73 genome sequence and structural similarity of its transforming gene product to that of Rous sarcoma virus. Nature 298:205-208.

19. LeBeau, M.M., C.A. Westbrook, M.O. Diaz, and J.D. Rowley (1984) Evidence for two distinct c-src loci on human chromosome 1 and 20. Nature 312:70-71.

20. Marth, J.D., R. Peet, E.G. Krebs, and R.M. Pelmutter (1985) A lymphocyte-specific protein-tyrosine kinase gene is rearranged and overexpressed in the murine T cell lymphoma LSTRA. Cell 43:393-404.

21. Masuko, T., K. Sugahara, M. Kozono, S. Otuki, T. Akiyama, T. Yamamoto, K. Toyoshima, and T. Hashimoto (1989) A murine mono-

clonal antibody that recognizes an extracellular domain of the human c-erbB-2 protooncogene product. Jpn. J. Cancer Res. 80:10-14.

22. Mori, S., T. Akiyama, Y. Morishita, I. Sugawara, K. Toyoshima, and T. Yamamoto (1989) c-erbB-2 gene product, a membrane protein commonly expressed on human fetal epithelial cells. Laboratory Investigation (in press).

23. Nishizawa, M., K. Semba, T. Yamamoto, and K. Toyoshima (1985) Human c-fgr gene does not contain coding sequence for actin-like protein. Jpn. J. Cancer Res. (Gann) 76:155-159.

24. Nishizawa, M., K. Semba, M.C. Yoshida, T. Yamamoto, M. Sasaki, and K. Toyoshima (1986) Structure, expression, and chromosomal location of the human c-fgr gene. Molec. Cell. Biol. 6:511-517.

25. Rudd, C.E., J.M. Trevillyan, J.D. Dasgupta, L.L. Wong, and S.F. Schlossman (1988) The CD4 receptor is complexed in detergent lysates to a protein-tyrosine kinase (pp58) from human T lymphocytes. Proc. Natl. Acad. Sci., USA 85:5190-5194.

26. Segatto, O., C.R. King, J.H. Pierce, P.P. Difore, and S. Aaronson (1988) Different structural alterations upregulate in vitro tyrosine kinase activity and transforming potency of the erbB-2 gene. Molec. Cell. Biol. 8:5570-5574.

27. Semba, K., N. Kamata, K. Toyoshima, and T. Yamamoto (1985) A v-erbB-related protooncogene, c-erbB-2, is distinct from the c-erbB-1/epidermal growth factor-receptor gene and is amplified in a human salivary gland adenocarcinoma. Proc. Natl. Acad. Sci., USA 82:6497-6501.

28. Semba, K., Y. Yamanashi, M. Nishizawa, J. Sukegawa, M.C. Yoshida, M. Sasaki, T. Yamamoto, and K. Toyoshima (1985) Location of the c-yes gene of the human chromosome and its expression in various tissues. Science 227:1038-1040.

29. Semba, K., M. Nishizawa, N. Miyajima, M.C. Yoshida, J. Sukegawa, Y. Yamanashi, M. Sasaki, T. Yamamoto, and K. Toyoshima (1986) yes-related protooncogene, syn, belongs to the protein-tyrosine kinase family. Proc. Natl. Acad. Sci., USA 83:5459-5463.

30. Semba, K., M. Nishizawa, H. Sato, S. Fukushige, M.C. Yoshida, M. Sasaki, K. Matsubara, T. Yamamoto, and K. Tokoshima (1988) Nucleotide sequence and chromosomal mapping of the human c-yes-2 gene. Jpn. J. Cancer Res. (Gann) 79:710-717.

31. Stern, D.F., and M.P. Kamps (1988) EGF-stimulated tyrosine phosphorylation of p185[neu], a potential model for receptor interactions. EMBO J. 7:995-1001.

32. Strebhardt, K., J.I. Mullins, C. Bruck, and H. Rubsamen-Waigmann (1977) Additional member of the protein-tyrosine kinase family: The src and lck-related protooncogene c-tkl. Proc. Natl. Acad. Sci., USA 84:8778-8782.

33. Sukegawa, J., K. Semba, Y. Yamanashi, M. Nishizawa, N. Miyajima, T. Yamamoto, and K. Toyoshima (1987) Characterization of cDNA clones for the human c-yes gene. Molec. Cell. Biol. 7:41-47.

34. Takeya, T., and H. Hanafusa (1983) Structure and sequence of the cellular gene homologous to the RSV src gene and the mechanism for generating the transforming virus. Cell 32:881-890.

35. Tsuda, H., S. Hirohashi, Y. Shimosato, T. Hirota, H. Yamamoto, N. Miyajima, K. Toyoshima, T. Yamamoto, J. Yokota, T. Yoshida, H. Sakamoto, M. Terada, and T. Sugimura (1989) Correlation between long-term survival in breast cancer patients and amplification of two putative oncogene-coamplification units; hst-1/int-2 and c-erbB-2/-ear-1. Cancer Res. (in press).

36. Veillette, A., M.A. Bookman, E.M. Horak, and J.B. Bolen (1988) The CD4 and CD8 T cell surface antigens are associated with the internal membrane tyrosine-protein kinase p56^{lck}. Cell 55:301-308.
37. Vornaova, A.F., and B.M. Sefton (1986) Expression of a new tyrosine protein kinase is stimulated by retrovirus promoter insertion. Nature 319:682-685.
38. Yamamoto, T., T. Nishida, S. Kawai, T. Ooi, and K. Toyoshima (1983) The erbB gene of avian erythroblastosis virus is a member of the src gene family. Cell 35:71-78.
39. Yamamoto, Y., S. Hori, M. Yoshida, T. Kishimoto, K. Inoue, T. Yamamoto, and K. Toyoshima (1989) Expression of a protein-tyrosine kinase, p56^{lyn}, is preferable in hematopoietic cells and is associated with HTLV-1 production. Proc. Natl. Acad. Sci., USA (in press).
40. Yamamoto, T., S. Ikawa, T. Akiyama, K. Semba, N. Nomura, N. Miyajima, T. Saito, and K. Toyoshima (1986) Similarity of protein encoded by the human c-erbB-2 gene to epidermal growth factor receptor. Nature 319:230-234.
41. Yamamoto, T., N. Kamata, H. Kawano, S. Shimizu, T. Kuroki, K. Toyoshima, K. Rikimaru, N. Nomura, R. Ishizaki, I. Pastan, S. Gamou, and N. Shimizu (1986) High incidence of amplification of the epidermal growth factor receptor gene in human squamous carcinoma cell lines. Cancer Res. 46:414-416.
42. Yamanashi, Y., F. Fukushige, K. Semba, J. Sukegawa, N. Miyajima, K. Matsubara, T. Yamamoto, and K. Toyoshima (1987) The yes-related cellular gene, lyn, encodes a possible tyrosine kinase similar to p56^{lck}. Molec. Cell. Biol. 7:237-243.
43. Yokota, J., T. Yamamoto, K. Toyoshima, M. Terada, T. Sugimura, H. Battifora, and M.J. Cline (1986) Frequent amplification of the c-erbB-2 oncogene in human adenocarcinomas. Lancet i:765-767.
44. Yokota, J., T. Yamamoto, N. Miyajima, K. Toyoshima, N. Nomura, H. Sakamoto, T. Yoshida, M. Terada, and T. Sugimura (1988) Genetic alterations of the c-erbB-2 oncogene occur frequently in tubular adenocarcinomas of the stomach and are often accompanied by amplification of the v-erbA homolog. Oncogene 2:283-287.
45. Yoshida, M.C., H. Satoh, M. Sasaki, K. Semba, T. Yamamoto, and K. Toyoshima (1986) Regional localization of a novel yes-related protooncogene, syn, on human chromosome 6 at band q21. Jpn. J. Cancer Res. (Gann) 77:1009-1061.
46. Yoshida, M.C., M. Sasaki, K. Mise, K. Semba, M. Nishizawa, T. Yamamoto, and K. Toyoshima (1985) Regional mapping of the human proto-oncogene c-yes-1 to chromosome 18q21.3. Jpn. J. Cancer Res. (Gann) 76:559-562.
47. Zhou, D., H. Battifora, J. Yokota, T. Yamamoto, and M.J. Cline (1987) Association of multiple copies of the c-erbB-2 oncogene with spread of breast cancer. Cancer Res. 47:6123-6125.

INHIBITION OF THE GENOTOXICITY OF 3-AMINO-1-METHYL-5H-PYRIDO[4,3-b]INDOLE (Trp-P-2) IN DROSOPHILA BY CHLOROPHYLL

Tomoe Negishi, Sakae Arimoto, Chiharu Nishizaki, and Hikoya Hayatsu

Faculty of Pharmaceutical Sciences
Okayama University
Tsushima, Okayama 700, Japan

INTRODUCTION

Green-yellow vegetables have been suggested to be anticarcinogenic, based on epidemiological evidence (9). Experimental studies have shown that among the components of vegetables, vitamins and fibers are anticarcinogenic (for a review, see Ref. 6). Since chlorophyll is a main component of green vegetables (3), it is important to evaluate its antimutagenic and anticarcinogenic potency. With bacterial mutagenicity assays, antimutagenic activities of chlorophyll and chlorophyllin have been observed in several laboratories (6). In this chapter, we wish to report the antigenotoxic effect of chlorophyll and chlorophyllin against the activities of heterocyclic amines, particularly of 3-amino-1-methyl-5H-pyrido[4,3-b]indole (Trp-P-2), as demonstrated in Drosophila test systems. In these experiments, reagents are administered to the larvae by feeding; obviously, this test system is a better model than bacterial tests for the human situation, especially so when the mutagen tested is a diet-related one.

MATERIALS AND METHODS

Chlorophyll, which was prepared from Chlorella vulgaris E25 grown in the dark, was obtained from Osaka Seiken Co. (Osaka); the preparation contained chlorophyll a and b in a molar ratio of 7:3. Chlorophyllin Cu-Na salt was purchased from Nakarai Chemicals (Kyoto). Trp-P-2 and 2-aminodipyrido[1,2-a:3',2'-d]imidazole (Glu-P-2) were generous gifts from Dr. M. Nagao of the National Cancer Center Research Institute, Tokyo. 2-Amino-3,8-dimethylimidazo[4,5-f]quinoxaline (MeIQx) was obtained through the carcinogen stock program supported by a Grant-in-Aid for Cancer Research from the Ministry of Health and Welfare, Japan.

The Drosophila wing spot test, originally developed by Würgler and co-workers and used by Yoo et al. (11) for the mutagenicity test of food-pyrolysate mutagens, was carried out according to the method described previously (7,8). Drosophila melanogaster strains used in the wing spot

341

test were kindly supplied by Dr. S. Kondo of Osaka University. We used the heterozygous progeny prepared by mating the virgin females (y; mwh jv) with males (y; Dp(1;3)sc^{J4}, flr/TM1, Me). The Drosophila used in the in vivo rec-assay, which had been developed by Fujikawa (4), were the recombination-deficient strain (rec$^-$; C(1)Dx, yf/scZw$^{+(TE)}$ mei-9^a mei-41^{DS}) and the recombination-proficient strain (rec+; C(1)Dx, yf/scZw^{+TE}). The rec$^-$ strain was obtained from Dr. K. Fujikawa of Takeda Chemical Co., Osaka, and the rec$^-$ strain from Dr. H. Ryo of Osaka University. In these tests, the third instar larvae were fed the reagents; the feeding method is detailed in our recent publication (7).

The spectroscopic studies were carried out by use of a Hewlett-Packard 8450A UV/VIS spectrophotometer.

RESULTS AND DISCUSSION

The genotoxicity in the wing spot test is detected by phenotypic alterations of adult fly wing hairs, alterations resulting from mutations in the somatic cell chromosomes of the larvae (5,10). An administration of Trp-P-2 at a dose of 2 mg per bottle induced both small single spots (one or two mutant hairs in a clone) with a frequency of 1.20 spots per wing (the control value, 0.29), and large single spots (three or more mutant hairs in a clone) with a frequency of 0.30 spots per wing (control, 0.03). Twin spots (the two different phenotype mutants co-existing in a clone) were also induced with a frequency of 0.07 per wing (control, 0.02). These values are in agreement with those previously reported by Yoo et al. (11) for the genotoxicity of Trp-P-2 in this system. When 200 mg chlorophyll was co-administered with Trp-P-2 to these larvae, the values of mutant clone numbers per wing decreased: small single spots to 0.28, large single spots to 0.15, and twin spots to 0.01. The decrease in small single spots represents a complete inhibition, because the frequency of spontaneous induction was 0.29. Chlorophyllin was also effective in reducing the frequencies of the mutant clone formation. In Fig. 1, we show the dose dependence of these inhibitions. Both chlorophyll and chlorophyllin seem to be less effective in suppressing the large single spot formation than in suppressing the small singles. The reason for this difference is not clear.

We observed the antigenotoxic activity of chlorophyllin against Trp-P-2 using another Drosophila test system, i.e., the recombination-deficient strain-specific genotoxicity test (4). In this assay, the number of adult male flies, which are recombination-deficient, will decrease if larvae are treated with a genotoxic compound, while the number of adult female flies, which are recombination-proficient, will be unaffected. With this test, one can assess the genotoxicity of a compound by simply observing the decrease in the male-to-female ratio. We found that Trp-P-2 was again genotoxic to the Drosophila larvae as assayed by this system, and that chlorophyllin was effective in preventing the toxicity; the sex ratio decrease induced by Trp-P-2 was abolished by co-administration of chlorophyllin, with the inhibitory effect being dependent on the dose of chlorophyllin. Genotoxicities of Glu-P-2 and MeIQx were also suppressed by chlorophyllin, as demonstrated by this assay. Control experiments were performed in these tests in which rec$^+$ male flies were treated in place of rec$^-$ males, and the mutagens and the modifying agents were always without effect on the sex ratios. The details of these results will be published elsewhere.

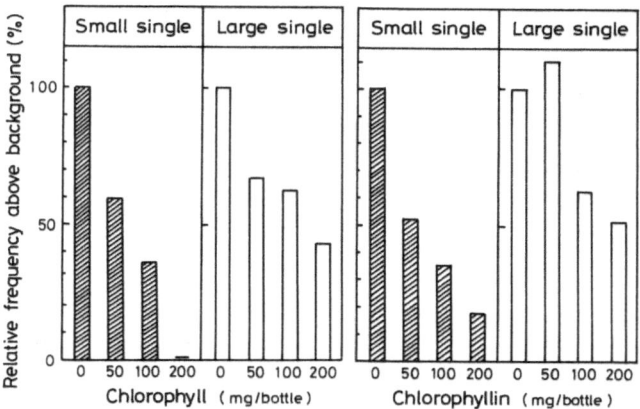

Fig. 1.   Inhibition of Trp-P-2 genotoxicity by chlorophyll and chlorophyllin.   Relative frequency (RF) was calculated as follows:

$$\frac{RF}{(\%)} = \frac{\text{spots/wing (Trp-P-2 + pigment)} - \text{spots/wing (solvent)}}{\text{spots/wing (Trp-P-2)} - \text{spots/wing (solvent)}} \times 100$$

Trp-P-2 dose/bottle was 2 mg.   The solvent was distilled water.

In the Ames test, chlorophyllin suppressed Trp-P-2-mediated mutagenicity (2).   Now we found that in the bacterial test, chlorophyll and
chlorophyllin can inhibit the direct-acting mutagenicity of N-hydroxylated
Trp-P-2, an activated form of Trp-P-2 (see also Ref. 1).   A finding relevant to the mechanism of these inhibitions is the following: we obtained
evidence showing that chlorophyll and chlorophyllin can make complexes
with Trp-P-2.   First, an equimolar mixture of chlorophyllin (chlorophyll)
and Trp-P-2 showed an absorption spectrum different from that of a calculated sum of spectra for the individual components.   This can be seen
from the difference spectrum shown in Fig. 2.   Second, chlorophyllin

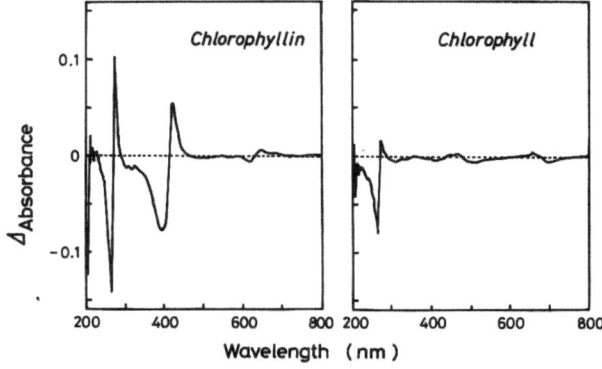

Fig. 2.   Difference spectra for mixtures of Trp-P-2 and chlorophyll/-
chlorophyllin.   The solutions contained Trp-P-2 and/or pigment
(20 μM each) in 0.1 M sodium phosphate buffer at pH 7.4.

covalently linked to Sepharose can efficiently adsorb Trp-P-2, whereas Sepharose itself does not adsorb Trp-P-2. These observations suggest that chlorophyllin and chlorophyll can trap Trp-P-2 by forming a complex. We detected no degradation of Trp-P-2 in the presence of these pigments, a fact indicating that degradation of Trp-P-2 is unlikely to be the cause of the inhibition. Glu-P-2 and MeIQx were also adsorbable to chlorophyllin-Sepharose (the details will be published elsewhere). Therefore, a straightforward mechanism of these inhibitions is that Trp-P-2 and other heterocyclic amines become no longer available to organisms due to formation of the chlorophyll complex.

ACKNOWLEDGEMENTS

This work is supported by a Grant-in-Aid for Research in Priority Area, Cancer-Bioscience, from the Ministry of Education, Science and Culture, Japan. We thank Nissan Science Foundation for support.

REFERENCES

1. Arimoto, S., and H. Hayatsu (1989) Role of hemin in the inhibition of mutagenic activity of 3-amino-1-methyl-5H-pyrido[4,3-b]indole (Trp-P-2) and other aminoazaarenes. Mutat. Res. (in press).
2. Arimoto, S., Y. Ohara, T. Namba, T. Negishi, and H. Hayatsu (1980) Inhibition of the mutagenicity of amino acid pyrolysis products by hemin and other biological pyrrole pigments. Biochem. Biophys. Res. Commun. 92:662–668.
3. Chipchase, M.I.H. (1961) Chemical components in plant tissues. In Biochemists' Handbook, C. Long, ed. E. & F.N. Spon, Ltd., London, pp. 1032–1033.
4. Fujikawa, K. (1988) A novel methodology of chemical mutagenicity assays in Drosophila melanogaster. J. Food Hyg. Soc. Japan 29:115–214.
5. Graf, U., F.E. Würgler, A.J. Katz, H. Frei, H. Juon, C.B. Hall, and P.G. Kale (1984) Somatic mutation and recombination test in Drosophila melanogaster. Environ. Mutag. 6:153–188.
6. Hayatsu, H., S, Arimoto, and T. Negishi (1988) Dietary inhibitors of mutagenesis and carcinogenesis. Mutat. Res. 202:429–446.
7. Negishi, T., S. Arimoto, C. Nishizaki, and H. Hayatsu (1989) Inhibitory effect of chlorophyll on the genotoxicity of 3-amino-1-methyl-5H-pyrido[4,3-b]indole (Trp-P-2). Carcinogenesis 10:145–149.
8. Negishi, T., K. Negishi, H. Ryo, S. Kondo, and H. Hayatsu (1988) The genotoxicity of N^4-aminocytidine in the Drosophila wing spot test. Mutagenesis 3:11–13.
9. Phillips, R.L. (1975) Role of life-style and dietary habits in risk of cancer among Seventh-Day Adventists. Cancer Res. 35:3515–3522.
10. Würgler, F.E., and E.W. Vogel (1986) In vivo mutagenicity testing using somatic cells of Drosophila melanogaster. In Chemical Mutagens, Vol. 10, F.J. de Serres, ed. Plenum Press, New York, pp. 1–72.
11. Yoo, M.-A., H. Ryo, T. Todo, and S. Kondo (1985) Mutagenic potency of heterocyclic amines in the Drosophila wing spot test and its correlation to carcinogenic potency. Jpn. J. Cancer Res. (Gann) 76:468–473.

INHIBITION OF TOBACCO-SPECIFIC NITROSAMINE 4-(METHYL-NITROSAMINO)-1-(3-PYRIDYL)-1-BUTANONE (NNK)-INDUCED LUNG TUMORS AND DNA METHYLATION IN F344 RATS AND A/J MICE BY PHENETHYL ISOTHIOCYANATE

M.A. Morse, S.S. Hecht, and F.L. Chung

Section of Nucleic Acid Chemistry
Division of Chemical Carcinogenesis
American Health Foundation
Valhalla, New York 10595

INTRODUCTION

4-(Methylnitrosamino)-1-(3-pyridyl)-1-butanone (NNK) (Fig. 1) is the most potent carcinogenic nitrosamine so far found in tobacco and to-bacco smoke (3). It induces lung, nasal cavity, liver, and pancreatic tumors in F344 rats; nasal cavity and lung tumors in hamsters; and lung tumors in mice (5). The organ-specific effect of NNK in the induction of lung tumors in all species tested supports its possible role in the devel-opment of lung cancer among smokers. Therefore, it is of great impor-tance to discover compounds, either synthetic or dietary-related, which can counteract the carcinogenic action of NNK.

Our previous studies demonstrated that pretreatment of rats with a diet containing phenyl isothiocyanate (PITC), benzyl isothiocyanate (BITC), or phenethyl isothiocyanate (PEITC) (Fig. 1) resulted in reduced metabolic demethylation of NNK in hepatic microsomes as well as a de-crease in hepatic DNA methylation by NNK (2). These results suggested that these aromatic isothiocyanates are potential inhibitors of NNK car-cinogenesis. Recently, Wattenberg demonstrated inhibition of forestomach tumors by diethylnitrosamine in A/J mice pretreated with benzyl isothio-cyanate (10). In this study we examined the effects of PEITC on NNK tumorigenesis in F344 rats, and of PITC, BITC, and PEITC on NNK tumorigenesis in A/J mice. We also studied their effects on NNK-induced DNA methylation in liver, lung, and nasal cavity of rats and in lung of mice, all target tissues of NNK tumorigenesis.

RESULTS

Table 1 shows the tumor incidences in lung, liver, and nasal cavity

345

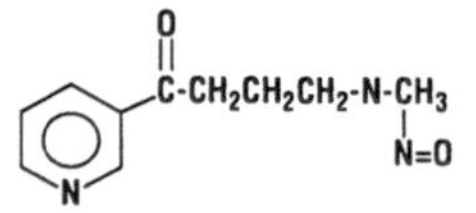

4-(methylnitrosamino)-1-(3-pyridyl)-1-butanone (NNK)

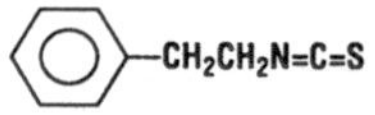

phenethyl isothiocyanate (PEITC)

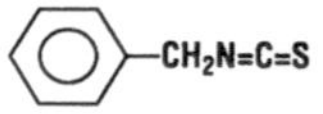

benzyl isothiocyanate (BITC)

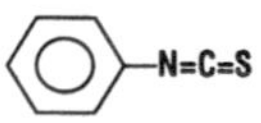

phenyl isothiocyanate (PITC)

Fig. 1.   Carcinogens found in tobacco and tobacco smoke.

in F344 rats upon treatment with NNK, NNK + PEITC, or PEITC. The
bioassay was terminated after 2 yr. The tumor incidences induced by
NNK alone were within the expected range, based on previous bioassay
results (4). Rats fed PEITC before and during NNK treatments developed only 43% lung tumor incidence compared to 80% in the group fed control diets. However, the PEITC diet did not alter the incidences of
tumors induced by NNK in liver or nasal cavity.

Tab. 1.   Incidence of lung, liver, and nasal cavity tumors after treatment with NNK, NNK + PEITC, and PEITC.

Treatment[a]	No. of rats	Lung			Liver			Nasal cavity		
		Adenoma	Carcinoma	Total	Adenoma	Hepatocellular carcinoma	Total	Benign[b]	Malignant[c]	Total
1. NNK	40	8	24	32 (80)	12	3	15 (38)	8	3	11 (28)
2. NNK + PEITC	40	5	12[d]	17 (43)[e]	9	5	14 (35)	6	1	7 (18)
3. PEITC	20	0	0	0 (0)	4	2	6 (28)	0	1	1 (5)
4. Control	20	1	0	1 (5)	3	1	4 (20)	0	1	1 (5)

Number of rats with tumors (%)

[a]Male F344 rats at 8 wk of age were randomized into four groups. Groups 2 and 3 were fed PEITC diets (3 µmol/g
diet) ad libitum for 21 wk while groups 1 and 4 were given only NIH-07 diets. After the first wk of feeding, NNK
(1.76 mg/kg b.w.) was administered to groups 1 and 2 by s.c. injection three times weekly for 20 wk. The experiment was terminated after 104 wk. Gross lesions and representative samples of all major organs were processed for
microscopic examination.
[b]Squamous cell papillomas, transitional-cell papillomas, polyps.
[c]Squamous cell carcinoma.
[d]One animal had squamous cell carcinoma and 11 had adenocarcinoma.
[e]P<0.05 compared to NNK group.

To determine the effects of PEITC diets on formation of DNA adducts by NNK, we chose to use experimental conditions analogous to those used in the bioassay. Table 2 shows the effects of 2 weeks' feeding of PEITC diets on DNA methylation in liver, lung, and nasal mucosa of rats. The levels of 7-methylguanine (7-mGua) were not affected in DNA of the liver and nasal mucosa of rats fed this diet. In lung, however, 7-mGua was reduced from 10.4 to 5.9 μmol/mol guanine, a reduction of nearly 50%.

In the A/J mouse bioassay, we examined the effects of PEITC and its homologues PITC and BITC on NNK-induced lung adenomas. Table 3 shows that a single i.p. administration of NNK at a dose of 10 μmol/mouse resulted in a 100% incidence of pulmonary adenomas with a multiplicity of 10.7 tumors/mouse in just 16 wk. The 5 μmol daily dose (20 μmol total) of PEITC did not significantly reduce the proportion of mice that developed pulmonary adenomas, but resulted in an approximate 70% reduction in tumor multiplicity. The 25 μmol daily dose (100 μmol total) of PEITC resulted in a 70% reduction of the percentage of mice that developed tumors and a nearly complete inhibition of tumor multiplicity. However, pretreatment with BITC or PITC at 5 μmol/da resulted in no significant change in the percentage of mice that developed tumors and in tumor multiplicity. It should be noted that both BITC and PITC proved too toxic to be tested at a daily dose of 25 μmol.

In an effort to relate the effects of PEITC, BITC, and PITC on NNK lung tumorigenicity to in vivo NNK:DNA adduct formation, the effects of these isothiocyanates on NNK-induced O^6-methylguanine (O^6-mGua) in A/J lung DNA were investigated. The same dosing regimen employed in the pulmonary adenoma assays was used in the O^6-mGua assays. Table 4 shows that, 6 hr after NNK administration, the 5 μmol daily dose of PEITC resulted in an 87% reduction of O^6-mGua levels, while the 25 μmol daily dose of PEITC yielded O^6-mGua levels that were undetectable. However, neither BITC nor PITC pretreatment resulted in significant reduction in O^6-mGua levels. These data clearly showed that the effects of the isothiocyanates on NNK-induced O^6-mGua formation were in good agreement with their effects on NNK lung tumorigenicity.

Tab. 2. DNA methylation in NNK-treated rats fed control or PEITC diets.

Diet[a]	7-mGua		
	Lung	Liver	Nasal mucosa
Control	10.4 ± 1.3[b]	20.6 ± 0.9[b]	21.5[c]
PEITC	5.9 ± 0.6[d]	22.8 ± 0.7	31.5

[a]Male F344 rats were fed control or test diets containing 3 μmol/g of diet PEITC for 2 wk. Beginning on day 11 of feeding, [^{3}H-CH$_3$]NNK was administered s.c. daily at a dose of 0.6 mg/kg b.w. for four consecutive days. Four hr after the last NNK dosing, rats were sacrificed and tissue DNA was isolated for analysis of 7-mGua. Numbers are expressed as μmol/mol guanine.
[b]Mean ± S.E. of 6 rats.
[c]Mean of two pooled preparations (2-3 rats/pool).
[d]P<0.05 compared to values in NNK group.

Tab. 3.   Effects of isothiocyanates on NNK-induced pulmonary adenomas
          in A/J mice.

Pretreatment[a]	Daily dose (μmol)	Number of mice	Weight at sacrifice (g)	% of mice with tumors	Tumors/mouse[b]
1. None	---	30	23.8	100	$10.7^1 \pm 0.8$
2. PEITC	5	18	23.0	89	$2.6^2 \pm 0.4$
3. PEITC	25	20	24.2	30^c	$0.3^3 \pm 0.1$
4. BITC	5	20	24.4	100	$7.6^1 \pm 0.5$
5. PITC	5	20	23.5	100	$9.5^1 \pm 1.2$

[a]Groups of 20-30 female A/J mice were administered corn oil or isothiocyanates by gavage daily for four consecutive days. Two hr after the final gavaging, a single dose of NNK (10 μmol/mouse) was administered i.p. Sixteen weeks after NNK administration, mice were sacrificed and pulmonary adenomas were quantitated.
[b]Mean ± S.E. Means bearing different superscripts under the tumors/mouse heading are statistically different (P<0.05) from one another as determined by analysis of variance followed by Newman-Keuls' ranges test.
[c]Significantly (P<0.01) less than that of group 1 as determined by the Chi-square test.

DISCUSSION

These data represent the first demonstration of inhibition of NNK tumorigenesis by any compound (6,8). While PEITC effectively reduced the incidence of lung tumors in rats, it had no effects on NNK-induced liver or nasal cavity tumors in rats. Several factors could account for the selective inhibition of lung tumors by PEITC in rats. These include effects on the tissue disposition of NNK, the detoxification of NNK, and/or the specific effects of PEITC on cytochrome P450 isozymes or other enzymes responsible for NNK activation.

The preferential formation and persistence of O^6-mGua in specific rat lung cells following NNK treatment suggests that DNA methylation is important in NNK-induced lung tumorigenesis (1). DNA methylation was reduced in the lungs of rats and mice administered PEITC. However, it was not affected in the lungs of mice fed PITC and BITC. These effects are consistent with the effects of the isothiocyanates on tumorigenicity induced by NNK in both species. These results clearly suggest that the protective effect of PEITC against NNK-induced lung tumorigenicity is due to its ability to inhibit DNA methylation by NNK. Since there is a good correlation between NNK tumorigenicity and DNA methylation, inhibition of DNA methylation could be used as a means of screening potential inhibitors of NNK carcinogenicity.

PEITC is a product of hydrolysis of gluconasturtiin, which is commonly found in turnips and rutabagas (9). In view of widespread human exposure to these compounds, it is important to demonstrate the inhibition of NNK lung tumorigenicity by a naturally-occurring dietary constituent. However, caution must be taken in relating the effects of a single compound in a specific diet to that of an entire diet, since there are other compounds present in the diet that could have opposing effects (7). The present study demonstrates that both rats and mice fed PEITC develop

Tab. 4. Effects of isothiocyanates on NNK-induced O^6-methylguanine formation in A/J mouse lung.

Pretreatment[a]	Daily dose (µmol)	µmol O^6-mGua/mol guanine[b]
1. None	---	$30.9^1 \pm 5.9$
2. PEITC	5	$3.9^2 \pm 1.2$
3. PEITC	25	N.D.[c]
4. BITC	5	$26.1^1 \pm 6.7$
5. PITC	5	$29.7^1 \pm 4.4$

[a]Groups of 5 mice were administered corn oil or isothiocyanates by gavage for four consecutive days. Two hr after the final gavaging, NNK was administered i.p. at a dose of 10 µmol/mouse. Mice were sacrificed at 6 hr after NNK administration. DNA was isolated from lung and hydrolyzed in 0.1 N HCl for 60 min. O^6-mGua was analyzed by strong cation exchange HPLC and fluorescence detection.
[b]Mean ± S.E. of 4-5 mice. Means that bear different superscripts are statistically different (P<0.05) from one another as determined by analysis of variance followed by Newman-Keuls' ranges test.
[c]Not detected.

considerably fewer NNK-induced lung tumors than controls. The goals of our future studies are to develop inhibitors of better efficacy via structure-activity studies and, ultimately, to test them in high-risk groups such as heavy smokers.

REFERENCES

1. Belinsky, S.A., C.M. White, T.R. Devereux, J.A. Swenberg, and M. Anderson (1987) Cell selective alkylation of DNA in rat lung following low dose exposure to the tobacco specific carcinogen 4-(methylnitrosamino)-1-(3-pyridyl)-1-butanone. Cancer Res. 47:1143-1148.
2. Chung, F.-L., M. Wang, and S.S. Hecht (1985) Effects of dietary indoles and isothiocyanates on N-nitrosodimethylmethylamine and 4-(methylnitrosamino)-1-(3-pyridyl)-1-butanone α-hydroxylation and DNA methylation in rat liver. Carcinogenesis 6:539-543.
3. Hecht, S.S., N. Trushin, A. Castonguay, and A. Rivenson (1986) Comparative tumorigenicity and DNA methylation in F344 rats by 4-(methylnitrosamino)-1-(3-pyridyl)-1-butanone and N-nitrosodimethylamine. Cancer Res. 46:498-502.
4. Hoffmann, D.E., A. Rivenson, S. Amin, and S.S. Hecht (1984) Dose-response study of the carcinogenicity of tobacco-specific N-nitrosamines in F344 rats. J. Cancer Res. Clin. Oncol. 108:81-86.
5. International Agency for Research on Cancer (1985) Evaluation of the carcinogenic risk of chemicals to humans: Tobacco habits other than smoking; betel-quid and areca-nut chewing; and some related nitrosamines. IARC Monographs 37:209-224.
6. Morse, M.A., S.S. Hecht, and F.-L. Chung (1989) Effects of aromatic isothiocyanates on tumorigenicity, O^6-methylguanine formation, and metabolism of the tobacco-specific nitrosamine 4-(methylnitrosamino)-1-(3-pyridyl)-1-butanone in A/J mouse lung. Cancer Res. (in press).

7. Morse, M.A., C.-X. Wang, S.G. Amin, S.S. Hecht, and F.-L.
    Chung (1988) Effects of dietary sinigrin or indole-3-carbinol on O^6-
    methylguanine-DNA-transmethylase   activity   and   4-(methylnitros-
    amino)-1-(3-pyridyl)-1-butanone-induced DNA methylation and tumor-
    igenicity in F344 rats. Carcinogenesis 9:1891-1895.
8. Morse, M.A., C.-X. Wang, G.D. Stoner, S. Mandal, P.B. Conran,
    S.G. Amin, S.S. Hecht, and F.-L. Chung (1989) Inhibition of 4-
    (methylnitrosamino)-1-(3-pyridyl)-1-butanone-induced DNA   adduct
    formation and tumorigenicity in the lung of F344 rats by dietary
    phenethyl isothiocyanate. Cancer Res. 9:549-553.
9. Tookey, H.L., C.H. Van Etten, and M.E. Daxenbichler (1980) Glu-
    cosinolates. In Toxic Constituents of Plant Foodstuffs, 2nd ed.,
    I.E. Liener, ed. Academic Press, Inc., New York, pp. 103-142.
10. Wattenberg, L.W. (1987) Effects of benzyl isothiocyanate adminis-
    tered shortly before diethylnitrosamine or benzo[a]pyrene on pul-
    monary and forestomach neoplasia in A/J mice. Carcinogenesis
    8:1971-1973.

PROTEOLYTIC ACTIVATION OF UmuD and MucA PROTEINS

FOR SOS MUTAGENESIS

Toshikazu Shiba, Hiroshi Iwasaki, Atsuo Nakata,
and Hideo Shinagawa

Department of Experimental Chemotherapy
Research Institute for Microbial Diseases
Osaka University
3-1 Yamadaoka, Suita
Osaka, Japan 565

ABSTRACT

SOS mutagenesis in Escherichia coli requires the functions of the umuD,C genes, or their functional analogues mucA,B derived from a plasmid pKM101, and the recA gene. However, mere derepression of these SOS genes does not increase the ability of the cell to perform mutagenesis. Activation of RecA protein to a form (RecA*) that mediates cleavage of the LexA repressor is required for mutagenesis. We present evidence that UmuD and MucA are proteolytically processed by RecA* and that the processed products are the active forms involved in mutagenesis.

INTRODUCTION

Exposure of the Escherichia coli cell to agents that damage DNA or interfere with DNA replication results in the induction of a diverse set of physiological responses, collectively termed the SOS response, that include an increased capacity to form mutagenesis (8,14,15). It has been shown that the products of umuD, umuC, or their plasmid-derived analogues mucA, mucB, and recA are required for mutagenesis induced by radiation or chemical agents (SOS mutagenesis) (3,7,13,14). The expression of the SOS genes including umuD,C, mucA,B, and recA is repressed by the lexA gene product (14). However, derepression of these SOS genes, such as in the lexA(Def) cell, does not increase the mutagenic activity of the cell (1,4,7,16), indicating that the derepression of the SOS genes is not sufficient for SOS mutagenesis. Ennis et al. (4) found that mutagenic activity was high only with conditions of the cell that activate the proteolytic function of RecA. These findings suggest that for SOS mutagenesis, RecA* is needed to perform a function in addition to the inactivation of LexA to derepress the SOS genes.

In this work, we examined the possibility that UmuD and MucA are proteolytically processed by RecA* and that the processed products are the active forms for mutagenesis.

RESULTS AND DISCUSSION

RecA*-Dependent Cleavage of UmuD and MucA Proteins

To identify UmuD or MucA protein in the cell, we performed the immunoblot analysis using antisera against LacZ-UmuD or LacZ-MucA hybrid protein. We found that the 17-kDa intact UmuD protein was proteolytically cleaved to a 14-kDa protein in a wild-type strain only after the cells were exposed to UV or mitomycin C (12). In a lexA(Def) strain, which expresses the SOS-regulated genes constitutively, only the intact UmuD protein was observed without mitomycin C treatment, and both the intact UmuD protein and the processed product were detected after mitomycin C treatment. However, the processing of UmuD protein was not observed in a recA/lexA(Def) strain with and without mitomycin C treatment. Similarly to UmuD protein, the 18-kDa MucA protein was processed to a 15-kDa protein by RecA*. The proteolytic processing was observed only in the cells that were active in mutagenesis. These results suggest that the intact UmuD or MucA protein is converted to the 14- to 15-kDa protein by proteolytic cleavage, which is mediated by RecA*, and that the processing is required for SOS mutagenesis.

Cleavage of UmuD or MucA does not require functional UmuC or MucB, since the cleavage was observed in umuC mutants or in the absence of the mucB gene.

Proteolytic Cleavage of Mutant UmuDs and MucAs

Homology of UmuD and MucA proteins to LexA and λ repressors suggests that the cleavage site by RecA* is likely the Cys^{24}-Gly^{25} bond of UmuD and the Ala^{25}-Gly^{26} bond of MucA (11). We constructed by site-directed mutagenesis the plasmids which carry mutant umuD or mucA genes that encode UmuDs or MucAs with replacements in the amino acids flanking the putative cleavage site. We examined whether the mutant types of UmuD and MucA are cleavable by RecA* and whether these mutants are capable of inducing mutation.

In the case of the UmuD mutants (6), changing the wild-type sequence of Cys^{24}-Gly^{25} to Ala^{24}-Gly^{25} sequence, which is the putative cleavage site of MucA (11), did not affect the proteolytic processing of the mutant UmuD and the phenotype of UV mutagenesis; while changing it to Ala^{24}-Asp^{25} sequence, which is the sequence of mutant LexA3(Ind⁻) (9), or to Cys^{24}-Asp^{25}, which is derived from Cys^{24} of the wild-type UmuD and the Asp^{85} residue of the LexA3(Ind⁻) cleavage site, made the UmuDs resistant to cleavage by RecA* and defective in UV mutagenesis. Nohmi et al. also constructed similar umuD mutants and showed that these mutants had reduced ability of UV mutagenesis (10).

In the case of mucA mutants, changing the wild-type sequence of Ala^{25}-Gly^{26} to Cys^{25}-Gly^{26}, which is the sequence for the putative cleavage site of UmuD (11), did not affect the cleavage of MucA and UV mutagenesis. However, in the two mutants in which the cleavage sequence was changed to Ala^{25}-Asp^{26} (sequence of mutant LexA3[Ind⁻])

or changed to Cys^{25}-Asp^{26} (Cys^{24} of wild-type UmuD and Asp^{85} of the LexA3[Ind$^-$] cleavage site), the cleavage products were not detected, and these two mutants were defective in UV mutagenesis.

The umuD or the mucA mutations that encode the products cleavable by RecA* were proficient in UV mutagenesis, and those that encode the noncleavable products were deficient. These results suggest that UmuD and MucA are converted to the active form for mutagenesis by proteolytic processing mediated by RecA*.

However, one exception was observed. A mucA mutant in which Ala^{25}-Gly^{26} was changed to Thr^{25}-Gly^{26} (sequence of mutant λ cI$_{AT111}$[Ind$^-$]) (5) was as proficient as the wild-type gene in UV mutagenesis, but the cleavage of the MucA by RecA* was not detected. This mutant MucA protein with Thr at the cleavage site might take a conformation that is active for mutagenesis without proteolytic processing.

The Cleavage Products of UmuD and MucA Are the Active Form for SOS Mutagenesis

The possibility that the cleavage reaction itself is required for mutagenesis and that the cleavage is a way to inactivate the protein was not completely ruled out. We constructed plasmids which carried a mutant umuD or mucA gene that encoded the putative cleavage product lacking the amino-terminal peptide of the intact protein. Since the umuD$^-$ strain carrying either of the plasmids was fully proficient in UV mutagenesis, we concluded that the cleavage product of UmuD (6) or MucA is the active form for UV mutagenesis. Essentially the same conclusion for UmuD was drawn independently by Nohmi et al. (10).

Role of Activated Form of UmuD and MucA in SOS Mutagenesis

Since cleavage removes the N-terminal 24 and 25 residues from UmuD and MucA proteins, respectively, the activation process might involve unmasking the active sites of the proteins which were masked by presence of these peptides in the nascent proteins. Burckhardt et al. (2) have shown that UmuD protein autocatalytically cleaves itself at alkaline pH and that the cleavage product is identical to the RecA*-cleaved product. This work suggests that UmuD protein is a potential protease which is activated at alkaline pH or by interaction with RecA*. We propose that UmuD and MucA are activated by proteolytic cleavage to active proteases which in turn cleave a component of DNA replication machinery and reduce replication fidelity. The potential candidate of the target protein might be DnaQ protein (ε-subunit of the DNA polymerase III holoenzyme) that has proofreading activity. Activation of proteases by proteolytic processing is common among many mammalian proteases such as pepsin and trypsin.

REFERENCES

1. Blanco, M., G. Herrera, P. Collado, J. Rebollo, and L.M. Botella (1982) Influence of RecA protein on induced mutagenesis. Biochimie 64:633-636.
2. Burckhardt, S.E., R. Woodgate, R.H. Scheurmann, and H. Echols (1988) The UmuD mutagenesis protein of Escherichia coli: Overproduction, purification and cleavage by RecA. Proc. Natl. Acad. Sci., USA 85:1811-1815.

3. Elledge, S.J., and G.C. Walker (1983) Proteins required for ultra-
    violet light and chemical mutagenesis: Identification of the products
    of the umuC locus of Escherichia coli. J. Molec. Biol. 164:175-192.
4. Ennis, D.G., B. Fisher, S. Edmiston, and D.W. Mount (1985) Dual
    role for Escherichia coli RecA protein in SOS mutagenesis. Proc.
    Natl. Acad. Sci., USA 82:3325-3329.
5. Gimble, F.S., and R.T. Sauer (1986) λ Repressor inactivation:
    Properties of purified ind⁻ proteins in the autodigestion and RecA-
    mediated cleavage reactions. J. Molec. Biol. 192:39-47.
6. Iwasaki, H., T. Shiba, A. Nakata, and H. Shinagawa (1988) Activa-
    tion of UmuD protein for SOS mutagenesis by proteolytic processing
    mediated by RecA*. In Mechanisms and Consequences of DNA Dam-
    age Processing, E.C. Friedberg and P.C. Hanawalt, eds. Alan R.
    Liss, Inc., New York, pp. 461-469.
7. Kato, T., and Y. Shinoura (1977) Isolation and characterization of
    mutants of Escherichia coli deficient in induction of mutations by
    ultraviolet light. Molec. Gen. Genet. 156:121-131.
8. Little, J.W., and D.W. Mount (1982) The SOS regulatory system of
    Escherichia coli. Cell 29:11-22.
9. Markham, B.E., J.W. Little, and D.W. Mount (1981) Nucleotide se-
    quence of the lexA gene of Escherichia coli K-12. Nucleic Acids
    Res. 9:4149-4161.
10. Nohmi, T., J.R. Battista, L.A. Dodson, and G.C. Walker (1988)
    RecA-mediated cleavage activates UmuD for mutagenesis: Mechanistic
    relationship between transcriptional derepression and posttransla-
    tional activation. Proc. Natl. Acad. Sci., USA 85:1816-1820.
11. Perry, K.L., S.J. Elledge, B.B. Mitchell, L. Marsh, and G.C.
    Walker (1985) umuDC and mucAB operons whose products are re-
    quired for UV light- and chemical-induced mutagenesis; UmuD, MucA
    and LexA proteins share homology. Proc. Natl. Acad. Sci., USA
    82:4331-4335.
12. Shinagawa, H., H. Iwasaki, T. Kato, and A. Nakata (1988) RecA
    protein-dependent cleavage of UmuD protein and SOS mutagenesis.
    Proc. Natl. Acad. Sci., USA 85:1806-1810.
13. Shinagawa, H., T. Kato, T. Ise, K. Makino, and A. Nakata (1983)
    Cloning and characterization of the umu operon responsible for
    inducible mutagenesis In Escherichia coli. Gene 23:167-174.
14. Walker, G.C. (1984) Mutagenesis and inducible responses to deoxy-
    ribonucleic acid damage in Escherichia coli. Microbiol. Rev.
    48:60-93.
15. Witkin, E.M. (1976) Ultraviolet mutagenesis and inducible DNA repair
    in Escherichia coli. Bacteriol. Rev. 40:869-907.
16. Witkin, E.M., and T. Kogoma (1984) Involvement of the activated
    form of RecA protein in SOS mutagenesis and stable DNA replication
    in Escherichia coli. Proc. Natl. Acad. Sci., USA 81:7539-7543.

ANALYSIS OF BLEOMYCIN-RESISTANT DNA SYNTHESIS IN CELLS

FROM AN INHERITED HUMAN DISORDER, ATAXIA TELANGIECTASIA

Asao Noda and Toshiharu Matsumura

Meiji Institute of Health Science
540 Naruda, Odawara 250, Japan

INTRODUCTION

For a better understanding of the effects of environmental and genetic factors on the molecular mechanisms in human cancer development, the study of cancer prone diseases has provided some important knowledge. An autosomal recessive hereditary disorder, ataxia telangiectasia (AT), shows a high incidence of cancer(s), especially of the lymphatic system (1). It is particularly important to note that heterozygotes of AT, which are estimated to be as frequent as 0.68 to 7.7% of the human population, have an increased risk of cancer (9).

Normal cells show a transient inhibition of DNA synthesis after exposure to a certain dose of X-ray irradiation, or of bleomycin, but can survive afterward. On the other hand, cells from affected individuals (AT homozygotes) show little inhibition of DNA synthesis after the same exposure, but then die out afterward. This transient inhibition of DNA synthesis in normal cells is explained by the inhibition of DNA replicon initiation (2,8). We have been speculating that when normal cells are X-ray irradiated or treated with bleomycin, specific gene expressions are induced, and then resulting RNA transcripts, proteins, or their metabolites act as inhibitors of DNA synthesis (Fig. 1). AT cells may lack this process or its ultimate product. In this chapter we report the results of the investigation as to whether or not the putative inhibitor is of protein nature, and introduce our current approaches to the molecular cloning of the gene responsible for the protein.

THE EFFECTS OF CYCLOHEXIMIDE ON DNA SYNTHESIS

If the putative inhibitor is of protein nature, it is expected that the protein synthesis inhibitor, cycloheximide, inhibits the synthesis of the protein in normal cells, and that the cycloheximide-treated normal cells show bleomycin-resistant DNA synthesis like AT cells. When normal cells were treated with cycloheximide 30 min before bleomycin treatment, they, in fact, showed bleomycin-resistant DNA synthesis, apparently mimicking AT cells (Fig. 2). However, the treatment with cycloheximide alone par-

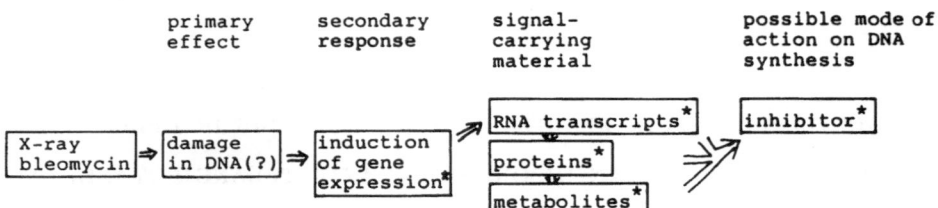

Fig. 1.   Hypothetical mechanisms for the damage-induced inhibition of
           DNA synthesis.   Items marked with * can possibly be modified
           in AT cells.

tially inhibited DNA synthesis both in normal cells and in AT cells.   A
sucrose density gradient study showed that cycloheximide inhibits DNA
replicon initiation (7).   Since bleomycin treatment is known to cause inhibition of DNA replicon initiation in normal cells (2), the results indicate
that cycloheximide acts on the same target in spite of the fact that the
actions of these two agents on DNA synthesis are apparently different.

Although these results did not provide evidence that a hypothetical
factor or process in normal cells that is induced after treatment with
bleomycin is of protein nature, they suggest that cycloheximide inhibits
the synthesis of a protein(s) with a short half-life, which is indispensable
for the cells to enter S phase (as reviewed in Ref. 7).   Furthermore, the
treatment with cycloheximide did not reverse the bleomycin-inhibition of
DNA synthesis in normal cells.

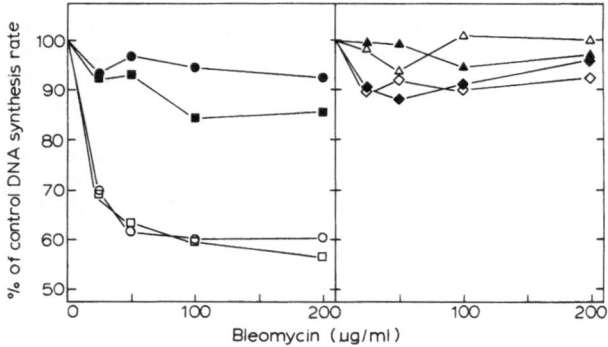

Fig. 2.   Relative rates of DNA synthesis in normal cells (IMR90, N21OS)
           and AT cells (AT1OS, AT4BI) as functions of the bleomycin
           concentration.   Open symbols: Cells were treated with bleo-
           mycin for 30 min, incubated in fresh medium without bleomycin
           for 30 min, and then pulse-labeled by ^{3}H-thymidine for 20 min
           for determination of the rate of DNA synthesis.   Closed sym-
           bols: Cycloheximide was added (5 µg/ml) 30 min before bleo-
           mycin treatment, and the cells were then incubated in the pres-
           ence of cycloheximide until pulse-labeling.   A value is shown in
           the figure as the percentage of the rate of DNA synthesis in
           cells without bleomycin as a control.   IMR90: (○, ●); N21OS
           (□, ■); AT1OS (◇, ◆); AT4BI (△; ▲).

DIFFERENTIAL HYBRIDIZATION SCREENING OF THE MESSAGES EXPRESSED IN NORMAL CELLS AND AT CELLS AFTER BLEOMYCIN TREATMENT

In due course, we have tried to detect specific transcripts that are expressed in normal cells, but not in AT cells, after bleomycin treatment. The cDNA library was generated from bleomycin-treated normal (IMR90) (6) cells and propagated in Escherichia coli. These E. coli colonies were grown on nitrocellulose filters on top of agar plates, and replica filters were made. These filters were then hybridized with a mixture of ^{32}P-labeled single-stranded cDNA probes that were prepared from total poly(A)$^+$ RNAs of either bleomycin-treated normal cells (IMR90) or bleomycin-treated AT cells (AT1OS) (Fig. 3).

If bleomycin-treated AT1OS cells (3) show altered gene expressions from that of·bleomycin-treated IMR90 cells, we could detect them by comparing these filters after autoradiography. Arrows in Fig. 3 indicate the prospective colonies in which specific transcript is not expressed (A) or overexpressed (B) in bleomycin-treated AT cells. Figure 4 shows the autoradiogram of this hybridization screening in which arrows indicate candidate colonies. Each of them represents a specific transcript which is expressed at a reduced level in bleomycin-treated AT1OS cells.

After 5 rounds of screening of about 3 x 10^5 E. coli clones, two colonies which gave positive hybridization signals when probed with bleomycin-treated IMR90 cell cDNA, but not when probed with bleomycin- treated AT1OS cell cDNA, were obtained. In addition, four colonies which

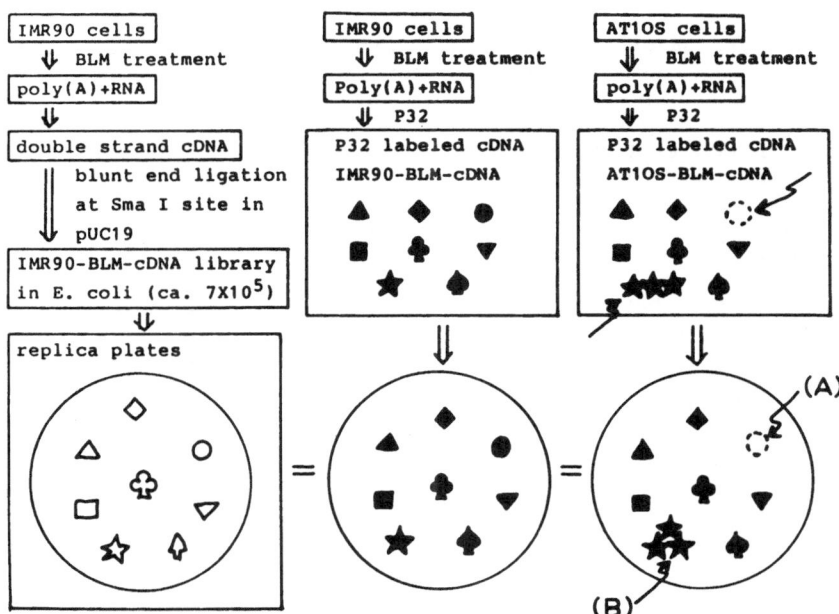

Fig. 3.   Methods of hybridization screening for those RNA transcripts which are expressed differently in IMR90 and AT1OS cells after bleomycin treatment.

IMR90 probe   AT1OS probe

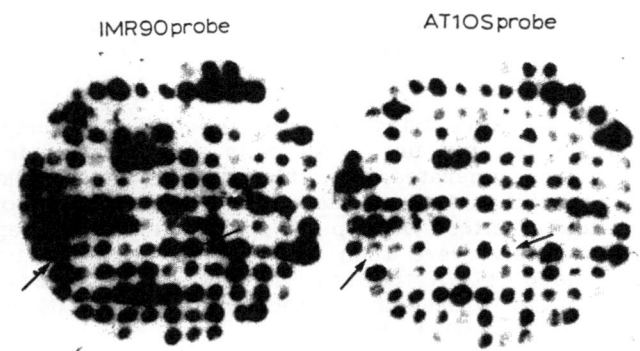

Fig. 4.   Candidate colonies of differential hybridization screening (see
           text for explanation).

were strongly positive when probed with bleomycin-treated AT1OS cell
cDNA, but weakly positive when probed with bleomycin-treated IMR90 cell
cDNA, were also obtained.   A northern blot hybridization study showed
that the RNAs hybridized to the former two cDNA clones were expressed
abundantly in bleomycin-treated normal cells but little in bleomycin-
treated AT cells.   The RNAs hybridized to the latter four cDNA clones
were expressed strongly in AT cells and weakly in normal cells without
regard to bleomycin treatment.

DISCUSSION AND FUTURE ASPECTS

There are many environmental and cellular substances that are
known to protect somatic mutations and neoplastic transformation.   When
cycling cells are exposed to mutagens and/or carcinogens, they normally
show a specific response, i.e., a transient inhibition of DNA synthesis
and delay in cell cycle progression in the G1 and G2 phases.   Cells may
have several protective mechanisms to ensure genetic fidelity in mitotic
division (10).   It is speculated, therefore, that AT cells may have a defect in one of these protective mechanisms which is recognized as a structural abnormality in DNA, and transiently arrests to enter S phase.

Abnormalities have been noted in certain protein synthesis in AT
cells, such as γ-endonuclease, primer activating enzyme (1), fibronectin
(5), and actin-related protein (4), although they were not fully characterized.   We believe that approaches with this particular cancer-prone
disorder, AT, provide some insight into antimutagenic mechanisms,
especially the control of DNA synthesis in association with potentially
mutagenic events in cellular DNA.

ACKNOWLEDGEMENTS

We are grateful to Dr. M. Ikenaga (Kyoto University), Dr. Y.
Fujiwara (Kobe University), the Coriell Institute for Medical Research,
and the Japanese Cancer Research Resources Bank for providing the cell
strains used in this study.

REFERENCES

1. Bridges, B.A., and D.G. Harnden, eds. (1982) Ataxia-telangiec-tasia, John Wiley and Sons, 402 pp.
2. Cramer, P., and R.B. Painter (1981) Bleomycin-resistant DNA synthesis in ataxia telangiectasia cells. Nature 291:671-672.
3. Ikenaga, M., M. Midorikawa, J. Abe, and T. Mimaki (1983) The sensitivities to radiations and radiomimetic chemicals of cells from patients with ataxia telangiectasia. Jpn. J. Human Genet. 28:1-10.
4. McKinnon, P.J., and L.A. Burgoyne (1985) Evidence for the existence of an actin-derived protein in ataxia telangiectasia lympho-blastoid cell lines. Expt. Cell Res. 158:413-422.
5. Murnane, J.P., and R.B. Painter (1983) Altered protein synthesis in ataxia telangiectasia fibroblasts. Biochemistry 22:1217-1222.
6. Nichols, W.W., D.G. Murphy, V.J. Cristofalo, L.H. Toji, A.E. Greene, and S.A. Dwight (1977) Characterization of a new human diploid cell strain, IMR-90. Science 196:60-63.
7. Noda, A. (1988) Replicon initiation in normal human cells and ataxia telangiectasia cells: Its differential inhibition by cycloheximide and bleomycin. Cell Biol. Int. Rep. 12:943-950.
8. Painter, R.B., and B.R. Young (1980) Radiosensitivity in ataxia telangiectasia: A new explanation. Proc. Natl. Acad. Sci., USA 77:7315-7317.
9. Swift, M., P.J. Reitnauer, D. Morrell, and C.L. Chase (1987) Breast and other cancers in families with ataxia-telangiectasia. N. Engl. J. Med. 316:1289-1294.
10. Weinert, T.A., and L.H. Hartwell (1988) The RAD9 gene controls the cell cycle response to DNA damage in Saccharomyces cerevisiae. Science 241:317-322.

TUMOR DOSE-RESPONSE STUDIES WITH AFLATOXIN B_1

AND THE AMBIVALENT MODULATOR INDOLE-3-CARBINOL:

INHIBITORY VERSUS PROMOTIONAL POTENCY

R.H. Dashwood, A.T. Fong, J.D. Hendricks, and G.S. Bailey

Department of Food Science and Technology
Oregon State University
Corvallis, Oregon 97331

ABSTRACT

Indole-3-carbinol (I3C), a natural compound from cruciferous vegetables, inhibits aflatoxin B_1 (AFB_1) carcinogenesis in trout when administered prior to and during carcinogen exposure, but also promotes it in the same species when given after AFB_1 initiation. To provide quantitative potency information for these opposing activities, detailed tumor dose-response studies were performed with AFB_1 (10-400 ppb) and I3C (0-4,000 ppm). In a plot of (logit) percent tumor response vs log AFB_1 exposure, the results generated a series of parallel AFB_1 dose-response curves. Increasing I3C doses displaced these curves, respectively, toward higher and lower AFB_1 doses in the inhibition and promotion studies. Similar potencies were observed over the dose range 0-1,500 ppm I3C; the 50% promotion and inhibition (P_{50} and I_{50}) values were 1,000 vs 1,400 ppm I3C, respectively. Differences in the protocols used in the two studies suggest that the inhibitory activity of I3C is more likely to supersede promotion under human exposure conditions.

INTRODUCTION

The human diet contains many compounds which inhibit the carcinogenic process (4,6,7). However, the overall protective effect which such compounds may exert, and their potential for deliberate chemoprevention, remains difficult to ascertain because unreasonably high doses often are employed in order to elicit an inhibitory effect. Moreover, in some experimental systems, dietary "anticarcinogens" may exhibit promotional activity. For example, indole-3-carbinol (I3C) is a naturally-occurring compound found in cruciferous vegetables (e.g., cabbage and broccoli) which inhibits aflatoxin B_1 (AFB_1) carcinogenesis in trout when administered prior to and during carcinogen exposure (3,5), but which also promotes it in the same species when given after AFB_1 initiation (1). Since human exposure to such dietary agents may be virtually unavoidable, it is important to provide detailed quantitative potency information for these opposing activities (inhibition vs promotion) in order to assess risk vs benefit. Thus, we have undertaken the following tumor studies:

(a) Inhibitory potency. In vivo dose-response analyses were made in trout pre-exposed to 0, 1,000, 2,000, 3,000, or 4,000 ppm I3C in the diet for 4 wk and subsequently treated for 2 wk with the same level of I3C together with dietary AFB_1 in the dose-range 10-320 ppb ($\sim$10,000 animals, Tab. 1). Details of this study have appeared in a recent paper (3).

(b) Promotional potency. Trout (4,800 animals, Tab. 1) were pre-treated with AFB_1 (12.5-400 ppb, 30 min bath exposure) and subsequently fed 0, 750, 1,500, or 2,000 ppm I3C in the diet for 24 wk.

Tab. 1. Tumor dose-response studies with aflatoxin B_1 and indole-3-carbinol. Reprinted from Ref. 3, with permission.

INHIBITORY POTENCY STUDIES			PROMOTIONAL POTENCY STUDIES		
Pre-treatment (ppm I3C)	Post-treatment (ppb AFB1)	*Percent Tumor response	Pre-treatment (ppb AFB1)	Post-treatment (ppm I3C)	*Percent Tumor response
0	10	10.7(3.7)	50	0	40.4(7.6)
0	20	32.2(3.3)	100	0	44.7(1.5)
0	40	55.6(4.2)	200	0	58.7(7.2)
0	80	68.4(7.5)	400	0	77.1(2.2)
1000	10	14.7(6.1)	50	750	45.1(5.7)
1000	20	21.1(4.5)	100	750	53.2(1.9)
1000	40	35.2(6.5)	200	750	65.5(3.6)
1000	80	57.6(6.7)	400	750	84.2(3.5)
2000	20	4.9(1.0)	25	1500	64.7(4.4)
2000	40	12.4(1.7)	50	1500	69.9(3.4)
2000	80	22.5(7.7)	100	1500	75.8(7.2)
2000	160	42.6(7.9)	200	1500	88.6(1.2)
3000	20	1.3(1.2)	12.5	2000	55.2(5.4)
3000	40	2.0(0.8)	25	2000	66.7(9.1)
3000	80	3.1(1.9)	50	2000	67.9(2.6)
3000	160	5.6(1.2)	100	2000	71.5(7.8)
4000	40	0.7(0.5)			
4000	80	1.0(0.8)			
4000	160	5.9(2.0)			
4000	320	7.0(2.3)			

*Tumor data are means(+/- SD) from triplicate groups of ~100 animals at each AFB1-I3C level. Plotted as logit percent tumor response vs. log AFB1 concentration, a series of parallel curves was produced in both the inhibition and promotion study. I3C-mediated shifts in these curves were calculated in terms of 'percent inhibition' and 'percent promotion', as shown in Figs. 1a and 1b, respectively.

RESULTS

Inhibitory Potency

Tumor dose-response data from the inhibition studies are presented in Tab. 1. Plotted as percent tumor response (logit scale) vs log ppb AFB_1 in diet, a series of 5 parallel straight-line curves was evident (see Fig. 2 in Ref. 3). Each increase in I3C dose further offset the position of the logit curve toward higher AFB_1 dose. From these results, the percent inhibition of AFB_1 tumorigenicity was calculated and plotted as a function of I3C dose (Fig. 1a). Figure 1a serves to highlight the fact that inhibition occurred in a dose-dependent manner, and while complete inhibition was not achieved over the I3C dose-range tested, it was suppressed by 95% at the top dose of 4,000 ppm. Interpolating from Fig. 1a, the dose producing 50% inhibition of AFB_1 tumorigenicity (I_{50}) was 1,400 ppm I3C.

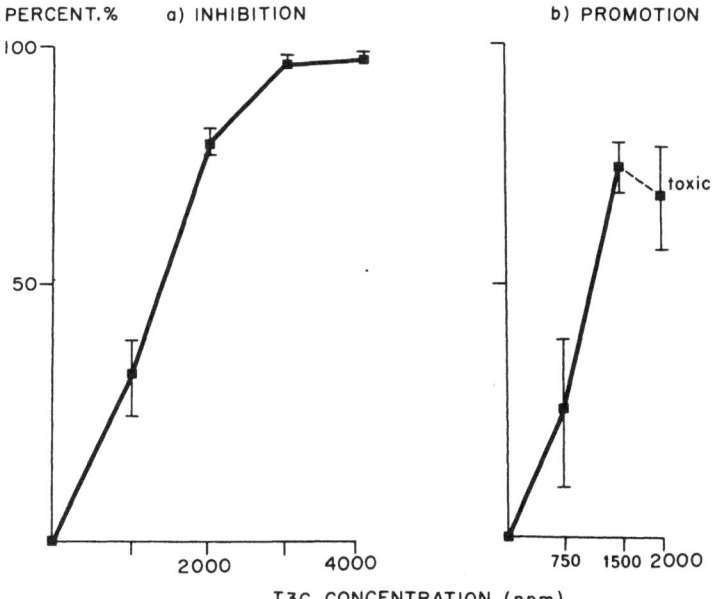

Fig. 1. Percent (a) inhibition and (b) promotion of AFB_1 tumor incidence versus I3C concentration in diet. Data points represent mean ± S.D. and were calculated from tumor dose-response curves using the formulae:

Percent inhibition = 100 (1 − TDx_0 / TDx_i)

Percent promotion = 100 (1 − TDx_i / TDx_0)

where "x" is the tumor incidence being compared (e.g., 50%), TDx_0 is the dose of AFB_1 needed to give that tumor incidence in the 0 ppm (control) group, and TDx_i is the dose of AFB_1 required to produce that tumor incidence in the group receiving "i" ppm I3C. Reprinted from Ref. 3, with permission.

Promotional Potency

Data from the promotion studies are presented in Tab. 1. When plotted in an analogous fashion to the inhibition data (logit tumor response vs log AFB_1 dose), a series of parallel curves also was produced (not shown). The results for groups receiving I3C were displaced horizontally toward lower AFB_1 dose. As a result of toxicity observed at the highest I3C dose, 2,000 ppm I3C failed to elicit the greatest promotional response in these studies (Fig. 1b). From Fig. 1b, the concentration of I3C producing 50% promotion (the P_{50} value) was ∿1,000 ppm I3C, and the maximum promotional response was ∿75% at the I3C dose of 1,500 ppm.

DISCUSSION

Inhibitory Versus Promotional Potency

The data presented in Fig. 1 indicate that, using the experimental protocols described in these studies, the inhibitory and promotional potencies of I3C were similar at doses <1,500 ppm. This is reflected in the shape of the dose-response curves in the dose range 0-1,500 ppm I3C, and in the 50% inhibition versus promotion (I_{50} vs P_{50}) values of 1,400 ppm and 1,000 ppm, respectively. At I3C doses >1,500 ppm, some discrepancy was observed due to toxicity (Fig. 1b), and this probably reflects the fact that in the inhibition studies trout received a total of 6 wk exposure to I3C compared with 24 wk postinitiation treatment in the promotion study.

These observations highlight an important point, namely, that the inhibitory and promotional potencies of I3C are highly dependent upon the exposure protocols employed. In this context the data in Fig. 1 may be considered misleading, since they imply approximate equivalence of the promotion and inhibition activities--that is, no overall benefit from human exposure to I3C in terms of chemoprevention. The promotion studies reported here, by design, were aimed at evaluating a possible "worst-case scenario" for I3C, i.e., continuous postinitiation exposure to relatively high dietary concentrations.

To provide data on the effects of various postinitiation treatments, including those which attempt to mimic likely human exposures, detailed tumor studies have been established in trout pretreated with AFB_1 and subsequently given I3C either immediately or at various times after terminating AFB_1 exposure (1, 3, 6, 9 mo) and fed either continuously or for various lengths of time (3, 6, 9 mo, or alternating months or weeks, or Mondays and Thursdays of each week only). Preliminary results indicate that continuous I3C exposure is not a prerequisite for enhancement, although exposure is required over an extended period of time (e.g., 2 days a week for several months; ms. in prep.). Since one week of I3C pretreatment is sufficient for anticarcinogenic activity in trout (unpubl. observ.), these data collectively suggest that the inhibitory activity of I3C probably supersedes promotion under human exposure conditions.

Low-Dose Thresholds

Previous studies of I3C inhibition toward AFB_1-DNA binding (2,3) produced a potency curve that was linear through the origin at I3C doses <1,500 ppm, suggesting that no significant threshold exists for I3C protection. Each data point in Fig. 1. was determined with precision using large numbers of animals (triplicate groups of approximately 100 animals at 4 AFB_1 dose-levels), but only one data point is available in the range of possible human exposure (1,000 ppm in Fig. 1a and 750 ppm in Fig. 1b). A large tumor study currently is in progress aimed at providing detailed information in the range 0-1,000 ppm I3C. This should yield important insight into the question of low-dose thresholds and the implied message from the DNA-binding studies (2,3) that even low dietary of I3C, such as those found in the human diet, may be protective.

ACKNOWLEDGEMENTS

Supported by PHS grants CA34742, EH03850, and ES00210.

REFERENCES

1. Bailey, G.S., J.D. Hendricks, D.W. Shelton, J.E. Nixon, and N.E. Pawlowski (1987) Enhancement of carcinogenesis by the natural anti-carcinogen indole-3-carbinol. J. Natl. Cancer Inst. 78:931-934.
2. Dashwood, R.H., D.N. Arbogast, A.T. Fong, J.D. Hendricks, and G.S. Bailey (1988) Mechanisms of anti-carcinogenesis by indole-3-carbinol: Detailed in vivo DNA binding dose-response studies after dietary administration with aflatoxin B_1. Carcinogenesis 9:427-432.
3. Dashwood, R.H., D.N. Arbogast, A.T. Fong, C. Pereira, J.D. Hendricks, and G.S. Bailey (1989) Quantitative interrelationships between aflatoxin B_1 carcinogen dose, indole-3-carbinol anti-carcinogen dose, target organ DNA adduction and final tumor response. Carcinogenesis 10:175-181.
4. Fiala, E.S., B.S. Reddy, and J.H. Weisburger (1985) Naturally occurring anticarcinogenic substances in foodstuffs. Ann. Rev. Nutr. 5:295-321.
5. Nixon, J.E., J.D. Hendricks, N.E. Pawlowski, C. Pereira, R.O. Sinnhuber, and G.S. Bailey (1984) Inhibition of aflatoxin B_1 carcinogenesis in rainbow trout by flavone and indole compounds. Carcinogenesis 5:615-619.
6. Shankel, D.M., P.E. Hartman, T. Kada, and A. Hollaender (1987) Synopsis of the First International Conference on Antimutagenesis and Anticarcinogenesis: Mechanisms. Environ. Mutagen. 9:87-103.
7. Wattenberg, L.W. (1983) Inhibition of neoplasia by minor dietary constituents. Cancer Res. (Suppl.) 43:2448s-2453s.

ANTIMUTAGENESIS IN YEAST BY SODIUM CHLORIDE, POTASSIUM CHLORIDE, AND SODIUM SACCHARIN

Kenneth R. Parker[1] and R.C. von Borstel[1,2]

[1]Department of Genetics
University of Alberta
Edmonton, Alberta, Canada T6G 2E9

[2]Basel Institute for Immunology
Grenzacherstrasse 487, CH-4058 Basel, Switzerland

ABSTRACT

Aqueous salt solutions containing NaCl, KCl, $MgCl_2$, Na_2SO_4, $CaCl_2$, NH_4Cl, or sodium saccharin are mutagenic in yeast when logarithmic growth of cells is interrupted by exposure to a 0.5-2.0 M salt solution. Stationary-phase cells are not mutated by this treatment. When placed in an enriched medium with the salt, the stationary-phase cells grow after a prolonged lag period. The compounds tested (NaCl, KCl, and sodium saccharin), under conditions in which growth in medium can take place, exhibit an antimutagenic response as measured by the compartmentalization test.

The antimutagenic action of salt solutions in yeast is concentration-dependent. Unlike the mutagenic action of these compounds, which approximates an osmolality-dependent response, the antimutagenic action seems to be correlated with toxicity as measured by growth rate reduction at increasing concentrations of the compounds. For example, sodium saccharin and NaCl exhibit almost identical osmolalities; however, 0.3 M sodium saccharin reduces the growth rate much more than does 0.3 M NaCl. At these same molar concentrations, the spontaneous mutation rate for histidine prototrophy is, for the control, 6.2×10^{-8} mutations/cell/-generation, 3.5×10^{-8} with 0.3 M NaCl, and 1.7×10^{-8} with 0.3 M sodium saccharin.

INTRODUCTION

Ionic salts are mutagenic (7) and clastogenic (1,2,6) to eukaryotic cells when the cells are exposed at sufficiently high concentrations. At these concentrations, a large osmotic gradient across the cell membrane results in loss of water from the cytosol, with possible disruption of membranes, which permits entry of extracellular solutes. It is not obvious why osmotic shock is mutagenic or clastogenic, and several hypotheses have been suggested, such as effects of excess ions on chromosomal proteins (1), effects on enzymes involved in DNA replication and repair (5), effects on microtubules (2), effects on topoisomerases (3), and disruption of lysosomes that might harbor nucleolytic enzymes (9). Moreover, DNA supercoiling under osmotic stress appears to regulate certain genes (4); some of these genes may be implicated in mutational lesions or their repair.

Yeast cells exhibit mutagenic and lethal effects at somewhat higher salt concentrations (as low as 44 mg/ml NaCl) than those reported to be clastogenic for mammalian cells (8-16 mg/ml NaCl). A further decrease in salt concentration (to 30-35 mg/ml NaCl) depresses the mutation frequency in yeast below the normal spontaneous level measured in medium that had no salt added (8). In order to confirm accurately the antimutagenic responses noted at exposures below 40 mg/ml of NaCl, we used the multicompartmented fluctuation test (11) to measure mutation rates in a cell population undergoing growth in a medium containing salts of varying concentrations.

MATERIALS AND METHODS

Strain

All experiments reported here were carried out with the haploid strain XV185-14C of the yeast Saccharomyces cerevisiae. Its genotype is as follows:

$$\underline{MATa}\ \underline{ade2\text{-}1}\ \underline{arg4\text{-}17}\ \underline{his1\text{-}7}\ \underline{hom3\text{-}10}\ \underline{lys1\text{-}1}\ \underline{trp5\text{-}48}\quad,$$

The alleles ade2-1, arg4-17, lys1-1, and trp5-48 are mutants with chain-terminating codons of the UAA variety; because ade2-1, arg4-17, and lys1-1 are specifically suppressible by mutants of tyrosine-tRNA, revertants arise as a consequence of three types of transversions. The allele his1-7 is a missense mutation revertible by intragenic internal missense suppression attributable to base substitutions which can be of either transitions or transversions. The allele hom3-10 is a +1 frameshift mutation revertible by a -1 frameshift, or by a +1 frameshift in the anticodon of a suppressing tRNA.

Measurement of the Mutation Rates

A multicompartmented fluctuation test (11) is used to measure the antimutagenic effects of salt on growing cells. The growth medium contains all of the auxotrophic requirements for growth, but the extent of growth is controlled by a limiting concentration of histidine (in this case, either 0.20 or 0.40 μg/ml of histidine). Stationary-phase cells are inoculated into the medium at approximately 3,000 cells/ml. One-ml aliquots are dispensed into 400 to 1,000 wells in 24-well tissue culture plates. The plates are then sealed in plastic bags and incubated at 26°C for at least 10 da. At that time, 20 wells that do not contain histidine prototrophs are sampled to determine the average number of cells per well. Calculation of mutation rates is based on the zeroth component of a Poisson distribution.

Tab. 1. Reversion rates of strain XV185-14C of Saccharomyces cerevisiae in growth medium with 0.2 M or 0.3 M salt (26°C).

	Histidine (μg/ml)	Generation Time (hours)	Saturation Density cells/ml (X10^{-6})	his1-7 → HIS$^+$ N[1]	N_0[2]	Reversion Rate (X 10^8)
Control	0.2	2.5	1.58	502	408	6.5
	0.4	2.1	3.68	502	317	6.2
NaCl						
0.2M	0.2	3.2	1.54	480	419	4.3
0.3M	0.2	3.1	1.31	500	434	3.5
NaSaccharin						
0.2M	0.2	9.0	0.79	467	427	5.4
0.3M	0.4	7.6	1.48	517	489	1.7
KCl						
0.3M	0.2	3.0	1.66	528	432	4.5

[1] N is the total number of compartments with growing cells.
[2] N_0 is the number of compartments without mutant clones.

RESULTS AND DISCUSSION

Antimutagenicity

In every case, the mutation rate for cells grown in salt-supplemented medium is less than the control spontaneous reversion rates without added solutes (Tab. 1). As the molarity of the salt is increased, the mutation rate is depressed even further. In a further experiment in which the molarity of the NaCl was increased to 1.0, the mutation rate declined to 1.2×10^{-8} histidine prototrophs/cell/generation (data not shown).

Salt-induced mutagenicity by osmotic shock to logarithmic-phase cells exhibits a pattern of mutation induction correlated quite closely with the ion concentration of the salt (7). Interestingly enough, 1 M NaCl is mutagenic for logarithmic-phase cells under conditions of osmotic shock but antimutagenic when cells are grown in the same medium. Somehow the stationary-phase cells become "conditioned" to the higher molarities of salt (cf. Ref. 4); cell replication ensues and mutations take place at a reduced rate.

One possible explanation for the decline in mutation rate in the presence of an antimutagen is that error-free repair operates more efficiently under conditions of mitotic inhibition (10). For example, rad9 mutants fail to arrest growth in the G2 portion of the cell cycle in response to radiation damage; consequently, the repair is less efficient (12). Nevertheless, it can be seen in Tab. 1 that there is no correlation between the reduction of the generation time and the depression in reversion rate.

The response of sodium saccharin is of special interest because it has a strong effect on generation time at concentrations of 0.2 M and 0.3 M. Whereas at the lower molarity the reversion rate is near that of the control, at the higher molarity the decline in reversion rate is marked. Therefore, different salts at the same molarities are antimutagenic to differing degrees. In support of this observation, we have measured the reversion rate for histidine prototrophy with KCl at 1.0 M. The rate was found to be 2.3×10^{-8} revertants/cell/generation. Thus, at molarities of both 0.1 M and 0.2 M, NaCl is slightly more antimutagenic than KCl.

The physical nature of the anion or cation therefore affects both the generation time and the reduction in mutation rate, and these are not necessarily correlated. It may depend on differing biological uptake mechanisms or on differing sensitivities of intracellular targets.

ACKNOWLEDGEMENTS

The authors are grateful to Nicole Schoepflin and Cynthia Baker for typing and editing the manuscript. The research was supported by a strategic grant from the Natural Sciences and Engineering Research Council of Canada. The Basel Institute for Immunology was founded and is supported by F. Hoffmann-La Roche, Ltd., Basel, Switzerland.

REFERENCES

1. Ashby, J., and M. Ishidate, Jr. (1986) Clastogenicity in vitro of the Na, K, Ca, and Mg salts of saccharin and of magnesium chloride: Consideration of significance. Mutat. Res. 163:63-73.
2. Galloway, S.M., D.A. Deasy, C.L. Bean, A.R. Kraynak, M.J. Armstrong, and M.O. Bradley (1987) Effects of high osmotic strength on chromosome aberrations, sister-chromatid exchanges and DNA strand breaks, and the relation to toxicity. Mutat. Res. 189:15-25.
3. Gaulden, M.E. (1987) Hypothesis: Some mutagens directly alter specific chromosomal proteins (DNA topoisomerase II and peripheral proteins) to produce chromosome stickiness, which causes chromosome aberrations. Mutagenesis 2:357-365.
4. Higgins, C.F., C.J. Dorman, D.A. Stirling, L. Waddell, I.R. Booth, G. May, and E. Bremer (1988) A physiological role for DNA supercoiling in the osmotic regulation of gene expression in Salmonella typhimurium and Escherichia coli. Cell 52:569-584.
5. Iliakis, G., P.E. Bryant, and F.Q.H. Ngo (1985) Independent forms of potentially lethal damage fixed in plateau-phase Chinese hamster cells by post-irradiation treatment in hypertonic salt solution or araA. Radiat. Res. 104:329-345.
6. Ishidate, Jr., M., T. Sofuni, K. Yoshikawa, M. Hayashi, T. Nohmi, M. Sawada, and A. Matsuoka (1984) Primary mutagenicity screening of food additives currently used in Japan. Food Chem. Toxicol. 22:623-636.
7. Parker, K.R., and R.C. von Borstel (1987) Base-substitution and frameshift mutagenesis by sodium chloride and potassium chloride in Saccharomyces cerevisiae. Mutat. Res. 189:11-14.
8. Parker, K.R., and R.C. von Borstel (1988) Salt-induced mutagenic and antimutagenic effects in yeast. Abstracts of the Nineteenth Annual Meeting of the Environmental Mutagen Society, Charleston, South Carolina, March 27-31, 1988. Environ. Molec. Mutag. 11(Suppl. 11):81.
9. Rowe, A.W., and L.L. Lenny (1980) Cryopreservation of granulocytes for transfusion: Studies on human granulocyte isolation, the effect of glycerol on lysosomes, kinetics of glycerol uptake and cryopreservation with dimethyl sulfoxide and glycerol. Cryobiology 17:198-212.
10. Von Borstel, R.C. (1986) The relation of activation and inactivation to antimutagenic processes. In Antimutagenesis and Anticarcinogenesis Mechanisms, D.M. Shankel, P.E. Hartman, T. Kada, and A. Hollaender, eds. Plenum Press, New York, pp. 39-43.
11. Von Borstel, R.C., K.T. Cain, and C.M. Steinberg (1971) Inheritance of spontaneous mutability in yeast. Genetics 69:17-27.
12. Weinert, T.A., and L.H. Hartwell (1988) The RAD9 gene controls the cell cycle response to DNA damage in Saccharomyces cerevisiae. Science 241:317-322.

THE USE OF DROSOPHILA AS AN IN VIVO SYSTEM

TO STUDY MODIFIERS OF CHEMICAL MUTAGENESIS

C. Ramel, H. Cederberg, J. Magnusson, and L. Romert

Department of Genetic and Cellular Toxicology
Wallenberg Laboratory, University of Stockholm
S-106 9 Stockholm, Sweden

INTRODUCTION

In order to study the mechanisms of modifiers of mutagenesis, in vitro systems are essential. However, bacteria, cell cultures, and other in vitro systems cannot reflect the complicated interactions taking place in intact animals. This is not the least true when it comes to the generation and effects of free radicals. The ubiquitous occurrence of oxygen radicals in aerobic organisms has enforced a complex array of defense mechanisms, which makes it difficult to extrapolate data from in vitro to in vivo systems. On the other hand, experimental work with mammals is expensive and time-consuming, and therefore there is a need for a simple and fast in vivo system to study the modifying effects of various agents on mutagenic and carcinogenic compounds. Drosophila constitutes a useful system in this context. The enzymatic detoxification system of Drosophila is essentially similar to the ones in mammals. The mixed-function oxygenase system operates efficiently with respect to phenobarbital-induced P450, while the metabolism of polyaromatic hydrocarbons by P448 differs somewhat quantitatively from that in mammals (3).

The use of Drosophila in genetic toxicology has been based previously almost entirely on germ cell mutagenicity. In later years somatic systems have been developed; they have turned out to be very sensitive, and they furthermore show a better correlation with carcinogenicity than do germ cell systems (6). These somatic systems have been named SMART, for somatic mutation and recombination tests, as they cover both of these genetic endpoints. We have been using a SMART system that was developed by Würgler and co-workers in Switzerland (2). The system is based on mutations and recombination involving two loci in the third chromosome, which act on wing hairs--multiple wing hair (mwh) and flare (flr). Larvae are treated with test substances, and spots of mutated cells are scored in wings of adult flies under the microscope. Each wing contains about 25,000 cells. We have used this system to study antimutagens and other modifiers of mutagenicity acting at different levels. The treatment was performed by mixing the chemicals in the substrate in accordance with a previously used method (1,2,6).

INDIRECT MUTAGENS

The effect in this somatic system of an indirect mutagen, which requires activation by P450, can be illustrated by dimethylnitrosamine (DMN). In accordance with mammalian systems, phenobarbital (PB) caused a significant increase in the effect of DMN, evidently through an enzyme induction (Fig. 1). Furthermore, ethylalcohol, which inhibits the activation of DMN in mammals, caused a significant decrease in the effect of DMN in this system also.

Conjugation with glutathione (GSH) is an efficient detoxification mechanism which is mediated by glutathione transferase. An opposite effect of GSH can, however, also occur in some cases. It was shown on Salmonella that 1,2-dichloroethane (DCE) and the corresponding bromo compound are activated to potent mutagens, half mustards, by conjugation to GSH (4). In the SMART system, DCE was an efficient mutagen. The GSH conjugation was manipulated by buthionine sulfoximine (BSO), which depletes the GSH pool, and by PB, which induces glutathione transferases. The result of combined treatments was according to expectation. BSO decreased the effect of DCE, while PB increased it (Fig. 2). The influence of the different compounds on GSH and on GSH- transferases was measured by HPLC chromatography, and by a spectrophotometric assay, respectively. BSO caused a depletion of GSH to less than 10%, while PB caused a significant increase of GSH transferases. Butylated hydroxyanisole (BHA) has been shown to act as an efficient anticarcinogen through the induction of GSH-transferases and the increased formation of GSH (5). In Drosophila, however, no effect of BHA either on GSH-transferases or on GSH could be seen.

INHIBITOR OF 3-AMINOBENZAMIDE

3-Aminobenzamide is an inhibitor of poly(ADP)-ribosylation. Poly-(ADP-ribose) is involved in mutations, transformation, and DNA repair, but the actual mechanism and function of poly(ADP)-ribosylation are not clear. In the SMART system, 3-aminobenzamide had no significant effect on the spontaneous frequency of spots, but it significantly increased

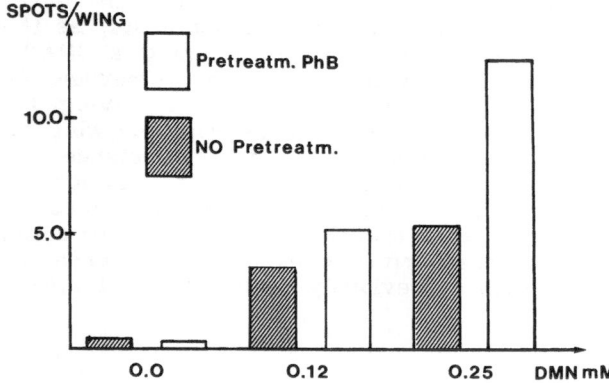

Fig. 1.   The effect of 1,000 ppm of phenobarbital (PhB) on the mutagenicity of dimethylnitrosamine (DMN) in the wing spot system.

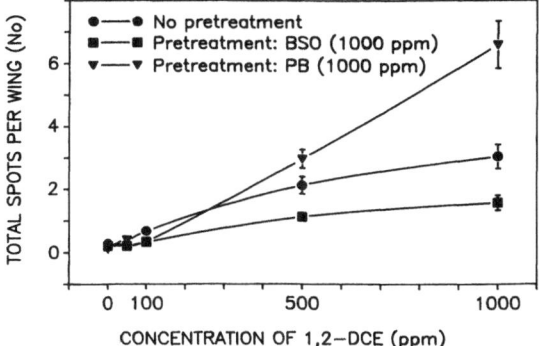

Fig. 2.   Mutagenicity of 1,2-dichloroethane (1,2-DCE) in the wing spot
           system with and without pretreatment with buthionine sulfoxi-
           mine (BSO) or phenobarbital (PB).

spots induced by methyl methanesulfonate (MMS) (Tab. 1). We have not
yet found out whether this effect of 3-aminobenzamide operates on recombination or on mutation. It could, however, be mentioned that the vast
majority of spots induced by MMS itself are caused by somatic recombination.

BLEOMYCIN

Finally, we have made a series of experiments with bleomycin (BLM),
which is a well-known radiomimetic compound that acts by the generation
of free radicals (1). BLM caused a significant and dose-dependent increase in spots, and that effect was drastically increased by increasing
the oxygen content of the air to 70% (Fig. 3).

It is of interest to notice that the effect of BLM was increased also
by post-treatment with oxygen. BLM has a rather unique mechanism of
action on DNA. It intercalates in DNA and generates oxygen radicals in

Tab. 1.   The effect of 500 ppm 3-aminobenzamide on the mutagenicity of
           10 ppm MMS in the wing spot system.

| Treatment | Benzamide (500 ppm) | | Spots/wing |
	Pretreatment	Co-treatment	
Control	−	−	0.3
MMS (10 ppm)	−	−	1.7
MMS	−	+	6.5
MMS	+	+	7.6
MMS	+	−	2.1
−	−	+	0.3
−	+	+	0.2
−	+	−	0.3

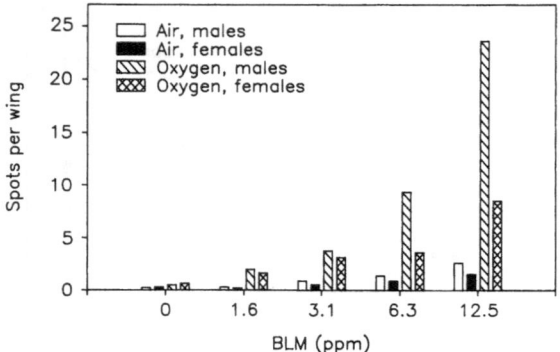

Fig. 3.   Mutagenicity of bleomycin (BLM) in the wing spot system with and without co-treatment with 80% oxygen.

direct connection with DNA.   The intercalation of BLM to DNA and the generation of oxygen radicals evidently can be separated by a fairly long time interval.   The fact that BLM acts in the immediate vicinity of DNA may explain the fact that our attempts to interfere with the mutagenicity by various antioxidants and inhibitors of the defense mechanisms were not successful (1).

In conclusion, experiments with combined treatments of mutagens and mutation modifiers in Drosophila have shown that the somatic mutation and recombination test is a valuable model system by which to study interaction between agents in mutation processes.

ACKNOWLEDGEMENTS

This work has been supported by grants from the National Swedish Environmental Protection Board, the Swedish Work Environment Fund, and the Erik Philip Sörensens Fund for Promotion of Genetic and Humanistic Scientific Research.

REFERENCES

1.   Cederberg, H., and C. Ramel (1989) Modifications of the effect of bleomycin in the somatic and recombination test in Drosophila melanogaster. Mutat. Res. (in press).
2.   Graf, U., F.E. Würgler, A.J. Katz, H. Frei, H. Juon, C.B. Hall, and P.G. Kale (1984) Environ. Mutag. 6:157-188.
3.   Hallstrom, I., and R. Grafström (1981) The metabolism of drugs and carcinogens in isolated subcellular fractions from Drosophila melanogaster. II. Enzyme induction and metabolism of benzo(a)py-rene. Chem.-Biol. Interact. 34:145-159.
4.   Rannug, U., A. Sundvall, and C. Ramel (1978) The mutagenic effect of 1,2-dichloroethane on Salmonella typhimurium. I. Activation through conjugation with glutathione in vitro. Chem.-Biol. Interact. 20:1-16.
5.   Wattenberg, L.V. (1980) Inhibitors of chemical carcinogens. J. Environ. Pathol. Toxicol. 3:35-52.

6. Würgler, F.E., and E.W. Vogel (1986) In vivo mutagenicity testing using somatic cells of Drosophila melanogaster. In Chemical Mutagens, Vol. 10, F.J. de Serres, ed. Plenum Press, New York, pp. 1-59.

THE MOUSE MUTANT, "WASTED": TISSUE-SPECIFIC RADIATION SENSITIVITY AND HEMATOPOIETIC CELL LINEAGES

Hideo Tezuka

National Institute of Genetics
Mishima, Shizuoka-ken 411, Japan

INTRODUCTION

Mice homozygous for the autosomal recessive mutation "wasted" (wst/wst) show pathological changes in the central nervous and lymphoid systems, and exhibit increases in the frequency of both spontaneous and gamma-irradiation-induced chromosomal aberrations (mainly chromatid-type). Such abnormalities are similar to those seen in ataxia telangiectasia (AT), one of the human hereditary diseases with chromosomal instability. Thus the wasted mouse has been isolated and proposed as an animal model for AT by Shultz et al. (4). The mouse die, mostly at 28 da of age.

My previous reports (3,5) show that the biological effects of the wst mutation are tissue-specific and age-dependent as assessed by changes in organ weight per body weight ratio. A high induction of chromosomal aberrations following treatment with ionizing radiation was observed in an age-dependent manner in the wasted-mouse bone marrow (3,5). In contrast, no marked tissue specificity for radiation sensitivity has been reported in human AT (1). In the present study, tissue specificity of sensitivity to ^{137}Cs gamma-ray irradiation was investigated in the mutant by the following two endpoints.

THE INDUCTION OF CHROMOSOMAL ABERRATIONS

In Vivo Studies

Figure 1 shows that the frequency of total cells with aberrations in the wasted-mouse bone marrow increased significantly (P<0.01), approximately two-fold over that in the littermate control, at almost every dose examined. In spermatogonia, the frequency increased linearly with radiation doses for both mutant and control groups, but no difference was observed between them.

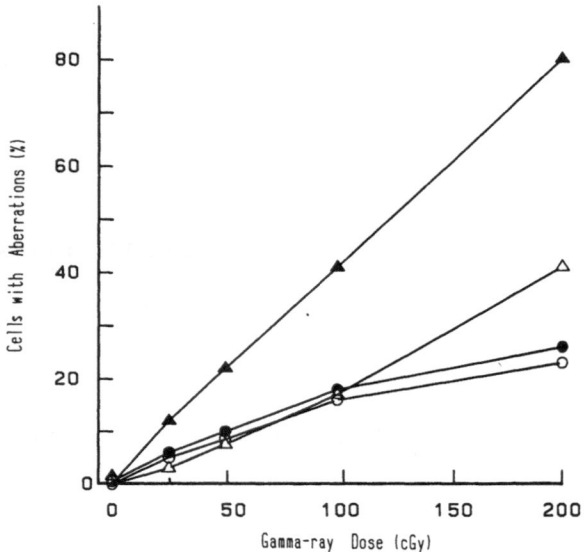

Fig. 1.   Tissue-specific induction of chromosomal aberrations in wasted
           mice (in vivo). Groups of animals were sacrificed 24 hr after
           the whole body irradiation, at 26 or 27 da of age. The animals
           received 4 mg colchicine/kg treatment for 3-4 hr before sacri-
           fice. Open triangles, control mice bone marrow; closed tri-
           angles, wasted mice bone marrow; open circles, control mice
           spermatogonia; closed circles, wasted mice spermatogonia.

In Vitro Studies

Since cells of erythropoietic lineage are the major part of bone marrow, the sensitivity of erythropoietin-stimulated bone marrow cells in culture was studied in vitro as a function of age. Animals were investigated
at two different ages: 21-23 da and 26-27 da.

As shown in Tab. 1, a significant, two-fold increase in the induced
frequency of aberrations was observed at the later age in the wasted-
mouse bone marrow when compared with the control. The increase was
age-dependent. The result coincides well with those shown in the in vivo
studies previously reported (5). However, no similar difference was observed in fibroblastic cells from either combination of groups. The difference in the induced frequency of aberrations per cell between bone
marrow and fibroblasts may be due to each cell's specificity--for example,
different cell cycling time.

CHANGES IN SURVIVAL OF IRRADIATED CELLS IN CULTURE

Figure 2 shows results of the survival study of CFU-E (colony forming unit erythroid, erythropoietic progenitor cells) and CFU-C (colony
forming unit in culture, progenitor cells to granulocytes and macrophages). At 22 da of age, no difference was observed between the sen-

Tab. 1. Tissue-specific and age-dependent induction of chromosomal aberrations in the wasted mouse (in vitro).

Age (days)	Genotype	Induced aberrations per cell	
		Bone marrow	Fibroblasts
21-23	+/	1.62	0.39
	wst/wst	1.35	0.33
26-27	+/	1.50	0.42
	wst/wst	2.86	0.48

Femur bone marrow cells from both groups of mice at both ages were harvested with FBS (fetal bovine serum), treated with ice-cold 0.83% NH$_4$Cl in 0.05M Tris, suspended in ice-cold culture medium of Eagle's MEM + 10% FBS, irradiated at 4°C, and incubated for 18 hr after the addition of 0.2 U erythropoietin/ml at 37°C in a humidified 5% CO$_2$ atmosphere. Tail fibroblastic cells at 2-3 passages were cultured in Eagle's MEM + 10% FBS, irradiated at 4°C, and incubated for 24 hr in the same CO$_2$ incubator. Irradiation dose was 200 cGy. Colcemid at 0.001 mM was added 0.5-1 hr before chromosome preparation. 100-200 cells were counted for each point. Wasted homozygotes were first recognized at 20-22 days of age by low body weight and ataxia, and later confirmed by body weight decrease and progressive paralysis. Final diagnosis was done by weighing spleens of the mice according to the previous report (5).

sitivity to radiation-induced killing of the wasted-mouse CFU-E and control cells. At 27 da of age, however, there was a four-fold increase of radiosensitivity in the wasted-mouse CFU-E (D$_{37}$:40 cGy) over that in the control mouse cells (D$_{37}$:155 cGy). These data correspond well with those shown in Tab. 1 in in vitro studies on the induction of chromosomal aberrations. Survival curves in both groups were biphasic, which indicates the existence of at least two subpopulations of different radiation sensitivity.

In the case of CFU-C, however, no difference or, rather, a radiation-resistant tendency was observed in the wasted-mouse CFU-C when compared with the control (at 22 da of age, difference was significant at 100 and 200 cGy at P<0.01). Also, no difference was observed in the survival experiment of fibroblastic cells in culture between these two groups of mice (2).

The above-mentioned results suggest the following:

a) The effect of the wasted mutation is tissue- and even cell-specific. Since this is unlike human AT, the wasted mouse may not be an exact model animal for this human disease, but may be a good model animal for studying the relation between radiation sensitivity, DNA repair, and cell differentiation.

b) Radiation sensitivity in the wasted-mouse bone marrow is observed in CFU-E and not in CFU-C in an age-dependent fashion, suggesting that development of radiation sensitivity may be based on the differentiation of erythropoietic cell lineage itself.

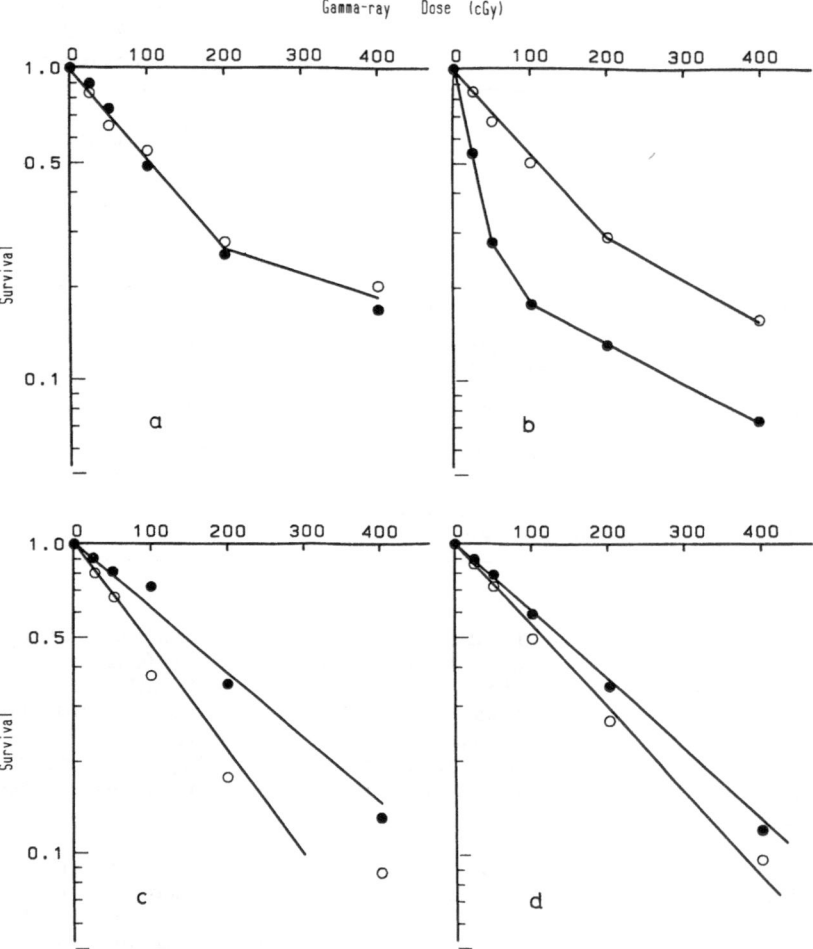

Fig. 2.   Cell-specific and age-dependent induction of radiation sensitivity (in vitro cell survival). Femur bone marrow cells were prepared in a similar way to the methods mentioned in Tab. 1.
Cells were incubated in plastic dishes with grids (d=35 mm, Lux
#5217) in alpha-MEM (Gibco) containing 30% FBS, 0.8% methyl
cellulose, 2.5 U erythropoietin/ml, 0.2 mM alpha-thioglycerol,
and 1% deionized bovine serum albumin at 37°C in a CO_2 incubator. Erythropoietin, TOYOBO Epo 301, is a crude product
including activity of colony-stimulating factor. Numbers of
CFU-E colonies were counted after 2 da, and those of CFU-C
colonies counted after 9 da. (a) and (b), CFU-E from mice at
22 and 27 da of age; (c) and (d), CFU-C from mice at 22 and
27 da of age, respectively. Open circles, control mice; closed
circles, wasted mice.

What is the wst gene mutation, and how is it responsible for these phenomena and the wasted-mouse clinical symptoms such as ataxia and progressive body weight loss? Why does the wasted mouse die so fast? One of the important keys to open the door of the mystery of life may be hidden in this mutant.

ACKNOWLEDGEMENTS

The author is grateful to Drs. D. Ayusawa, T. Inoue, T. Seno, and the late T. Kada for their discussions and support. This work was supported in part by a grant-in-aid for scientific research from the Ministry of Education of Japan.

REFERENCES

1. Bridges, B.A., and D.G. Harnden, eds. (1982) Ataxia-Telangiectasia, a Cellular and Molecular Link between Cancer, Neuropathology, and Immune Deficiency, John Wiley and Sons, New York.
2. Inoue, T., K. Aikawa, H. Tezuka, T. Kada, and L.D. Shultz (1986) Effect of DNA-damaging agents on isolated spleen cells and lung fibroblasts from the mouse mutant "wasted," a putative animal model for ataxia telangiectasia. Cancer Res. 46:3979-3982.
3. Inoue, T., H. Tezuka, T. Kada, K. Aikawa, and L.D. Shultz (1986) The mouse mutant "wasted," an animal model for ataxia telangiectasia. In Antimutagenesis and Anticarcinogenesis Mechanisms, D.M. Shankel, P.E. Hartman, T. Kada, and A. Hollaender, eds. Plenum Press, New York, pp. 323-335.
4. Shultz, L.D., H.O. Sweet, M.T. Davisson, and D.R. Coman (1982) "Wasted," a new mutant of the mouse with abnormalities characteristic of ataxia telangiectasia. Nature 297:402-404.
5. Tezuka, H., T. Inoue, T. Noguchi, T. Kada, and L.D. Shultz (1986) Evaluation of the mouse mutant "wasted" as an animal model for ataxia telangiectasia. I. Age-dependent and tissue-specific effects. Mutat. Res. 161:83-90.

ANTIMUTAGENIC EFFECTS OF TUMOR PROMOTERS--

CO-MUTAGENIC EFFECTS OF CO-CARCINOGENS

Rudolf Fahrig

Fraunhofer-Institut für Toxikologie und Aerosolforschung
Abteilung Genetik
3000 Hannover 61, Federal Republic of Germany

According to the generally accepted initiator (= carcinogen)-promoter model of carcinogenesis (6), promoter substances are responsible for expression of the tumorigenic genotype.

In tests of the initiator-promoter model (6), a promoting agent is applied repeatedly after a single subthreshold dose of an initiating carcinogen. In testing the co-carcinogenicity model, two agents are administered simultaneously. A test is positive when higher tumor incidences occur than either agent would produce alone. Not all tumor promoters are co-carcinogens, and vice versa (8).

The approach of our work was to investigate the influence of tumor promoters and co-carcinogens upon mutagenicity and recombinogenicity of genotoxic substances, using yeast strain MP1 (2). Saccharomyces cerevisiae is a suitable organism in which to study mechanisms which result in the expression of a heterozygous recessive gene. Previous short-term experiments have shown that in stationary phase cells chromosome loss can be induced neither by tumor promoters and co-carcinogens alone, nor in combination with a carcinogen (3). In such experiments, tumor promoters were neither mutagenic nor recombinogenic when given alone; only a few co-carcinogens were weakly mutagenic (2,3).

Figure 1 illustrates by means of a few typical examples that tumor promoters and co-carcinogens are genetically active when given in combination with a mutagen. The activity is independent of the mutagen/carcinogen used, but specific for a given co-carcinogen or tumor promoter. Substances which are only toxic do not alter the genetic effects of a mutagen (3).

The following substances have dose/effect-curves similar to that of hydroquinone:

pyrogallol butylated hydroxytoluene*
catechol   β-estradiol*
cholic acid   diethylstilbestrol*
Cl$_5$-phenoxyphenol   testosterone*
sodium arsenite

The following substances have dose/effect-curves similar to that of phenol:

d-limonene   chrysarobin
coumarin   8-hydroxyquinoline
phenobarbital   p-dimethylaminoazobenzol
lithocholic acid   butylated hydroxytoluene
acridine orange   diethylstilbestrol
harman   testosterone
mezerein

The following substances have dose/effect-curves similar to that of n-decane:

TPA   deoxycholic acid
n-dodecane   hyodeoxycholic acid
1-decanol   iodoacetic acid
gossypol   biphenyl

Although for many substances the carcinogenicity data are incomplete (literature not shown), it is apparent that there are similarities between the occurrence of specific genetic effects and specific effects in carcinogenicity tests:

1.   Co-carcinogens are co-mutagenic.

2.   Tumor promoters which are anticarcinogenic if given simultaneously with the carcinogen are co-recombinogenic and antimutagenic.

3.   Tumor promoters which can be transformed into co-carcinogens can reverse their genetic effects from co-recombinogenicity to co-mutagenicity upon metabolic activation.

4.   Substances which are tumor promoters as well as co-carcinogens are also co-recombinogens and co-mutagens.

In the carcinogenicity experiment, the anticarcinogenic effect of tumor promoters cannot be distinguished from a simple descarcinogenic (4) effect (inactivation of carcinogens by whatever means) without further

---

*Only in the presence of activating S-9 mix.

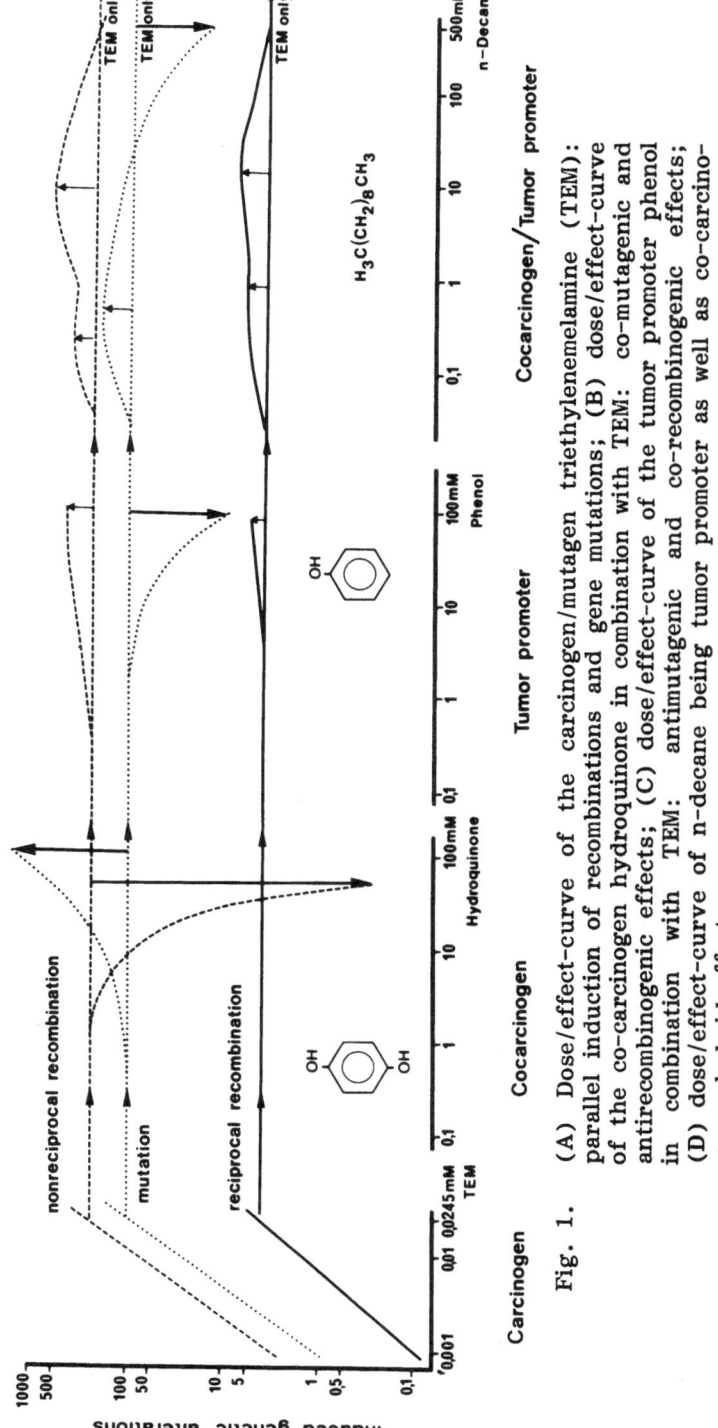

Fig. 1. (A) Dose/effect-curve of the carcinogen/mutagen triethylenemelamine (TEM): parallel induction of recombinations and gene mutations; (B) dose/effect-curve of the co-carcinogen hydroquinone in combination with TEM: co-mutagenic and antirecombinogenic effects; (C) dose/effect-curve of the tumor promoter phenol in combination with TEM: antimutagenic and co-recombinogenic effects; (D) dose/effect-curve of n-decane being tumor promoter as well as co-carcinogen: hybrid effects.

biochemical analyses. In the genetic tests, however, the simultaneous increase in mutations and decrease in recombinations is indicative of interference with DNA repair processes.

The relationships between the effects of substances in carcinogenicity tests and in genetic experiments do not prove that there is a causal
connection between the two processes, but they offer at least plausible
explanations for hitherto conflicting results in carcinogenicity experiments. For example, the anticarcinogenic effects of tumor promoters, like
phenol (8), limonene (1,7), and phenobarbital (5) in co-carcinogenicity
experiments can simply be attributed to the antimutagenic effect of this
type of tumor promoter. Thus, the genetic effects observed may be relevant to the carcinogenic process.

REFERENCES

1. Elegbede, J.A., C.E. Elson, A. Qureshi, M.A. Tanner, and M.N.
    Gould (1984) Inhibition of DMBA-induced mammary cancer by the
    monoterpene d-limonene. Carcinogenesis 5:661-664.
2. Fahrig, R. (1984) Genetic mode of action of cocarcinogens and tumor
    promoters in yeast and mice. Molec. Gen. Genet. 194:7-14.
3. Fahrig, R. (1987) Effects of bile acids on the mutagenicity and re-
    combinogenicity of triethylene melamine in yeast strains MP1 and
    D61.M. Arch. Toxicol. 60:192-197.
4. Kada, T. (1982) Mechanisms and genetic implications of environmen-
    tal antimutagens. In Environmental Mutagens and Carcinogens, T.
    Sugimura, S. Kondo, and H. Takebe, eds. University of Tokyo
    Press, Tokyo, and Alan R. Liss, Inc., New York, pp. 355-359.
5. Peraino, C., R.J.M. Fry, and E. Staffeldt (1971) Reduction and en-
    hancement by phenobarbital of hepatocarcinogenesis induced in the
    rat by 2-acetylaminofluorene. Cancer Res. 31:1500-1513.
6. Rous, P., and J.G. Kidd (1941) Conditional neoplasms and sub-
    threshold neoplastic states. A study of tar tumors in rabbits. J.
    Exp. Med. 73:365-389.
7. Sparnins, V.L., and L.W. Wattenberg (1985) Effects of citrus fruit
    oils on glutathione S-transferase (GST) activity and benzo(a)pyrene
    (BP)-induced neoplasia. Carcinogenesis 26:123.
8. Van Duuren, B.L. (1976) Tumor promoting and co-carcinogenic
    agents in chemical carcinogenesis. ACS Monogr. 173:24-51.

ANTIMUTAGENIC EFFECTS OF DIMETHYL SULFOXIDE ON METABOLISM AND GENOTOXICITY OF BENZENE IN VIVO

William W. Au,[1] Wagida A. Anwar,[2] Elie Hanania,[3]
and V.M. Sadagopa Ramanujam[1]

[1]Department of Preventive Medicine and Community Health
[3]Department of Human Biological Chemistry and Genetics
University of Texas Medical Branch
Galveston, Texas 77550

[2]Department of Community, Environmental
    and Occupational Medicine
Faculty of Medicine
Ain Shams University
Cairo, Egypt

INTRODUCTION

By understanding the mechanism of action of hazardous agents, it may be possible to utilize specific procedures to block the action of these agents in order to achieve antimutagenic and anticarcinogenic effects. We have conducted a study to investigate the genotoxic activity of benzene in vivo and to evaluate a potential antimutagenic procedure to block the activity of benzene.

The metabolism of benzene has been reported to be mediated by hydroxyl radicals (5-7,14), and dimethyl sulfoxide (DMSO) is a known hydroxyl radical scavenger (9-11). Therefore, we have conducted our study by using DMSO to block metabolism of benzene and by correlating the different metabolites with the expression of genotoxicity in treated mice. Details of our study are presented elsewhere (1), and a summary of our work is presented here.

MATERIALS AND METHODS

Adult ICR mice were randomly assigned (5 mice per group) to negative control, solvent control (olive oil), positive control (benzene 440 mg/kg b.w.), and exposure groups. The exposure groups included mice exposed to different concentrations of DMSO (1.25%, 3.75%, and 12.5% alone) or to benzene (440 mg/kg) plus DMSO. The combined exposure groups (benzene plus DMSO) included dose-response groups which were

exposed to benzene and, 1 hr later, to 1.25, 3.75, or 12.5% DMSO, and
time-response groups which were exposed to benzene and, 1, 3, or 5 hr
later, to 12.5% DMSO.

Different groups of mice were housed immediately in separate clean
metabolic cages containing food and drinking water. The urine from each
group of 5 mice was collected in glass tubes maintained in ice baths for
24 hr and then frozen at -70°C. At the time of analysis, urine samples
were thawed to room temperature and the exact volume of urine collected
from each group was measured. For the isolation and identification of
metabolites, the method of Gad-El Karim et al. (3) was used.

Mice were sacrificed at 30 hr after treatment. For the combined
treatment group, mice were sacrificed at 30 hr after exposure to benzene,
irrespective of exposure time to DMSO. Slides were prepared from bone
marrow samples according to Schmid (12), coded, stained, and scored by
the same observer for the presence of micronuclei (MN) in the polychromatic erythrocytes (PCE).

RESULTS

Micronucleus Test

The data summarized in Tab. 1 show that the micronuclei frequencies
in negative controls, in solvent controls, and in mice treated with various
concentrations of DMSO are low (0-2.8 MN/1,000 PCE). Mice exposed to
benzene showed a significant increase in MN frequencies (48.8 ± 5.6/1,000
PCE) over the control groups. In the presence of DMSO, the free-radical
scavenger produced a significant positive dose-dependent and an inverse
time-dependent reduction of benzene-induced MN frequencies (Tab. 1).
However, when DMSO was administered to mice at 5 hr after benzene exposure, the reduction effect was not detectable.

Tab. 1. Induction of micronuclei in polychromatic erythrocytes (PCE) of
mice after exposure to benzene (BZ) with or without dimethyl
sulfoxide (DMSO).

DMSO concentration	Micronuclei frequency/1,000 PCE			
	Without BZ	With BZ		
		1 hr	3 hr	5 hr
0%	0.8 ± 0.4	48.8 ± 5.6	-	-
1.25%	0 ± 0	29.4 ± 10.9*	-	-
3.75%	1.0 ± 0.3	20.0 ± 7.6*	-	-
12.50%	0.8 ± 0.2	2.6 ± 0.7*	3.4 ± 0.8*	36.2 ± 12.1

*Significantly different from mice treated with BZ alone, P<0.01.

The micronuclei frequency for olive oil control is 2.8 ± 1.0. All data are
shown as a mean frequency from five mice ± S.E.M.

Urinary Metabolites

As shown in Tab. 2, phenol is the major metabolite found in vivo after exposure to benzene. This observation is consistent with data from the literature (8,13). Other metabolites detected by us are trans-trans muconic acid (MA), hydroquinone (HQ), and catechol. Our metabolite data suggest that DMSO affects predominantly the metabolism of benzene toward the formation of MA and phenol. The quantity of these two metabolites in urine is altered by the presence of DMSO in a positive dose-dependent and an inverse time-dependent manner.

Correlation Between Induction of Micronuclei and Presence of Benzene Metabolites

The micronuclei and metabolite data are plotted on the same graph (Fig. 1) to show the dose- and time-dependent changes of these two measured parameters. Linear regression analysis was performed to assess the relationship between the quantity of the benzene metabolites and the MN frequencies. It was found that changes in MA quantity are much better correlated with alterations in MN frequencies ($p < 0.007$) than with total and conjugated phenol ($p < 0.03$). On the other hand, there is no correlation between HQ or catechol concentration and MN frequency ($p > 0.5$ and $p > 0.33$, respectively).

CONCLUSION

Our study shows that the presence of phenol and MA is best correlated with the induction of MN; therefore, our observation suggests that they are the genotoxic metabolites of benzene in vivo. Since MA is the oxidized form of trans-trans muconaldehyde in urine, it is believed that trans-trans muconaldehyde is responsible for genotoxicity in vivo instead

Tab. 2. Benzene metabolites in urine of mice exposed to benzene (BZ) with or without dimethyl sulfoxide (DMSO).

| Treatment | Urinary benzene metabolites (mg/kg b.w.) | | | | |
| | Trans-trans muconic acid | Total phenol | Glucuronide conjugated | | |
			Phenol	Hydroquinone	Catechol
Negative control	0	0	0	0	0
Olive oil control	0	0	0	0	0
DMSO 1.25%	0	0	0	0	0
3.75%	0	0	0	0	0
12.50%	0	0	0	0	0
BZ 440 mg/kg	15.7	57.2	27.0	7.1	8.0
BZ + DMSO 1.25%, 1 hr	13.0	52.8	20.0	3.3	2.4
3.75%, 1 hr	8.8	49.0	19.6	5.4	4.7
12.50%, 1 hr	3.1	34.0	17.3	3.3	4.3
12.50%, 3 hr	7.4	51.1	20.3	6.2	4.1
12.50%, 5 hr	11.4	52.6	22.6	4.3	4.1

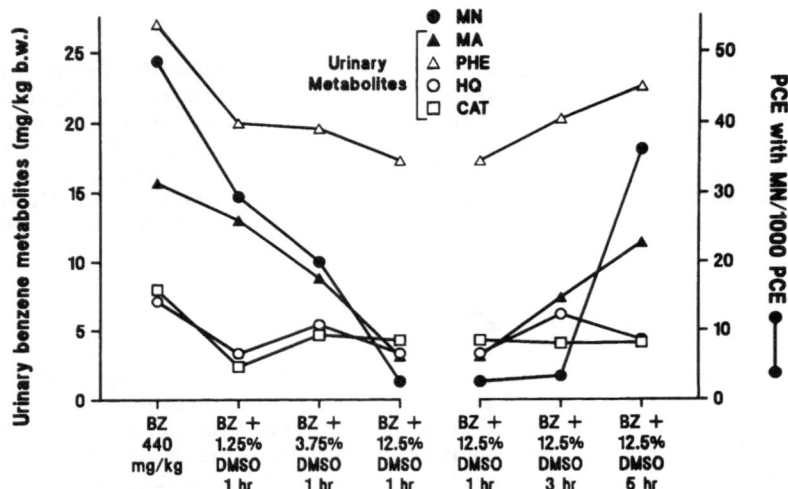

Fig. 1. Dose- and time-dependent changes in bone marrow micronuclei
(MN) and urinary benzene metabolites in mice exposed to ben-
zene with or without DMSO. PCE = polychromatic erythrocytes,
MA = muconic acid, PHE = phenol, HQ = hydroquinone, and
CAT = catechol.

of MA. Our study also demonstrated that the mutagenic and potentially
carcinogenic activity of benzene can be blocked by a free-radical scavenger, DMSO.   Therefore, the potential use of free-radical scavengers to
protect workers exposed to benzene may be considered.

ACKNOWLEDGEMENTS

This study is funded by the Fogarty International Fellowship, number 1 FO5 TW 03793-01 BI-5 (12). The presentation of this paper in the
symposium is partially funded by a travel award from the organizing committee.

REFERENCES

1. Anwar, W.A., W.W. Au, M.S. Legator, and V.M. Sadagopa
Ramanujam (1989) Effect of dimethyl sulfoxide on the genotoxicity
and metabolism of benzene in vivo. Carcinogenesis 10:441-445.
2. Dean, B.J. (1985) Recent findings on the genetic toxicology of
benzene, toluene, xylenes and phenols. Mutat. Res. 154:153-181.
3. Gad-El Karim, M.M., V.M. Sadagopa Ramanujam, and M.S. Legator
(1985) Trans-trans muconic acid, an open chain urinary metabolite of
benzene in mice. Quantification by high pressure liquid chromatography. Xenobiotica 15:211-221.
4. Gad-El Karim, M.M., V.M. Sadagopa Ramanujam, and M.S. Legator
(1986) Correlation between the induction of micronuclei in bone marrow by benzene exposure and the excretion of metabolites in urine of
CD-1 mice. Toxicol. Appl. Pharm. 85:464-477.

5. Griffiths, J.C., G.F. Kalf, and R. Snyder (1986) The metabolism of benzene and phenol by a reconstituted purified phenobarbital-induced rat liver mixed function oxidase system. Adv. Exp. Med. Biol. 197:213-222.
6. Ingelman-Sundberg, M., and A.L. Hagbjork (1982) On the significance of the cytochrome P450-dependent hydroxyl radical mediated oxygenation mechanism. Xenobiotica 12:673-686.
7. Johansson, I., and M. Ingelman-Sundberg (1983) Hydroxyl radical mediated cytochrome P450-dependent metabolic activation of benzene in microsomes and reconstituted enzyme systems from rabbit liver. J. Biol. Chem. 258:7311-7316.
8. Kalf, G.F., G.B. Post, and R. Snyder (1987) Solvent toxicology: Recent advances in the toxicology of benzene, the glycol ethers and carbon tetrachloride. Ann. Rev. Pharmacol. Toxicol. 27:399-427.
9. Littlefield, L.G., E. Joiner, E.L. Frome, and S.P. Colyer (1987) Modulation of radiation-induced chromosome aberrations by DMSO--An OH radical scavenger. Environ. Mutag. 9(8):61.
10. Reuvers, A.P., C.L. Greenstock, J. Borsa, and J.D. Chapman (1973) Studies on the mechanism of chemical radioprotection by dimethyl sulphoxide. Int. J. Radiat. Biol. 24:533-536.
11. Sasaki, M.S., and S. Matsubara (1977) Free radical scavenging in protection of human lymphocytes against chromosome aberration formation by gamma ray irradiation. Int. J. Radiat. Biol. 32:439-445.
12. Schmid, W. (1975) Micronucleus test. Mutat. Res. 31:9-15.
13. Snyder, R., S.L. Longacre, C.M. Witmer, and J.J. Kocsis (1982) Metabolic correlates of benzene toxicity. In Biological Reactive Intermediates II, R. Snyder, D.V. Park, J.J. Kocsis, D.J. Jollow, G.G. Gibson, and C.M. Witmer, eds. Plenum Press, New York.
14. Trush, M.A., M.J. Reasor, M.E. Wilson, and K. Van Dyke (1984) Oxidant-mediated electronic excitation of imipramine. Biochem. Pharmacol. 33:1401-1410.

POSSIBLE ANTITUMOR PROMOTER IN THE GLANDULAR STOMACH:

CALCIUM CHLORIDE

C. Furihata and T. Matsushima

Department of Molecular Oncology
Institute of Medical Science
University of Tokyo
Tokyo 108, Japan

ABSTRACT

Studies were made on the inhibitory effects of $CaCl_2$ and 13-cis-retinoic acid on induction of replicative DNA synthesis (RDS) in the pyloric mucosa of male F344 rats by the glandular stomach tumor promoter, NaCl. RDS in the pyloric mucosa showed a maximum of about a ten-fold increase 17 hr after administration of 3.3 M NaCl and returned to the control level 48 hr after the administration of NaCl. Administration of 400 mM $CaCl_2$ 1 hr before NaCl resulted in 60-80% inhibition of the increase in RDS 4-48 hr after NaCl administration. Administration of 20 to 400 mM $CaCl_2$ 1 to 2 hr before NaCl caused dose-dependent inhibition of the increase in RDS 17 hr after NaCl administration, with 400 mM $CaCl_2$ causing 80-100% inhibition. Administration of 400 mM $CaCl_2$ 1 hr before NaCl also decreased the histological damage of the surface epithelial cells induced by NaCl. Administration of 13-cis-retinoic acid at doses of 10 µg-10 mg/kg body weight did not inhibit the increase in RDS in the pyloric mucosa that was induced by NaCl. These results suggest that $CaCl_2$, but not 13-cis-retinoic acid, inhibits tumor promotion in the pyloric mucosa of rat stomach.

INTRODUCTION

Previously, we found that all six glandular stomach tumor promoters examined [NaCl (3), taurocholate, glyoxal, catechol, $K_2S_2O_5$, and formaldehyde] and all five glandular stomach carcinogens examined [4-nitroquinoline 1-oxide, N-methyl-N'-nitro-N-nitrosoguanidine, N-nitroso-N-methylurethane, N-propyl-N'-nitro-N-nitrosoguanidine, and N-ethyl-N'-nitro-N-nitrosoguanidine (2)] induced about a ten-fold increase in RDS in the proliferative zone of the pyloric mucosa of F344 male rats, with maxima 16-24 hr after their administration, and also a 20- to 100-fold increase in ornithine decarboxylase (ODC) activity in the pyloric mucosa of F344 rat stomach, with maxima 4-24 hr after their administration (2,3). However, ethanol, which, like NaCl, induces acute ulcers in the stomach mucosa of rats, did not induce RDS or ODC in their pyloric mucosa (2) and did not enhance stomach carcinogenesis. Moreover, nonglandular stomach carcinogens, such as 2-acetylaminofluorene, dimethylnitrosamine, and 3-amino-1-methyl-5H-pyrido[4,3-b]indole, did not stimulate RDS or ODC in the pyloric mucosa of rats (2).

These glandular stomach tumor promoters and glandular stomach carcinogens induced erosion of the surface epithelial cells in the pyloric mucosa of F344 rats (3). This tissue damage was rapidly repaired, and during its repair, ODC activity was induced and RDS in the proliferative zone of the pyloric mucosa was stimulated. As this RDS was thought to be the end point of these sequential phenomena, in the present study we tested $CaCl_2$ and 13-cis-retinoic acid for ability to inhibit induction of RDS by the glandular stomach tumor promoter NaCl in the pyloric mucosa.

MATERIALS AND METHODS

Groups of 5 male Fischer rats (F344/DuCrj: Charles River Japan, Inc., Kanagawa), 7 to 8 wk old, were given $CaCl_2$ in 1 ml of distilled water or 13-cis-retinoic acid in 0.1 ml of dimethylsulfoxide (DMSO) by gastric tube, and after a suitable interval were given 3.3 M NaCl in 1 ml of distilled water by gastric tube. Control rats were given distilled water or DMSO only. The following day, RDS was determined by in vitro organ culture in the presence of tritiated thymidine ([^{3}H]dThd, ∿80 Ci/m mole, New England Nuclear, Boston, Massachusetts) as described previously (2). Briefly, the stomach was removed and the pyloric mucosa was cut up with scissors and chopped into small pieces (1 x 1 x 1 mm) with a razor. Then the minced tissue was cultured in the presence of [^{3}H]dThd (10 µCi/ml) at 37°C for 2 hr. The DNA fraction was extracted from the tissue, and incorporation of [^{3}H]dThd into DNA was determined in a liquid scintillation counter. The DNA content of the DNA fraction was determined with 3,5-diaminobenzoic acid (Tokyo Kasei Kogyo Co., Tokyo) with calf thymus DNA (Worthington Biochemical Co., Freehold, New Jersey) as a standard. All values shown are means for five individual rats.

RESULTS

As seen in Fig. 1, RDS in the pyloric mucosa of rats showed a maximum increase to 15-fold the initial value 17 hr after administration of 3.3 M NaCl, and returned to the control level 48 hr after NaCl administration. However, administration of 400 mM $CaCl_2$ 1 hr before NaCl resulted in 60 to 80% inhibition of the increase in RDS in the pyloric mucosa 4 to 48 hr after NaCl administration. Fig. 2 shows that administration of 20 to 400 mM $CaCl_2$ 1 hr before NaCl caused dose-dependent inhibition of the increase in RDS in the pyloric mucosa 17 hr after administration of NaCl. Administration of 400 mM $CaCl_2$ 1 to 2 hr before administration of NaCl resulted in 80% inhibition of the increase in RDS in the pyloric mucosa, but its administration 3 hr before NaCl resulted in only 30% inhibition of the increase in RDS in the pyloric mucosa (Fig. 3). Fig. 4a shows the appearance of untreated rat pyloric mucosa. At 30 min after administration of 3.3 M NaCl, the surface epithelial cells were damaged and became detached from the mucosa (Fig. 4b). Administration of 400 mM $CaCl_2$ 1 hr before NaCl decreased NaCl-induced damage of the surface epithelial cells (Fig. 4c). Only slight damage of the pyloric mucosa was seen 90 min after administration of 400 mM $CaCl_2$ alone, as a control (Fig. 4d). As seen in Fig. 5, administration of 13-cis-retinoic acid,

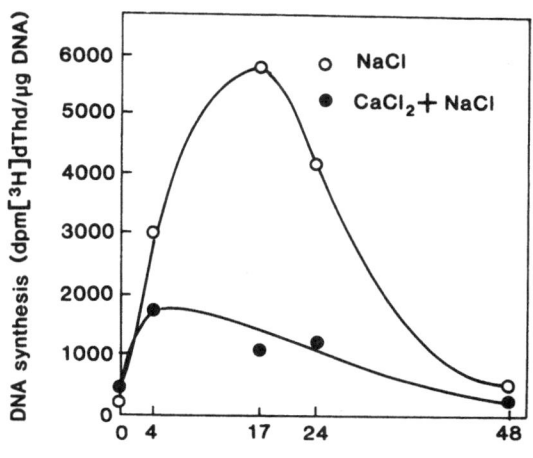

Fig. 1.   Inhibition by CaCl₂ of induction of RDS in the pyloric mucosa 4-48 hr after NaCl administration.   Rats were given 1 ml of 3.3 M NaCl with (●) or without (o) 1 ml of 400 mM CaCl₂.

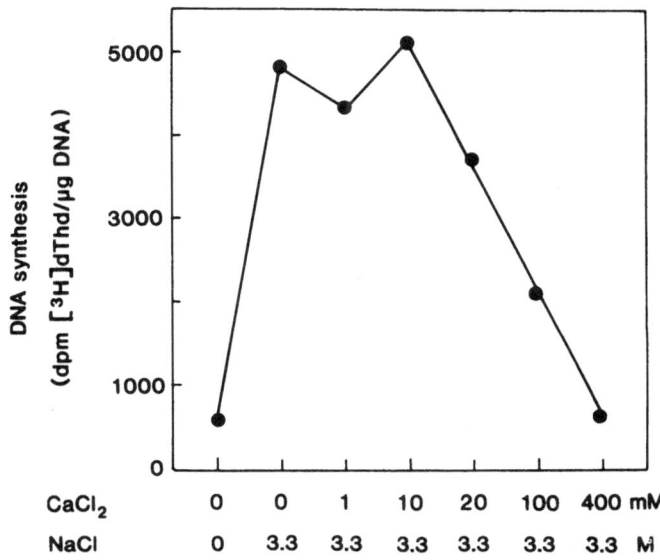

Fig. 2.   Dose-dependent inhibition by CaCl₂ of induction of RDS in the pyloric mucosa 17 hr after NaCl administration.

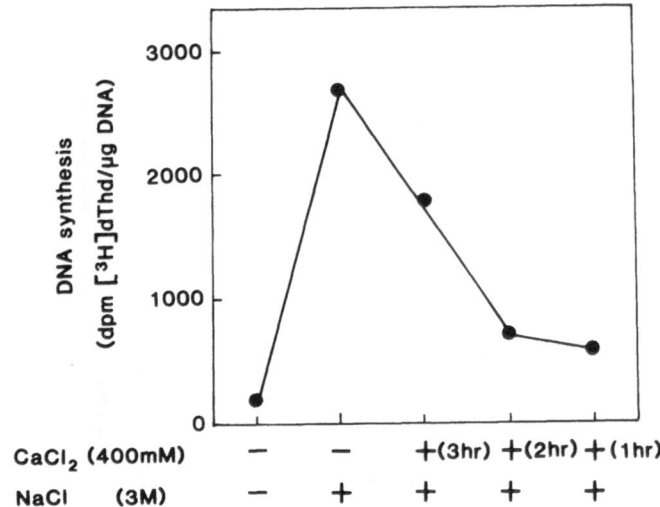

Fig. 3.   Effects of the time between $CaCl_2$ and NaCl administrations on
inhibition of induction of RDS in the pyloric mucosa.

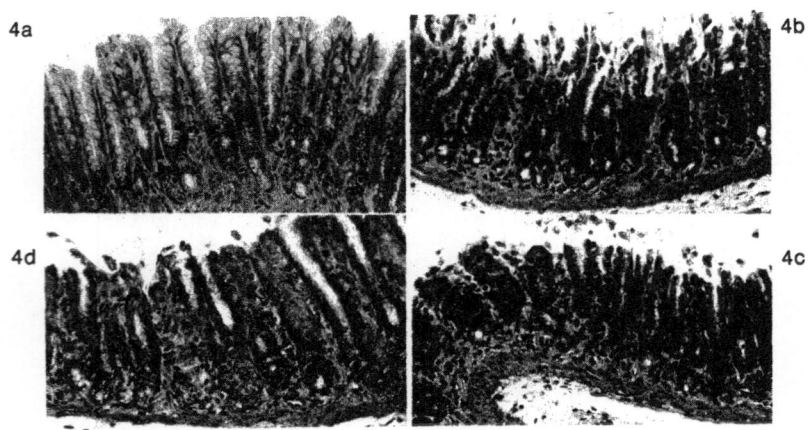

Fig. 4.   (a) Pyloric mucosa of an untreated control rat.   (b) Pyloric
mucosa of a rat 30 min after administration of 3.3 M NaCl.
(c) Pyloric mucosa of a rat given 400 mM $CaCl_2$ 1 hr before
NaCl and killed 30 min after NaCl administration.   (d) Pyloric
mucosa of a rat 1.5 hr after administration of 400 mM $CaCl_2$
only.

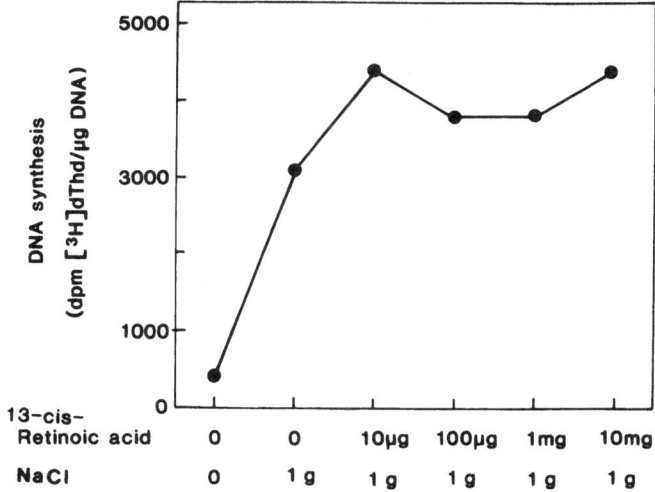

Fig. 5.   Absence of inhibition by 13-cis-retinoic acid of induction of RDS by NaCl in the pyloric mucosa.

vitamin A, at doses of 10 µg to 10 mg/kg body weight before administration of 1 ml of 3.3 M NaCl did not inhibit the increase in RDS in the pyloric mucosa 17 hr after NaCl administration.

DISCUSSION

In the present study we found that administration of aqueous $CaCl_2$ solution to rats before administration of aqueous NaCl solution inhibited induction by the latter of RDS in the pyloric mucosa. $CaCl_2$ seemed to protect the mucosal tissue from damage by NaCl. Similar results have been obtained in rat colon. That is, Ca salts were found to inhibit increase in the mitotic index in the rat colon mucosa induced by bile acid, a colon tumor promoter (1). Moreover, Ca salts also decreased tissue damage induced by bile acid in the colon mucosa (4). Ca-free medium is used to isolate cells from organs for primary culture. Calcium may protect the cell surface structure from breakage. The mechanism of its protective effect against tissue damage requires study.

REFERENCES

1. Bird, R.P., R. Schneider, D. Stamp, and W.R. Bruce (1986) Effect of dietary calcium and cholic acid on the proliferative indices of murine colonic epithelium. Carcinogenesis 7:1657-1661.
2. Furihata, C., S. Yoshidi, Y. Sato, and T. Matsushima (1987) Induction of ornithine decarboxylase and DNA synthesis in rat stomach mucosa by glandular stomach carcinogens. Jpn. J. Cancer Res. (Gann) 78:1363-1369.
3. Furihata, C., Y. Sato, M. Hosaka, T. Matsushima, F. Furukawa, and M. Takahashi (1984) NaCl induced ornithine decarboxylase and DNA synthesis in rat stomach mucosa. Biochem. Biophys. Res. Commun. 121:1027-1032.

4. Wargovich, M.J., V.W.S. Eng., H.L. Newmark, and W.R. Bruce
    (1983) Calcium ameliorates the toxic effect of deoxycholic acid on
    colonic epithelium. Carcinogenesis 4:1205-1207.

REVERSION OF TRANSFORMED NIH 3T3 CELLS TO FLAT CELLS BY INHIBITORS OF POLY(ADP-RIBOSE) POLYMERASE

Minako Nagao, Michie Nakayasu, Hiroshi Shima,
Shizu Aonuma, and Takashi Sugimura

Carcinogenesis Division
National Cancer Center Research Institute
Tsukiji, Chuo-ku, Tokyo 104, Japan

INTRODUCTION

Poly(ADP-ribose) polymerase catalyzes the formation of poly(ADP-ribose) from NAD in the presence of histones and DNA (5). Many nuclear proteins have been found to be acceptors for poly(ADP-ribose) (1). Studies using various poly(ADP-ribose) polymerase inhibitors have suggested the involvement of poly(ADP-ribose) polymerase in DNA repair, sister chromatid exchanges, DNA replication, and cell differentiation (1). Recently, coumarin, a DNA-site inhibitor of poly(ADP-ribose) polymerase, was found to prevent tumorigenesis of oncogene-transformed rat fibroblasts without affecting the levels of transcripts and translation products from the activated c-Ha-ras (6). The inhibitory effect of a poly(ADP-ribose) polymerase inhibitor, 3-aminobenzamide, on the integration of exogenous DNA (3) and on the modification of transforming efficiency in an oncogene- specific manner (2) was reported.

We found a new function of poly(ADP-ribose) polymerase inhibitors (4). NIH 3T3 cells transformed with exogenous oncogenes reverted to flat and nontumorous cells by losing exogenous transforming genes. Various poly(ADP-ribose) polymerase inhibitors showed similar effects. Transformants yielded by transfecting activated Ha-ras, Ki-ras, N-ras, c-raf, and ret-II all reverted to flat cells after treatment with benzamide. We analyzed the mechanisms of flat reversion.

RESULTS

Induction of Flat Cells by Benzamide

An NIH 3T3 transformant cell, al-1, yielded by transfecting activated genomic c-Ha-ras of human bladder carcinoma T24 cells (hc-Ha-ras^{T24}), was used after cloning. Approximately 30 cells of the clone were plated on a 10-cm Petri dish; on day 2, benzamide was added to a final concentration of 5 mM and the cells were cultured further for an additional 14 da. Eight colonies developed. All colonies were composed of a mixed cell population, i.e., transformed cells and flat cells, the morphology of the latter being the same as normal NIH 3T3 cells. Many flat cell clusters, which were surrounded by transformant cells, were observed within a colony, as shown in Fig. 1A. All colonies that developed in the nontreated control plate were composed of only transformed cells. These results strongly suggest that benzamide induced reversion of transformed cells to normal cells.

This phenomenon was also confirmed with two other independent clones of al-1 cells. The cells obtained after treatment with benzamide were cloned by limiting dilution. Clones with transformed cell morphology (T) and flat revertant cell morphology (F) were obtained. Five clones of

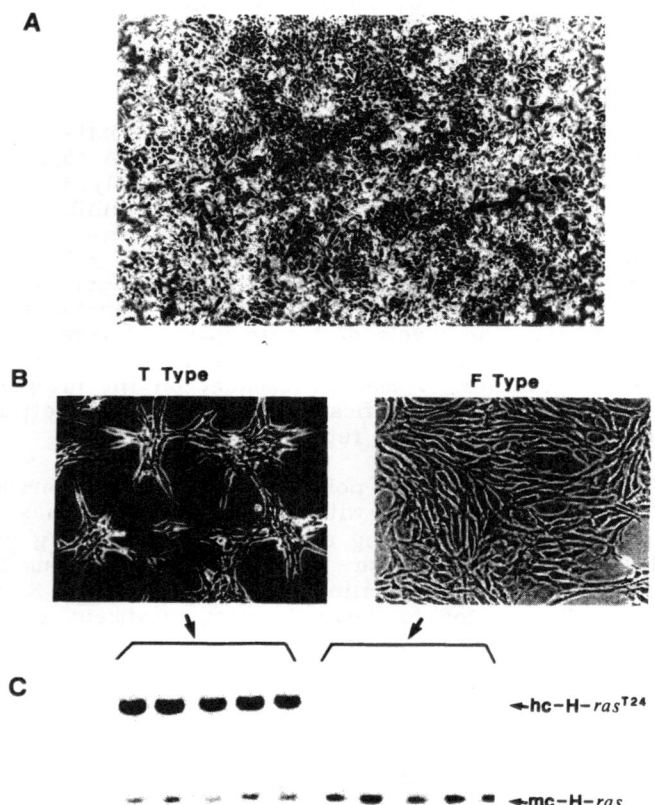

Fig. 1.   Analysis of cell morphology and exogenous hc-Ha-ras^{T24} sequence after cell cloning. (A) Enlarged section of a colony of an al-1 cell clone cultured with 5 mM benzamide for 2 wk. Arrows show clusters of flat revertant cells. (B) Cell morphology of cloned T-type and F-type cells. (C) Exogenous hc-Ha-ras^{T24} and endogenous mc-Ha-ras sequences in T-type and F-type cells.

each cell morphology (T_1-T_5 and F_1-F_5) were subjected to further analysis. The doubling times of T_1 cells and F_1 cells were 18 and 30 hr in the absence of benzamide, and 36 and 48 hr in the presence of benzamide, respectively. Thus, T_1 cells were shown to have a growth advantage over F_1 cells. This suggests that benzamide induced F-type cells with an efficiency that overcame the growth advantage of T-type cells.

Loss of Exogenous hc-Ha-ras^{T24}

Presence of the transforming gene, hc-Ha-ras^{T24}, in flat revertant cells was examined by Southern blot analysis. As shown in Fig. 1, T-type cells retained hc-Ha-ras^{T24}. The ratio of the hybridization band density of hc-Ha-ras^{T24} to endogenous mouse Ha-ras (mc-Ha-ras) was 4:1 in the untreated al-1 cell clones. This hybridization intensity did not change in T-type cells, indicating that complete copies of hc-Ha-ras^{T24} were retained. However, no hybridization band of hc-Ha-ras^{T24} was found in any of the five independent F clones, as illustrated in Fig. 1. Also, in these cases, the endogenous mc-Ha-ras was not affected by benzamide treatment. From these data, it is clear that benzamide induced the complete loss of amplified hc-Ha-ras^{T24} from al-1 cells, resulting in the conversion of T-type cells to F-type cells. F-type as well as T-type cells were morphologically stable for at least 3 mo in the absence of benzamide.

T_1 cells, as well as an untreated clone of al-1 cells, produced tumors in nude mice 3 da after transplantation of 4×10^6 cells. T_1 cell DNA yielded transformant NIH 3T3 cells by transfection. Flat revertants (clones F_1-F_3) did not induce tumors. No transformant was induced by transfecting F_1-F_3 DNA, as expected.

Induction of Flat Revertants by Various Poly(ADP-ribose) Polymerase Inhibitors

We examined whether or not F-type cell induction was a common property of poly(ADP-ribose) polymerase inhibitors, using a low dose of benzamide and various types of inhibitors. About 90 cells of an al-1 clone were plated on a 10-cm Petri dish and cultured for 12 da in the presence of the various compounds. The results are summarized in Tab. 1. The colonies were classified into three morphological groups. Group A colony did not contain any flat cells, group B colony contained less than 50% flat cells, and group C colony contained more than 50% flat cells. Benzamide induced flat revertants even at 1 mM, and the population of flat revertants increased dose-dependently. The NAD$^+$-site inhibitors of poly(ADP-ribose) polymerase, 2-, 3-, and 4-aminobenzamide, and nicotinamide, induced flat cells. A DNA-site inhibitor, coumarin, and a newly found inhibitor, luminol, were also good inducers of flat cells. The noninhibitory analogues of benzamide, benzoic acid, and 3-aminobenzoic acid induced only a slight increase in the appearance of group C colony. The loss of exogenous hc-Ha-ras^{T24} was also observed with luminol, without affecting endogenous mc-Ha-ras (data not shown).

Effect of Benzamide on Transformant Cells with Other Oncogenes

Four transformed cell lines obtained by transfecting activated oncogenes, rat Ki-ras, human N-ras, human c-raf, and human ret-II, were used for examining effects of benzamide on the exogenous oncogenes. From each cell line, three clones were obtained, and 2-3×10^4 cells of

Tab. 1.   Flat cell induction by poly(ADP-ribose) polymerase inhibitors.

Compounds	mM	Morphologies of colonies (%)			Plating efficiency
		A	B	C	(%)
Benzamide	1.0	70.0	21.3	8.6	73.0
	2.5	36.5	38.5	25.0	54.8
	5.0	8.2	27.4	64.4	27.0
2-Aminobenzamide	5.0	52.0	28.3	19.7	64.1
3-Aminobenzamide	5.0	65.8	17.1	17.1	54.1
4-Aminobenzamide	1.0	63.9	24.1	12.0	87.0
Nicotinamide	5.0	60.3	31.8	7.9	83.9
Coumarin	1.0	33.3	41.4	25.3	73.3
Luminol	0.25	54.1	35.9	10.0	77.4
	1.0	4.4	16.8	78.8	41.9
Benzoic acid	1.0	89.0	3.6	7.3	20.4
	2.5	100	0	0	1.9
3-Aminobenzoic acid	1.0	80.6	16.4	5.0	61.1
	5.0	68.4	28.4	3.2	35.2
Untreated		94.3	3.9	1.8	83.9

Approximately 90 al-1 cells were plated on a 10 cm Petri dish and treated with the above compounds for 12 days. Group A, colony without any flat cells; Group B, colony consists of less than 50% of flat cells; Group C, colony consists of more than 50% flat cells. Three plates were used for each dose of each compound.

each clone were plated on 10-cm Petri dishes and cultured for 2 wk in the presence of 5 mM benzamide with passage. The characteristic rat Ki-ras bands observed at 1.7, 3.9, and 11 kb in the EcoRI digest were attenuated after treatment with benzamide. Similarly, human N-ras bands observed at 1.6, 2.6, and 3.8 kb in the PstI digest, human c-raf bands observed at 7.4 kb in the PstI digest, and human ret-II sequence observed at 7.0 kb in the EcoRI digest were attenuated after a 2-wk culture in the presence of 5 mM benzamide (data not shown).

DISCUSSION

     In this study we found that poly(ADP-ribose) polymerase inhibitors can eliminate exogenous transforming genes, irrespective of the properties of transforming gene products. The ras family genes, raf and ret-II, which encode G-proteins, a serine/threonine kinase, and probably a tyrosine kinase, respectively, were all eliminated. All these exogenous transforming genes are originated from human or rat tumors and are amplified in the transformants. Thus, these gene losses might be due to amplification and exogenous and/or species differences. The chromatin structures might be disturbed around the DNAs where exogenous DNA are integrated, and the disturbed region might be susceptible to poly(ADP-ribose) polymerase inhibitors which are potent inducers of sister chromatid exchanges. We do not know whether there are some mechanisms for recognizing species-specific sequences such as the Alu repetitive sequence in humans.

We found a loss of c-myc amplification in granulocyte-differentiated HL-60 cells induced by benzamide (Nagao et al., ms. in prep.). The effect of the poly(ADP-ribose) polymerase inhibitors on integrated viral sequences remains to be studied.

REFERENCES

1. Althaus, F.R., and C. Richter (1987) ADP-Ribosylation of proteins. Molec. Biol. Biochem. Biophys. 37:45-58.
2. Diamond, A.M., C.J. Der, and J.C. Schwartz (1989) Alterations in transformation efficiency by the ADPRT-inhibitor 3-aminobenzamide are oncogene specific. Carcinogenesis 10:383-385.
3. Farzaneh, F., G.N. Panayotou, L.D. Bowler, B.D. Hardas, T. Broom, C. Walther, and S. Shall (1988) ADP-Ribosylation is involved in the integration of foreign DNA into the mammalian cell genome. Nucl. Acids Res. 16:11319-11326.
4. Nakayasu, M., H. Shima, S. Aonuma, H. Nakagama, M. Nagao, and T. Sugimura (1988) Deletion of transfected oncogenes from NIH3T3 transformants by inhibitors of poly(ADP-ribose) polymerase. Proc. Natl. Acad. Sci., USA 85:9066-9070.
5. Sugimura, T. (1973) Poly(adenosine diphosphate ribose). Prog. Nucl. Acid Res. Molec. Biol. 13:127-151.
6. Tseng, A., W.M.F. Lee, E. Kirsten, A. Hakam, J. McLich, K. Buki, and E. Kun (1987) Prevention of tumorigenesis of oncogene-transformed rat fibroblast with DNA site inhibitors of poly(ADP-ribose) polymerase. Proc. Natl. Acad. Sci., USA 84:1107-1111.

INACTIVATION OF MUTAGENIC HETEROCYCLIC AND ARYL AMINES

BY LINOLEIC ACID 13-MONOHYDROPEROXIDE AND METHEMOGLOBIN

Tetsuta Kato, Takuya Takahashi, and Kiyomi Kikugawa

Tokyo College of Pharmacy
1432-1 Horinouchi, Hachioji
Tokyo 192-03, Japan

INTRODUCTION

Mutagenic and carcinogenic heterocyclic amines are generated during pyrolysis of proteins and are present in processed foodstuffs (7). Peroxidation of lipids takes place during processing of foodstuffs. It is thus important to investigate the effect of the free radical species generated during lipid peroxidation on the mutagenic heterocyclic and aryl amines. We have investigated the effect of the radical species generated by interaction of linoleic acid 13-monohydroperoxide (LOOH) with methemoglobin (MetHb) (2,3,4,6) on the mutagenicity of 3-amino-1,4-dimethyl-5H-pyrido[4,3-b]indole (Trp-P-1), 3-amino-1-methyl-5H-pyrido[4,3-b]indole (Trp-P-2), and 2-aminofluorene (2AF) to Salmonella typhimurium TA98. We report here that the radical species inactivated these heterocyclic and aryl amines rather than activated them.

RESULTS

Effect of LOOH and LOOH/MetHb on the Mutagenicity of Trp-P-1 and Trp-P-2 to Salmonella typhimurium TA98

Trp-P-1 (2.5 µg) was preincubated with various doses of LOOH (∿100 µg) at 37°C for 20 min. Analysis of the mixtures using HPLC did not show any significant loss of the mutagen. The mixtures were then incubated with S. typhimurium TA98 strain. Depending on the LOOH dose, mutagenicity to the bacteria developed, but the number of His[+] revertant colonies did not exceed 300 (1.2% of the colonies developed by 2.5 µg Trp-P-1 against TA98 with S-9 mix).

When the mutagenicity of untreated Trp-P-1 (0.25 µg) was tested in the presence of S-9 mix, it produced 2,575 His[+] revertant colonies. Trp-P-1 treated with increasing doses of LOOH showed significantly decreased mutagenicity to the bacteria in the presence of S-9 mix; 40% of the mutagenicity was lost with 100 µg of LOOH. Because HPLC analysis revealed that the conversion of Trp-P-1 by S-9 mix was little affected by

LOOH (∿100 μg), the activity of S-9 may not be destroyed by LOOH. The active metabolite produced by S-9 mix might be partially converted into a nonactive form(s) by LOOH. It seems likely that LOOH hardly reacted with Trp-P-1 but transformed its active metabolite into a nonactive form(s). A similar effect of LOOH was observed on the development of the mutagenicity of Trp-P-2.

Trp-P-1 (2.5 and 0.25 μg) was preincubated with mixtures of LOOH (10 μg) and various amounts of MetHb (∿1,000 μg) at 37°C for 20 min; the LOOH dose employed here hardly affected the mutagenicity of Trp-P-1. Analysis of the mixtures using HPLC revealed that the amount of the mutagen was dramatically decreased at the MetHb dose of 20 μg: a 55% decrease for 2.5 μg of Trp-P-1 and a more than 95% decrease for 0.25 μg of Trp-P-1 (Fig. 1A, solid line). This decrease was inhibited at higher doses of MetHb. The extensive loss of Trp-P-1 may be due to the conversion of the mutagen by the radical species (LOO· or LO·) generated by the interaction of LOOH and MetHb. The inhibition of the loss by a large amount of MetHb may be due to scavenging of the radical species by MetHb. Trp-P-1 treated with LOOH/MetHb at the MetHb dose of 20 μg showed no significant mutagenicity (Fig. 1A, dotted line), indicating that the mutagen was converted into some nondirect-acting form(s) by the radical species.

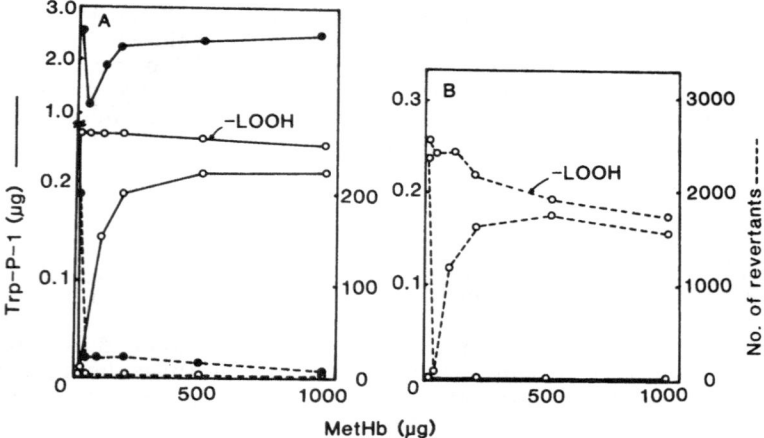

Fig. 1. Effect of LOOH/MetHb on the decrease and the mutagenicity of Trp-P-1 in the absence (A) and presence (B) of S-9 mix. Trp-P-1 [2.5 μg (●) or 0.25 μg (o)] was preincubated with LOOH (10 μg) and the indicated amount of MetHb at 37°C for 20 min. (A) The amount of Trp-P-1 in the preincubation mixtures was determined by HPLC (——). Mutagenicity of the mixtures was measured on incubation with Salmonella typhimurium TA98 at 37°C for 20 min (----). (B) The preincubation mixtures were incubated with S-9 mix at 37°C for 20 min, and the amount of Trp-P-1 was determined (——). The mutagenicity of the preincubation mixtures was assayed by incubation with the bacteria and S-9 mix (----). The effect of MetHb alone is also indicated. All the data shown are the mean values of more than two experiments.

When the mutagenicity of Trp-P-1 treated with LOOH/MetHb was tested in the presence of S-9 mix, the mutagenicity was completely lost at the MetHb dose of 20 μg, and this decrease was suppressed at higher doses of MetHb (Fig. 1B, dotted line). This decrease of mutagenicity corresponded to the loss of mutagen (Fig. 1A, solid line). Thus, Trp-P-1 was effectively degraded by the radical species into a nonmutagenic compound(s) that could not be activated in the presence of S-9 mix. The effect of LOOH/MetHb on the loss and the development of the mutagenicity of Trp-P-2 (2.5 and 0.25 μg) was similar to that in the case of Trp-P-1.

Effect of LOOH and LOOH/MetHb on the Mutagenicity of 2-Aminofluorene to Salmonella typhimurium TA98

2-Aminofluorene (5 μg) was incubated with various doses of LOOH (∿100 μg). The incubation mixtures showed no decrease of the mutagen and no increase in the mutagenicity. Thus, 2AF did not react with LOOH. When the mutagenicity of untreated 2AF (5 μg) was tested in the presence of S-9 mix, it produced 1,120 His$^+$ revertant colonies. Treatment of 2AF with increasing doses of LOOH markedly increased the mutagenicity in the presence of S-9 mix; treatment with LOOH (100 μg) increased the mutagenicity by two-fold. It seems likely that the active metabolite produced by S-9 mix was converted into a more active form(s) by reaction with LOOH.

2-Aminofluorene (5 μg) was treated with a mixture of LOOH (10 μg) and various doses of MetHb (∿1,000 μg). The effect of the mutagen was decreased depending on the MetHb doses; the maximal decrease (60%) occurred at the MetHb doses of 20-100 μg, and a slight decrease at the higher doses of MetHb. 2-Aminofluorene treated with LOOH/MetHb showed a significant increase in the mutagenicity in the absence of S-9 mix, indicating that a certain active form(s) was produced during the loss of the mutagen, but the maximal number of His+ revertant colonies was not more than 200 at the MetHb doses of 20-100 μg (18% of the colonies developed by 5 μg 2AF against TA98 with S-9 mix).

When the mutagenicity of 2AF treated with LOOH/MetHb was tested in the presence of S-9 mix, the mutagenicity was lowered to 60% at the MetHb dose of 100 μg. It is likely that LOOH/MetHb degraded the mutagen into a nonactive form(s) rather than activated it. In contrast to the effective activation of 2AF by LOOH, the active species generated from LOOH/MetHb must have destroyed the mutagen.

DISCUSSION

In the present study, the direct effect of the radical species formed during lipid oxidation on the mutagenic heterocyclic and aryl amines was investigated. The radical species from LOOH/MetHb (2,3,4,6) effectively destroyed the heterocyclic amine mutagens Trp-P-1 and Trp-P-2, and the aryl amine mutagen 2AF. The radical species generated from LOOH/MetHb may be the LOO· or LO· radical, which is the common species formed during lipid oxidation. The potency of the mixture of LOOH/MetHb depended on the amount of MetHb. In every experiment, a catalytic amount of MetHb (20-100 μg) was most effective to degrade the mutagens. It has been previously shown that the potency of the radical species (LOO· or

LO·) was greatly lowered with an increasing amount of MetHb (1,5). The loss of potency caused by MetHb has been considered to be due to the radical scavenging property of the porphyrin ring of the MetHb (1,6).

It may be postulated that treatment of heterocyclic and aryl amine mutagens with the radical species formed during lipid oxidation can effectively transform the mutagens into nonactive forms that cannot be activated by S-9 mix. Mutagenic and carcinogenic heterocyclic amines are generally produced and present in cooked foodstuffs (7). The heterocyclic amines in the foodstuffs can be partly destroyed by free radical species, when lipid oxidation simultaneously takes place during cooking of the food.

REFERENCES

1. Beppu, M., M. Nagoya, and K. Kikugawa (1986) Role of heme compounds in the erythrocyte membrane damage induced by lipid hydroperoxide. Chem. Pharm. Bull. 34:5063-5070.
2. Hamberg, M. (1975) Decomposition of unsaturated fatty acid hydroperoxides by hemoglobin: Structures of major products of 13L-hydroperoxy-9,11-octadecadienoic acid. Lipids 10:87-92.
3. Hawco, F.J., C.R. O'Brien, and P.J. O'Brien (1977) Singlet oxygen formation during hemoprotein catalyzed lipid peroxide decomposition. Biochem. Biophys. Res. Commun. 76:354-361.
4. Kikugawa, K., T. Sasahara, and T. Kurechi (1983) Oxidation of sesamol dimer by active species produced in the interaction of peroxide and hemoglobin. Chem. Pharm. Bull. 31:591-599.
5. Kikugawa, K., T. Nakahara, Y. Taniguchi, and M. Tanaka (1985) Chromogenic determination of lipid hydroperoxides by sesamol dimers. Lipids 20:475-481.
6. O'Brien, P.J. (1969) Intercellular mechanisms for the decomposition of a lipid peroxide. I. Decomposition of a lipid peroxide by metal ions, heme compounds, and nucleophiles. Can. J. Biochem. 47:485-492.
7. Sugimura, T., and S. Sato (1983) Mutagens-carcinogens in foods. Cancer Res. 43(Suppl.):2415-2421.

ESTABLISHMENT OF A HIGHLY REPRODUCIBLE TRANSFORMATION ASSAY OF A Ras-TRANSFECTED BALB 3T3 CLONE BY TREATMENT WITH PROMOTERS

Kiyoshi Sasaki,[1] Hiroshi Mizusawa,[2]
Motoi Ishidate,[2] and Noriho Tanaka[1]

[1]Laboratory of Cell Toxicology, Department of Cell Biology
Hatano Research Institute, Food and Drug Safety Center
729-5 Ochiai, Hadano, Kanagawa 257, Japan

[2]Laboratory of Cell Development, Division of Mutagenesis
National Institute of Hygienic Sciences
1-18-1 Kami-Yoga, Setagaya-ku, Tokyo 158, Japan

INTRODUCTION

In the detection of cancer-inducing agents, it is important to establish appropriate screening systems, not only for initiators but also for promoters. Almost all the initiators can be detected by short-term tests using bacteria, such as the Ames test, or cultured mammalian cells, such as cytogenetic tests and drug-resistant mutation tests (4). In recent years, some studies indicate that active ras genes act as initiators in two-stage carcinogenesis in vivo as well as in vitro (1,2,3,7). We cloned v-Ha-ras-transfected BALB 3T3 cells (Bhas 42) by co-transfection with pSV2-neo genes. Bhas 42 cells were found to be sensitive to contact inhibition, but also to undergo a drastic transformation by treatment with 12-O-tetradecanoylphorbol-13-acetate (TPA) (7). These results suggest that Bhas 42 cells are initiated cells in two-stage transformation. Therefore, we tried to apply Bhas 42 cells to screening of promoters by using the two-stage transformation assay.

MATERIALS AND METHODS

Bhas 42 cells were cloned as described above (Fig. 1) (7). Bhas 42 cells were inoculated at a density of 100 cells together with 10^4 BALB 3T3 cells per dish, treated with TPA for 2 wk, and scored for the number of transformed foci at 6 wk after inoculation. Transformation frequency was expressed as the number of foci per plating efficiency of Bhas 42 cells. As the currently used method, 10^4 BALB 3T3 cells were plated and exposed to 3-methylcholanthrene (MCA) for 72 hr, then treated with TPA, and scored for the number of foci (6).

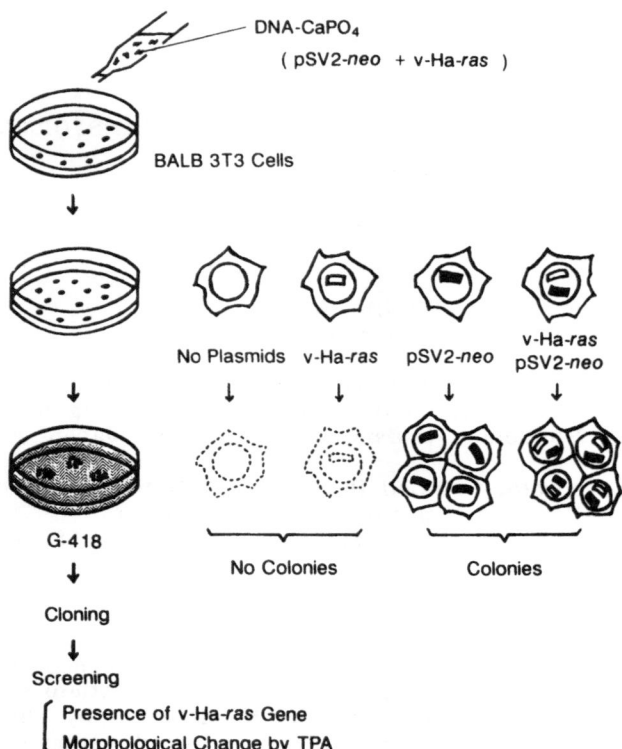

Fig. 1.   Strategy for isolation of v-Ha-ras transfected BALB 3T3 cells.

RESULTS

Characterization of Bhas 42 Cells

        The doubling time of Bhas 42 cells was about 24 hr.   The proliferation of these cells ceased at the confluent state after several days, with a 1.5-fold increase in cell density of BALB 3T3 cells (Fig. 2).   Bhas 30 cells (v-Ha-ras-transfected), which showed transformed morphology, were kept growing for more than 13 da (Fig. 2).

        Bhas 42 cells in the arrested state showed a flat shape and formed a monolayer as is the case in BALB 3T3 cells (Fig. 3A).   However, when Bhas 42 cells were exposed to TPA, a dramatic transformation was observed, as shown by the criss-cross, multilayered, and fibroblastic morphology (Fig. 3B).   These morphological changes of Bhas 42 cells occurred within 36 hr after the addition of TPA.   On the other hand, only the increase in cell density was observed in control BALB 3T3 cells (Fig. 3C, D).   These results imply that Bhas 42 cells are the initiated cells in two-stage transformation (7).

Bhas 42 Transformation Assay System

        In the standard transformation assay, each MCA-induced transformed focus appeared with various morphological types, and each focus was

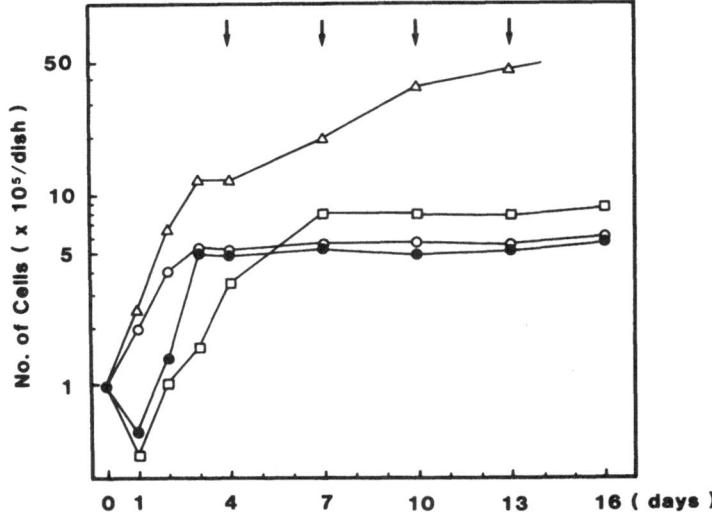

Fig. 2. Comparison of cell growth among various BALB 3T3 clones transfected with exogenous DNAs. (●) BALB 3T3; (○) BALB 3T3 transfected with pSV2-neo; (□) Bhas 42; (△) Bhas 30 cells. Arrows indicate the timing of medium changes.

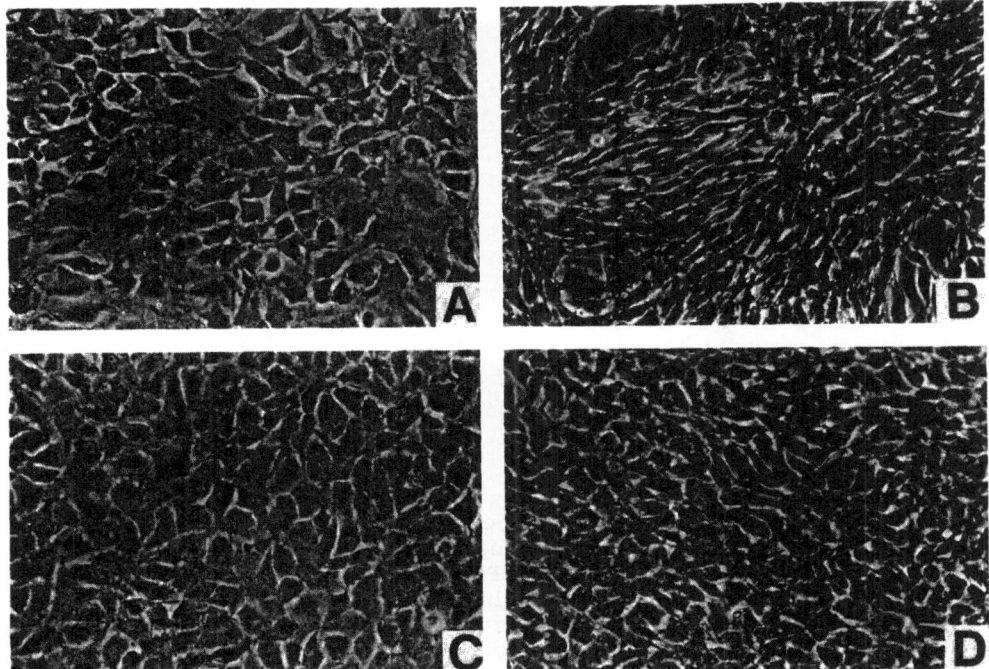

Fig. 3. Morphological alterations of quiescent cells at 5 da after addition of TPA (1,000 ng/ml). (A) Bhas 42 cells (control) without TPA; (B) Bhas 42 cells with TPA; (C) BALB 3T3 cells (control) without TPA; (D) BALB 3T3 cells with TPA.

difficult to judge without the criteria described below (10). In the control experiment, transformed foci were not observed in BALB 3T3 cells with or without TPA treatment (Fig. 4A, B). Without TPA treatment, foci of Bhas 42 cells were not induced after 6 wk of culture (Fig. 4C). However, after treatment with TPA, the majority of foci of Bhas 42 cells displayed a typical transformed phenotype having the following properties: basophilic, multilayered, criss-crossed, and spindle-shaped morphology (Fig. 4D). Therefore, the Bhas 42 system was easy to score objectively for the number of transformed foci.

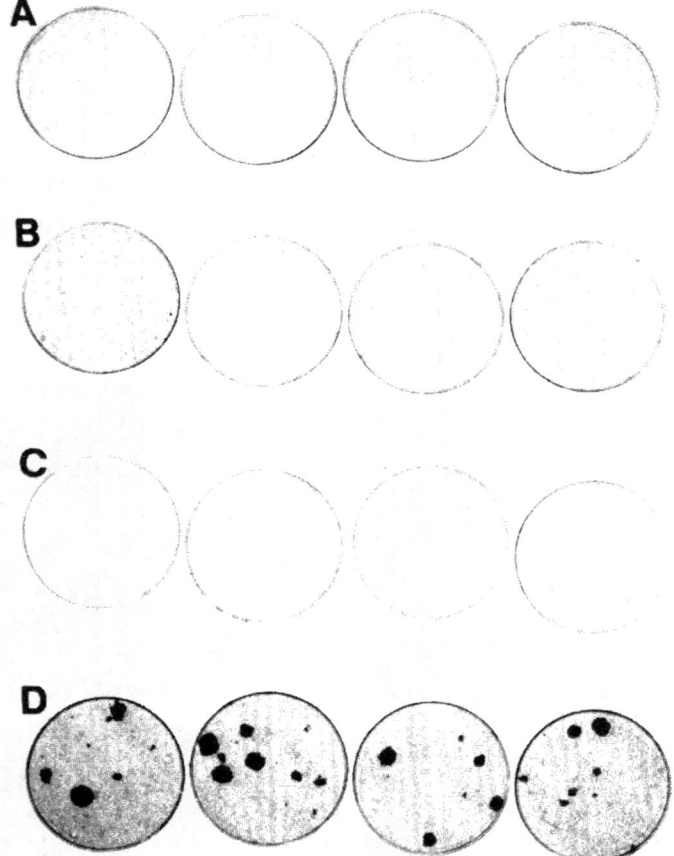

Fig. 4.   Induction of transformed foci of Bhas 42 cells. Row A: Dishes plated with 100 BALB 3T3 cells and 10^4 BALB 3T3 cells together, and cultured without TPA treatment. Row B: Dishes plated as described in A, and cultured with TPA treatment. Row C: Dishes plated with 100 Bhas 42 cells and 10^4 BALB 3T3 cells together, and cultured without TPA treatment. Row D: Dishes plated as described in C, and cultured with TPA treatment. All of the cells were cultured for 6 wk. TPA was treated at a concentration of 1,000 ng/ml for 2 wk.

The frequency of chemically-induced transformation is dependent on batches of serum, not only in BALB 3T3 cells but also in other cells (5,8,9). On the contrary, the transformation frequency of Bhas 42 cells was constant in the sense that about 20% of the cells formed transformed foci by treatment with TPA (1,000 ng/ml, for 2 wk) in several batches of fetal calf serum.

The transformation frequency of Bhas 42 cells was dependent on concentrations (1-1,000 ng/ml) and treatment times (3-14 da) of TPA. This frequency was highly reproducible in each experiment as compared with the MCA-induced experiment.

DISCUSSION

We established a highly reproducible transformation assay system using Bhas 42 cells. The Bhas 42 system had an advantage over the initiator-induced transformation assay system at the following three points: (1) objective criteria of transformed foci; (2) no influence of batches of serum; and (3) high reproducibility of the transformation frequency in each experiment. The present study shows that the Bhas 42 transformation assay system is useful for screening of promoters and antipromoters.

REFERENCES

1. Brown, K., M. Quintanilla, M. Ramsden, I.B. Kerr, S. Young, and A. Balmain (1986) v-ras genes from Harvey and BALB murine sarcoma viruses can act as initiators of two-stage mouse skin carcinogenesis. Cell 46:447-456.
2. Dotto, G.P., L.F. Parada, and R.A. Weinberg (1985) Specific growth response of ras-transformed embryo fibroblasts to tumour promoters. Nature 318:472-475.
3. Hsiao, W.-L.W., S. Gattoni-Celli, and I.B. Weinstein (1984) Oncogene-induced transformation of C3H 10T1/2 cells is enhanced by tumor promoters. Science 226:552-555.
4. International Commission for Protection Against Environmental Mutagens and Carcinogens (1985) Guide to Short-term Tests for Detecting Mutagenic and Carcinogenic Chemicals (Environmental Health Criteria 51), World Health Organization, Geneva.
5. Oshiro, Y., P.S. Balwierz, and C.E. Piper (1982) Selection of fetal bovine serum for use in the C3H/10T1/2 CL8 cell transformation assay system. Environ. Mutag. 4:569-574.
6. Sasaki, K., K. Chida, H. Hashiba, N. Kamata, E. Abe, T. Suda, and T. Kuroki (1986) Enhancement by 1-α-25-dihydroxyvitamin D_3 of chemically induced transformation of BALB 3T3 cells without induction of ornithine decarboxylase or activation of protein kinase C. Cancer Res. 46:604-610.
7. Sasaki, K., H. Mizusawa, and M. Ishidate (1988) Isolation and characterization of ras-transfected BALB/3T3 clone showing morphological transformation by 12-O-tetradecanoyl-phorbol-13-acetate. Jpn. J. Cancer Res. (Gann) 79:921-930.
8. Schuman, R.F., R.J. Pienta, J.A. Poiley, and W.B. Lebherz III (1979) Effect of fetal bovine serum on 3-methylcholanthrene-induced transformation of hamster cells in vitro. In Vitro 15:730-735.

9.   Sivak, A., M.C. Charest, L. Rodenko, D.M. Silveira, I. Simons, and A.M. Wood (1980) BALB/c-3T3 as target cells for chemically induced neoplastic transformation. In Advances in Modern Environmental Toxicology, Vol. 1, Mammalian Cell Transformation by Chemical Carcinogens, N. Mishra, V. Dunkel, and M. Mehlman, eds. Senate Press, Princeton Junction, New Jersey, pp. 133-180.

10. Working Group (1985) Recommendations for experimental protocols and for scoring transformed foci in BALB/c 3T3 and C3H 10Tl/2 cell transformation. In Transformation Assay of Established Cell Lines: Mechanisms and Application (IARC Scientific Publications No. 67), T. Kakunaga and H. Yamasaki, eds. International Agency for Research on Cancer, Lyon, pp. 207-219.

ANTIMUTAGENIC EFFECT OF umuD MUTANT PLASMIDS: ISOLATION AND CHARACTERIZATION OF umuD MUTANTS REDUCED IN THEIR ABILITY TO PROMOTE UV MUTAGENESIS IN ESCHERICHIA COLI

Takehiko Nohmi,* John R. Battista, Toshihiro Ohta,
Vivien Igras, William Sun, and Graham C. Walker

Department of Biology
Massachusetts Institute of Technology
Cambridge, Massachusetts 02139

INTRODUCTION

umuD and umuC are essential genes for UV and most chemical muta-genesis in Escherichia coli (4,15,18). Both umuD and umuC mutants are virtually nonmutable with ultraviolet (UV) and a variety of chemicals (6,17). The umuD and umuC genes are organized as an operon (4,15) and encode proteins of 15.0 and 47.7 kilodaltons (kDa), respectively (7,12). The umuDC operon is repressed by the LexA protein (1,4) and regulated as a part of SOS response of E. coli (10,18,19). SOS response occurs by activated RecA mediating the proteolytic cleavage of a bond between Ala^{84} and Gly^{85} of LexA (9), apparently facilitating a conditional autodigestion of LexA (16).

UmuD shares sequence homology with LexA and the repressors of bacteriophages lambda, 434, P22, and ϕ80 (3,12). This similarity led us to hypothesize that UmuD is cleaved at a homologous site between Cys^{24} and Gly^{25} in a RecA-dependent fashion during SOS induction. Cleavage was recently demonstrated in vivo and in vitro (2,14), and we have shown that the resulting C-terminal polypeptide (12 kDa polypeptide, UmuD*) is necessary and sufficient for the role of UmuD in UV mutagene-sis (11). In this paper, we have isolated and sequenced a set of umuD mutants reduced in their ability to promote UV mutagenesis in order to better define the structural determinants required for the RecA-mediated cleavage.

---

*Present address: Division of Mutagenesis, Biological Safety Research Center, National Institute of Hygienic Sciences, 1-18-1 Kamiyoga, Seta-gaya-Ku, Tokyo 158, Japan.

ISOLATION AND SEQUENCE OF umuD MUTANTS REDUCED IN THEIR ABILITY TO PROMOTE UV MUTAGENESIS

The umuD mutants were generated by mutagenizing pGW2020 (11), a multicopy plasmid carrying wild-type umuD, with hydroxylamine. The mutagenized plasmids were introduced into AB1157 umuD44 strain, and transformants which showed a less UV-mutable phenotype were identified by screening. Candidates were further characterized by quantitative mutagenesis assay after transfer of the plasmid DNA into a fresh umuD44 background.

Out of about 5,000 transformants screened, we isolated 15 independent mutant plasmids putatively carrying umuD mutations. The umuD44 strains harboring these mutant plasmids were significantly less mutable (0.4 to 25%) than those carrying their wild-type parent, pGW2020. To identify the mutational change in these mutants, the entire region of the umuD gene and its promoter region were then sequenced. All the mutations that we found resulted in the alteration of a single base pair, and all except one were missense mutations. Some of the independent mutations had the same sequence changes, so that the collection contains a total of 11 different mutations, affecting ten residue sites (Fig. 1). Nearly half of the mutations (5/11) were localized in the region between Cys^{24} and Ala^{30} close to where the cleavage site of UmuD is located. No mutations were found in the promoter region. When MucA, LexA, lambda

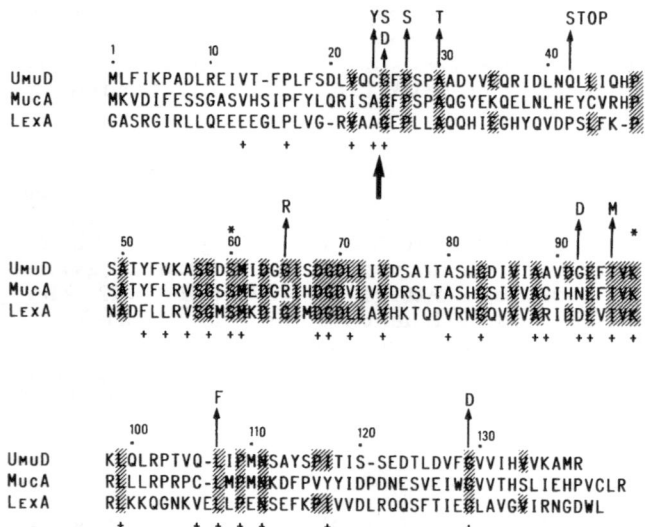

Fig. 1.   Location of mutations in the UmuD protein and homology with MucA and LexA. The small arrows (⟶) indicate the mutational changes of UmuD. Residues conserved between UmuD and LexA are highlighted (////////). + indicates the residues conserved between LexA and cI repressor of lambda phage. The large arrow (⟹) indicates the site of cleavage of LexA mediated by RecA protein. * indicates the residue of Ser^{60} or Lys^{97}. Numbers indicate the position of amino acid residues from N-terminus of UmuD.

cI repressor, and UmuD are aligned, 17 residues are conserved. Five of the sites of umuD mutations (Gly^{25} to Ser, Gly^{25} to Asp, Thr^{95} to Met, Leu^{107} to Phe, Gly^{129} to Asp) affect one of these highly conserved positions, and three mutations (Pro^{27} to Ser, Ala^{30} to Thr, Gly^{65} to Arg) affect residues that are conserved between UmuD and LexA.

The changes at the cleavage site (Cys^{24} to Tyr, Gly^{25} to Ser, Gly^{25} to Asp) abolished the ability of UmuD to function in UV mutagenesis by more than 97%, while the changes of Pro^{27} to Ser and Ala^{30} to Thr caused less severe reduction (about 85%) of the ability of UmuD to function in UV mutagenesis. The mutant whose change (Gly^{65} to Arg) is located around Ser^{60} showed about 5% remaining ability compared to the wild-type UmuD, whereas the mutants (Gly^{92} to Asp, Thr^{95} to Met) close to Lys^{97} had 15% and 25% remaining ability to promote UV mutagenesis. The mutation of Leu^{107} to Phe caused about 85% reduction of the ability of UmuD. The mutation whose change (Gly^{129} to Asp) is close to the C-terminal end had the strongest effect on the function of UmuD, and a derivative of the umuD44 strain harboring this plasmid had only 0.4% remaining UV mutability compared to that having wild-type umuD plasmid. The mutant with the nonsense mutation (Gln^{42} to Stop) had no ability to promote UV mutagenesis. Lin and Little (8) have previously reported that lexA mutations randomly generated with hydroxylamine and formic acid were clustered in three regions, namely, around the cleavage site and around two residues, Ser^{119} and Lys^{156}, which correspond to Ser^{60} and Lys^{97} of UmuD. Our umuD mutations were, however, located not only in the three regions but also in the region which is close to the C-terminal end of UmuD.

THE MUTANT umuD PROTEINS ARE NONCLEAVABLE

To investigate whether the mutant UmuD proteins are cleavable in a RecA-dependent fashion, we have labeled the cellular proteins of an AB1157 derivative harboring the plasmids that carry the umuD mutations after UV irradiation, and have analyzed them by SDS polyacrylamide gel electrophoresis followed by fluorography. The pulse-labeling experiments indicated that all the mutant UmuD proteins, except for the mutant harboring the nonsense mutation, were induced and accumulated their uncleaved form after UV irradiation, whereas the wild-type UmuD was rapidly processed to UmuD*. Although most of the mutant UmuD proteins showed almost the same mobility as the wild-type UmuD protein on the gel, certain of the mutant UmuD proteins exhibited irregular mobility. The mutant UmuD having a change of Gly^{65} to Arg or Thr^{95} to Met appeared to migrate faster than the wild type, and the mutant having a change of Gly^{92} to Asp appeared to migrate slower than the wild type. Examination of radiolabeled proteins in a recA1 background using the maxicell technique (13) confirmed that these bands having irregular mobility also represented the uncleaved forms of the mutant UmuD proteins.

From these results, we concluded that all these missense mutants of UmuD were not cleaved efficiently, if at all, following irradiation. As mentioned above, none of the missense mutants of UmuD completely abolished UV mutability. It is possible that the uncleaved UmuD has some limited capacity to participate in UV mutagenesis, or that some proteolysis is occurring which cannot be detected by the methodology we have employed to examine cleavage.

THE MUTANTS OF UmuD HAD ANTIMUTAGENESIS EFFECTS
ON UV-INDUCED AND SPONTANEOUS MUTAGENESIS

   We have introduced the plasmids carrying the various mutant umuD
alleles into a umuD$^+$ strain and compared their UV mutability with the
mutability of one harboring pGW2020. The introduction of plasmids which
have the mutation around the cleavage site (Cys^{24} to Tyr, Gly^{25} to Ser,
Gly^{25} to Asp) or the C-terminal end (Gly^{129} to Asp) reduced UV mutability of the umuD$^+$ strain by more than 90%. The most effective mutation
was the change of Gly^{25} to Asp, which reduced the mutability of the
umuD$^+$ strain by 98%. All mutants except for that giving rise to Gln^{42} to
Amber exhibited dominance. The plasmid carrying mutation of Thr^{95} to
Met had the weakest effect and reduced the mutability by 30%.

   The recA730 mutation causes constitutive expression of SOS functions
and increases spontaneous mutation frequency (19). We have introduced
the plasmids carrying the various umuD mutations into a recA730 background and tested whether the introduction of the plasmids affects the
spontaneous mutation frequency of this strain. The plasmids carrying a
mutation of Cys^{24} to Tyr or Gly^{129} to Asp had strong effects, and they
reduced the spontaneous mutation frequency by more than 75%. The
plasmids carrying mutations of Gly^{65} to Arg and Thr^{95} to Met showed
moderate effects, and they reduced the frequency by 70% and 60%, respectively. The mutation of Leu^{107} to Phe had weak effects, and it reduced the mutation frequency by 40%.

   These results indicate that the mutants of umuD are dominant to
chromosomal umuD$^+$ and that the degree of dominance depends on the site
of mutation. The mutants of UmuD might titrate out another component
essential to mutagenesis such as wild-type UmuD (if UmuD, like LexA, is
a dimer), UmuC, or RecA, thereby interfering with induced and spontaneous mutagenesis. Overexpression of some mutant polypeptides can dis-
rupt the activity of the wild-type protein (5). These dominant negative
mutants in genes encoding essential proteins involved in mutagenesis
could be a most effective antimutagen, i.e., a specific inhibitor of UV and
spontaneous mutagenesis in E. coli.

REFERENCES

1. Bagg, A., C.J. Kenyon, and G.C. Walker (1981) Inducibility of a
    gene product required for UV and chemical mutagenesis in Escheri-
    chia coli. Proc. Natl. Acad. Sci., USA 78:5749-5753.
2. Burckhardt, S.E., R. Woodgate, R.H. Scheuermann, and H. Echols
    (1988) UmuD mutagenesis protein of Escherichia coli: Overproduc-
    tion, purification, and cleavage by RecA. Proc. Natl. Acad. Sci.,
    USA 85:1811-1815.
3. Eguchi, Y., T. Ogawa, and H. Ogawa (1988) Cleavage of bacterio-
    phage φ80 CI repressor by RecA protein. J. Molec. Biol. 202:565-
    574.
4. Elledge, S.J., and G.C. Walker (1983) Proteins required for ultra-
    violet light and chemical mutagenesis: Identification of the products
    of the umuC locus of E. coli. J. Molec. Biol. 164:175-192.
5. Herskowitz, I. (1988) Functional inactivation of genes by dominant
    negative mutations. Nature 329:219-222.

6. Kato, T., and Y. Shinoura (1977) Isolation and characterization of mutants of Escherichia coli deficient in induction of mutations by ultraviolet light. Molec. Gen. Genet. 156:121-131.

7. Kitagawa, Y., E. Akaboshi, H. Shinagawa, T. Horii, H. Ogawa, and T. Kato (1985) Structural analysis of the umu operon required for inducible mutagenesis in Escherichia coli. Proc. Natl. Acad. Sci., USA 82:4336-4340.

8. Lin, L., and J.W. Little (1988) Isolation and characterization of non-cleavable (Ind⁻) mutants of the LexA repressor of Escherichia coli K-12. J. Bacteriol. 170:2163-2173.

9. Little, J.W., S.H. Edmiston, L.Z. Pacelli, and D.W. Mount (1980) Cleavage of the Escherichia coli lexA protein by the recA protease. Proc. Natl. Acad. Sci., USA 77:3225-3229.

10. Little, J.W., and D.W. Mount (1982) The SOS regulatory system of Escherichia coli. Cell 29:11-22.

11. Nohmi, T., J.R. Battista, L.A. Dodson, and G.C. Walker (1988) RecA-mediated cleavage activates UmuD for mutagenesis: Mechanistic relationship between transcriptional derepression and posttranslational activation. Proc. Natl. Acad. Sci., USA 85:1816-1820.

12. Perry, K.L., S.J. Elledge, B.B. Mitchell, L. Marsh, and G.C. Walker (1985) umuDC and mucAB operons whose products are required for UV light- and chemical-induced mutagenesis: UmuD, MucA, and LexA protein share homology. Proc. Natl. Acad. Sci., USA 82:4331-4335.

13. Sancar, A., R.P. Wharton, S. Seltzer, B.M. Kacinski, N.D. Clarke, and W.D. Rupp (1981) Identification of the uvrA gene product. J. Molec. Biol. 148:45-62.

14. Shinagawa, H., H. Iwasaki, T. Kato, and A. Nakata (1988) RecA protein-dependent cleavage of UmuD protein and SOS mutagenesis. Proc. Natl. Acad. Sci., USA 85:1806-1810.

15. Shinagawa, H., T. Kato, T. Ise, K. Makino, and A. Nakata (1983) Cloning and characterization of the umu operon responsible for inducible mutagenesis in Escherichia coli. Gene 23:167-174.

16. Slilaty, S.N., and J.W. Little (1987) Lysine-156 and serine-119 are required for LexA repressor cleavage: A possible mechanism. Proc. Natl. Acad. Sci., USA 84:3987-3991.

17. Steinborn, G. (1978) Uvm mutants of Escherichia coli deficient in UV mutagenesis. I. Isolation of uvm mutants and their phenotypical characterization in DNA repair and mutagenesis. Molec. Gen. Genet. 165:87-93.

18. Walker, G.C. (1984) Mutagenesis and inducible responses to deoxyribonucleic acid damage in Escherichia coli. Microbiol. Rev. 48:60-93.

19. Witkin, E.M. (1976) Ultraviolet mutagenesis and inducible DNA repair in Escherichia coli. Bacteriol. Rev. 40:869-907.

20. Witkin, E.M., J.O. McCall, M.R. Volkert, and I.E. Wermundsen (1982) Constitutive expression of SOS functions and modulation of mutagenesis resulting from resolution of genetic instability at or near the recA locus of Escherichia coli. Molec. Gen. Genet. 185:43-50.

ANTITUMOR-PROMOTING ACTIVITY OF SESQUITERPENE

ISOLATED FROM AN HERBAL SPICE

Takeshi Matsumoto[1] and Harukuni Tokuda[2]

[1]Research Center
Daicel Chemical Industries, Ltd.
Aboshi-ku, Himeji 671-12, Japan

[2]Department of Microbiology
Faculty of Medicine
Kyoto University
Yoshida, konoe-cho, Sakyo-ku, Kyoto 606, Japan

INTRODUCTION

Recent studies on Epstein-Barr virus (EBV) activation in human lymphoblastoid cell latently infected Raji cells revealed a considerable overlapping between the inhibitors of EBV activation and the antitumor promoters of mouse skin carcinogenesis (5). Based on this parallel relationship, we have widely searched for possible antitumor promoters from edible plants, using the short-term in vitro assay of EBV activation in Raji cells (4). In the course of the survey, our interest has focused upon herbal spices commonly used in Japanese food, and we found that the crude methanol extract of one herbal spice, water pepper (Polygonum hydropiper), effectively inhibited EBV activation.

Water pepper has long been used as a hot-tasting spice in China, Japan, and Europe. The sprout of water pepper, called "mejiso" or "benitade" in Japanese, is a well-known relish for "sashimi," and the seed was sometimes used as a substitute for pepper in Europe. Water pepper is also used as a folk medicine against tumors (scirrhous, hydropic, and edematous tumors), uterine fibromas, and malignant ulcers (3).

In this paper we report the isolation of the active components from the sprout of water pepper, monitoring them by the inhibitory activity on EBV activation and the in vivo effects of the components on mouse skin. We also discuss here the in vitro and/or in vivo activities of a variety of natural or synthetic aldehydes, since the aldehyde group in the active components appears to be essential for the antitumor-promoting activity.

MATERIALS AND METHODS

Bioassay of EBV Activation (In Vitro Test)

The inhibition of EBV activation was assayed using the same method as described previously (4). The cells were incubated for 48 hr in a medium containing n-butyric acid, 12-O-tetradecanoylphorbol-13-acetate (TPA), and test compounds. The expression of EBV was observed by indirect immunofluorescence and compared to that of control without test compounds.

Tumor Promotion Test (In Vivo Test)

The method employed was described in detail previously (5). Mice were initiated with DMBA and were treated with the test compound 1 hr prior to each promotion with TPA. The incidence of papillomas was observed weekly for 20 wk.

Isolation of the Inhibitors

Isolation of the inhibitors was monitored by the bioassay of EBV activation. Fresh sprouts of Polygonum hydropiper L. were extracted with methanol with sonification. The crude methanol extract was dispersed into water, and the aqueous suspension was extracted with chloroform, ethyl acetate, and n-butanol. The chloroform fraction showed potent inhibition and was fractionated by silica gel column chromatography and/or reversed-phase (C18) liquid chromatography to give two sesquiterpenes, polygodial and warburganal, as active components.

Binding Assay

The binding assay was performed by a modified Blumberg's method (1). The inhibition of [^{3}H]TPA binding by TPA, teleocidin B, and warburganal was tested. The mouse skin particulate was first incubated with various concentrations of inhibitors for 30 min at 4°C. Four pmol of [^{3}H]TPA was then added and further incubated for 3 hr at 4°C. Radioactivity was measured in the usual manner. All values were corrected for quenching, and nonspecific binding of [^{3}H]TPA was determined in the presence of large excess amounts of unlabeled TPA.

RESULTS AND DISCUSSION

The crude methanol extract from the sprout of P. hydropiper inhibited 80% of the induction of EBV-EA (EBV early antigen) on Raji cells at 100 μg/ml. This has now led to the isolation of potent inhibitors, sesquiterpene dialdehydes, of EBV activation. The inhibitors were identified as polygodial (1) and warburganal (2) by the direct comparison of the spectral data with those of authentic compounds (2).

The inhibitory effects of these dialdehydes on EBV activation and viabilities of Raji cells are shown in Tab. 1. Both dialdehydes exhibited strong activities at 100 (mol ratio/TPA), especially warburganal, which showed remarkable activity even at 10 (mol ratio/TPA). (No EBV-EA-inducing activity was observed at the 10 μg/ml of the dialdehydes.) The inhibitory activity of warburganal is stronger than that of retinoic acid, which is known as a typical antitumor promoter (6).

Tab. 1.   Inhibitory effects of various aldehydes on TPA-induced EBV activation.

Sample	Concentration (mol ratio/TPA) **		
	1000	100	10
	% to control (% viability)		
TPA (32 pmol) 100 *			
Aldehydes			
Hexadienal	0 (10.0)	91.7(80.0)	100
Hexenal	0 (00)	83.3(70.0)	100
Cyclohexanecarbox aldehyde	83.3(70.0)	100	100
Glyceraldehyde	39.2(70.0)	91.3	100
n-Butyraldehyde	73.0(70.0)	100	100
n-Valeraldehyde	80.0(70.0)	100	100
n-Hexylaldehyde	50.0(70.0)	92.3	100
n-Heptylaldehyde	50.0(70.0)	76.9	100
n-Octylaldehyde	75.0(60.0)	87.5	100
n-Nonylaldehyde	37.5(60.0)	41.7	100
n-Decylaldehyde	29.2(70.0)	62.5	100
n-Dodecylaldehyde	28.4(60.0)	65.7	100
Benzaldehyde	56.5(60.0)	100	100
Phenylacetaldehyde	52.0(60.0)	100	100
Salicylaldehyde	15.2(70.0)	74.1	91.2
p-Anisaldehyde	79.8(80.0)	86.3	100
α-Naphtaldehyde	54.5(70.0)	88.8	100
trans-Cinnamaldehyde	88.9(60.0)	100	100
Furfral	100 (70.0)	100	100
Sesquiterpene dialdehyde			
Warburganal	0 (70.0)	15.8	50.8
Polygodial	0 (60.0)	50.5	90.8

 * Positive control
 ** Relative ratio as compared to TPA (32 pmol)

    The inhibitory effects of warburganal and polygodial on TPA-induced tumor promotion are shown in Fig. 1 and 2.   They delayed the formation of papillomas, and markedly reduced both the rate of papilloma-bearing mice and the number of papillomas per mouse.   Thus, these sesquiterpene dialdehydes were found to be strong antitumor promoters, in which warburganal was more effective than polygodial.

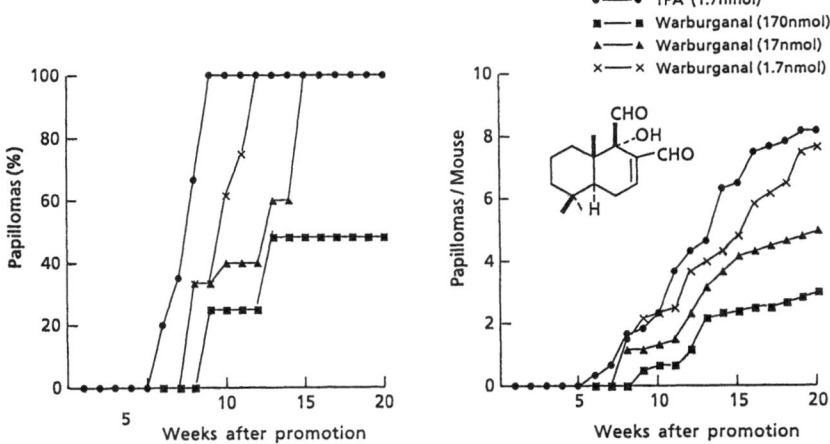

Fig. 1.   Inhibition of TPA-induced tumor promotion by warburganal.

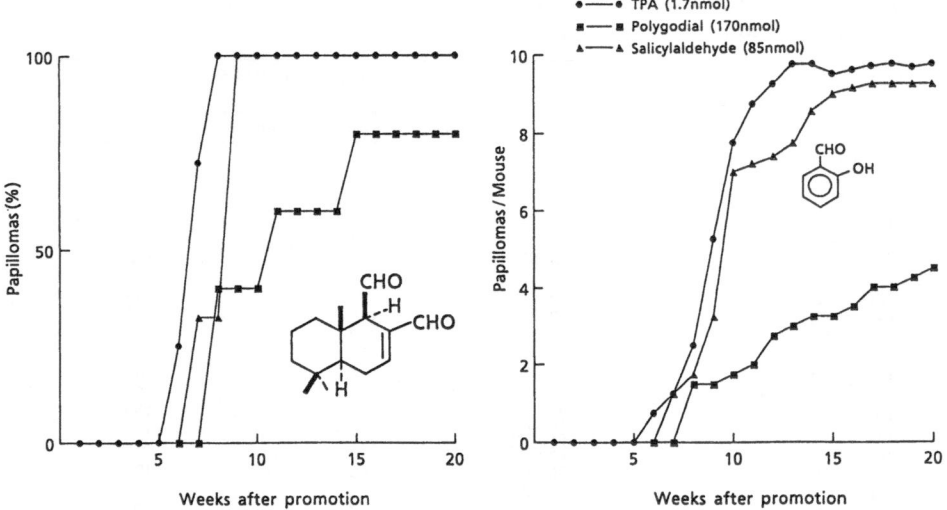

Fig. 2.   Inhibition of TPA-induced tumor promotion by polygodial and salicylaldehyde.

As shown in Fig. 3, warburganal did not inhibit the specific binding of [^{3}H]TPA; therefore, it is suggested that the dialdehydes act on some event(s) after binding of TPA to the receptor(s) of mouse skin.

To find the structural requirement for EBV-EA induction, 20 aldehydes (natural or synthetic) were studied by the in vitro assay, and the results are listed in Tab. 1.   Among the saturated acyclic monoaldehydes, nonyl-, decyl-, and dodecylaldehydes exhibited more potent activities than the shorter-chain aldehydes.   Generally, the saturated acyclic or aromatic monoaldehydes preserved high viability of the cells;   whereas

Fig. 3.   Specific binding of the inhibitors to mouse skin particulate.

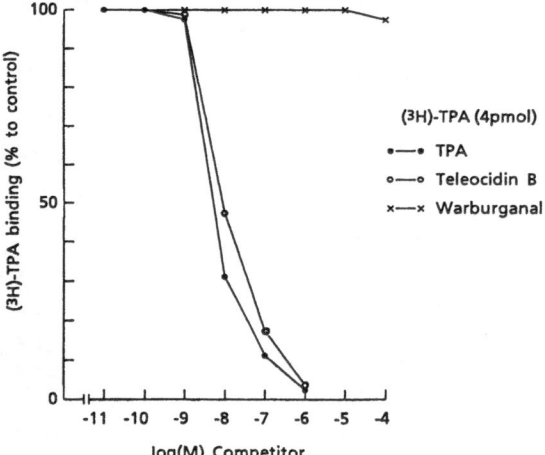

unsaturated acyclic monoaldehydes showed strong cytotoxicity against the Raji cells. Trans-cinnamaldehyde had no activity even at 1,000 (mol ratio/TPA), and this implies that the enal moiety alone is not sufficient. Moreover, it should be noted that the inhibitory activity of salicylaldehyde is stronger than that of benzaldehyde. This result seems to correspond to the data shown for the sesquiterpene dialdehydes, and suggests that a hydroxyl group in close proximity to the aldehyde moiety may enhance the inhibitory activity. It is more important, however, that none of the aldehydes listed in Tab. 1 is comparable to the natural dialdehydes in the inhibition of EBV activation or of tumor promotion in vivo (Fig. 2). These results suggested that the functionality of both the enal moiety and the other aldehyde group is essential for the inhibition of tumor promotion on mouse skin.

The expressed juice of the freshly gathered green leaves of water pepper has been used as a dip sauce when we eat Japanese traditional river fish, called "ayu," as a grilled fish which contains highly mutagenic materials from heating protein during cooking. Various naturally-occurring substances, especially of plant origin, are known to arrest the promotion of tumor growth by carcinogens. In the present study, data show a possible relationship between an herbal spice and prevention of the occurrence of cancer.

REFERENCES

1. Dunphy, W.G., K.B. Delclos, and P.M. Blumberg (1980) Characterization of specific binding of (^{3}H)phorbol-12,13-dibutyrate and (^{3}H)phorbol-12-myristate-13-acetate to mouse brain. Cancer Res. 40:3635-3641.
2. Fukuyama, Y., T. Sato, I. Miura, and Y. Asakawa (1985) Drimane-type sesqui- and norsesquiterpenoids from Polygonum hydropiper. Phytochemistry 24:1521-1524.
3. Hartwell, J.L. (1982) Plant Used Against Cancer, Quarterman Publications, Inc., Massachusetts, pp. 477-478.
4. Ito, Y., S. Yanase, J. Fujita, T. Harayama, M. Takashima, and H. Imanaka (1981) A short-term in vitro assay for promoter substances using human lymphoblastoid cells latently infected with Epstein-Barr virus. Cancer Lett. 13:29-37.
5. Tokuda, H., H. Ohigashi, K. Koshimizu, and Y. Ito (1986) Inhibitory effects of ursolic and oleanolic acid on skin tumor promotion by 12-O-tetradecanoylphorbol-13-acetate. Cancer Lett. 33:279-285.
6. Verma, A.K., T.J. Slaga, R.W. Wertz, G.C. Mueller, and R.K. Boutwell (1980) Inhibition of skin tumor promotion by retinoic acid and its metabolite 5,6-epoxy-retinoic acid. Cancer Res. 40:2367-2371.

INHIBITORY EFFECTS OF CHLOROGENIC ACID, RESERPINE, POLYPRENOIC ACID (E-5166), OR COFFEE ON HEPATOCARCINOGENESIS IN RATS AND HAMSTERS

Takuji Tanaka, Akiyoshi Nishikawa, Hiroto Shima,
Shigeyuki Sugie, Tokuro Shinoda, Naoki Yoshimi,
Hitoshi Iwata, and Hideki Mori

Department of Pathology
Gifu University School of Medicine
40 Tsukasa-machi, Gifu City 500, Japan

ABSTRACT

Four different experiments were performed in order to examine the modifying effects of chlorogenic acid (CA), reserpine, polyprenoic acid (E-5166), and coffee on chemical carcinogenesis in rats or hamsters. Experiment 1: The numbers of hyperplastic liver cell foci and the incidence of colon tumors in male and female Syrian golden hamsters given a single intravenous injection of methylazoxymethanol (MAM) acetate and then fed the diet containing 0.025% CA for 24 wk were significantly lower than those of hamsters given MAM acetate alone. Experiment 2: The incidence of altered hepatocellular foci in female ACI/N rats given N-2-fluorenylacetamide (FAA, 0.02% in diet) for 10 wk and reserpine (weekly subcutaneous injections, 1 µg/g body weight) during or after (17 wk) FAA exposure was significantly lower than that of rats given FAA alone. Experiment 3: The number of hepatocellular foci in male ACI/N rats given 0.02% FAA diet for 13 wk and E-5166 by gavage (40 mg/kg body weight, 3 times/wk) for 16 wk after the end of FAA exposure was significantly smaller than that in rats given FAA diet alone. Experiment 4: Incidences of liver tumors and hepatocellular foci of rats given concurrent dietary administration of aminopyrine (0.01%) and sodium nitrite (0.1%) and coffee solution as a drinking water for 630 da were significantly lower than those of rats given aminopyrine and sodium nitrite. Thus, the tested compounds had inhibitory effects on chemical carcinogenesis in liver or colon.

INTRODUCTION

Chemoprevention studies, which focus on the inhibition of carcinogenesis by chemical agents, are based on the concept that certain chemicals, in particular synthetic agents, as well as certain naturally-occurring products found in various foods may inhibit carcinogenesis (25). Until now, certain chemopreventive agents have proved effective on chemical carcinogenesis in different organs. However, the mechanisms of inhibition of these compounds have not been well established, although Wattenberg recently tried to classify them into three main categories: "compounds preventing formation of carcinogen from precursor compounds," "blocking agents," and "suppressing agents" (26). In the present study, four different experiments were performed in order to examine the modifying effects of several chemicals, including a plant constituent, a drug, newly synthesized polyprenoic acid, and coffee, on chemical carcinogenesis.

EXPERIMENT 1: EFFECT OF CHLOROGENIC ACID ON LIVER AND COLON CARCINOGENESIS INITIATED WITH METHYLAZOXYMETHANOL ACETATE

Chlorogenic acid (CA) is a phenolic compound which is widely distributed as a plant constituent (19). This phenylpropanoid has been found to have antioxidant effects (6) and to be a potent inhibitor of many enzyme activities (4). CA has a protective effect on chemical carcinogenesis in mice (9), although some genotoxic effects were reported (22). An experiment was undertaken to examine the modifying effect of CA on liver and colon carcinogenesis in hamsters.

A total of 98 Syrian golden hamsters (Japan Clea Inc., Tokyo, Japan) of both sexes, 2 mo old, were used. Animals were divided into four groups and treated as shown in Fig. 1. All animals were killed and autopsied in week 24 of the experiment. All tissues, including gastrointestinal tract and liver, were prepared by the conventional method for histopathological examination. Colon neoplasms and liver lesions were diagnosed according to the criteria described by Emminger and Mohr (5) and those described by Stewart et al. (21), respectively.

The results are summarized in Tab. 1 and 2. The numbers of hyperplastic liver cell foci in male and female hamsters given MAM acetate and CA were significantly smaller than those in hamsters given MAM acetate alone. The combined incidence of total large intestinal tumors in male and female hamsters, and the combined incidence of colon adenocarcinomas or the incidence of the carcinomas in male or female animals of the group given MAM acetate and CA, were significantly lower than those in hamsters given MAM acetate alone. These results indicate an inhibitory effect of CA on MAM acetate-induced carcinogenesis. The present results appear to support the report by Lasca (9) showing the protective effect of CA on benzo(a)pyrene-induced tumorigenesis in mice. Stich et al. (22) reported some genotoxic effects of CA. However, in the present study, neither neoplastic nor preneoplastic changes were present in hamsters treated with CA alone for 24 wk. This may suggest that CA has no clear carcinogenic potency. The mechanism of inhibitory effect of CA on MAM acetate-induced carcinogenesis may be related to antioxidative effects or inhibition of microsomal enzyme activities.

EXPERIMENT 2: EFFECT OF RESERPINE ON LIVER CARCINOGENESIS INDUCED BY N-2-FLUORENYLACETAMIDE

Reserpine is a naturally-occurring alkaloid isolated from the roots of Rauwolfia serpentina, and is widely used as a medicine, e.g., as an antihypertensive drug or as a tranquilizer (23). Recently, some

PROTOCOL —— Exp. 1

Carcinogen: MAM acetate (Ash Stevens, Milwaukee, WI, U.S.A.)

Test compound: Chlorogenic acid (Sigma Chem. Co., St. Louis, MO, U.S.A)

Animal: A total of 98 Syrian golden hamsters (49 males and 49 females), 8 wks old

Experimental duration: 24 weeks

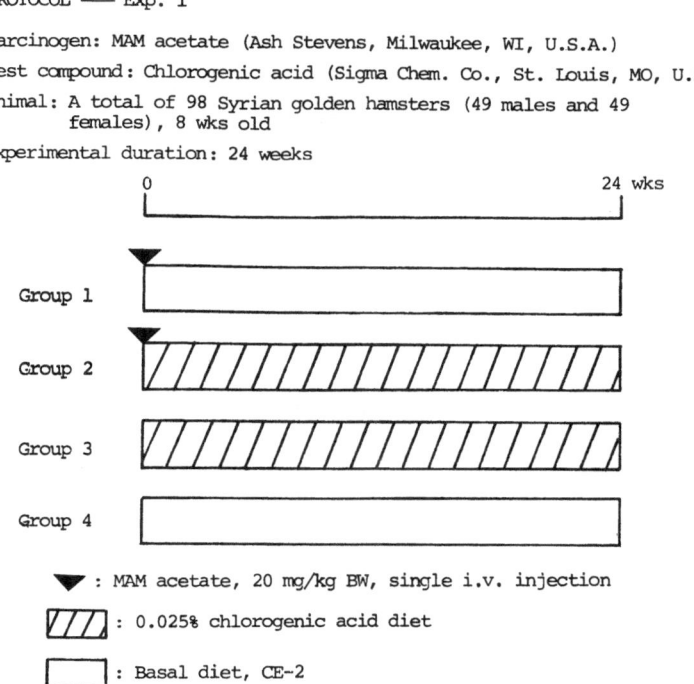

▼ : MAM acetate, 20 mg/kg BW, single i.v. injection

▨ : 0.025% chlorogenic acid diet

☐ : Basal diet, CE-2

Fig. 1.   Protocol for Experiment 1:   Effect of chlorogenic acid on liver and colon carcinogenesis initiated with methylazoxymethanol acetate.

epidemiological studies have reported a positive correlation between long-term use of reserpine and the occurrence of mammary cancer (7,16).   The modifying effects of reserpine on carcinogenesis have been reported (12, 27).   However, the results of its effects have not been consistent.   In this experiment, the modifying effect of reserpine was examined in the rat model for an early stage of hepatocarcinogenesis caused by N-2-fluorenyl-acetamide (FAA).

A total of 72 of an inbred strain of female ACI/N rats, maintained in our laboratory, were used.   At 6 wk of age, rats were transferred to the holding room and randomized into 5 groups.   The experimental schedule is shown in Fig. 2.   Animals were killed 27 wk after the start of the experiment.   Prior to killing, all rats were subcutaneously injected with 125 mg iron/kg body weight (Ferrimicrodex brand, 100 mg elemental iron/ml) 3 times/wk for 2 wk.   Liver tissues were stained with hematoxylin and eosin and exposed to the Prussian blue technique for iron.   The results are summarized in Tab. 3.   The incidence of altered hepatocellular foci in rats given FAA and reserpine simultaneously for 10 wk was significantly smaller than that in rats exposed to FAA alone.   Similarly, the incidence of the foci in the group given reserpine for 16 wk after discontinuation of FAA exposure was also significantly smaller than that in the group given FAA alone.   No hepatocellular foci were seen in rats given reserpine alone for 27 wk and in untreated controls.

Tab. 1. Tumor incidence of the liver and colon of hamsters in Experiment 1.

Group no.	Treatment	No. of effective animals ($\male$ / $\female$)[a]	No. of animals with liver tumors ($\male$ / $\female$)			No. of animals with colon tumors (no. of tumors) : $\male$ / $\female$		
			NN[b]	BDAD	HG	Total	AD	ADC
1	MAM acetate	20(10/10)	1(0/1)	4(2/2)	2(1/1)	10(17): 6(9)/4(8)	6(7): 4(5)/2(2)	8(10): 4(4)/4(6)
2	MAM acetate + CA	24(12/12)	0	3(0/3)	0	3[c](3): 2(2)/1(1)	3(3): 2(2)/1(1)	0[d]: 0[e]/0[e]
3	CA	20(10/10)	0	0	0	0	0	0
4	No treatment	30(15/15)	0	0	0	0	0	0

[a]Survived until the termination of the experiment (24 weeks).

[b]NN = neoplastic nodule; BDAD = bile duct adenoma; HG = hemangioma; AD = adenoma; ADC = adenocarcinoma.

[c-e]Significantly different from group 1 by Fisher's exact probability test ([c]$P<0.008$, [d]$P<0.0008$ and [e]$P<0.03$).

These results indicate an inhibitory effect of reserpine on the early stage of FAA-induced hepatocarcinogenesis. The reasons for the inhibitory effect of reserpine on the liver carcinogenesis with its exposure after the FAA treatment are not clear. It is known that reserpine administration produces enzymatic activation in the livers of rodents (23,28). The anticarcinogenic action of reserpine demonstrated in this study, with its exposure together with FAA, may be attributed to some detoxification process depending on such biological activity. Reserpine has also been reported to function as a calcium channel antagonist (11). Such an effect of the chemical might be related to an enhancement of the phenotypic reversion of the altered cells through alteration of cellular conjugation.

Tab. 2. Number of altered hepatocellular foci in hamsters of Experiment 1.

Group no.	Treatment	No. of hamsters examined	Total no. of hepatocellular foci and each cell type[a]	No. of foci/cm^2
1	MAM acetate	Male 10	282: E 182 (65%) C 18 (6%) B 82 (29%)	10.0 ± 3.8[b]
		Female 10	285: E 195 (68%) C 6 (2%) B 84 (30%)	11.2 ± 4.7
2	MAM acetate + CA	Male 11	110: E 74 (67%) C 6 (5%) B 30 (28%)	5.3 ± 2.3[c]
		Female 11	138: E 97 (70%) C 6 (4%) B 35 (26%)	6.4 ± 3.0[c]

[a]E = eosinophilic cell foci; C = clear cell foci; B = basophilic cell foci.

[b]Mean $\pm$ SD

[c]Significantly different from group 1 by Student's t-test ($P<0.01$).

PROTOCOL —— Exp. 2

Carcinogen: FAA (Nakarai Chem., Kyoto, Japan)

Test compound: Reserpine (Diichi Pharm. Co., Tokyo, Japan)

Animal: A total of 72 female ACI/N rats, 6 weeks old

Experimental duration: 27 weeks

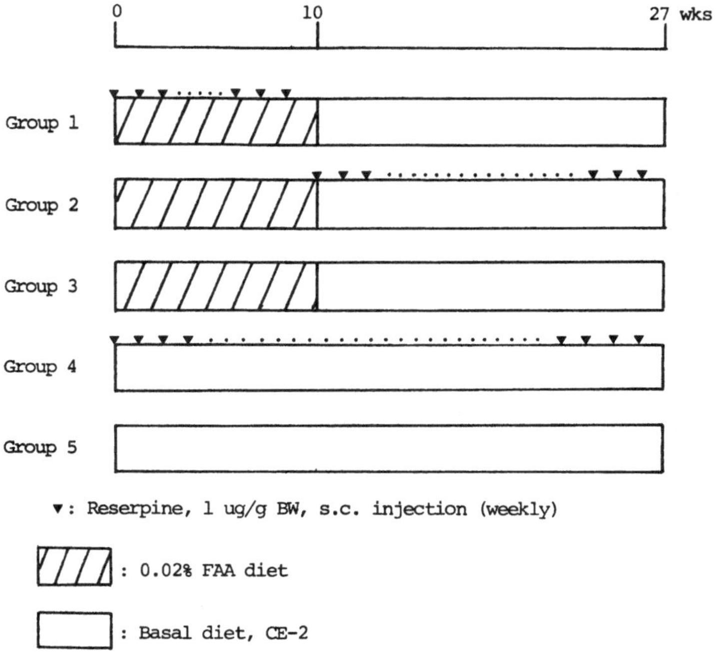

▼: Reserpine, 1 ug/g BW, s.c. injection (weekly)

▨ : 0.02% FAA diet

☐ : Basal diet, CE-2

Fig. 2. Protocol for Experiment 2: Effect of reserpine on liver carcinogenesis induced by N-2-fluorenylacetamide.

Furthermore, reserpine has been reported to affect cellular mitosis and
differentiation (1). The chemical is also known to depress [^{3}H]thymidine
incorporation into cellular DNA (10). These biological functions may also
relate to the inhibitory effect on hepatocarcinogenesis.

EXPERIMENT 3: EFFECT OF POLYPRENOIC ACID (E-5166) ON HEPATOCARCINOGENESIS INDUCED BY N-2-FLUORENYLACETAMIDE

Current studies on vitamin A and its derivatives (retinoids) have
shown that these compounds have inhibitory effects on tumor development
in various organs (20). Thus, the chemicals are of interest as possible
chemopreventive agents (20). Recently, Muto and Moriwaki have found a
new polyprenoic acid, 3,7,11,15-tetramethyl-2,4,6,10,14-hexadecapentae-
noic acid (E-5166), which has retinoid properties and is less toxic than
other synthetic retinoids (14). Some inhibitory effects of E-5166 on
spontaneous occurrence of hepatoma in C_3H/HeNCrJ strain mice and on
induction of rat liver neoplasms by 3'-methyl-4-dimethylaminoazobenzene

Tab. 3.   Effect of reserpine on the incidence of altered liver cell foci in rats of Experiment 2.

Group no.	Treatment	No. of rats examined (initial/final)	Percent of liver weight	No. of foci (/cm^2)
1	FAA + reserpine[a]	15/15	3.07 ± 0.22[b]	1.51 ± 0.58[c]
2	FAA/reserpine[a]	15/15	3.29 ± 0.54	1.51 ± 0.62[c]
3	FAA/basal diet	15/14	3.28 ± 0.21	11.46 ± 3.13
4	Reserpine[a] alone	17/17	3.13 ± 0.15	0
5	No treatment	10/10	3.10 ± 0.21	0

[a]Reserpine (1 μg/g body weight) was given by sc injection once a week for 10 weeks in group 1, for 16 weeks in group 2 and for 27 weeks in group 4.

[b]Mean ± SD

[c]Significantly different from group 3 by Student's t-test (P<0.001).

have been observed during long-term administration (14).   In the present experiment, we examined the effects of E-5166 during carcinogen exposure or at the promotion state on rat hepatocarcinogenesis by FAA.

A total of 74 male ACI/N rats, 6 wk old, were divided into 6 groups.   All animals of these groups were fed a basal diet, CE-2 (Japan Clea, Inc., Tokyo) containing 0.02% FAA for 13 wk, and a summary chart of additional treatments is presented in Fig. 3.   To identify hepatocellular lesions, hepatic siderosis was produced by subcutaneous injection of iron dextran as for Exp. 2.

The results are shown in Tab. 4 and 5.   Rats tolerated well the assigned treatments, including E-5166.   The numbers of altered liver cell foci in rats of group 1 and in rats of group 2 were almost the same, indicating that E-5166 had no effect at the stage of carcinogen exposure. However, the number of foci in group 4 was significantly smaller than that in group 3.   These results suggest some anticarcinogenic activity of E-5166, possibly involving the phenotypic expression of the preneoplastic foci.   Furthermore, the number of altered foci in rats of group 6 was also significantly smaller than that in rats of group 5.   The incidence of neoplastic nodules of the liver in group 6 at the end of the experiment was also lower than in group 5.

The results indicate that the polyprenoic acid had no effect during carcinogen exposure.   Conversely, E-5166 had an antipromoting activity. This is in agreement with reports postulating that inhibitory effects of retinoids on tumor development are mainly due to their antipromoting activities in the case of skin papilloma (2).   Weak enhancing effects of retinoids on chemical carcinogenesis have been reported in other experimental models (17).   It appears that they involve multiple factors such as the type of retinoid, the duration of exposure, and the concentration used in the animal models.   It has been proved that E-5166 has an affinity for the cellular retinoid-binding protein, F-type or cellular retinoic acid-binding protein (15).   This may be one of the factors eliciting the inhibitory effect on chemical hepatocarcinogenesis.

PROTOCOL —— Exp. 3

Carcinogen: FAA (Nakarai Chem., Kyoto, Japan)

Promoter: Phenobarbital (Nakarai Chem.)

Test compound: E-5166 (3,7,11,15-tetramethyl-2,4,6,10,14-
                       hexadecapentaenoic acid, Eisai Co., Tokyo)

Animal: A total of 74 male ACI/N rats, 6 weeks old

Experimental duration: 29 weeks old

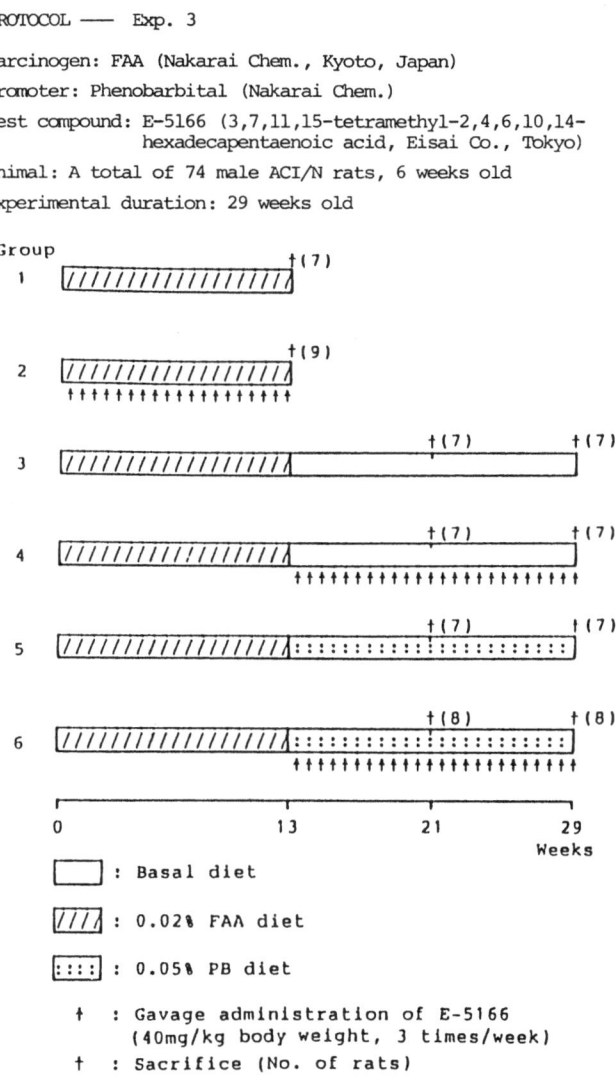

Fig. 3.   Protocol for Experiment 3:   Effect of polyprenoic acid (E-5166)
           on hepatocarcinogenesis induced by N-2-fluorenylacetamide.

EXPERIMENT 4: EFFECT OF COFFEE ON LIVER CARCINOGENESIS
INDUCED BY AMINOPYRINE AND SODIUM NITRITE

       Epidemiological studies have reported a positive relationship between
coffee consumption and occurrence of cancer (18).   Experimental investigations have also provided some evidences to support them (13).   Co-carcinogenic potency of coffee constituents through a catalyzing effect on N-
nitrosamine formation has also been suggested (3).   The results in Exp. 1
have shown an anticarcinogenic effect of CA, a major constituent of coffee, in a hamster model.   Furthermore, roles of coffee constituents for

Tab. 4.   Relative liver weight of rats in Experiment 3.

Group no.	No. of rats examined	Relative liver weight per 100 g of body weight (mean ± SD)		
		13 wk[a]	21 wk	29 wk
1	7	6.4 ± 0.1	-	-
2	9	6.1 ± 0.7	-	-
3	14	-	5.4 ± 0.2	5.2 ± 0.2
4	14	-	5.7 ± 0.6	5.6 ± 0.4[b]
5	14	-	7.0 ± 0.7[c]	7.4 ± 0.7[d]
6	16	-	6.9 ± 0.3	7.1 ± 0.7

[a]wk: weeks at sacrifice.
[b-d]Significantly different from group 3 by Student's t-test
   ([b]$P < 0.05$, [c]$P < 0.01$ and [d]$P < 0.001$).

carcinogen detoxification have been recently suggested (8).   Thus, it can be said that experimental results of coffee or coffee constituents relating to carcinogenesis are not consistent, and the effects of coffee on chemical carcinogenesis are still in question.   In this experiment, using concurrent administration of aminopyrine and sodium nitrite, the effect of long-term coffee drinking on nitrosamine formation or nitrosamine-induced liver tumorigenesis was examined in rats.

A total of 60 female Sprague-Dawley rats, 4 wk old, which were supplied by Japan Clea, Inc., were used.   The animals were divided into 5 groups.   The experimental protocol is schematized in Fig. 4.   On termination of the experiment (630 da), all animals were killed and autopsied.   The rats that survived more than 600 da were handled as effective numbers in this study, and the results obtained on these animals are

Tab. 5.   Incidence of hepatocellular foci and neoplastic nodules of rats in Experiment 3.

Group no.	No. of rats examined	No. of hepatocellular foci /cm² (mean ± SD)			No. of rats with neoplastic nodules of the liver (total no. of the nodules)		
		13 wk[a]	21 wk	29 wk	13 wk	21 wk	29 wk
1	7	32.5 ± 5.1	-	-	0	-	-
2	9	31.5 ± 9.1	-	-	0	-	-
3	14	-	29.3 ± 4.8	15.4 ± 2.8	-	1(1)	2(2)
4	14	-	25.5 ± 3.0	12.3 ± 2.5[b]	-	3(3)	1(2)
5	14	-	34.0 ± 5.2	20.0 ± 3.3[b]	-	1(1)	6(9)
6	16	-	30.9 ± 3.6	11.5 ± 6.9[c]	-	2(2)	0[d]

[a]wk: weeks at sacrifice.
[b]Significantly different from group 3 by Student's t-test ($p < 0.05$).
[c]Significantly different from group 5 by Student's t-test ($p < 0.05$).
[d]Significantly different from group 6 by Fisher's exact probability test ($P < 0.002$).

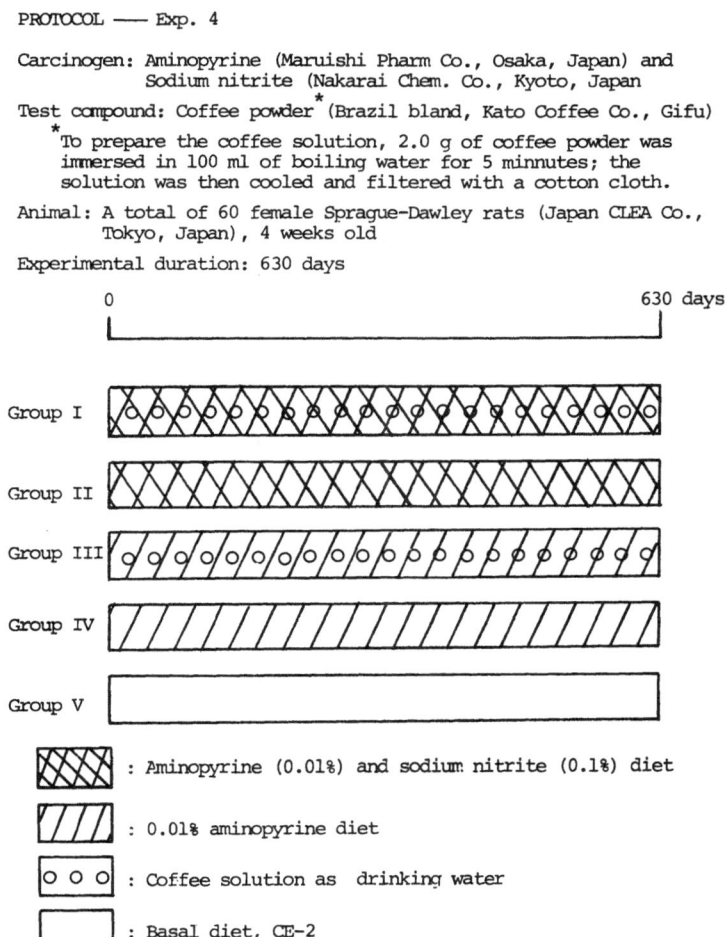

PROTOCOL —— Exp. 4

Carcinogen: Aminopyrine (Maruishi Pharm Co., Osaka, Japan) and
            Sodium nitrite (Nakarai Chem. Co., Kyoto, Japan

Test compound: Coffee powder* (Brazil bland, Kato Coffee Co., Gifu)

   *To prepare the coffee solution, 2.0 g of coffee powder was
   immersed in 100 ml of boiling water for 5 minnutes; the
   solution was then cooled and filtered with a cotton cloth.

Animal: A total of 60 female Sprague-Dawley rats (Japan CLEA Co.,
        Tokyo, Japan), 4 weeks old

Experimental duration: 630 days

0   630 days

Group I

Group II

Group III

Group IV

Group V

: Aminopyrine (0.01%) and sodium nitrite (0.1%) diet

: 0.01% aminopyrine diet

: Coffee solution as drinking water

: Basal diet, CE-2

Fig. 4.   Protocol for Experiment 4:   Effect of coffee on liver carcinogenesis induced by aminopyrine and sodium nitrite.

summarized in Tab. 6 and 7.   The incidences of liver tumors (liver cell
adenoma, carcinoma, and hemangioendothelial sarcoma) and hepatocellular
foci (basophilic cell foci) of rats surviving more than 600 da in group 1
were significantly lower than those of rats in group 2.   No liver tumors
were detected in groups 3, 4, and 5.

These results indicate some properties of nitrosamine-induced hepatocarcinogenesis in induction of preneoplastic hepatocellular lesions as well
as liver neoplasms, and suggest an inhibitory effect of coffee on hepatocarcinogenesis in this rat model.   The inibitory effect of coffee is considered to relate to either of two causative possibilities: (a) inhibition on
the process of nitrosamine formation, and (b) inhibition on the process of
hepatocarcinogenesis ·by nitrosamine (formed by aminopyrine and sodium
nitrite).   Lam et al. (8) stated that some coffee constituents, such as

Tab. 6.   Incidence of tumors in rats of Experiment 4.

Group no.	Treatment	No. of rats (surviving)	No. of rats with liver tumors				No. of rats with other tumors
			Total (%)	NN[a]	HCC	Hemangioendo-thelial sarcoma	
1	Aminopyrine + sodium nitrite + coffee	9	2(22)[b]	2	0	0	5MT
2	Aminopyrine + sodium nitrite	9	7(78)	5	1	1	7MT
3	Aminopyrine + coffee	7	0	0	0	0	3MT
4	Aminopyrine	8	0	0	0	0	3MT, 1OT
5	Control	10	0	0	0	0	3MT

[a]NN = neoplastic nodule; HCC = hepatocellular carcinoma; MT = mammary tumor; OC = ovarian cancer.

[b]Significantly lower than the number of group 2 by Fisher's exact probability test ($P < 0.03$).

kahweol palmitate and cafestol palmitate, increase glutathione S-transferase, effecting carcinogen detoxification. Wattenberg has classified the inhibitory effect against carcinogenesis of plant materials, such as green coffee beans generating a detoxification response, as a "blocking agent" of inhibitors (25). The result that incidence of the basophilic foci, which appear in high incidence in aged rats with different strains (24), in the group given aminopyrine and coffee was lower than that in the group given aminopyrine alone suggests that coffee has inhibited spontaneous occurrence of these hyperplastic lesions in rats.

Tab. 7.   Incidence of altered liver cell foci in rats of Experiment 4.

Group no.	Treatment	No. of rats examined	No. of rats with liver cell foci			Mean no. of foci/cm^2		
			Total (%)	E[a] (%)	B (%)	Total	E	B
1	Aminopyrine + sodium nitrite + coffee	9	6(67)	6(67)	3(33)[b]	1.43	1.22	0.22[b]
2	Aminopyrine + sodium nitrite	9	8(89)	5(56)	7(78)	3.98	1.30	2.68
3	Aminopyrine + coffee	7	3(43)	3(43)	2(29)	0.56	0.48	0.09[c]
4	Aminopyrine	8	7(88)	3(38)	5(63)	1.53	0.18	1.28
5	Control	10	6(60)	3(30)	5(50)	0.94	0.12	0.96

[a]E = eosinophilic cell foci; B = basophilic cell foci.

[b]Significantly different from group 2 by Mann-Whitney U-test ($P < 0.03$).

[c]Significantly different from group 4 by Mann-Whitney U-test ($P < 0.03$).

REFERENCES

1. Birnbaum, L., T. Sapp, and J. Moore (1976) Effects of reserpine, epidermal growth factor, and cyclic nucleotide modulator on epidermal mitosis. J. Invest. Dermatol. 66:313-318.
2. Bollag, W. (1974) Therapeutic effect of an aromatic retinoic acid analog on chemically induced skin papillomas and carcinomas of mice. Eur. J. Cancer 10:731-737.
3. Challis, B.C., and C.D. Bartlett (1975) Possible cocarcinogenic effects of coffee constituents. Nature 254:532-533.
4. Das, M., D.R. Bickers, and H. Mukhtar (1984) Plant phenols as in vitro inhibition of glutathione S-transferase(s). Biochem. Biophys. Res. Commun. 127:427-433.
5. Emminger, A., and U. Mohr (1982) Tumours of the oral cavity, cheek pouch, salivary glands, oesophagus, stomach and intestines. In Pathology of Tumours in Laboratory Animals. Vol. III: Tumours of the Hamster, V.S. Turusov, ed. IARC, Lyon, pp. 45-68.
6. Hayase, F., and H. Kato (1984) Antioxidative compounds of sweet potatoes. J. Nutri. Sci. Vitaminol. 30:37-46.
7. Heinonen, O.P., S. Shapiro, L. Touminen, and M.I. Turunen (1974) Reserpine use in relation to breast cancer. Lancet 2:675-677.
8. Lam, L.K.T., V.L. Sparnins, and L.W. Wattenberg (1982) Isolation and identification of kahweol palmitate and cafestol palmitate as active constituents of green coffee beans that enhance glutathione S-transferase activity in the mouse. Cancer Res. 42:1193-1198.
9. Lasca, P. (1983) Protective effects of ellagic acid and other plant phenols on benzo(a)pyrene-induced neoplasia in mice. Carcinogenesis 4:1651-1653.
10. Lewis, P.D., A.J. Patel, G. Béndek, and R. Balasz (1977) Effect of reserpine on cell proliferation in the developing rat brain: A quantitative histological study. Brain Res. 129:1229-1308.
11. Login, I.S., A.M. Judd, M.J. Cronin, T. Yamamoto, and R.M. Macleod (1985) Reserpine is a calcium channel antagonist in normal and GH3 rat pituitary cells. Am. J. Physiol. 252:E15-19.
12. Lupulescu, A. (1984) Reserpine and carcinogenesis: Inhibition of carcinoma formation in mice. J. Natl. Cancer Inst. 71:57-62.
13. Mori, H., and I. Hirono (1977) Effect of coffee on carcinogenicity of cycasin. Br. J. Cancer 35:369-371.
14. Muto, Y., and H. Moriwaki (1984) Antitumor activity of vitamin A and its derivatives. J. Natl. Cancer Inst. 73:1389-1393.
15. Muto, Y., H. Moriwaki, and M. Omori (1981) In vitro binding affinity of novel synthetic polyprenoid (polyprenoic acids) to cellular retinoid-binding proteins. Gann 72:974-977.
16. Ross, R.K., A. Paganini-Hill, M.D. Krilo, V.R. Gerkins, B.E. Henderson, and M.C. Pike (1984) Effects of reserpine on prolactin levels and incidence of breast cancer in postmenopausal women. Cancer Res. 44:3106-3108.
17. Schroder, E.W., and P.H. Black (1980) Retinoids: Tumor preventers or tumor enhancers? J. Natl. Cancer Inst. 65:671-674.
18. Snowdon, D.A., and R.L. Phillips (1984) Coffee consumption and risk of fatal cancers. Am. J. Publ. Health 74:820-823.
19. Sondenheimer, E. (1964) Chlorogenic acids and related depsides. Bot. Rev. 30:677-712.
20. Sporn, M.B., and D.L. Newton (1979) Chemoprevention of cancer with retinoids. Fed. Proc. 38:2528-2534.

21. Stewart, H.L., G.M. Williams, C.H. Keysser, L.S. Lombard, and R.J. Montali (1980) Histologic typing of liver tumors of the rat. J. Natl. Cancer Inst. 64:177-206.

22. Stich, H.F., M.P. Rosin, C.H. Wu, and W.D. Powrie (1981) A comparative genotoxicity study of chlorogenic acid (3-O-caffeloquinic acid). Mutat. Res. 90:201-212.

23. Stizel, R.E. (1977) The biological fate of reserpine. Pharmacol. Rev. 28:179-205.

24. Ward, J.M. (1982) Background data and variations in tumor rates of control in rats and mice. Prog. Exp. Tumor Res. 26:241-258.

25. Wattenberg, L. (1981) Inhibition of chemical carcinogenesis. In Cancer: Achievements, Challenges. and Prospects for the 1980's, J.H. Burchenal, ed. Grune and Stratton, New York, pp. 517-539.

26. Wattenberg, L.W. (1985) Chemoprevention of cancer. Cancer Res. 45:1-8.

27. Welch, C.W., and J. Meites (1970) Effects of reserpine on development of 7,12-dimethylbenzanthracene-induced mammary tumors in female rats. Experientia 26:1133-1134.

28. Wu, J.M. (1982) The influence of reserpine on nitrogen metabolizing enzymes in chick liver. Biochim. Biophys. Acta 715:57-62.

ANTIMUTAGENIC STRUCTURE MODIFICATION OF QUINOLINE:

FLUORINE-SUBSTITUTION AT POSITION-3

Masatsugu Kamiya, Yoko Sengoku, Kazuhiko Takahashi,
Kohfuku Kohda, and Yutaka Kawazoe

Faculty of Pharmaceutical Sciences
Nagoya City University
Tanabedori, Mizuho-ku, Nagoya 467, Japan

INTRODUCTION

The present study concerns the effect of substituents (halogens and methyl) on the mutagenicity of hepatotumorigenic quinoline in Salmonella typhimurium TA100 in the presence of an S-9 mix. The aim of this study is to determine whether quinoline is deprived of its mutagenic property through the inhibition of mutagenic metabolism by introducing a substituent such as fluorine. In addition, this study attempts to confirm the mutagenic activation process of quinoline.

MATERIALS AND METHODS

The metabolic activation pathway of quinoline has previously been studied based mainly on the structure of metabolites and the property of the DNA-bound adduct (8,15-17). In this study, quinoline and its 3-, 5-, 6-, and 8-fluoro derivatives were synthesized and tested for mutagenicity in TA100 in the presence of an S-9 mix consisting of the S-9 obtained from rat liver [induced with phenobarbital and 5,6-benzoflavone (Oriental Yeast Co., Tokyo)], KCl, $MgCl_2$, glucose-6-phosphate (Glu-6-P), Glu-6-P dehydrogenase, NADPH, NADH, and isotonic phosphate buffer (pH 7.4). The mutagenicity assay was carried out after preincubation for 20 min with the chemicals (18). TA100 is a tester strain sensitive to the mutagenicity of quinoline. In 3-fluoroquinoline, the F atom is located on the pyridine moiety involved in the genotoxic metabolism (15-17), whereas that of the 5-, 6-, and 8-fluoroquinolines is located on the benzene moiety which is presumed to be the site involved in the detoxification process. In addition, some chloro, bromo, and methyl derivatives were also tested. It is noted here that none of the derivatives examined induced revertants without an S-9 mix.

RESULTS

Mutagenicity of Fluoroquinolines

The mutagenicity of fluoroquinolines in the presence of the S-9 mix, in comparison to that of quinoline, is shown in Fig. 1. 5-Fluoroquinoline was more potently mutagenic than quinoline itself; the mutagenicity of 8-fluoroquinoline was a little less potent than that of quinoline. The 6-fluoro derivative was weakly mutagenic. 3-Fluoroquinoline was not mutagenic up to a dose of 10 μmol/plate.

Mutagenicities of 2-, 3-, 4-, 5-, 6-, and 8-Chloroquinolines and 3- and 6-Bromoquinolines

With regard to the chloro derivatives, 4-chloroquinoline proved to be weakly mutagenic, whereas 2- and 3-chloroquinolines were nonmutagenic, as shown in Fig. 2. 2-Chloroquinoline was previously reported to be nonmutagenic and noncarcinogenic (5,14). The 5- and 6- chloroquinolines were more potently mutagenic than quinoline, although they were more cytotoxic. The 8-chloro isomer was mutagenic and severely cytotoxic. For the bromo derivatives (data not shown), 3-bromoquinoline was deficient in mutagenicity, whereas the 6-bromo derivative proved to be mutagenic. It is to be noted that, contrary to the results obtained with the fluoroquinolines, all the chloro and bromo derivatives examined were cytotoxic, even in a low dose range, regardless of the presence or absence of the S-9 mix. The mechanism of enhanced cytotoxicity of chloro and bromo derivatives is now under investigation.

Mutagenicity of Methylquinolines

The mutagenicities of methylquinolines are shown in Fig. 3. These derivatives were previously tested in a similar assay system, in which ATP was used as an additional co-factor (2). There were no significant discrepancies in the relative mutagenic potencies of any of these compounds. 4-Methylquinoline was the most mutagenic, followed by 7-methylquinoline and quinoline. 3-Methyl and 6-methyl derivatives were somewhat less mutagenic, while 2-methylquinoline was less active than all the other methylquinolines.

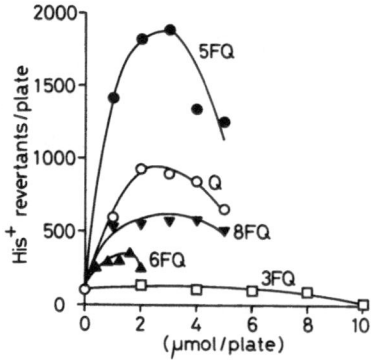

Fig. 1.   Mutagenicity of fluoroquinolines in <u>Salmonella typhimurium</u> TA100.

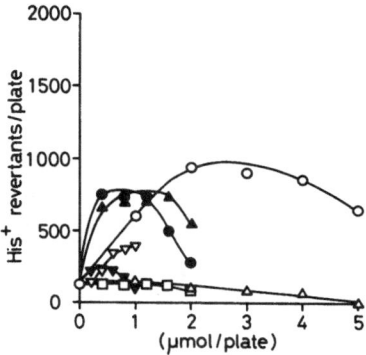

Fig. 2.   Mutagenicities of chloroquinolines in S. typhimurium TA100.
o, quinoline (Q); Δ, 2ClQ;   , 3ClQ; ▽, 4ClQ; ●, 5ClQ; ▲, 6ClQ;
▽, 8ClQ.

DISCUSSION

It is a widely accepted concept that a halogen atom substituted on
the aromatic ring blocks the oxidative genotoxic metabolism at the site of
the halogen-substitution in polycyclic aromatic hydrocarbons (1,4,6,9,11-
13). It may be suggested that introducing a halogen atom at a putative
site in the molecule will deprive the molecule of any genotoxic effects.
Among halogens, F atom is sometimes ignored in the interactions with biomolecules, including enzymes. The steric disturbance of the F atom may
be minimal because of its small van der Waals radius (1.35 Å), which is
close to that of hydrogen (1.20 Å). Therefore, F-substitution might possibly deprive the molecule of genotoxicity without affecting some other
biological activities of the parent molecule.

The present study reveals that 3-fluoro, 2- and 3-chloro, and 3-
bromo derivatives of potently mutagenic quinoline were deficient in mutagenicity in S. typhimurium TA100, whereas all the derivatives carrying a
halogen atom at the 4-, 5-, 6-, and 8-positions were mutagenic. These
results provide further support for our earlier proposals (15-17) that

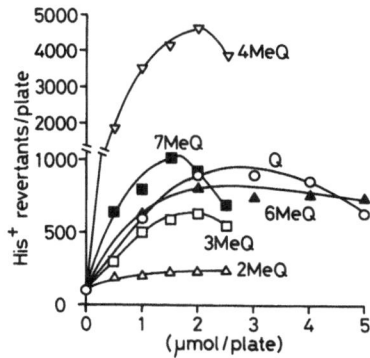

Fig. 3.   Mutagenicities of methylquinolines in S. typhimurium TA100.

position-2 and position-3 are involved in the oxidative metabolic activation leading to genotoxicity, and that the oxidation of the benzene moiety is the major process of detoxification metabolism. Thus, when the DNA-quinoline adduct, which was formed by treatment of DNA with ^{3}H-labeled quinoline in the presence of an S-9 mix, was exposed to acid- or alkaline-treatment, the entire amount of ^{3}H-label was recovered as a form of ^{3}H-labeled 3-hydroxyquinoline. This result suggested that the ultimate structure of genotoxic quinoline may be the 3,4-dihydro-3,4-epoxide of quinoline or, alternatively, the 2,3-dihydro-2,3-epoxide of the 1,4-hydrate of quinoline, provided that the ultimate reactant is an arene oxide, as is well known for carcinogenic aromatic hydrocarbons.

The present study seems to eliminate the former possibility that position-4 can be involved in the oxidative metabolism leading to the genotoxic ultimate structure, since 4-chloroquinoline is definitely mutagenic. In addition, looking at the mutagenicities of methylquinolines, an extraordinary increase in the mutagenicity of the 4-methyl isomer may suggest that position-4 cannot be involved in the oxidative activation (7,20, 21). A bulky methyl group at position-4 might protect the adjacent 2,3-dihydro-epoxide from a detoxifying attack of epoxide hydrolase and/or glutathione S-transferase, resulting in an enhancement of the mutagenicity. This type of enhancement effect is known for a bay region methyl group on tumorigenicity of polycyclic aromatic hydrocarbons (1,3, 10). The 2-methyl isomer is very weakly mutagenic, suggesting that this position may possibly be involved in DNA binding, as shown in Scheme 1. It is known that methyl-substitution at the site involved in the epoxide formation is known to partially reduce the mutagenicity (7,20,21).

In conclusion, the present study, together with our previous studies (15-17), reveals that the mutagenic metabolite is produced through microsomal oxidation of the pyridine moiety but not the benzene moiety. Also, the 2,3-epoxide of 1,4-hydrate of quinoline may be the most probable structure of the mutagenic principle among all the other possible arenic epoxides (19), although direct evidence is not yet available. In addition, it is worth noting that the introduction of a halogen atom, especially a fluorine, onto position-3 can deprive the quinoline molecule of its mutagenic property.

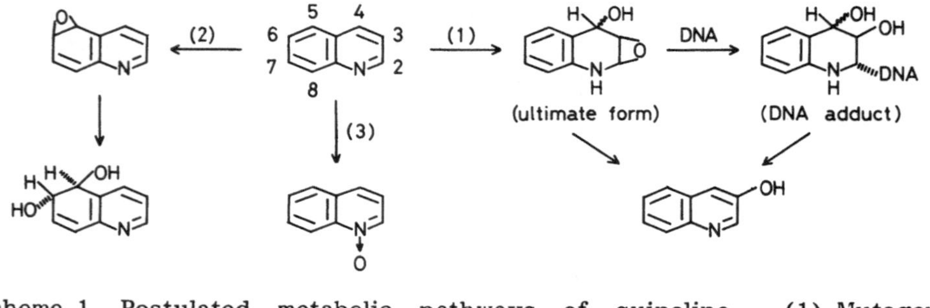

Scheme 1. Postulated metabolic pathways of quinoline. (1) Mutagenic process; (2) detoxification by P448-oxidases; (3) detoxification by P450-oxidases. Absolute configurations of metabolites have not yet been examined.

REFERENCES

1. Diamond, L., K. Cherian, R.G. Harvey, and J. DiGiovanni (1984) Mutagenic activity of methyl- and fluoro-substituted derivatives of polycyclic aromatic hydrocarbons in a human hepatoma cell-mediated assay. Mutat. Res. 136:65-72.
2. Dong, M., I. Schmertz, E. LaVoie, and D. Hoffmann (1978) Aza-arenes in the respiratory environment: Analysis and assays for mutagenicity. In Polynuclear Aromatic Hydrocarbons, Vol. 3, Jones and Freudenthal, eds. Raven Press, New York, pp. 97-108.
3. Hecht, S.S., S. Amin, K. Huie, A.A. Melikian, and R.G. Harvey (1987) Enhancing effect of a bay region methyl group on tumorigenicity in newborn mice and mouse skin of enantiomeric bay region diol epoxides formed stereoselectively from methylcrysenes in mouse epidermis. Cancer Res. 47:5310-5315.
4. Hecht, S.S., E.J. LaVoie, V. Bedenko, L. Pingaro, S. Katayama, D. Hoffmann, D.J. Sardella, E. Boger, and R.E. Lehr (1981) Reduction of tumorigenicity and of dihydrodiol formation by fluorine substitution in the angular rings of dibenzo(a,i)pyrene. Cancer Res. 41:4341-4345.
5. Hirao, K., Y. Shinohara, H. Tsuda, S. Fukushima, M. Takahashi, and N. Ito (1976) Carcinogenic activity of quinoline on rat liver. Cancer Res. 36:329-335.
6. Hubenman, E., and T.J. Slaga (1979) Mutagenicity and tumor-initiating activity of fluorinated derivatives of 7,12-dimethylbenz(a)anthracene. Cancer Res. 39:411-414.
7. Kinoshita, T., M. Konieczny, R. Santella, and A.M. Jeffrey (1982) Metabolism and covalent binding to DNA of 7-methylbenzo(a)pyrene. Cancer Res. 42:4032-4038.
8. LaVoie, E.J., E.A. Adams, A. Shigematsu, and D. Hoffmann (1983) On the metabolism of quinoline and isoquinoline: Possible molecular basis for differences in biological activities. Carcinogenesis 4:1169-1173.
9. LaVoie, E.J., L. Tulley-Freiler, V. Bedenko, and D. Hoffmann (1983) Mutagenicity of substituted phenanthrenes in Salmonella typhimurium. Mutat. Res. 116:91-102.
10. Melikian, A.A., S. Amin, K. Huie, S.S. Hecht, and R.G. Harvey (1988) Reactivity with DNA bases and mutagenicity toward S. typhimurium of methyl-chrysene diol epoxide enantiomers. Cancer Res. 48:1781-1787.
11. Miller, E.C., and J.A. Miller (1960) The carcinogenicity of fluoro derivatives of 10-methyl-1,2-benzanthracene. I. 3- and 4-monofluoro-substitution. Cancer Res. 20:133-137.
12. Miller, J.A., E.C. Miller, and G.C. Finger (1953) On the enhancement of the carcinogenicity of 4-dimethylaminoazobenzene by fluoro-substitution. Cancer Res. 13:93-97.
13. Miller, J.A., E.C. Miller, and G.C. Finger (1957) Further studies on the carcinogenicity of dyes related to 4-dimethylaminoazobenzene. The requirement for an unsaturated 2-position. Cancer Res. 17:387-398.
14. Nagao, M., T. Yahagi, Y. Seino, T. Sugimura, and N. Ito (1977) Mutagenicities of quinoline and its derivatives. Mutat. Res. 42:335-342.
15. Tada, M., K. Takahashi, and Y. Kawazoe (1982) Dependence of specific metabolism of quinoline on the inducer of microsomal mixed-function oxidases. In Microsome, Drug Oxidation and Drug Toxicity, R. Sato and R. Kato, eds. Japan Sci. Soc. Press, Tokyo, p. 517.

16.   Tada, M., K. Takahashi, and Y. Kawazoe (1982) Metabolites of quin-
      oline, a hepatocarcinogen, in a subcellular microsomal system.
      Chem. Pharm. Bull. 30:3834-3837.
17.   Tada, M., K. Takahashi, Y. Kawazoe, and N. Ito (1980) Binding of
      quinoline to nucleic acid in a subcellular microsomal system. Chem.-
      Biol. Interact. 29:257-266.
18.   Takahashi, K., T. Kaiya, and Y. Kawazoe (1987) Structure-muta-
      genicity relationship among aminoquinolines, aza-analogues of naph-
      thylamine, and their N-acetyl derivatives. Mutat. Res. 187:191-197.
19.   Tullis, D.L., and S. Banerjee (1984) Covalent hydration: A possi-
      ble mechanism for aza-arene carcinogenesis. Cancer Lett. 23:241-
      244.
20.   Yang, S.K. (1982) The absolute stereochemistry of the major trans-
      dihydrodiol enantiomers formed from 11-methylbenz(a)anthracene by
      rat liver microsomes. Drug Metab. Dispos. 10:205-211.
21.   Yang, S.K., M.W. Chou, H.B. Weems, and P.D. Fu (1979) Enzymat-
      ic formation of an 8,9-diol from 8-methylbenz(a)anthracene. Bio-
      chem. Biophys. Res. Commun. 90:1136-1141.

MUTAGENICITY AND ANTIMUTAGENICITY OF THAI MEDICINAL PLANTS

Wannee Rojanapo, Anong Tepsuwan,
and Pongpan Siripong*

Biochemistry and Chemical Carcinogenesis Section
*Medicinal Plant Research Section
Research Division
National Cancer Institute
Bangkok 10400, Thailand

ABSTRACT

Crude extracts and partially purified as well as purified fractions were prepared from three Thai medicinal plants, namely, Acanthus ebracteatus Vahl, Plumbago indica Linn, and Rhinacanthus nasuthus Kurz, and then tested for their mutagenic and antimutagenic potentials using the Salmonella/microsome mutagenicity test. All fractions tested were not mutagenic toward either strain TA98 or TA100 whether tested in the presence or absence of S-9 mix. Interestingly, however, various fractions-- especially those extracted by organic solvents such as petroleum ether, hexane, and chloroform, as well as some purified compounds from these plants--could strongly inhibit the mutagenicity of aflatoxin B_1 (AFB_1), an indirect mutagen, when tested in the presence of S-9 mix but not that of 2-(2-furyl)-3-(5-nitro-2-furyl)acrylamide (AF-2), which does not require metabolic activation for its mutagenicity. Furthermore, these fractions could markedly inhibit the activity of rat liver aniline hydroxylase, which is one of the cytochrome-P450-mediated reactions. These results therefore suggest that these Thai medicinal plants contain an antimutagen(s) which inhibits chemical mutagenesis by inhibiting the enzyme activities necessary for activation of indirect mutagens/carcinogens. Identification as well as anticarcinogenicity of purified compounds of these plants are being investigated in our laboratory.

INTRODUCTION

Acanthus ebracteatus Vahl, Rhinacanthus nasuthus Kurz, and Plumbago indica Linn are among the Thai plants widely used as folklore medicines to treat various diseases, i.e., dermatitis, inflammation, leprosy, lymphangitis, and bacterial, fungal, and parasitic infections, as well as some forms of cancer (7). Acanthus ebracteatus Vahl has been shown to reduce the mortality rate and the spleen size in leukemic mice (10).

However, various crude extracts have been found to have no cytotoxicity against some cancer cell lines such as KB cells (P. Picha, unpubl. observ.).

Plumbago indica Linn has also been used to treat cancer. Nonpolar solvent extracts have been reported to exhibit remarkable antitumor activity in vitro (8). Interestingly, plumbagin, a major constituent, has been reported to inhibit the growth of fibrosarcoma in rats induced by methylcholanthrene (4). The antitumor activity of R. nasuthus Kurz has also received much attention. Various crude extracts from R. nasuthus have been found to possess strong antitumor activity toward cancer cell lines (6).

All the above results indicate that these Thai medicinal plants or their constituents may have antitumor activity. However, these products have never been evaluated for either mutagenic or carcinogenic potentials, although they are widely used in this country. Together with an increasing finding of antimutagens in plants (3), it is therefore of great interest to study both the mutagenic and the antimutagenic activities of some commonly used Thai medicinal plant ingredients. In this communication, therefore, we report the mutagenicity and antimutagenicity of various extracts prepared from A. ebracteatus Vahl, R. nasuthus Kurz, and P. indica Linn, as well as their possible mechanism of antimutagenicity.

PREPARATION OF THAI MEDICINAL PLANT EXTRACTS

Parts of A. ebracteatus Vahl (stem), P. indica Linn (root), and R. nasuthus Kurz (root) were dried and then extracted sequentially with hexane (or petroleum ether in some cases), chloroform, and then methanol. Petroleum ether or hexane extracts were further purified by column, and thin-layer chromatographies and, finally, partially purified fractions as well as pure compounds were obtained and subjected to mutagenicity and antimutagenicity testing.

MUTAGENIC POTENTIAL OF THAI MEDICINAL PLANT EXTRACTS

Various medicinal plant extracts were tested for their mutagenic potentials toward Salmonella typhimurium TA98 and TA100 using the standard Salmonella/microsome mutagenicity test with slight modification (1,5,9). Results (Tab. 1) revealed that either crude extracts or purified fractions from these three plants were not mutagenic to both strains of S. typhimurium whether tested in the presence or absence of PCB-induced S-9 mix, although the tests were performed up to 5 mg per plate, or until reaching the toxic level. These results indicate that these Thai medicinal plants do not contain compounds mutagenic toward S. typhimurium TA98 or TA100.

ANTIMUTAGENICITY OF THAI MEDICINAL PLANT EXTRACTS

Test for antimutagenicity of plant extracts was performed, using the Ames test, both in the presence and absence of S-9 mix in which AFB_1 and AF-2 were used as standard mutagens. Most fractions obtained from all three plants, particularly those extracted by petroleum ether, hexane,

Tab. 1.   Mutagenicity of plant extracts on Salmonella typhimurium TA98 and TA100.

Plant extract	Amount (µg)	No. of His$^+$ Revertants/plate			
		TA 98		TA 100	
		−S-9 mix	+S-9 mix	−S-9 mix	+S-9 mix
A. ebracteatus Vahl (A)					
A-P	50	39	40	229	239
	500	49	49	246	265
	5,000	36	57	218	271
A-P-3	25	24	40	159	268
	125			183	183
	250	27	37	210	214
A-C	50	35	51	239	259
	500	44	60	222	244
	5,000	37	52	268	271
A-M	50	36	45	239	216
	500	48	39	248	240
	5,000	40	33	231	245
P. indica Linn (P)					
P-H	50	43	45	269	293
	500	10	31	301	229
	5,000	K	K	K	K
P-H-3	12.5	37	41	181	260
	25.0	44	30	218	221
	50.0	K	38	K	247
R. nasuthus Kurz (R)					
R-H	20	41	47	204	251
	100	51	46	220	282
	500	50	34	204	238
R-C	50	43	50	208	262
	500	36	42	211	202
	5,000	19	32	225	206
R-M	50	37	42	178	232
	500	50	47	198	251
	5,000	31	41	224	242
Solvent control:					
DMSO	100 µl	35	45	218	241
Positive controls:					
AF-2	0.02	N.D	N.D	706	N.D
	0.2	765	N.D	N.D	N.D
AFB$_1$	0.03	N.D	1,378	N.D	2,099

Results are means of 2 separate experiments. S-9 mix contained 100 µl/plate of PCB-induced rat liver S-9 fraction and NADPH-regenerating system as used by Maron and Ames (5).

Abbreviations :
   C : chloroform extract ;   H : hexane extract
   M : methanol extract   ;   P : petroleum ether extract
 H-3 : purified fraction from hexane extract
 P-3 : purified fraction from petroleum ether extract

and chloroform, could strongly inhibit the mutagenicity of AFB$_1$, an indirect mutagen, toward S. typhimurium TA100 when tested in the presence of S-9 mix (Fig. 1). As expected, inhibitory effects of purified fractions were higher than those of the original hexane or petroleum extract. On the other hand, extracts from all three plants did not inhibit, or only slightly inhibited, the mutagenicity of AF-2, which is a direct mutagen, when tested in the absence of S-9 mix (Fig. 2). In addition, methanol extracts of all three plants tested did not inhibit the mutagenicity of either AFB$_1$ or AF-2 (Fig. 1 and 2).

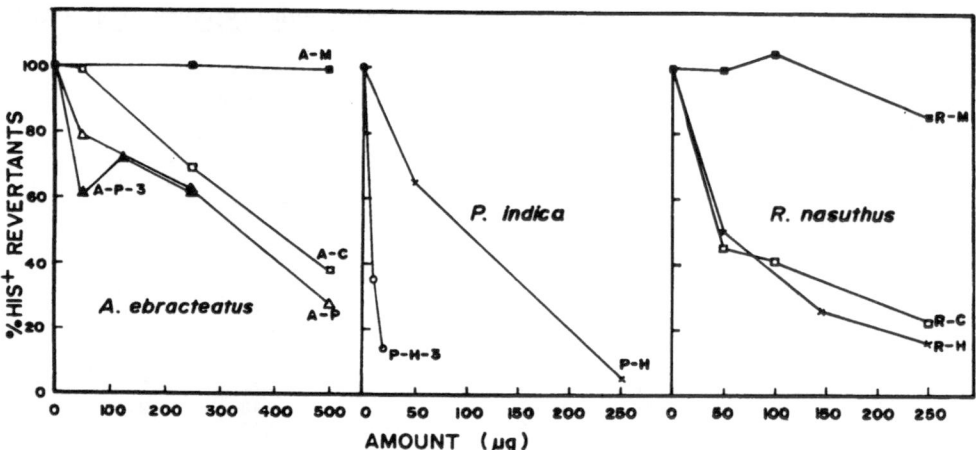

Fig. 1.   Inhibition of plant extracts on the mutagenicity of AFB$_1$ toward
           <u>S. typhimurium</u> TA100. The test was performed in the pres-
           ence of S-9 mix, and 0.03 µg of AFB$_1$ was used as standard
           mutagen (100% His$^+$ revertants represents 2,150 His$^+$ revert-
           ants/plate). For abbreviations, see Tab. 1.

        Our results demonstrate that these three Thai medicinal plants con-
tain a compound(s), probably a nonpolar one(s), capable of inhibiting the
mutagenic activity of indirect mutagens/carcinogens.

INHIBITORY EFFECT OF PLANT EXTRACTS ON ACTIVITY
OF RAT LIVER ANILINE HYDROXYLASE

        The effect of medicinal plant extracts on the activity of rat liver
aniline hydroxylase, which is one of the cytochrome P450-mediated enzyme
reactions, was determined in order to shed some light on the possible
mechanism of their inhibitory effect on the mutagenicity of indirect mutagens.   In this study, the PCB-induced S-9 fraction was used as the
source of aniline hydroxylase, and the activity was determined by the
method described by Carpenter et al. (2).   The results shown in Fig. 3
indicated that all plant extracts including purified fractions, but not
methanol extracts, could also inhibit the activity of this enzyme.   These
results were in parallel with those on the inhibition of mutagenicity
described above, both qualitatively and quantitatively.

DISCUSSION

        Results in the present study demonstrate that commonly used Thai
medicinal plants, namely, <u>A</u>. <u>ebracteatus</u> Vahl, <u>P</u>. <u>indica</u> Linn, and <u>R</u>.
<u>nasuthus</u> Kurz, do not contain mutagenic compounds but do contain an
antimutagen(s) capable of inhibiting the mutagenicity of only the indirect
mutagens.   All fractions possessing the antimutagenicity were also found
to inhibit the activity of rat liver aniline hydroxylase, which is an enzyme
in the drug-metabolizing enzyme or mixed-function oxygenase system.   We
therefore conclude that the mechanism by which these antimutagens inhibit

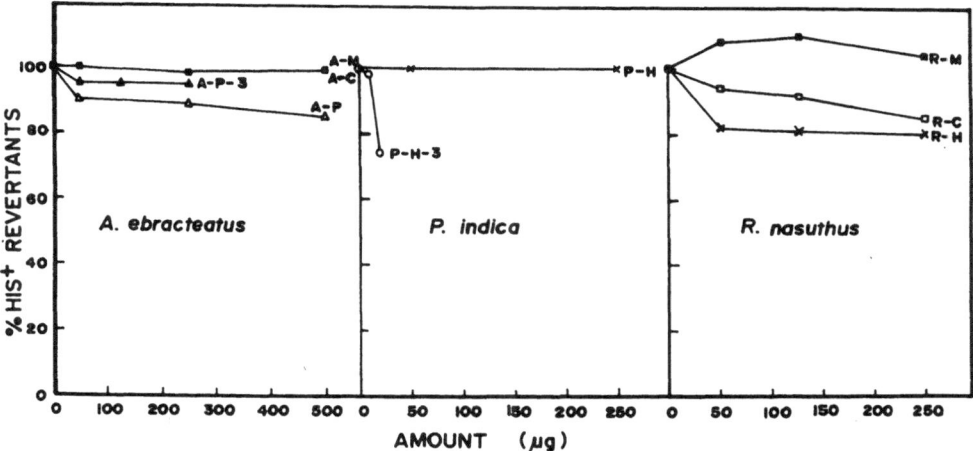

Fig. 2.   Inhibition of plant extracts on the mutagenicity of AF-2 toward S. typhimurium TA100. The test was performed in the absence of S-9 mix, and 0.02 μg of AF-2 was used as standard mutagen (100% His⁺ revertants represents 841 His⁺ revertants/plate). For abbreviations, see Tab. 1.

the mutagenicity of indirect mutagens may be the inhibition of the enzyme activities necessary for the metabolic activation of indirect mutagens/carcinogens to their active metabolites.

Some other plant extracts, e.g., from lettuce and string beans, as well as from a number of natural chemicals, have also been reported to reduce the mutagenicity of the indirect mutagens such as benzo(a)pyrene

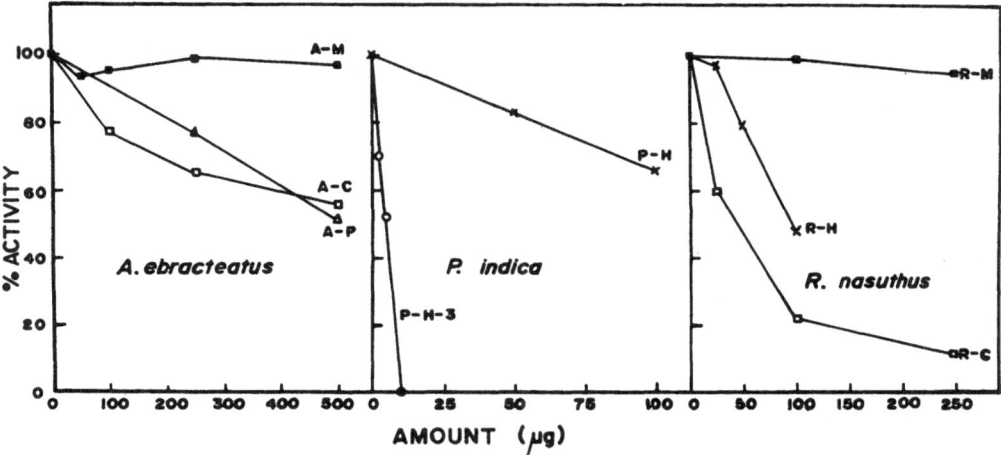

Fig. 3.   Inhibitory effect of plant extracts on the activity of rat liver aniline hydroxylase. The test was performed as described in the text, and 100% activity represents 0.267 nmole p-aminophenol formed/min/mg protein. For abbreviations, see Tab. 1.

and cigarette smoke condensate (11). The antimutagenicity was therefore suggested to be due to the interaction between antimutagen and enzyme(s) in the liver homogenate (11). The antimutagens found in this study are probably nonpolar compounds, since they are soluble only in nonpolar solvents. Some of them have already been purified and are now being investigated for their chemical structure and anticarcinogenicity.

ACKNOWLEDGEMENTS

The authors would like to thank Mr. Seni Raruen for drawing the illustrations, Miss Chintana Soodprasert and Mr. Sathaporn Boonprasert for excellent technical assistance, and Miss Panporn Srimanop for typing the manuscript. This work was partly supported by the Association for International Cancer Research and by the World Health Organization.

REFERENCES

1. Ames, B., J. McCann, and E. Yamasaki (1975) Methods for detecting carcinogens and mutagens with the Salmonella/mammalian microsome mutagenicity test. Mutat. Res. 31:347-364.
2. Carpenter, M.P., and C.N. Howard (1974) Vitamin E, steroids, and liver microsomal hydroxylations. Ann. J. Clin. Nutr. 27:966-979.
3. Kada, T., T. Inoue, and Y. Shirasu (1985) Antimutagens and their modes of action. In Antimutagenesis and Anticarcinogenesis Mechanisms, D.M. Shankel, P.E. Hartman, T. Kada, and A. Hollaender, eds. Plenum Press, New York, pp. 185-196.
4. Krishnaswamy, M., and K.K. Purushothaman (1980) Plumbagin: A study of its anticancer, antibacterial and antifungal properties. Indian J. Exp. Biol. 18:876-877.
5. Maron, D.M., and B.N. Ames (1983) Revised methods for the Salmonella mutagenicity test. Mutat. Res. 113:173-215.
6. Picha, P., M. Rienkijkarn, K. Preechanukool, et al. (1987) Ten antitumor screening models and their natural product testing. In Abstract of the First Princess Chulabhorn Science Congress 1987 International Congress on Natural Products, Bangkok, Thailand, December 10-13, 1987, p. 31.
7. Pongboonrod, S. (1971) Plants in Thailand: Properties of Thai and Foreign Folklore Medicine, Fuang Aksorn Press, Bangkok.
8. Rienkijkarn, M., P. Picha, P. Siripong, and K. Preechanukool (1984) Anticancer activity of a Plumbaginaceae plant (Chettamunphlo-eng-daeng) in vitro. Abstract of the Eighth Annual Meeting of Medical Technologists Association of Thailand, April 24-26.
9. Rojanapo, W., P. Kupradinum, A. Tepsuwan, S. Chutimataewin, and M. Tanyakaset (1986) Carcinogenicity of an oxidation product of p-phenylenediamine. Carcinogenesis 7:1997-2002.
10. Srivatanakul, P., and L. Naka (1981) Effect of Acanthus illicifolius Linn in treatment of leukemic mice. Thai Cancer J. 7:89-93.
11. van der Hoeven, J.C.M. (1986) Occurrence and detection of natural mutagens and modifying factors in food products. In Diet, Nutrition and Cancer, Y. Hayashi et al., eds. Japan Science Society Press, Tokyo/VNU Science Press, Utrecht, pp. 119-137.

INFLUENCE OF POTENTIAL ANTIOXIDANTS ON FREE-RADICAL DAMAGE OF LYMPHOCYTES

B. Binková, J. Topinka, and R.J. Šrám

Psychiatric Research Institute
181 03 Prague 8, Czechoslovakia

INTRODUCTION

Oxidative damage in biological systems can occur due to the presence of high levels of superoxide and hydroxyl free radicals and hydrogen peroxide, which can oxidize susceptible target molecules in living cells. They initiate the process of membrane lipid peroxidation and react also with DNA, causing modifications of the molecular structure of the genetic material. Gradual cellular degeneration and damage may be the consequence of free-radical action. To understand these processes may be important from the point of view of aging. Living cells contain three species of protective enzymes involved in the detoxification of free radicals--superoxide dismutase, catalase, and glutathion peroxidase. An important role is played also by natural or synthetic antioxidants. It is believed that the nootropic effect of some drugs may be just the result of their antioxidant activity (2,5,7).

This is the reason why the present paper deals with the protective influence of vitamins (α-tocopherol and riboflavin) and nootropics (pyritinol, centrophenoxine, and 7-methoxytacrine) on the oxidative damage in human peripheral lymphocytes from healthy volunteers.

MATERIALS AND METHODS

Chemicals

D,L-α-Tocopherol (α-TP), Serva, Federal Republic of Germany (FRG); riboflavin (RBF), Serva, FRG; centrophenoxine (CPH), Cetrexin, Spofa, Czechoslovakia; pyritinol (PYR), Encephabol, Polfa, Poland; 9-amino-7-methoxy-1,2,3,4-tetrahydroacridinium lactate (7MX), VUFB, Prague, Czechoslovakia.

Blood Samples

The experiments were carried out on healthy volunteers of both sexes, with an average age of 43.1 yr (s.d. 8.0). The effect of all

453

compounds on lipid peroxidation (LPO) and unscheduled DNA synthesis (UDS) was tested in peripheral blood lymphocytes.

Sample Preparation

The lymphocytes were isolated on Ficoll 400-Verografin gradients (3). The content of the proteins in the suspension of isolated cells was determined according to Bradford (2). The samples from each donor were diluted in medium RPMI 1640 at a cell density of 0.3 mg of protein per ml and divided to several aliquots, in which molecular oxygen was activated by adding Fe^{2+} (2 μmol/l) and sodium ascorbate (30 μmol/l) or hydrogen peroxide (1 mmol/l). On the basis of preliminary experiments, the influence of the pretreatment with studied compounds in the final concentration of 200 μmol/l was investigated. All compounds were added before oxygen-activating systems. The last aliquots of samples served as controls without any induction of the molecular oxygen. After incubation of all samples for 30 min at 37°C, the lipid peroxidation and the oxidative damage to DNA were determined.

Lipid Peroxidation (LPO)

The LPO level in lymphocytes was determined by the modified thiobarbituric acid (TBA) assay (6). TBA-active products of lipid peroxidation were measured spectrophotometrically after extraction into butanol as a difference of the absorbances at 532 nm and 580 nm on a Specord M 40 (GDR). The obtained values were expressed as nanomoles MDA per mg of protein, using 1,1,3,3-tetraethoxypropane as a standard for all treated (T) and control (C-endogenous levels) samples. The ratio T/C gives information about in vitro-induced LPO levels in the suspension of lymphocytes under given conditions.

Unscheduled DNA Synthesis (UDS)

The level of the oxidative damage to DNA was estimated by the scintillometric measurement of in vitro UDS induced by the activation of molecular oxygen. Induction of UDS by the alkylating agent 1-methyl-3-nitro-1-nitrosoguanidine (MNNG) has been used as the positive control. DNA from cells was isolated according to Martin et al. (4). The incorporated radioactivity was measured on a Beckman LS 5801 liquid scintillation counter. The concentration of DNA was estimated by the diphenylamine method. Specific activity of samples was expressed as CPM/μg DNA and calculated for all treated (T) and control (C) samples. The ratio T/C gives information about the increased incorporation of the radiolabeled nucleoside [methyl-^{3}H]thymidine as the consequence of DNA damage (excision repair).

RESULTS AND DISCUSSION

In the preliminary experiments, the influence of the sodium ascorbate and hydrogen peroxide concentration on LPO and UDS levels was tested for a given Fe^{2+} concentration. It was found that for simultaneous induction of LPO and UDS, the optimal concentration is 30 μmol/l of Na-ascorbate and, in the case of H_2O_2, 1 mmol/l.

Table 1 summarizes the results with the lymphocytes treated by all compounds. The activation of molecular oxygen by the above-mentioned

Tab. 1. Effect of potential antioxidants (AO) in in vitro-induced LPO
         and UDS in peripheral lymphocytes.

AO (200 µmol/l)		Lipid peroxidation			Unscheduled DNA synthesis		
		C*	LPO T/C	LPO + AO T/C	C**	LPO T/C	LPO + AO T/C
α-TP	mean	1.05	2.50	1.40	61.5	1.61	1.26
	s.d.	0.19	0.83	0.45	15.9	0.51	0.38
	N=20						
RBF	mean	1.20	1.88	1.73	66.2	1.65	1.29
	s.d.	0.10	0.08	0.23	18.6	0.16	0.24
	N=5						
CPH	mean	1.10	1.81	1.67	53.5	1.34	1.23
	s.d.	0.07	0.32	0.26	11.2	0.10	0.12
	N=5						
PYR	mean	1.01	2.35	2.25	63.9	1.77	1.81
	s.d.	0.17	0.88	0.87	28.6	0.59	0.54
	N=13						
7-MX	mean	0.94	2.43	2.70	68.6	2.01	2.60
	s.d.	0.12	0.86	0.84	12.5	0.62	0.76
	N=10						

*nmol MDA per mg of proteins in the cell suspension.
**CPM per µg DNA.

C, control (endogenous) level of LPO and ^{3}H-thymidine incorporation; LPO, samples
with the activation of molecular oxygen; LPO + AO, the same as LPO but in the
presence of potential antioxidant (AO).

conditions induced a 2.5 times higher level of LPO and a 1.8 times increase of [^{3}H]thymidine incorporation into DNA compared to the control
(endogenous) levels. α-TP decreased the induced LPO by 74%, UDS by
43%; RBF suppressed LPO by 20%, UDS by 38%; and CPH suppressed LPO
by 17%, and UDS by 32%. On the other hand, nootropic PYR had no effect on LPO and UDS levels, and 7MX increased the LPO level by 19% and
UDS by 58% under the given conditions. The data obtained with RBF and
CPH were analyzed only on 5 samples. If a larger number of samples
were used, they could be observed for higher interindividual variability,
as was already seen with α-TP, PYR, and 7MX.

    It seems that efficient protective action against free radicals is re-
lated not only to their scavenging, but also to the stabilization of the
membrane lipid bilayers in the cells. It is obvious that the hydrophobic
isoprenoic chain with a hydrophilic chromanol nucleus of TP is structurally similar to the phospholipids. Thus, the molecules of α-TP may be
directly incorporated into lipid bilayer, and may affect the structural stability of the membrane. The antioxidant activity of the α-TP is explained
by the presence of the OH-group on the chromanol nucleus. From the
multifunctional effect of α-TP as a universal stabilizer of biological membranes, the advantages of α-TP are obvious.

    The results from RBF antioxidant activity indicate that it may be an
effective radical scavenger. This protective mechanism should, therefore,

be studied more thoroughly. CPH is expected to affect protein solubility in the brain, which was explained by its antioxidant activity (5). The absence of an effect of PYR on in vitro-induced LPO and UDS levels suggests that its nootropic activity may not be related to the scavenging of free radicals. Surprising results were observed with 7MX, which increased LPO and UDS levels. These results should, therefore, be resolved in the intact mammal.

In vitro-induced LPO and UDS in human peripheral lymphocytes seems to be a promising approach for the study of drug antioxidant activity.

REFERENCES

1. Bradford, M.M. (1976) A rapid and sensitive method for the quantitation of microgram quantities of protein utilizing the principle of protein-dye binding. Anal. Biochem. 72:248-254.
2. Fusek, J., J. Patočka, J. Bajgar, J. Bielavský, J. Herink, and V. Hrdina (1974) Pharmacology of 1,2,3,4-tetrahydro-9-amino-acridine. Activ. Nerv. Super. 16:226-228.
3. Harris, R., and E.O. Ukaejiofo (1970) Tissue typing using a routine one-step lymphocyte separation technique. Brit. J. Haematol. 18:229-235.
4. Martin, C.N., A.C. McDermid, and R.C. Garner (1978) Testing of known carcinogens and noncarcinogens for their ability to induce unscheduled DNA synthesis in HeLa cells. Cancer Res. 38:2621-2627.
5. Zs. Nagy, I., K. Nagy, V. Zs. Nagy, A. Kalmar, and E. Nagy (1981) Alterations in total content and solubility characteristics of proteins in rat brain and liver during ageing and centrophenoxine treatment. Exp. Geront. 16:229-240.
6. Ohkawa, H., N. Ohishi, and K. Yagi (1979) Assay for lipid peroxides in animal tissues by thiobarbituric acid reaction. Anal. Biochem. 95:352-358.
7. Pavlík, A., O. Benešová, and J. Pilar (1988) Nootropic effect of pyritinol: Cholinergic and free radical scavenger components. Psychopharmacology 96(Suppl.):241.

SPECIFICITY OF ANTIMUTAGENS AGAINST CHEMICAL MUTAGENS
IN MICROBIAL SYSTEMS

D.M. Shankel[1] and C.H. Clarke[2]

[1]University of Kansas
Lawrence, Kansas

[2]University of East Anglia
Norwich, England

ABSTRACT

Procedures have been developed which enable the study of antimutagenic specificity of certain antimutagenic chemicals against chemical mutagens/carcinogens. Modifications of the Ames Salmonella assay, the Bacillis subtilis rec assay of Kada and co-workers, and the Luria-Delbrück fluctuation test, along with procedures we have developed utilizing E. coli K12 strain ND160 developed by Dworkin, all are employed in these studies.

Using these procedures, a number of naturally-occurring compounds and/or their derivatives have been shown to produce antimutagenic specificity either against changes at different specific genetic loci or against activity of specific chemical mutagens such as nitrofurazone, ethyl methanesulfonate, or caffeine. Compounds that demonstrate this activity include cinnamaldehyde, chlorophyllin, an extract of Glycyrrhiza glabra, spermine, and mixtures of guanosine and cytidine. The data demonstrate that some antimutagens act specifically against spontaneous mutations, while others inhibit the development of chemically-induced mutations at specific loci. These results have potential application to the prevention of chemical toxicological damage.

INTRODUCTION

"Mutagenic specificity" was for many years a goal of mutation workers who envisioned the potential for producing specific desirable changes in genes. It is now clear that a level of "mutagenic specificity" does exist, and that some mutagens act only on specific areas of genes or specific base pair combinations. "Antimutagenic specificity" thus becomes a potentially achievable goal for those who visualize the potential for preventing toxicological damage due to the actions of specific mutagenic chemicals or physical agents such as ultraviolet- or X-irradiation.

MATERIALS AND METHODS

Procedures and Strains Employed

A modified Ames Salmonella microsomal assay was utilized with Salmonella typhimurium strains TA100 and TA1535. The standard Ames
protocol (5) was modified in the following ways: aliquots of cells were
treated for 30 min with appropriate dilutions of mutagens (as shown in
control experiments), then washed with pH 7.0 phosphate buffer; the
treated cells were mixed with the antimutagen in "top agar" and then
plated; survival (colony-forming ability) of diluted suspensions of the
treated cells was determined in the same manner.

A modified rec assay (4) was also employed. Bacillus subtilis strains
M45 rec⁻ and H17 rec⁺ (7) were utilized, and the standard procedure was
modified by adding the test antimutagen to the filter paper disk prior to
the addition of the mutagen.

A modified Luria-Delbrück fluctuation test (3) was also employed.
In these experiments, Escherichia coli K12 ND160 (1) was used. This
strain is lactose-requiring due to a polar lacZ frameshift mutation, is
melibiose-requiring due to the polarity of the lacZ frameshift extending
into the adjacent lacY gene, and is sensitive to a variety of agents such
as 2-deoxygalactose and 6-azauracil. Thus, it can be used to screen for
reversion and forward mutations. The standard Luria-Delbrück fluctuation test was modified by including series of tubes with nothing added to
detect levels of spontaneous mutations; with mutagen added to determine
frequencies of induced mutations; with only antimutagen added to evaluate
effects on spontaneous mutation frequencies; and with mutagen plus antimutagen added simultaneously to determine effects of antimutagen on frequency of induced mutations occurring during growth.

RESULTS

Ames Assays

Due to space limitations, we will refer to previously published data
which make the points we wish to make here. The modified Ames test we
employed can be utilized to determine actual mutation frequencies (2).
Using this procedure, we demonstrated that concentrations of cinnamaldehyde up to 25 µg/ml in the plating media produced a true antimutagenic
effect on mutation frequency; while concentrations above 25 µg/ml could
produce an "apparent antimutagenic effect" due to lethality which might
not be detected in the standard Ames test protocol.

Rec Assays

Glycyrrhiza glabra is a member of the licorice family of plants. The
extracts of this plant show good antimutagenic activity in both the Ames
assay systems (6) and the rec assay. Table 1 presents an abbreviated
summary of some results obtained in the rec assay.

Luria-Delbrück Fluctuation Tests

Table 2 presents summary results of a number of experiments to determine effects of selected antimutagens on three different loci in E. coli

Tab. 1. Effect of Glycyrrhiza glabra extract in the rec assay.

Mutagen	Amount of extract add (μg/disc)	Diameter of inhibition zone (mm)	
		H17 rec$^+$	M45 rec$^-$
EMS	0	1.3	15.6
EMS	100 (added with EMS)	0	4.0
EMS	100 (added pre-EMS)	0	3.3
DMSO (control)	0	0	0

K12 ND160. The mutagenic effects and/or antimutagenic effects of caf-feine, guanosine plus cytidine, or induction with the gratuitous inducer of the lac operon, IPTG (isopropyl-β-d-thiogalactoside), were evaluated. The results show that caffeine is mutagenic for Lac$^-$ to Lac$^+$ reversions, but that induction with IPTG is antimutagenic against both spontaneous and caffeine-induced mutations at this locus. They also show that caf-feine is antimutagenic against spontaneously-occurring azauracil-resistant mutations. Added guanosine plus cytidine is antimutagenic against spon-taneous Lac$^-$ to Lac$^+$ reversions, but appears to be mutagenic for azaura-cil resistance and to have no effect on fucose resistance frequencies.

DISCUSSION

The fact that organisms survive and preserve their genomes largely intact in an environment heavily laden with mutagens--sunlight and chem-icals, for example--is attributable to many factors. Fortunately, most cells possess repair systems which repair much of the genetic damage in-flicted on DNA. In addition, nature has provided a bountiful supply of antimutagens/anticarcinogens to counteract the effects of many of the genetic insults to which cells are exposed. The determination of the

Tab. 2. Luria-Delbrück fluctuation test results with Escherichia coli K12 ND160.

Components added to minimal E. glycerol ± B$_1$ growth tubes	Effects on numbers of mutants/tube for three mutations		
	Lac$^+$ mutants	AzaU^R mutants	Fuc^R mutants
A. Nothing added	Spontaneous	Spontaneous	NT*
Caffeine-500 μg/ml	Mutagenic	Antimutagenic	NT
IPTG-0.5 mM	Antimutagenic	Antimutagenic	NT
Caffeine + IPTG	Antimutagenic	No effect	NT
B. Nothing added	Spontaneous	Spontaneous	Spontaneous
Nitrofurazone-1 μg/ml	No effect	Mutagenic	No effect
Guanosine + cytidine (100 μg/ml)	Antimutagenic	Mutagenic	No effect
NFZ + G + C	No effect	Mutagenic	No effect

*Not tested.

specific antimutagens which counteract specific types of DNA-damaging agents could have significant implications for the prevention of toxicological damage occurring as a result of low-level long-term exposure or high-level short-term accidental exposure to genotoxic agents. We have provided several examples, using microbial systems, of how certain antimutagens counteract certain mutagens, or alternatively influence the expression of genetic damage at specific gene loci. Further studies should enable us to expand these findings into areas affecting specific human concerns.

REFERENCES

1. Clarke, C.H., and M.J. Wade (1975) Evidence that caffeine, 8-methoxypsoralen and steroidal diamines are frameshift mutagens for E. coli K12. Mutat. Res. 28:123-125.
2. de Silva, H., and D.M. Shankel (1987) Effects of the antimutagen cinnamaldehyde on reversion and survival of selected Salmonella tester strains. Mutat. Res. 187:11-19
3. Luria, S.E., and M. Delbrück (1943) Mutations of bacteria from virus sensitivity to virus resistance. Genetics 28:491-511.
4. Kada, T., Y. Sadaie, and Y. Sakamoto (1984) Bacillus subtilis repair test. In Handbook of Mutagenicity Test Procedures, B.J. Kilbey, M. Legator, W. Nichols, and C. Ramel, eds. Elsevier Science Publishers, Amsterdam, The Netherlands, pp. 13-31.
5. Maron, D., and B. Ames (1984) Revised methods for the Salmonella mutagenicity test. In Handbook of Mutagenicity Test Procedures, B.J. Kilbey et al., eds. Elsevier Science Publishers, Amsterdam, The Netherlands, pp. 93-140.
6. Mitscher, L.A., S. Drake, S.R. Gollapudi, J.A. Harris, and D.M. Shankel (1985) Isolation and identification of higher plant agents active in antimutagenic assay systems: Glycyrrhiza glabra. In Antimutagenesis and Anticarcinogenesis Mechanisms, D.M. Shankel, P.E. Hartman, T. Kada, and A. Hollaender, eds. Plenum Press, New York, pp. 153-165.
7. Sadaie, Y., and T. Kada (1976) Recombination-deficient mutants of Bacillus subtilis. J. Bacteriol. 125:489-500.

CLASSIFICATION OF MECHANISMS OF INHIBITORS

OF MUTAGENESIS AND CARCINOGENESIS

Silvio De Flora[1] and Claes Ramel[2]

[1]Institute of Hygiene and Preventive Medicine
University of Genoa
I-16132 Genoa, Italy

[2]Department of Genetic and Cellular Toxicology
Wallenberg Laboratory
University of Stockholm
S-106 91 Stockholm, Sweden

According to their mechanisms, inhibitors of mutagenesis and carcinogenesis can be classified into several categories and subcategories, as reported in Tab. 1. We refer to the article by De Flora and Ramel (2) and to the whole Mutation Research special issue (1) for details and examples of the proposed mechanisms.

REFERENCES

1. De Flora, S., ed. (1988) Role and Mechanisms of Inhibitors in Prevention of Mutation and Cancer. Mutat. Res. (special issue) 202: 277-446.
2. De Flora, S., and C. Ramel (1988) Mechanisms of inhibitors of mutagenesis and carcinogenesis. Mutat. Res. 202:285-306.
3. Kada, T., T. Inoue, and N. Namiki (1982) Environmental desmutagens and antimutagens. In Environmental Mutagenesis and Plant Biology, E.J. Klekowski, ed. Praeger, New York, pp. 137-151.
4. Ramel, C., U.K. Alekperov, B.N. Ames, T. Kada, and L.W. Wattenberg (1986) Inhibitors of mutagenesis and their relevance to carcinogenesis. Report by ICPEMC Expert Group on Antimutagens and Desmutagens. Mutat. Res. 168:7-65.
5. Wattenberg, L.W. (1981) Inhibitors of chemical carcinogens. In Cancer: Achievements, Challenges and Prospects for the 1980s, J.H. Burchenal and H.F. Oettgen, eds. Grune and Stratton, New York, pp. 517-540.

Tab. 1.   Classification of inhibitors of mutagenesis and carcinogenesis by
           mechanism.   Reprinted from Ref. 2, with permission.

1. Inhibitors of mutagenesis acting extracellularly[a]

    1.1. Inhibiting the uptake of mutagens or of their precursors

        1.1.1. Hindering their penetration
               1.1.1.1. Into the organism
               1.1.1.2. Into cells
        1.1.2. Favoring their removal

    1.2. Inhibiting the endogenous formation of mutagens

        1.2.1. Inhibiting the nitrosation reaction
        1.2.2. Modifying the microbial intestinal flora

    1.3. Deactivating mutagens

        1.3.1. By physical reaction
        1.3.2. By chemical reaction
        1.3.3. By enzymatic reaction

2. Inhibitors of mutagenesis acting intracellularly[b]

    2.1. Modulators of metabolism[c]

        2.1.1. Inhibiting cell replication
        2.1.2. Favoring sequestration of mutagens in nontarget cells
        2.1.3. Inhibiting the activation of promutagens
        2.1.4. Inducing the detoxifying mechanisms

    2.2. Blocking reactive molecules[c]

        2.2.1. Reacting with electrophiles
               2.2.1.1. By chemical reaction
               2.2.1.2. By enzymatic reaction
        2.2.2. Scavenging reactive oxygen species
        2.2.3. Protecting nucleophilic sites of DNA

    2.3. Modulators of DNA replication or repair[d]

        2.3.1. Increasing the fidelity of DNA replication
        2.3.2. Favoring the repair of DNA damage
        2.3.3. Inhibiting error-prone repair pathways

3. Inhibitors acting on initiated or neoplastic cells[e]

    3.1. Modulators of tumor promotion

        3.1.1. Inhibiting genotoxic effects
        3.1.2. Scavenging free radicals
        3.1.3. Inhibiting cell proliferation
        3.1.4. Inducing cell differentiation
        3.1.5. Modulating signal transduction

    3.2. Modulators of tumor progression

        3.2.1. Inhibiting genotoxic effects
        3.2.2. Acting on hormones or growth factors
        3.2.3. Modulating signal transduction
        3.2.4. Acting on the immune system
        3.2.5. Physical, chemical, or biological antineoplastic agents

[a]Stage 1 inhibitors according to Ramel et al. (4). Those acting in
vitro outside target cells have also been referred to as desmutagens
by Kada et al. (3).
[b]Stage 2 inhibitors according to Ramel et al. (4).
[c]Blocking agents according to Wattenberg (5).
[d]Bioantimutagens or antimutagens in strict sense according to Kada et
al. (3).
[e]Suppressing agents according to Wattenberg (5).

ANTIMUTAGENIC EFFECTS OF CHLOROPHYLLIN

G. Bronzetti, A. Galli, and C. Della Croce

Istituto di Mutagenesi e Differenziamento
C.N.R.
Pisa 56100, Italy

INTRODUCTION

It has been suggested that the majority of human cancers are caused by environmental exposures, but little is known about the effective environmental and occupational risks of cancer. It is likely that the process of mutagenesis and the intricate balance between mutagenesis and antimutagenesis are involved in aging, evolution, and other fundamental life processes (15).

Man is exposed to many risks: the air we breathe, the food we eat, the water we drink, and the places where we work may be contaminated by toxic substances or additives, and there are also many natural carcinogens such as sunshine, smoke, and so on. It is also known that living for one year in Los Angeles is equivalent to a person smoking one cigarette (5). There are many studies on the cooking process; it is clear that this process produces a variety of toxic substances. In fact, in roast beef we find tryptophan pyrolysate (16); from alcohol derives acetaldehyde (2); in coffee we find highly genotoxic methylglioxal (4). The fatty acids that are oxidized during the cooking process form several toxic substances (16). Since a vast number of chemical mutagens are now detected in our environment and especially in our diet, it is important to know how antimutagens interfere with their mutagenic effects (1).

For our purpose we have used sodium-copper-chlorophyllin (CuChlNa), a known food additive (E141) (Fig. 1) that is also used in gastrointestinal medicine (6) and for acceleration of wound healing (8).

There are many metallochlorophyllins: they are derivatives of chlorophyll in which the chelated metal, magnesium, is replaced by other metals such as copper, cobalt, or iron (7). It was shown that metallochlorophyllins can protect against the toxic effects of metabolites produced when the hepatic microsomal system degrades exogenous agents (6).

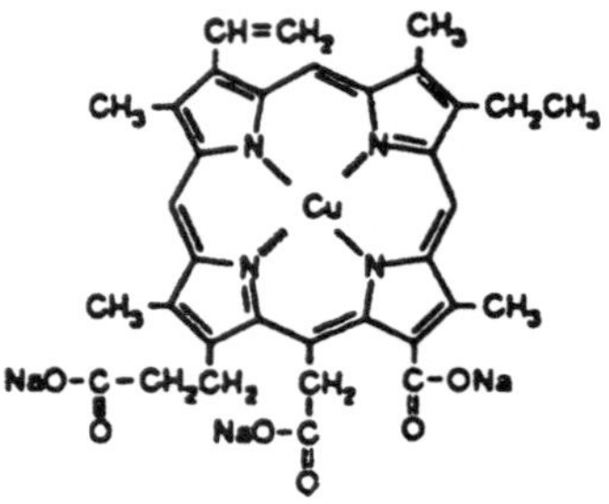

Fig. 1.   Chemical composition of chlorophyllin.

Little is known about sodium-copper-chlorophyllin. It possesses antioxidant activity, and in addition it is an antimutagenic agent. We know that sodium-copper-chlorophyllin has a protective effect on peroxidative damage because it stabilizes membranes (11).

Sato et al. have shown, through spectral analysis, that chlorophyllin is an antioxidant (12-13). This antioxidative action can be assigned to two methylic esters, derivative of chlorophyllin, and perhaps is not due to chlorophyllin itself (14).

Chlorophyllin was tested as an antimutagenic agent against environmental and dietary complex compounds; its activity is heat-stable, and it is not toxic to Salmonella or to yeast (Saccharomyces cerevisiae) (9).

Sodium-copper-chlorophyllin has an antioxidative effect on the lipid peroxidation in vivo, probably due to its action as a radical scavenger (11). The few published studies on the action of chlorophyllin in vitro concern prokaryotes. Chlorophyllin was shown to inhibit the genotoxicity induced by the derivatives of tryptophan and glutamate when assayed by the Ames test. A single paper concerning the eukaryotic system (Neurospora crassa) showed that chlorophyllin inhibits the mutagenicity of aflatoxin B_1 (AFB$_1$) (10).

Throughout our experiments we propose to elucidate the mechanism by which chlorophyllin exerts its antimutagenic effects. We have tested the effect of chlorophyllin on gene conversion and point reverse mutation induced by physical and chemical mutagens on the D7 strain of S. cerevisiae. The physical mutagen was X-rays; the chemicals were ethidium bromide (EtBr) and styrene oxide (SO). The effect of sodium-copper-chlorophyllin has also been compared with the activity of ascorbic acid, a known antioxidant and "scavenger" of free radicals.

MATERIALS AND METHODS

Saccharomyces cerevisiae D7 strain was obtained from Dr. F.K. Zimmermann. With this strain we can detect, simultaneously, mitotic gene conversion (GC) at the trp5 locus, point reverse mutation of the mutant allele ilv-92, and mitotic recombination between the centromere and the ade2 locus. Mitotic crossing over can be detected visually as pink and red twin-sectored colonies, which are due to the formation of homozygous cells of the genotype ade2-40/ade2-40 (deep red) and ade2-119/ade2-119 (pink) from the originally heteroallelic cells, which form white colonies.

Mitotic gene conversion can be detected by the appearance of tryptophan-nonrequiring colonies on selective media. Mutation can be followed by the appearance of isoleucine-requiring colonies on selective media (17).

Suspension Test

From a stationary culture that has a low spontaneous frequency of GC and PM, stored at 4°C, we draw about 500 million cells that are incubated, in the presence of the substances that we test, for 2 hr at 37°C. Then the cells are seeded on petri dishes and incubated for 3 or 4 da at 30°C before the number of colonies are counted (3).

X-Irradiation was supplied by a Siemens stabilipon X-ray source operating at 200 kV and 20 mA, with a distance from the focus of 25 cm. The doses were about 26 krad. (The estimated dose-rate was 800 rad/min as determined by a Simplex universal dosimeter.) Irradiation was performed in the cold (4°C) to avoid repair mechanisms.

Statistical Analysis

Results were analyzed using the Student's "t" test.

RESULTS

The results we have obtained are shown in the following tables.

In Tab. 1 we see that the increases in both mitotic gene conversions and in point reverse mutations induced by styrene oxide significantly decrease in the presence of either ascorbic acid or chlorophyllin.

In Tab. 2 it is clear that chlorophyllin protects against genetic damages produced by EtBr.

Tab. 1. The influence of chlorophyllin (Chl) and ascorbic acid (AA) on mitotic gene conversions and point reverse mutations induced by ethydium bromide (EtBr) on the D7 strain of $\underline{S}$. $\underline{cerevisiae}$.

	Colonies (% survival)	Convertants TRP^+ counted	Convertants/10^5	Revertants ILV^+ counted	Revertants/10^6
I. Control	11,365 (100)	1,013	0.89 ± 0.13	309	0.27 ± 0.05
II. SO (2 mM)	7,750 (68)	36,775	47.45 ± 7.68 (***)	4,255	5.49 ± 0.49 (***)
III. Chl (30 mM)	10,365 (91)	1,225	1.18 ± 0.15	385	0.37 ± 0.08
IV. AA (100 mM)	9,355 (82)	885	0.94 ± 0.11	280	0.30 ± 0.05
V. SO + AA	6,670 (59)	3,345	5.01 ± 0.51 (***)	805	1.21 ± 0.11 (***)
VI. SO + Chl (5 mM)	5,790 (51)	12,250	21.15 ± 4.56 (***)	1,440	2.49 ± 0.64 (***)
VII. SO + Chl (10 mM)	6,125 (54)	12,550	20.48 ± 2.09 (***)	1,520	2.48 ± 0.48 (***)
VIII. SO + Chl (20 mM)	6,050 (53)	8,205	13.56 ± 1.46 (***)	1,070	1.77 ± 0.46 (***)

Frequencies, means of five independent experiments, are expressed as colonies on selective medium/total colonies survived ± SD.
***P<0.001

Tab. 2. The influence of chlorophyllin (Chl) and ascorbic acid (AA) on mitotic gene conversions and point reverse mutations induced by styrene oxide (SO) on the D7 strain of <u>S</u>. <u>cerevisiae</u>.

	Colonies (% survival)	Convertants TRP$^+$ counted	Convertants/10^5	Revertants ILV$^+$ counted	Revertants/10^6
I. Control	11,365 (100)	1,013	0.89 ± 0.13	309	0.27 ± 0.05
II. EtBr (0.5 mM)	6,450 (57)	2,162	3.35 ± 0.79 (***)	445	0.69 ± 0.16 (**)
III. Chl (30 mM)	10,365 (91)	1,225	1.18 ± 0.15	385	0.37 ± 0.08
IV. AA (100 mM)	9,355 (82)	885	0.94 ± 0.11	280	0.30 ± 0.05
V. EtBr + AA	7,196 (63)	2,967	4.12 ± 0.93	341	0.47 ± 0.16
VI. EtBr + Chl (5 mM)	11,290 (99)	1,330	1.18 ± 0.10 (***)	330	0.29 ± 0.05 (**)
VII. EtBr + Chl (10 mM)	10,360 (91)	1,630	1.57 ± 0.22 (***)	515	0.49 ± 0.09 (*)
VIII. EtBr + Chl (20 mM)	8,175 (72)	1,160	1.42 ± 0.44 (***)	395	0.48 ± 0.17 (*)
IX. EtBr + Chl (30 mM)	9,600 (85)	1,045	1.08 ± 0.27 (***)	315	0.33 ± 0.09 (**)

Frequencies, means of five independent experiments, are expressed as colonies on selective medium/total colonies survived ± SD.
  *P<0.05
 **P<0.01
***P<0.001

Since cells treated with X-rays show (Tab. 3), in the presence of chlorophyllin, lower mutation and higher survival, it is probable that chlorophyllin acts as a free radical "scavenger."

DISCUSSION

In our experiments we have tested the antimutagenic activity of chlorophyllin against different physical and chemical mutagens. In our results we found that chlorophyllin interacts directly with epoxides such as styrene oxide.

Tab. 3. The influence of chlorophyllin (Chl) and ascorbic acid (AA) on mitotic gene conversions and point reverse mutations induced by X-rays on the D7 strain of <u>S</u>. <u>cerevisiae</u> (cells were irradiated in the presence of chlorophyllin).

	Colonies (% survival)	Convertants TRP$^+$ counted	Convertants/10^5	Revertants ILV$^+$ counted	Revertants/10^6
I. Control	7,607 (100)	816	1.07 ± 0.27	488	0.64 ± 0.09
II. Chl (30 mM)	7,599 (100)	804	1.05 ± 0.30	447	0.58 ± 0.11
III. X (26 Krad)	3,553 (46)	26,065	73.35 ± 2.49 (***)	8,685	25.11 ± 5.40 (***)
IV. X (26 Krad) + Chl (10 mM)	7,194 (94)	26,810	37.27 ± 0.82 (***)	9,525	13.24 ± 2.70 (*)

Frequencies, means of five independent experiments, are expressed as colonies on selective medium/total colonies survived ± SD.
  *P<0.05
***P<0.001

Ethidium bromide is a well-known intercalating agent which in aqueous solution exists in cation form. It is possible that ethidium bromide may form a complex with the chlorophyllin and is then no longer available for interaction with DNA.

Analyzing the results obtained in cells irradiated by X-rays, we can suggest the hypothesis that chlorophyllin may act as a "scavenger" of free radicals. It is postulated that chlorophyllin decreases the formation of free radicals by X-rays and it affects the formation of active intermediates from styrene oxide.

On the basis of results obtained, we can ascribe chlorophyllin to a larger class of desmutagens, in particular to those which do not interact with the stabilization function of the mutation and could therefore act both inside and outside the cell.

ACKNOWLEDGEMENTS

The authors wish to thank M. Minks for revising the manuscript; E. Morichetti, R. Vellosi, R. Del Carratore, R. Fiorio, and D. Rosellini for suggestions and operative assistance; and G. Cecchi for typing the manuscript.

REFERENCES

1. Ames, B.N. (1983) Dietary carcinogens and anticarcinogens (oxygen radicals and degenerative disease). Science 221:1256-1264.
2. Ames, B.N. (1984) Cancer and diet. Science 224:668-670; 757-760.
3. Bronzetti, G., C. Bauer, C. Corsi, C. Leporini, R. Nieri, and R. Del Carratore (1981) Genetic activity of vinylidene chloride in yeast. Mutat. Res. 89:179-185.
4. Bronzetti, G., C. Corsi, D. Del Chiaro, P. Boccardo, R. Vellosi, F. Rossi, M. Paolini, and G. Cantelli-Forti (1987) Methylglyoxal genotoxic studies and its effects "in vivo" on the microsomal monooxygenase system of the mouse liver. Mutagenesis 2(4):275-277.
5. Garfield, E. (1982) Risk analysis. Part 2. How we evaluate the health risks of toxic subtances in the environment. Current Contents 35:5-11.
6. Imai, K., T. Aimoto, M. Sato, K. Watanabe, R. Kimura, and T. Murata (1986) Effect of sodium metallochlorophyllins on the activity and components of microsomal drug metabolizing enzyme system in rat liver. Chem. Pharm. Bull. 34(10):4287-4293.
7. Kephart, J.C. (1955) Chlorophyll derivatives, their chemistry, commercial preparation and uses. Econ. Bot. 9:3-38.
8. Krasnikova, N.A. (1973) Proliferation of the ephythelium surrounding a skin wound in hairless mice exposed to sodium chlorophyllin. Byul. Eksp. Biol. Med. 76:99-102.
9. Ong, T. Man, W.Z. Whong, J. Stewart, and H.E. Brockman (1986) Chlorophyllin: A potent antimutagen against environmental and dietary complex mixtures. Mutat. Res. 173:111-115.
10. Robin, E. (1986) Inhibition of aflatoxin mutagenicity by chlorophyllin in Neurospora crassa. In Antimutagenesis and Anticarcinogenesis Mechanisms, D.M. Shankel, P.E. Hartman, T. Kada, and A. Hollaender, eds. Plenum Press, New York, pp. 575-576.

11. Sato, M., K. Konagai, R. Kimura, and T. Murata (1983) Effect of sodium copper chlorophyllin on lipid peroxidation. V. Effect on peroxidative damage of rat liver lysosomes. Chem. Pharm. Bull. 31(10):3665-3670.
12. Sato, M., K. Imai, R. Kimura, and T. Murata (1984) Effect of sodium copper chlorophyllin on lipid peroxidation. VI. Effects of its administration on mitochondrial and microsomal lipid peroxidation in rat liver. Chem. Pharm. Bull. 32(2):716-722.
13. Sato, M., K. Konagai, T. Kuwana, R. Kimura, and T. Murata (1984) Effect of sodium copper chlorophyllin on lipid peroxidation. VII. Effect of its administration on the stability of rat liver lysosomes. Chem. Pharm. Bull. 32(7):2855-2858.
14. Sato, M., I. Fujimoto, T. Sakai, T. Aimoto, R. Kimura, and T. Murata (1986) Effect of sodium copper chlorophyllin on lipid peroxidation. IX. On the antioxidative components in commercial preparations of sodium copper chlorophyllin. Chem. Pharm. Bull. 34(6):2428-2434.
15. Shankel, D.M., P.E. Hartman, T. Kada, and A. Hollaender, eds. (1986) Antimutagenesis and Anticarcinogenesis Mechanisms, Plenum Press, New York.
16. Sugimura, T., and S. Sato (1983) Carcinogenicity of mutagenic heterocyclic amines formed during the cooking process. Mutat. Res. 150:33-41.
17. Zimmermann, F.H., R. Kern, and H. Rasemberger (1975) A yeast strain for simultaneous detection of induced mitotic crossing over, mitotic gene conversion and reverse mutation. Mutat. Res. 28:381-388.

PARTICIPANTS, SPEAKERS, AND CHAIRMEN

Aikawa, K., National Institute of Animal Industry, Tsukuba Norindanchi, Ibaraki 305, JAPAN
Akiyama, Y., Japan Tobacco Inc., Toranomon, Minato-ku, Tokyo, JAPAN
Alain, Sarasin, CNRS, Villejuif, FRANCE
Alekperov, U., Azerbaijan Academy of Sciences, Baku, U.S.S.R.
Ando, N., Midori-Juji Co. Ltd., Kanzaki-gun, Hyogo, JAPAN
Ansher, S., U.S. Food and Drug Administration, Bethesda, Maryland, U.S.A.
Anwar, W.A., University of Texas Medical Branch, Galveston, Texas, U.S.A.
Aoki, K., Shihara Sangyo Co., Ltd., Kusatsu, Siga, JAPAN
Arimoto, S., Okayama University, Tsushima, Okayama, JAPAN
Ashida, H., Kobe University, Nada-ku, Kobe, JAPAN
Au, W.W., University of Texas Medical Branch, Galveston, Texas, U.S.A.

Baba, T., Daicel Chemical Industry, Himeji, Hyogo, JAPAN
Bailey, G.S., Oregon State University, Corvallis, Oregon, U.S.A.
Benedict, W.F., Childrens' Hospital of Los Angeles, California, U.S.A.
Bhilwade, H.N., Bhabha Atomic Research Centre, Bombay, INDIA
Binkova, B., Psychiatric Research Institute, Prague, CZECHOSLOVAKIA
Boone, C.W., National Cancer Institute, Bethesda, Maryland, U.S.A.
Brockman, H.E., Illinois State University, Normal Illinois, U.S.A.
Brockman, M.S., Illinois State University, Normal Illinois, U.S.A.
Brockman, T.L., Illinois State University, Normal Illinois, U.S.A.
Bronzetti, G.L., Instituto di Mutagenesi e Differenziamento, Pisa, ITALY

Caterina, T., Universita La Sapienza, Rome, ITALY
Charunut, S., Toxicology Division, Bangkok, THAILAND
Chauhan, P.S., Bhabha Atomic Research Centre, Bombay, INDIA
Chaveca, S., Av. Forcas Armadas, Lisbon, PORTUGAL
Chaveca, T., Av. Forcas Armadas, Lisbon, PORTUGAL
Cheruvanky, R., Indian Council of Medical Research, Hyderabad, Andhra Pradesh, INDIA
Chung, F.L., American Health Foundation, Valhalla, New York, U.S.A.
Cozzi, Renata, Universita la Sapienza, Rome, ITALY

Dashwood, R.H., Oregon State University, Corvallis, Oregon, U.S.A.
Davison, A.J., British Columbia Cancer Research Centre, Vancouver,
    British Columbia, CANADA
DeFlora, Silvio, University of Genoa, Genoa, ITALY
Devaki, N.S., Osmania University, Hyderabad, A.P., INDIA

Ebata, J., Osaka City University, Sumiyoshi-ku, Osaka, JAPAN
Ebitani, N. Shohoku Junior College, Atsugi, Kanagawa, JAPAN
Eisenstark, A., University of Missouri, Columbia, Missouri, U.S.A.
Ekitmoto, H., Nihon Kayaku Col., Ltd., Kita-ku, Tokyo, JAPAN
Endang, T.M., Institute of Physical & Chemical Research, Wako, Saitama,
    JAPAN
Esumi, H., National Cancer Center, Chuo-ku, Tokyo, JAPAN

Fahrig, R., Fraunhofer Institute fur Toxikologie, Hannover, FEDERAL
    REPUBLIC OF GERMANY
Friedberg, E.C., Stanford University, Stanford, California, U.S.A.
Fuchs, R.P.P., CNRS, Strasbourg, Cedex, FRANCE
Fujie, K., Osaka Women's University, Daisen-cho, Osaka, JAPAN
Fujikawa, K., Takeda Chemical Industries, Ltd., Yodogawa-ku, Osaka,
    JAPAN
Fujiki, H., National Cancer Center Research Institute, Chuo-ku, Tokyo,
    JAPAN
Fujiki, M., National Cancer Center Research Institute, Chuo-ku, Tokyo,
    JAPAN
Fujita, S., Chiba University, Yayoi-cho, Chiba, JAPAN
Fukatsu, F., Tokyo Institute of Technology, Yokohama, Kanagawa,
    JAPAN
Fukui, I., ICI Japan Limited, Chiyoda-ku, Tokyo, JAPAN
Furihata, C., University of Tokyo, Minato-ku, Tokyo, JAPAN
Furukawa, H., Laboratory of Environmental Science, Tenpaku-ku,
    Nagoya, JAPAN

Georgian, L., V. Babes Institute, Bucharest, ROMANIA
Ghaskadbi, S., M.A.C.S. Research Institute, Pune, M.S., INDIA
Glickman, B.W., York Universtiy, Toronto, Ontario, CANADA
Goncharova, R.I., U.S.S.R. Academy of Sciences, Minski, U.S.S.R.
Grigg, G.W., CSIRO, North Ryde, NSW, AUSTRALIA
Guevara, A.P., University of the Philippines, Diliman, Quezon City,
    PHILIPPINES

Hachiya, N., Akita University School of Medicine, Hondo chome, Akita,
    JAPAN
Hamasu, Y., Central Institute Nihon-Shinyaku, Minami-ku, Kyoto, JAPAN
Hara, Y., Food Research Laboratories, Fujieda, Shizuoka, JAPAN
Hattori, M., Institute of Japanese & Chinese Medicine, Sugitani, Toyama,
    JAPAN
Hayashi, H., Asahikasei-Kogyo Co., Chiyoda-ku, Tokyo, JAPAN
Hayatsu, H., Okayama University, Tsushima, Okayama, JAPAN
Hennig, E.E., Medical Academy, Warsaw, POLAND
Hiraga, Y., Mitsubishi-Kasei Institute, Inashiki-gun, Ibaraki, JAPAN
Hirayama, K., Mitsubishi Kasei Corporation, Yokohama, Kanagawa, JAPAN
Hirono, I., Fujita-Gakuen Health University, Toyoake, Aichi, JAPAN
Hirose, M., Nagoya City University, Mizuho-ku, Nagoya, JAPAN
Hoshino, H., Yamanashi Medical College, Nakaknoma, Yamanashi, JAPAN

Huang, M.-T., Rutgers University, Piscataway, New Jersey, U.S.A.
Iga, H., Utsunomiya University, Ichimine-cho, Utsunomiya, JAPAN
Ilino, T., University of Tokyo, Bunkyo-ku, Tokyo, JAPAN
Ikeda, Y., Mitsubishi-Kasei Co., Ltd., Inashiki-gun, Ibaraki, JAPAN
Ikehata, H., Kyoto University, Kyoto, JAPAN
Ikushima, T., Kyoto University, Sennan-gun, Osaka, JAPAN
Ikuta, Y., Mitsui Petrochemical Industries, Kuga-gun, Yamaguchi, JAPAN
Imanishi, H., Institute of Environmental Toxicology, Kodaira, Tokyo,
    JAPAN
Inoue, H., Mitsubishi Kasei Corporation, Midori-ku, Yokohama, JAPAN
Inoue, M., Chugai Pharmaceutical Co., Ltd., Toshima-ku, Tokyo, JAPAN
Inouye, T., Institute of Environmental Toxicology, Kodaira, Tokyo,
    JAPAN
Ishihara, Y., Nissan Chemical Ind. Ltd., Shiraoka, Saitama, JAPAN
Ito, N., Nagoya City University, Mizuho, Nagoya, JAPAN
Ito, Y., Institute of Kobe City, Chuo-ku, Kobe, JAPAN
Iwama, M., Tokyo University of Agriculture, Setagaya, Tokyo, JAPAN
Iwasaki, H., Osaka University, Suita City, Osaka, JAPAN
Iyatomi, A., Nihon Tokushu Noyaku Seizo, Hino-shi, Tokyo, JAPAN

Jain, A.K., K.G. Medical College, Lucknow, INDIA
Jenssen, D., University of Stockholm, Stockholm, SWEDEN

Kale, R.K., Jawaharlal Nehru University, New Delhi, INDIA
Kamiya, M., Nagoya City University, Mizuho-ku, Nagoya, JAPAN
Kanazawa, K., Kobe University, Nada-ku, Kobe, JAPAN
Kaneko, I., Riken Institute, Wako-shi, Saitama, JAPAN
Kato, R., Keio University, Shinjuku-ku, Tokyo, JAPAN
Kato, Ta., Science University of Tokyo, Shinzyuku-ku, Tokyo, HAPAN
Kato, Te., Tokyo College of Pharmacy, Hachioji, Tokyo, JAPAN
Kato, To. Institute of Environmental Toxicology, Kodaira, Tokyo, JAPAN
Kawai, K., Meijo University, Tenpaku-ku, Nagoya, JAPAN
Kawaji, H., Fuji Life Science Inc., Kobuchizawaw, Yamanashi, JAPAN
Kawakishi, S., Nagoya University, Tikusa-ku, Nagoya, JAPAN
Kawamura, H., Tsumura Pharmacological Institute, Inashiki-gun, Ibaraki,
    JAPAN
Kawazoe, Y., Nagoya City University, Mizuho, Nagoya, JAPAN
Kimura, Y., National Cancer Center Research Institute, Chuo-ku, Tokyo,
    JAPAN
Kitagawa, Y., Suntory Co., Mishima-gun, Osaka, JAPAN
Kobayashi, H., Shiseido Toxicological & Analytical Research Center,
    Kitaku, Yokohama, JAPAN
Kogiso, S., Sumitomo Chemical Co., Ltd., Takarazuka, Hyogo, JAPAN
Kojima, H., Nippon Menard Cosmetic Co., Ltd., Ogaki, Gifu, JAPAN
Komatsu, M., Kawauchi-Nagano, Osaka, JAPAN
Konishi, Y., Nara University, Kasiwara, Nara, JAPAN
Konoshima, T., Kyoto Pharmaceutical University, Yamashina-ku, Kyoto,
    JAPAN
Koyama, J., Kobe Women's College of Pharmacy, Kobe, Hyogo, JAPAN
Kritchevsky, D., The Wistar Institute, Philadelphia, Pennsylvania,
    U.S.A.
Kubiak, R., Polish Academy of Sciences, Krakow, POLAND
Kumari, M.V.R., Jawharlal Nehru University, New Delhi, INDIA
Kunkel, T.A., National Institute of Environmental Health Sciences,
    Research Triangle Park, North Carolina, U.S.A.

Kurashima, Y., National Cancer Center Research Institute, Chuo-ku, Tokyo, JAPAN

Kurihara, K., Tokyo Metropolitan Institute of Medical Science, Bunkyo-ku, Tokyo, JAPAN

Kurishita, A., Kyoto University, Sakyo-ku, Kyoto, JAPAN

Kuroda, K., Osaka City Institute of Public Helath and Environmental Sciences, Tennozi-ku, Osaka, JAPAN

Kuroda, Y., National Institute of Genetics, Mishima, Shizuoka, JAPAN

Kuroki, D., Utsunomiya University, Ichimine-cho, Utsunomiya, JAPAN

Kusano, T., Hiroshima Women's University, Minami-ku Ujina, Hiroshima, JAPAN

Kutuzawa, M., Kuraray Co., Ltd., Chuo-ku, Tokyo, JAPAN

Lang, Reiner, Schering AG Exp. Toxikologie, Berlin, WEST GERMANY

Laurent, C.P.A., Genetic Toxicology Laboratory, Tilman-Liege, BELGIUM

Laxminarayana, Osmania University, Hyderabad, INDIA

Lee, K.-H., Hokkaido University, Sapporo, Hokkaido, JAPAN

Lim-Sylianco, C., University of the Philippines, Diliman, Quezon City, PHILIPPINES

Linghua, Wang, Second Military Medical College, Shanghai, CHINA

Liu, P.K., Case Western Reserve University, Cleveland, Ohio, U.S.A.

Loprieno, N., Universita La Sapienza, Rome, ITALY

MacPhee, D.G., LaTrobe University, Bundoora, Victoria, AUSTRALIA

MacPhee, M.C., LaTrobe University, Bundoora, Victoria, AUSTRALIA

Maeda, S., Kobe University Medical School, Chuo-ku, Tokyo, JAPAN

Maki, H., Kyusyu University, Higashi-ku, Fukuoka, JAPAN

Maldonado, A.M., Institute of Physical and Chemical Research, Hirosawa, Saitama, JAPAN

Marchetti, G., University of Pisa, Pisa, ITALY

Mario, Fiore, Universita La Sapienza, Rome, ITALY

Mason, R.P., National Institute of Environmental Health Sciences, Research Triangle Park, North Carolina, U.S.A.

Masubuchi, Y., Chiba University, Yayoi-cho, Chiba, JAPAN

Matsuda, H., Hatano Research Institute, Hadano, Kanagawa, JAPAN

Matsui, S., Kyoto University, Otsu, Shiga, JAPAN

Matsumoto, T., Daicel Chemical Industries, Ltd., Himeji, Hyogo, JAPAN

Matsumura, T., Meiji Institute of Health Sciences, Odawara, Kanagawa, JAPAN

Matsushima, T., University of Tokyo, Minato-ku, Tokyo, JAPAN

Matsusita, H., National Institute of Public Health, Minato-ku, Tokyo, JAPAN

Matumoto, K., Institute of Environmental Toxicology, Kodaira, Tokyo, JAPAN

Minamoto, S., Ashiya University, Ashiya, Hyogo, JAPAN

Miwa, M., National Food Research Institute, Tsukuba, Ibaraki, JAPAN

Mizuno, M., Kobe University, Nada-ku, Kobe, JAPAN

Mochizuki, M., Kyoritsu College of Pharmacy, Minato-ku, Tokyo, JAPAN

Moon, R.C., Illinois Institute of Technology, Chicago, Illinois, U.S.A.

Mori, Yukio, Gifu Pharmaceutical University, Mitahora-higashi, Gifu, JAPAN

Morse, M.A., American Health Foundation, Valhalla, New York, U.S.A.

Moustacchi, E.E., Institut Curie Biologie, Paris, FRANCE

Muraoka, N., San-yo Gakuen Women's Junior College, Hirai, Okayama, JAPAN

Murata, K., Kumiai Chemical Industry Co., Ltd., Ogasa-gun, Shizuoka, JAPAN

Nagahama, A., Kikkoman Corporation, Noda, Chiba, JAPAN
Nagao, M., National Cancer Center Research Institute, Chuo-ku, Tokyo, JAPAN
Nakadate, T., Keio University, Shinjuku-ku, Tokyo, JAPAN
Nakagawa, Y., Tokyo Metropolitan Research Laboratory of Public Health, Shinjyuku-ku, Tokyo, JAPAN
Nakamura, S., Osaka Prefecture Institute of Public Health, Higashinari, Osaka, JAPAN
Nakamura, Y., University of Shizuoka, Oshika, Shizuoka, JAPAN
Namiki, M., Meito-ku, Nagoya, JAPAN
Narui, K., Ajinomoto, Co., Inc., Yokohama, Kanagawa, JAPAN
Nasu, K., University of Shizuoka, Hamamatsu, Shizuoka, JAPAN
Nasuno, S., Kikkoman Corporation, Noda, Chiba, CHINA
Natake, M., Kobe University, Nada-ku, Kobe, JAPAN
Negishi, T., Okayama University, Tushima, Okayama, JAPAN
Neriishi, K., Radiation Effects Research Foundation, Minami-ku, Hiroshima, JAPAN
Nestmann, E.R., Cyanamid Canada, Inc., Willowdale, Ontario, CANADA
Nishimura, S., National Cancer Center Research Institute, Chuo-ku, Tokyo, JAPAN
Nishino, H., Kyoto Prefectural University of Medicine, Kamigyoku, Kyoto, JAPAN
Nishioka, H., Doshisha University, Kamigyo-ku, Kyoto, JAPAN
Nito, S., Tanabe Seiyaku Co., Ltd., Yodogawa-ku, Osaka, JAPAN
Noda, A., Meiji Institute of Health Science, Odawara, Kanagawa, JAPAN
Nohmi, T., National Institute of Hygienic Sciences, Setagaya-ku, Tokyo, JAPAN

O'Brien, T.G., The Wistar Institute, Philadelphia, Pennsylvania, U.S.A.
Oba, H., Hamamatsu Health Junior College, Mikatahara-cho, Hamamtsu, JAPAN
Obana, H., Osaka Prefectural Institute of Public Health, Higashinari-ku, Osaka, JAPAN
Oda, Y., Osaka Prefectural Institute of Public Health, Higashinari-ku, Osaka, JAPAN
Odagiri, Y., Saitama Medical School, Iruma-gun, Saitama, JAPAN
Oguni, I., University of Shizuoka, Hamamatsu, Shizuoka, JAPAN
Ohara, Y., Okayama University, Tsushima-naka, Okayama, JAPAN
Ohara, Y., Ajonomoto Co., Inc., Yokohama, Kanagawa, JAPAN
Ohnishi, T., Nara Medical University, Kashihara, Nara, JAPAN
Ohnishi, Y., Univeristy of Tokushima, Kuramoto-cho, Tokushima, JAPAN
Ohnuki, H., Utsunomiya University, Ichimine-cho, Utsunomiya, JAPAN
Ohta, Ta., Science University of Tokyo, Shinjuku-ku, Tokyo, JAPAN
Ohta, To., Institute of Environmental Toxicology, Kodaira, Tokyo, JAPAN
Okamoto, S., Hoyu Co., Ltd., Higashi-ku, Nagoya, JAPAN
Okuyama, T., Meiji College of Pharmacy, Setagaya-ku, Tokyo, JAPAN
Osawa, T., Nagoya University, Chikusa, Nagoya, JAPAN

Palitti, F., Universita' Degli Studi della Tuscia, Viterbo, ITALY
Pariza, M.W., University of Wisconsin, Madison, Wisconsin, U.S.A.
Parker, K.R., University of Alberta, Edmonton, Alberta, CANADA
Perticone, P., Consiglio Nazionale delle Ricerche, Rome, ITALY
Peto, R., University of Oxford, Oxford, UNITED KINGDOM
Picha, P., National Cancer Institute, THAILAND
Pueyo, C., Universidad de Cordoba, Cordoba, SPAIN
Radu, Liliana, Victor Babes Institute, Bucharest, ROMANIA
Ramel, C., University of Stockholm, Stockholm, SWEDEN

Rao, A.R., Nehru University, New Delhi, INDIA
Rao, K.P., Osmania University, Hyderabad, INDIA
Renner, H.W., Federal Research Centre for Nutrition, Karlsruhe,
    FEDERAL REPUBLIC OF GERMANY
Richter, W.W., National Cancer Center Research Institute, Chuo-ku,
    Tokyo, JAPAN
Ricordy, R., Roma Universita, Rome, ITALY
Rojanapo, W., National Cancer Institute, Bangkok, THAILAND
Romert, L., University of Stockholm, Stockholm, SWEDEN
Rosin, M.P., British Columbia Cancer Research Center, Vancouver,
    British Columbia, CANADA
Ruggero, R., University La Sapienza, Rome, ITALY
Rukmini, C., National Institute of Nutrition, Hyderabad, INDIA
Ryo, H., Osaka University, Kita-ku, Osaka, JAPAN

Sadaie, Y., National Institute of Genetics, Mishima, Shizuoka, JAPAN
Sadhu, D.N., Osmania University, Hyderabad, INDIA
Sakai, A., National Institute of Hygienic Sciences, Setagaya-ku, Tokyo,
    JAPAN
Sakai, K., University of Chiba, Yayoi-cho, Chiba, JAPAN
Sakamoto, K., Food and Drug Safety Center, Hatano, Kanagawa, JAPAN
Sampedro, F., Hospital de la San Pedlo, Barcelona, SPAIN
Sasaki, K., Hatano Research Institute, Hatano, Kanagawa, JAPAN
Sasaki, M., Tokyo Metropolitan Research Laboratory of Public Health,
    Shinjuku-ku, Tokyo, JAPAN
Sasaki, Y.F., Institute of Environmental Toxicology, Kodaira, Tokyo,
    JAPAN
Sato, S., Toyama Institute of Health, Imizu-gun, Toyama, JAPAN
Sato, T., Gifu Pharmaceutical University, Mitahora-higashi, Gifu, JAPAN
Sato, Y., Kyoritu College of Pharmacy, Minato-ku, Tokyo, JAPAN
Sekiguchi, M., Kyushu University, Higashi-ku, Fukuoka, JAPAN
Sekijima, M., Sogo Biological and Medical Institute, Kawagoe, Saitama,
    JAPAN
Semba, M.S., Food & Drug Safety Center, Hatano, Kanagawa, JAPAN
Seno, T., National Institute of Genetics, Mishima, Shizuoka, JAPAN
Shahin, A., Aulnay-sous-Bois, FRANCE
Shahin, M.M., l'Oreal Research Laboratories, Aulnay-sous-Bois, FRANCE
Shankel, D.M., University of Kansas, Lawrence, Kansas, U.S.A.
Shephard, S.E., National Cancer Center Research Institute, Chuo-ku,
    Tokyo, JAPAN
Shiba, T., Osaka University, Suita, Osaka, JAPAN
Shibahara, T., Otsuka Pharmaceutical Co., Ltd., Kawauchi-cho,
    Tokushima, JAPAN
Shibuya, T., Hatano Research Institute, Hadano, Kanagawa, JAPAN
Shimada, H., Daiichi Seiyaku Co., Ltd., Edogawa-ku, Tokyo, JAPAN
Shimoi, K., University of Shizuoka, Oshika, Shizuoka, JAPAN
Shinohara, K., National Food Research Institute, Kannondai, Tsukuba,
    JAPAN
Shioya, S., Ajinomoto Co., Inc., Yokohama, Kanagawa, JAPAN
Shiragiku, T., Otsuka Pharmaceutical Co., Ltd., Kawauchi-cho,
    Tokushima, JAPAN
Shirasu, Y., Institute of Environmental Toxicology, Kodaira, Tokyo,
    JAPAN
Simic, D., University of Belgrade, Takovska, Beograd, YUGOSLAVIA

Simic, M.G., National Bureau of Standards, Gaithersburg, Maryland, U.S.A.
Sobels, F.H., University of Leiden, Leiden, THE NETHERLANDS
Sonoda, T., Chiba University, Chiba, JAPAN
Sorsa, Marja, Institute of Occupational Health, Topeliuksenkatu, Helsinki, FINLAND
Sporn, M.B., National Institute of Health, Bethesda, Maryland, U.S.A.
Sram, R.J., Psychiatric Research Institute, Prague, CZECHOSLOVAKIA
Standbridge, E., University of California, Irvine, California, U.S.A.
Stich, H.F., British Columbia Cancer Research Center, Vancouver, British Columbia, CANADA
Suga, K., Metropolitan Institute of Environmental Sciences, Toto-ku, Tokyo, JAPAN
Suganuma, M., National Cancer Center Research Institute, Chuo-ku, Tokyo, JAPAN
Sugimoto, K., Kikkoman Corporation, Noda, Chiba, JAPAN
Sugimura, T., National Cancer Center Research Institute, Chuo-ku, Tokyo, JAPAN
Sugiyama, C., Shiseido Toxicological & Analytical Research Center, Kita-ku, Yokohama, JAPAN
Suttajit, M., Chiang Mai University, Chang Mai, THAILAND
Suwa, Y., Suntory Co., Mishima-gun, Osaka, JAPAN
Suzuki, H., Chiba University, Chiba, JAPAN
Suzuki, J., University of Tokyo, Shinjyuku-ku, Tokyo, JAPAN
Suzuki, N., Chiba University, Inohana, Chiba, JAPAN

Taira, M., Mukogawa Women's University, Nishinomiya, Hyogo, JAPAN
Takahashi, K., Nagoya City University, Mizuho-ku, Nagoya, JAPAN
Takata, M., Meiji College of Pharmacy, Setagaya-ku, Tokyo, JAPAN
Takayama, K., National Cancer Center Research Institute, Chuo-ku, Tokyo, JAPAN
Takayama, S., Seiyaku Co., Ltd, Edogawa-ku, Tokyo, JAPAN
Takayama, Syo., National Cancer Center Research Institute, Chuo-ku, Tokyo, JAPAN
Takeda, K., Kyoritsu College of Pharmacy, Minato-ku, Tokyo, JAPAN
Takeo, T., ITO-EN Co., Haibara-gun, Shizuoka, JAPAN
Takeshita, T., Yamanashi Medical College, Nakakoma, Yamanashi, JAPAN
Tamai, K., Institute of Science Health, Hodogaya, Yokohama, JAPAN
Tamura, T., Asahikasei-Kogyo Co., Chiyoda-ku, Tokyo, JAPAN
Tanaka, M., JAPAN
Tanaka, N., Food & Drug Safety Center, Hadano, Kanagawa, JAPAN
Tanaka, T., Gifu University Medical School, Tsukasa-cho, Gifu, JAPAN
Tazima, Y., Kokubunji-shi, Tokyo, JAPAN
Terada, M., National Cancer Center Research Institute, Chuo-ku, Tokyo, JAPAN
Tezuka, H., National Institute of Genetics, Mishima, Shizuoka, JAPAN
Tokiwa, H., Fukuoka Environmental Research Center, Dazaifu, Fukuoka, JAPAN
Tokuda, H., Kyoto University, Sakyo-ku, Kyoto, JAPAN
Tomita, Isao, University of Shizuoka, Ozika, Shizuoka, JAPAN
Toyoda, Y., Suntory Co., Mishima-gun, Osaka, JAPAN
Troll, E., New York University Medical Center, New York, New York, U.S.A.
Troll, W., New York University Medical Center, New York, New York, U.S.A.

Tsuda, M., National Cancer Center Research Institute, Chuo-ku, Tokyo,
JAPAN
Tutikawa, K., National Institute of Genetics, Mishima, Shizuoka, JAPAN

Uda, Y., Utsunomiya University, Mine-cho, Utsunomiya, JAPAN
Udagawa, T., Tokyo Institute of Technology, Midori-ku, Yokohama,
JAPAN
Ueda, Y., National Cancer Center Research Institute, Chuo-ku, Tokyo,
JAPAN
Ueno, Y., Science University of Tokyo, Shinzyuku, Tokyo, JAPAN

Verma, A.K., University of Wisconsin, Madison, Wisconsin, U.S.A.
Vierling, Th., Beohringer Mannheim GmbH, Mannheim, FEDERAL
REPUBLIC OF GERMANY
von Borstel, P.A., University of Alberta, Edmonton, Alberta, CANADA
von Borstel, R.C., University of Alberta, Edmonton, Alberta, CANADA

Wakabayashi, K., National Cancer Center Research Institute, Chuo-ku,
Tokyo, JAPAN
Walker, G.C., Massachusetts Institute of Technology, Cambridge,
Massachusetts, U.S.A.
Wall, M.E., Research Triangle Institute, Research Triangle Park, North
Carolina, U.S.A.
Wang, L., Second Military Medical College, Shanghai, CHINA
Watanabe, K., Institute of Environmental Toxicology, Kodaira, Tokyo,
JAPAN
Watanabe, M., Institute of Environmental Toxicology, Kodaira, Tokyo,
JAPAN
Watanabe, M., Asahikasei-Kogyo Co., Chiyoda-ku, Tokyo, JAPAN
Waters, M.D., U.S. Environmental Protection Agency, Research Triangle
Park, North Carolina, U.S.A.
Wattenberg, L., University of Minnesota, Minneapolis, Minnesota, U.S.A.
Weisburger, J.W., American Health Foundation, Valhalla, New York,
U.S.A.
Wilson, C.M., Claire Wilson & Associates, Washington, DC, U.S.A.
Wirth, Peter, National Cancer Center Research Institute, Chuo-ku,
Tokyo, JAPAN
Wurgler, F.E., University of Zurich, Schwerzenbach, SWITZERLAND
Wurzner, H., Nestle Research Centre, Vevey, SWITZERLAND

Yagi, H., National Institute of Diabetes & Digestive and Kidney Diseases,
Bethesda, Maryland, U.S.A.
Yamamoto, S., Keio University, Shinjuku-ku, Tokyo, JAPAN
Yamamoto, T., University of Tokyo, Minato-ku, Tokyo, JAPAN
Yasukawa, K., Nihon University, Chiyoda-ku, Tokyo, JAPAN
Yasunaga, K., Mitsubishi Kasei Corporation, Midori-ku, Yokohama, JAPAN
Yatagai, F., Institute of Physical and Chemical Research, Wako-shi,
Saitama, JAPAN
Yoshida, Y.H., Tachikawa College of Tokyo, Akishima, Tokyo, JAPAN
Yoshitake, A., Sumitomo Chemical Co., Ltd., Konohana-ku, Osaka,
JAPAN
Yoshizawa, S., National Cancer Center Research Institute, Chuo-ku,
Tokyo, JAPAN
You, X., ITO-EN Co., Sagara-cho, Shizuoka, JAPAN